Catalogue of Indian Insects

THE EDITOR

Thomas Bainbrigge Fletcher (1878-1950) was an English entomologist. He worked in the Royal Navy; he became an expert on the "microlepidoptera" and was appointed as the second Imperial Entomologist in India, succeeding Harold Maxwell-Lefroy at the Imperial Agricultural Research Institute at Pusa. He was a meticulous naturalist and very careful on matters of systematics and taxonomic nomenclature.

He produced a list of publications on India Entomology and Catalogue of Indian Insects. He also wrote general works on entomology including Some South Indian Insects, and a Veterinary Entomology for India and hints on collecting and preserving insects. He was Fellow of the Linnean Society, the Royal Entomological Society, Zoological Society of London and a President of the Cotteswold naturalists field club.

Catalogue of Indian Insects

Volume 2

–Editor–

Thomas Bainbrigge Fletcher

2016

Daya Publishing House®

A Division of

Astral International Pvt. Ltd.

New Delhi – 110 002

Catalogue of Indian Insects (5 Volumes Set)

Publisher's note:

The book "***Catalogue of Indian Insects***" by Thomas Bainbrigge Fletcher was published in 1921. But presently complete set not available anywhere. We have tried to reproduced the complete valuable set from different sources in a form as close to the original possible. This historical reproduction fulfill this long felt needs of students, teachers and researchers engaged in the field of entomology, zoology and other related discipline.

Cataloging in Publication Data—DK
Courtesy: D.K. Agencies (P) Ltd. <docinfo@dkagencies.com>

Catalogue of Indian insects / editor, Thomas Bainbrigge Fletcher.

volumes cm
First published in 1921.
ISBN: 978-93-86071-07-1 (Vol. 2)
ISBN: 978-93-5130-922-2 (Set) (International Edition)

1. Insects—Catalogs and collections—India. I. Fletcher, T. B. (Thomas Bainbrigge), 1878-1950, editor.

QL468.2.C38 2016 DDC 595.70954 23

Published by : **Daya Publishing House®**
A Division of
Astral International Pvt. Ltd.
– ISO 9001:2008 Certified Company –
4760-61/23, Ansari Road, Darya Ganj
New Delhi-110 002
Ph. 011-43549197, 23278134
E-mail: info@astralint.com
Website: www.astralint.com

Printed at : **Replika Press Pvt. Ltd.**

डा. एस. अय्यप्पन
सचिव एवं महानिदेशक
Dr. S. AYYAPPAN
SECRETARY & DIRECTOR GENERAL

भारत सरकार
कृषि अनुसंधान और शिक्षा विभाग एवं
भारतीय कृषि अनुसंधान परिषद
कृषि एवं किसान कल्याण मंत्रालय, कृषि भवन, नई दिल्ली 110 001

GOVERNMENT OF INDIA
DEPARTMENT OF AGRICULTURAL RESEARCH & EDUCATION
AND
INDIAN COUNCIL OF AGRICULTURAL RESEARCH
MINISTRY OF AGRICULTURE AND FARMERS WELFARE
KRISHI BHAVAN, NEW DELHI 110 001
Tel.: 23382629; 23386711 Fax: 91-11-23384773
E-mail: dg.icar@nic.in

Foreword

With rapidly increasing human population that hardly matched by the rate of increase in agriculture production, the per capita availability of food and other products is bound to fall short. Therefore, increasing production per unit area is the most plausible alternative to sustain self sufficiency in food, fodder and fibre. Insects have got both positive and negative impacts in agriculture. Insects become pests when they interfere with our welfare, aesthetics or profits. A reduction or loss in yield and quality of crops is partly due to result of insect feeding activities and as vector/host to pathogen transmissions. On the other hand these small creatures serve us a lot as pollinating agents of many important agricultural crops, as biological controlling agents of pests and weeds as well as by producing honey, silk and many other materials essential to us.

This book 'Catalogue of Indian Insects' is divided into 5 volumes and covers 29 parts. This historical compilation is a dedicated effort to present the different aspects of the insects in a very concise and relevant manner. I believe that the students and scientist working in the area of entomology, agriculture, fisheries, zoology, botany and veterinary science will find this book as a source of useful and authentic information.

I appreciate the efforts of Daya Publishing House, A Division of Astral International (P) Ltd. New Delhi.

(Dr. S. Ayyappan)
Secretary (DARE) & Director General (ICAR)
Krishi Bhawan, New Delhi – 110001

Preface

There is a immense number of insect species on the planet. The class Insecta is a large and diverse as well as important ecologically. When compared to other groups, the number of insect species is much larger and represents more than three quarters of all known animal species. Most insects are either beneficial or harmless. Yet there are large numbers which are economically important as pests, pollinators, natural enemies and other beneficial insects like honey bees, silkworms, etc. Insects are fascinating to investigate, their classification is complex. It important to learn the main differences among insects so you can distinguish one group from another. Knowledge of insect classification is of primary importance in carrying out proper control procedures and quality pest management. Proper insect classification and identification is extremely important. If a beneficial or harmless insect is improperly identified, the resultant unnecessary and undesirable pesticide application usually disrupts natural control agents.

Insects deliver a host of ecological services fundamental to the survival of humankind. Insects play an important role in plant reproduction. Thousands of pollinator species have been identified and almost all of these are insects. Over 90 percent of the flowering plant species depend on pollinators. Insects play an equally vital role in waste biodegradation. They clean up dead plant matter, breaking down organic matter until it is fit to be consumed by fungi and bacteria. Insects also play a significant role in decomposing animal carcasses and manure. Insects have a large array of natural enemies, predators and parasitoids, keeping them under economic threshold levels. Due to large number of insects species it is necessary that these catalogued so as to enable locating the species into correct classification. One shall also know what are the species knows from a specific group. Especially in case of Indian Insects where large number of species are known. These are important Catalogue of Indian Insects as a compilation serves this purpose.

"Catalogue of Indian Insects" is divided into 5 volumes and covers 29 parts Volume 1 deals with Part I to Part IX. It covers Acrydidae, Culicidae, Bombyliidae, Trypetidae, Nitidulidae, Stapylinidae, Lasiocampidae, Amatidae and Zygaenidae. Volume 2 gives a detailed account of Part X to Part XVII. This volume covers Stephanidae, Brenthidae, Tabanidae,

Cicindelidae, Palpicornia, Cecidomyidae, Cosmopterygidae and Yponomeutidae. Volume 3 is devoted to Part XVIII i.e. Carabidae. Volume 4 covers in detail of Part XIX to Part XXV. This volume deals with Gyrinoidea, Alucitidae, Lycidae, Phalonidae and Chlidanotidae, Chalcidoidea, Evanidae and Thysanoptera. Volume 5 of the book describes the Part XXVI to Part XXIX. This volume deals with Serphoidea, Isoptera, Anthribidae and Asilidae.

This historical reproduction fulfill the long felt needs of students, teachers, researchers engaged in the field of entomology, agriculture, ecology, fisheries, forestry, paleontology, zoology, botany, plant science, medical and veterinary science, towards knowing insects.

We are extremely grateful to Dr. S. Ayyapppan, Secretary and Director General, Indian Council of Agricultural Reserarch, New Delhi for providing a foreword. The cooperation and support of Dr. V.V. Ramamurthy, Principal Scientist (Retd.), Division of Entomology, IARI, New Delhi is duly acknowledged.

All suggestions for future improvement for this book will be highly solicited.

Publisher

Contents

VOLUME 3

VOLUME 4

VOLUME 5

PART X – STEPHANIDÆ

Stephanidæ : Kieffer, Genera Insectorum, LXXVII, pp. 1-10, t. 1; Elliot, Proc. Zool. Soc., 1922, pp. 705-831, 8 figs. (Nov. 1922), [characters pp. 707-712; Key to Genera, p. 713].

1. Stephanus, Jurine.

Ichneumon, Fabricius, Ent. Syst. Suppl. p. 219 (1798). Type–*serrator,* Fb.

Bracon, Fabricius, Syst. Piez., p. 102 (1804).

Pimpla, Fabricius, ibidem, p. 112 (1804).

Stephanus, Panzer, Krit. Rev. I, p. 77 (1805); jurine, Nouv. Mèth. Hym. et Dipt. I, p. 91 (1807); Enderlein, Zoologische Anzeiger XXXIII, p. 473 (1905); Kieffer, Genera Insectorum, LXXVII. p. 3 (1908) (characters pp. 3-4, Tab., fig. 2); Elliot, Proc. Zool. Soc. London, p. 715 (1922) (Key to spp. pp. 715-718).

Xorides, Lamarck, Hist. Nat. Anim. S. Vert. IV, p. 135 (1817).

Megischus, Brullè, Hist. Nat. Ins. Hym. IV, p. 537 (1846).

Bothriocerus, Sichel, Ann. Soc. Ent. Fr. (3) VIII, p. 759 (1860).

Fænatopus, Smith, Journ. Proc. Linn. Soc. Lond. V, p. 58 (1861).

borneensis, Sauss.

Megischus borneensis, Sauss., Mission Pavie Indo-Chine, III, p. - 201, ♀♂ (1901).[1] Borneo.[2]

Stephanus borneensis, Elliot, P. Z. S. pp. 735-736 (1922).[2]

coronator, Fb.

Pimpla coronator, Fb., Syst. Piez., p. 118 (1804).[1]

Megischus viduus, Sm., Journ. Proc. Linn. Soc. Zool. V, p. 138, ♂ (1861).[2]

Megischus longicaudatus, Costa (Ach.) Rendie, Accad. Sc. Napoli, p. 271, ♀♂, (1866).[3]; Costa, (Ach.) Ann. 'Indes Orientales,[6 7] East India.[12]

Museo. Zool. Univ. Napoli, P. 81, t. IV, fig. l*a*, ♂ 1b ♀ (1866).[4] — Sarawak.[5, 12] Java.[11, 12]

Megischus insularis, Sm., Journ. Proc. Linn. Soc. Zool. II, p. 120, ♀♂ (1857).[5] — Bachian.[11, 12] Ceram.[11, 12]

Megischus coronator, Brullè, Hist. Nat Ins. Hymen., p. 538 (1846);[6] Sichel, Ann. Soc. Ent. France, V, ser. 4, 477, ♀♂ (1865).[7] — Amboine.[11, 12] kaissa.[2, 12] New Guinea.[6, 7, 11, 12] Australia.[11, 12]

Stephanus coronator, Boisd, Faun. Ent. (Voyage de la Astrolabe, 1832), p. 656, Tab. XII, fig. 7, ♀(1835)[8]; Schlett., Berlin. Entom. Zeits. XXXIII, p. 114, ♀♂ (1889)[9]; Enderl, Stett. Entom. Zeit. LXVIII p. 306, ♀♂ (1906)[10]; Kieffer, Genera Insectorum, LXXVII p. 4, Tab. f. 4 (1908)[11]; Elliot P. Z. S., pp. 738-739, figs. 2 and 3, ♀♂ (1922).[12] — Sumatra.[10, 11, 12] Manila.[7]

ceylonicus, Cam.

Stephanus ceylonicus, Cam., Journ. Straits Asiatic Soc. XXXIX, p. 102, (1903)[1]; Kieffer, Genera Insectorum, LXXVII. p. 4 (1908)[2]; Morl., Entomologist, L.p. 34 (1917)[3]; Elliot, P. Z. S., p. 735, ♀(1922).[4] — Ceylone.[2] Trincomali.[3, 4] Kandy.[3, 4] Galgamua.[3, 4]

ducalis, Westw.

Megischus, ducalis, Westw., Trans. Ent. Soc. (2) I, p. 229, ♀ (1850-51).[1] — India orientali.[1]

Stephanus ducalis, Schlett., Berlin. Entom. Zeits., XXXIII, p. 112, (1889), ♀[2]; Kieffer, Genera torum, LXXVII, p. 4, (1908)[3]; Elliot, P.Z.S., p. 740,♀♂ (1922).[4] — East India.[4]

Stephanus lucidus, Szèpl. Termès Füz, XXV, p. 532 (1902) ♂.[5] — Sumatra.[3, 4], Java.[4] Malacca.[3, 4] Celebes.[3, 4], Sumba.[3] Key Island.[2, 3, 4] Borneo.[4] New Guinea.[4] Pulu-Penang.[4] Bachian.[4]

hornianus, Enderl.

Stephanus hornianus, Enderl. Entomologische mitteilungen, I. No. 12, pp. 390-391, ♀♂, (1912).[1]; Elliot, P.Z.S., pp. 726-727,♀♂ (1922).[2] — Ceylon.[1, 2]

nigricans (Sichel) Schlett.

Megischus tarsatus, var, *nigricans* Sichel, Ann. Soc. Ent., France, V (Ser. 4), p. 476,♀ (1865).[1] — Ceylon.[1, 3, 4]

Stephanus nigricans, Schlett., Berl., Entom. Zeit. XXXIII, p. 109 (1889)[2]; Kieffer, Genera Insectorum, LXXVII, p. 4 (1906)[3]; Elliot, P. Z. S., p. 733, (1922).[4]

pilosus, Elliott.

Stephanus pilosus, Elliott, Entomologist, LIV, p. 256,♂ (1921); Elliott, P. Z. s. p. 744, ♂, (1922).[2] — Paleboehen.[1, 2] Ratoe.[1, 2]

Ruficeps, Sauss.

Megischus ruficeps, Sauss., Mission Pavie Indo-Chine, III, p. 201, ♀ (1901).[1] — Annam.[3]

Stephanus saussurei, Sehulz, Berl. Ent. Zeit. LI, p. 322 (1907)[2]; Kieffer, Genera Insectorum, LXXVII, p. 4 (1908).[3] — Cambodge.[3] Siam.[4]

Stephanus ruficeps, Elliot, P. Z. S., pp. 737-738 (1922).[4]

rugosus, Elliot.

Stephanus reugosus, Elliot, entomologist LIV, p. 256, ♂ (1921)[1]; Elliot P. Z. S., p. 745, ♂ (1922).[2] — Java.[1, 2]

sulcifrons, Schlett.

Megischus insidiator, Sm. Journ. Proc. Linn. Soc. Zool. VII, p. 7, ♂(1864).[1] — India.[2] Mindanao.[5]

Stephanus insidiator, Kieffer, Genera Insectorum LXXVII, p. 4 (1908).[2] — Philippines.[4] Mysol.[1]

Stephanus sulcifrons, Schlett, Berlin. Entom. Zeit. XXXIII, p. 110, ♀ (1889)[3]; Kièffer, Genera Insectorum, LXXVII, p. 4 (1908)[4]; Elliot, P.Z.S., pp. 724-745. ♀(1922).[5]

tarsatus, Sichel.

Megischus tarsatus, Sichel, Ann. Soc. Entom. France, V. Ser. 4, p. 476, (1865), t. X, figs. 4 and 5, ♀♂.[1] — Manila.[1, 4], Bachian.[4] (Molucca Islands.)

Stephanus tarsatus, Schlett, Berlin, Entom. Zeit., XXXIII, p. 107, ♀♂ (1889)[2]; Kieffer, Genera insectorum, LXXVII, p. 4 (1908)[3]; Elliot, P. Z. S., p. 729 (1922).[4] — Batjan.[3] (Philippines)

tortus, Morl.

Stephanus tortus, Morl. Entomologist, L, p. 84, ♀ (1919)[1]; Elliot, P. Z. S., p. 731, ♂ (1922).[2] — Nedunkernie.[1, 2] (Ceylon)

unicolor, Schlett.

Stephanus unicolor, Schlett., Berlin. Entom. Zeit. XXXIII,. 104, ♀ (1889)[1]; Kieffer, Genera Insectorum, LXXVII, p. 4 (1908)[2]; Elliott, P. Z. S. pp. 725-726, ♀ (1922).[3] — Mindanao.[3] (Philippines)[2]

2. Parastephanellus, Enderlein.

Stephanus (part) Westw. Theseur, Ent. Oxon., p. 126 (1874). — Type–*pygmæus,* Endrl.

Parastephanus (nec Hackel) Enderlein, Zool. Anzeig., p. 474 (1905).

Parastephanellus, Enderlein, Stett. Ent. Zeit. LXVII, p. 301 (1906); Kieffer, Genera Insectorum, LXXVII, p. 5 (Chatracters, p. 5, t. fig. 1), (1908); Elliot, P. Z. S., p. 746 (1922) (Key to spp. 747-748).

claripennis, Cam.

Megischus claripennis, Cam., Entomologist, XLIV, p. 56, ♀ (1911).[1] — Kuching.[1, 2]

parastephanellus claripennis, Elliot, P. Z. S., p. 758, ♀ (1922).[2]

eburneus, Morl.

Parastephanellus eburaneus Morl., Entomologist L, p. 35 (1917) ♀, ♂[1]; Elliot, P. Z. S. p. 752, ♀ (1922).[2] — Peradeniya[1] kandy.[1, 2] Pradenula.[2] (?Peradeniya.)

glaber, Elliot.

Parastephanellus glaber, Elliot, Entomologist, LII, pp. 129-130, ♀ (1919)[1]; Elliott, P. Z. S. p. 749, ♀ (1922).[2] — Singapore.[1, 2]

laevicollis, Morl.

Parastephanellus Laevieollis, Morl., Entomologist, L, p. 111, ♀, ♂ (1917)[1]; Elliot, P. Z. S. pp. 750-751, ♀ ♂ (1922).[2] — Ding-Ding Island.[1, 2] (Indian Ocean.)

maculifrons, Cameron.

**Megiseleus maculifrons,* Cam., Journ. Strait. Asiat. Soc. XXXVII, p. 100 (1902)[1]; Kieffer, Genera Insectorum, LXXVII, p. 5 (1908).[2] — Borneo.[2] New-guinea.[2] Sarawak.[3]

Parastephanellus maculifrons, Elliot, P. Z. S., pp. 751-752, ♂ (1922).[3]

(**Megiseleus* appears to be a printer's error for *Megischus.*)

martini, Stadlm.

Stephanus martini, Stadlm, Berl. Entom. Zeit. XXVII, p. 236, ♀♂ (1892).[1]

Parastephanellus martini, Kieffer, Genera Insectorum, LXXVII, p. 5 Tab. f. 9 (1908)[2]; Elliot, P. Z. S., p. 748, ♀, ♂ (1922).[3]

——————

Sumatra.[2, 3]

pallicitarsis, Cam.

Megischus palliditarsis, Cam., Entomologist, LXIV, pp. 56-57, ♀ (1911).[1]

Megischus reficollis, Cam., I. c. p. 57, ♂ (1911).[2]

Parastephanellus palliditarsis, Elliot, P. Z. S., pp. 756-757, ♀♂ (1922).[3]

——————

Kuching.[1, 2, 3]
(Borneo)

rubripictus, Elliott.

Stephanus damellicus, Enderl. (*nec* Wstw.) Arch. f. Naturg. Berlin, LXVII, p. 129, ♀♂ (1901).[1]

Parastephanellus rubripictus, Elliot, Entomologist, L, p. 129, ♀♂ (1917)[2]; Elliot, P. Z. S., pp. 759-760 ♀♂ 1922).[3]

——————

Borneo.[1, 2, 3]
Milne Bay.[1, 2, 3]
Bismarck Arch.[1, 2, 3]

scitus, Elliott.

Parastephaneelus scitus, Elliott, Entomologist, LII, p. 163, ♂ (1919)[1]; Elliott, P. Z. S. p. 757, ♀♂ (1922).[2]

Pusa.[1, 2], Ceylon.[1, 2]

——————

Ding-Dring Island.[1, 2], Boreno.[1, 2]
Australia.[1, 2]

3. Fænatopus, Smith.

Stephanus (Part) Westw., Ann Mag. Nat. Hist. VII, p. 538 (1841).

Megischus (Part) Westw., Trans. Ent. Soc. London, (2) I, p. 230 (1850-1851).

Fænatopus, Smith, Journ. Proc. Linn. Soc. Lond. V, p. 58 (1861); Enderlein, Zool. Anzeiger XXVIII, 1906; Kieffer, Genera insectorum, Westw. LXXVII, p. 6 (1908) (characters, p. 6, t. f. 15); Elliott, P. Z. S., p. 782 (1922) (Key to spp. 782-784).

Type–*indicus,* Westw.

indicus, Westw.

Stephanus indicus, Westw., Ann. Mag. Nat. His. VII, . 538 (1841)[1]; Schlett. Berl. Entom. Zeitz. XXXIII, p. 125, ♀♂ (1889).[2]

Ceylone.[5, 9]
Malabar.[1, 5, 9]
East India.[9]

Megischus indicus, Westw., Tr. Ent. Soc. Lond. I, ser. 2, p. 230 (1851).[3]

———
Philippines.[3, 5, 9]

Faenatipus ruficeps, Sm., Journ Proc. Linn. Soc. (Zool.) V, p. 58, ♀ (1861).[4]

Boreno.[5, 9]
Sumatra.[5, 7]

Fænatopus indicus, Kieffer, Genera Insectorum, LXXVII, p. 6 (1908)[5]; Enderl, Zool. Anzeiger XLII, p. 290 (1913)[6]; Stettin, Entom, Zeitung, LXVIII, p. 298 (1906)[7]; Morl., Entomologist, L, p. 36, ♀♂ (1917)[8]; Elliott, P. Z. S., p. 784-785, ♀♂ (1922).[9]

Celebes.[5, 9]
Sumbava.[9]
Mt. Matang.[8]
Makassar.[4]
Malay Archipelago.[8]

iridipennis, Elliott.

Fænatopu iridipennis, Elliott, Entomologist, LII, p. 30, ♀ (1911)[1]; Elliott, P. Z. S., p. 796-797, ♀♂ (1922).[2]

Dehra Dun.[1, 2]
Kangra. 4, 000 ft.[1, 2]

lacteipennis, Schlett.

Stephanus lacteipennis, Schlett., Berl. Entom. Zeits. XXXIII, p. 119,♀ (1889)[1]; Kieffer, Genera Insectorum, LXXVII, p. 6 (1908).[2]

———
Borneo.[2, 3]

Fænatopus lacteipennis, Elliott, P. Z. S., p. 790, ♀, (1922).[3]

longicauda, Elliott.

Fænatopus longicauda, Elliott, Entomologist, LII, p. 30, ♀ (1919)[1]; Eliott, P. Z. S., p. 794, ♀ (1922).[2]

Nilgiris, 3, 000 ft.[1, 2]

longicollis, Cam.

Megiseleius longicollis,* Cam., Journ. Straits Branch, Roy. Asia, Soc. XXXVII, p. 32, ♀ (1902).[1]

———
Sarawak.[2]

Fænatopus longicollis, Elliott, P. Z. S., pp. 790-791, ♀ (1922).[2]

ocellatus, Elliott.

Fænatopus ocellatus, Elliott, Entomologist, LII, p. 131, ♀ (1919)[1]; Elliott, P. Z. S. p. 784 (1922).[2]

———
Mt. Matang.[1, 2]

punctatus Elliott.

Fænatopus punctatus, Elliott, Entomologist, LII, p. 73, ♀ (1919)[1]; Elliott, P. Z. S. p. 794/7, ♀ (1922).[2]

Burma.[1, 2]

rugiceps, Elliott.

Fænatopus rugiceps, Elliott, Entomologist, LII, p. 29, ♀ (1919)[1]; Elliot P. Z. S. pp. 794-795, fig. 5 (1922).[2]

S. India.[1, 2]

---

* Appears to be a printer's error for *Megischus.)*

sulcaticollis, Enderlein.

Fænatopus indicus var. *sulcaticollis,* Enderlein, Stett. Ent. Zeit. LXVII, p. 298 (1906)[1]; Kieffer, Genera Insectorum, LXXVII, p. 6 (1908).[2] — Sumatra.[1, 2, 3]

Fænatopus sulcatiocllis, Elliott, P. Z. S., p. 787, ♀ (1922).[3]

sumbanus, Enderlein.

Fænatopus indicus, var. *sumbana,* enderlein, Stett. Ent. Zeit. LXVII, p. 298 (1906)[1]; Kieffer. Genera Insectourm, LXXVII, p. 6 (1908).[2] — Sumba Island.[1, 2, 3]

Fænatopus sumbanus, Enderlein, Zoologische Anzeiger, XLII, p. 209, ♀ (1913); Elliott, P. Z. S., pp. 785-786, ♀ (1922).[3]

Variidens, Elliott.

Fænatopus variidens, Elliott, entomologist, LIV, p. 257, ♀ (1921)[1]; Elliott, P. Z. S., p. 792, ♀ (1922).[2] — Paleboehen, [1, 2] Ratoe.[1, 2]

4. Diastephanus, Enderl.

Stephanus (Part) Schlett., Berl. Ent. Zeit., XXXIII, p. 121 (1889). — Type–*flavomaculatus,* Enderl.

Diastephanus, Enderl., Zool. Anzeig., XXXIII, p. 474, f. 6 (1905); kieffer, Genera Insectorum, LXXVII, p. 7 (1908) (Characters, p. 7, t. f. 16); Elliott, P. Z. S. p. 800 (1922). (Key to spp. pp. 800-802).

bilineatus, Elliott.

Diastephanus bilineatus, Elliott, entomologist, LII, p. 162, ♀ (1919)[1]; Elliott, P. Z. S., p. 805, ♀ (1922).[2] — Pusa.[1, 2] Chupra.[1, 2], (Bihar.)

carinifrons, Enderl.

Diastephanus carinifrons, Enderl., Stettin, Entom., Zeit., LXVII, p. 293, (1906)[1]; Kieffer, Genera Insectorum, LXXVII, p. 7 (1908)[2]; Elliott, P. Z. S., p. 811, ♀ (1922).[3] — Sumatra.[1, 2, 3]

dohrni, Enderl.

Diastephanus dohrni, Enderl. Stettin, Entom., Zeit. LXVII, pp. 291-293, ♀♂, (1906)[1]; Kieffer, Genera Insectorum, LXXVII, p. 7, (1908)[2]; Elliott, P. Z. S., p. 817, ♀ (1922).[3] — Sumatra.[1,2, 3,]

elegans, Elliott.

Diasterphansu elegans, Elliott, Entomologist, LII, p. 31, ♀ (1919)[1]; Elliott, P. Z. S., pp. 810-811, ♀ (1922).[2] — Singapore.[1, 2]

fasciatus, Szepl.

Fænatopus fasciatus, Szèpl., Termes. Füz. XXV, p. 529, ♂ (1902).[1]

Diastephanus Fasciatus, Enderl., Zool. Anzeig., XXXIII, p. 475 (1905)[2]; Kieffer, Genera Insectorum, LXXVII, p. 7 (1908)[3]; Elliott, P. Z. S., p. 826, ♂ (1922).[4]

———

Sumbava.[3, 4]

flavifrons, Elliott.

Diastephanus flavirons, Elliott, Entomologist, L, p. 131, ♂ (1917)[1]; Elliott, P. Z. S., p. 825, ♂ (1922).[2]

———

West Sarawak.[1]
Quop.[2]

flavonototus, Elliott

Diastephanus flavonotatus, Elliott, Entomologist, LII, p. 32, ♂ (1919)[1]; Elliot P. Z. S., pp. 826-827, ♂(1922).[2]

———

Kuching (Sarawak).[1, 2]

frontilinea, Morley.

Diastephanus frontilinea, Morl., Entomologist, L. p. 109, ♀, (1917)[1]; Elliott, P. Z. S., p. 813, ♀ (1922).[2]

Rangpur.[1]

fuscinervis, Cam.

Fænatopus fuscinervis, Cam. Journ. Straits Branch, Roy. Asiat. Soc. XXXIX, p. 101, ♂ (1903)[1]; Kieffer, Genera Insectorum, LXXVII, p. 6. (1908).[2]

Diastephanus fuscinervis, Elliott, P. Z. S., pp. 824-825, ♂ (1922).[3]

———

Borneo.[2]
Kuching (Sarawak).[3]

leucodontus, Schlett.

Stephanus leucodontus, Schlett., Berl. Ent. Zeit., XXXIII, p. 121, ♀ (1889).[1]

Diastephanus leucodontus, p. 475, (1905)[2]; Kieffer, Genera Insectorum, LXXVII, p. 7 (1908)[3]; Elliott, P. Z. S. p. 810, ♀ (1922).[4]

———

Sarawak.[3, 4]

maculifemur, Enderl.

Diastephanus maculifiermur, Enderl., Stettin, Entom. Zeit. LXVII, pp. 294-296, ♀ (1906)[1]; Kieffer, Genera Insectorum, LXXVII, p. 7 (1908)[2]; Elliott P. Z. S. p. 819, ♀ (1922).[3]

———

Perak (Malacca).[1, 2, 3]

nigripes, Szèpl.

Fænatopus nigripes Szèpl. Termes. Füz., XXV, p. 531 (1902).[1]

———

Sumbava.[1, 4]

Diastephanus nigripes, Enderl. Zool. Anzeig., XXXIII, p. 475 (1905)[2]; Kieffer, Genera Insectorum, LXXVII, p. 7, (1908)[3]; Elliott, P. Z. S., p 812, ♀ (1922).[4]

Pallescens, Schlett. ————

Stephanus pallescens, Schlett., Berl. Ent. Zeit. XXXIII, p. 123, ♀ (1889).[1] Philippines.[3, 4]

Diastephanus pallescens, Enderl. Zool. Anzeig., XXXIII, p. 475 (1905)[2]; Kieffer, Genera Insectorum, LXXVII, p. 7 (1908)[3]; Elliott, P. Z. S., pp. 803-804, ♀ (1922).[4]

parviceps, Enderl. ————

Diastephanus parviceps, Enderl., Stettin., entom, Zeit. LXVIII, ♀, pp. 296-298, ♀ (1906)[1]; Kieffer, Genera Insectorum, LXXVII, p. 7 (1908)[2]; Elliott, P. Z. S., p. 812, ♀ (1912).[3] Sumatra.[1, 2, 3]

quadridens, Elliott. ————

Diasterphanus quadridens, Elliott, Entomologist, LIII, p. 81, ♀ (1920)[1]; Elliott, P. Z. S., p. 818, ♀ (1922).[2] Indo-China.[1, 2]

sulcatus, Elliott. ————

Diastephans suleatus, Elliott, Entomologist, LIII, pp. 80-81, ♀ (1920)[1]; Elliott, P. Z. S. pp. 820-821, f. 7-8, ♀ (1922).[2] Indo-China.[1, 2]

PART XI – BRENTHIDAE.

CALODROMINÆ.

Calodromini, Kln., Ent. B1, 18, 1922, p. 161.

1. Calodromus Guérin (1832)

Type : *mellyi* Guér.

Guér. Mag. Zool. 1832, t. 34; Lacord. Gen. Col. VII, 1866, p 407; v. Schœnf. Gen. Ins. (Wytsman) 1908, p. 5; Cat. Col.(Junk) 7, 1910, p. 4.

mellyi Guér., Mag. Zool., t. 34 ♂[1]; Boh. Act. Holm. 1837, t. 6, f. 1-4; Schœnh., Gen. Curc. V. 184, p. 577[2]; Westw., Cab. Or. Ent. 1848, t. 15, f. 4-5; Kleine, Jl. F. M. S.: Mus. XIII. 1925.	Lakhimpur, Assam, Coromandel[1] Katha, Burma. Andamans Nicobars Ceylon, ----------- F. M. S. Malacca Sumatra, Borneo Philippines[2]

2. Cyphagogus Parry (1849).

Type : *whitei* Westw. (Philippines).

Parry, Trans. Ent. Soc. Lond. V, 1849, p. 182; Lacord., Gen. Col. VII, 1866, p. 410; v. Schœnf., Gen. Ins. (Wytsman) 1908, p. 5; Cat. Col. (Junk) 7, 1910, p. 4; Kln., Arch. Nat. 87, 1921, A. 6, p. 268; Beeson, Ind. For. Rec. XI, 1925, pp. 173-177 (Biol.).

Taphroderes Westw., Cab. Or. Ent. 1848, t. 15.

Calodromus J. Thoms., Arch. Ent. I, 1857, p. 119.

angusticeps Senna, Ann. S. Ent. Ital. 1893, p. 296, t. 2, f. 2[1]; Kln., Arch. Nat. 87, 1921, A. 6, p. 315, f. 8[2].	Nilgiris Java[2], Sumatra[1] Borneo
buccatus Klu., Ent. Mitt. 1916, 1-4, p. 9[1]; Arch. Nat. 87, 1921, A. 6, p. 311[2]; Jl. F. M. S. Mus. XIII, 1925; Beeson, Ind. For. Rec. XI, 1925, p. 175 (biol.).	Siwaliks Goalpara, Assam N. Toungoo, W. Salween, Burma Ceylon[2], Andamans

	Pahang, F. M. S. Perak, Singapore Sumatra[1], Java[2] Borneo[2]
concavus Kln., Ind. For Rec. XI, 1925, p. 124[1].	Penang[1]
confertulus Kln., Ind. For. Rec. XI, 1925, p. 125[1]; Beeson *t. c.* 1925, p. 175 (biol.).	Lakhimpur, Assam[1] Burma[1], Andamans
confidens Kln., 1924, Ind. For Rec. XI, 1925, p. 126[1].	Siam[1]
corporaali Kln., Tijidschr. Ent. X, 1917, p. 177[1]; Arch. Nat. 87, 1921, A. 6, p. 299, f. 12[2]; Beeson, Ind. For Rec. XI, 1925, p. 176, Pl. III, f. 1 (biol.); Gardner, *t. c.* 1925, p. 194 (descr. pupa.).	Siwaliks Pathri, Saharanpur, U.P., North Gujrat Nilgiris Nilambur, Madras Chikkaballapura Burma Dikoya, Ceylon
	Java,[1, 2], Borneo[2]
densepunctatus Kln., Ent. Mitt. V, 1916, 1-4, p. 4, t. II, f. 3[1]; Arch. Nat. 87, 1921, A. 6, p. 312, f. 10[2].	Singapore Borneo[1, 2]
eichhorni Kirsch., Mitt. Zoll. Mus. Dresd. I, 1875, p. 45[1]; Kln., Arch. Nat. 87, 1921, A. 6, p. 319, f. 15[2]; Beeson, Ind. For. Rec. XI, 1925, p. 176 (biol.); Kln., Jl. F. M. S. Mus., XIII, 1925.	Patkai Mts., Assam Singhbhum, B. & O. Burma[2], Andamans
	F. M. S., Siam, Mentawei, Malacca[1, 2] Sumatra[2], Borneo[2] Moluccas
fragosus Kln., Ind. For Rec. XI, 1925, p. 126.	Garo Hills, Assam, 3,500 - 3,900 ft.
gladiator Kln., Arch. Nat. 87, 1921, A. 6, p. 307, f. 3, 13[1]; *l. c.* A. 10, p. 22, f. 1[2]; Beeson, Ind. For rec. XI, 1925, p. 176 (biol.).	Sibsagar, Assam Perak Mentawei[1] Sumatra[1], Borneo[1, 2] Philippines[1]
	Sibsagar, Assam
incisus Senna, Bull. S. Ent. Ital. XXXIV, 1902, p. 153[1]; Kln., Arch. Nat. 87, A. 6, p. 318, f. 15, 23[2].	Nilgiris,
	Sumatra[1, 2] Java

longulus Senna, Leyd. Mus. XX, 1898, p. 52[1]; Kln., Arch. Nat. 87, 1921, A. 6, p. 308, f. 9, 19[2]; Kln., Jl. F. M. S. Mus., XIII, 1925. — F. M. S., Siam Java[1, 2], Ceram Batjan

modiglianii Senna, Ann. Mus. Civ. Stor. Nat. Gen. (2) XIII (XXXIII) 1893, p. 258; Kln., Jl. F. M. S. Mus. XIII, 1925. — Pahang, F. M. S.

obconiceps Senna, Bull. S. Ent. Ital. XXXIV, 1902, p. 154[1]; Kln., Arch. Nat. 87, 1921, A. 6, p. 315[2]; Beeson, Ind. For. Rec. XI, 1925, p. 177 (biol.). — Lakhimpur, Assam.

Malacca Sumatra[1, 2], Borneo

planifrons Kirsch., Mitt. Zool. Mus. Dresd. I, 1875, p. 46[1]; Kln., Arch. Nat. 87, 1921, A. 6, p. 305, f. 2, 14, 28, 29[2]; Beeson, Ind. For. Rec. XI, 1925, p. 177 (biol.). — Sibsagar, Assam Nilgiris

---

Siam, Malacca[1, 2] Sumatra[2], Borneo

ruflrostris Kln., Stett. E. Z. 1914, p. 162[1]; Arch. Nat. 87, 1921, A. 6, p. 304[2]. — Philippines[2] Ceylon[1, 2]

---

silvanus Senna, Bull. S. Ent. Ital. XXXIV, 1902, p. 154[1]; Kln., Arch. Nat. 87, 1921, A. 6, p. 310, f. 19[2]. — Shevaroys, Madras Coorg, East India

---

Perak, Mentawei[1, 2] Sumatra[1, 2]

simulator Senna, Bull. S. Ent. Ital. XXXIV, 1902, p. 155[1]; Kln., Arch. Nat. 87 1921, A. 6, p. 317, f. 7[2]; Beeson, Ind. For Rec. XI, 1925, p. 177 (biol.). — Lakhimpur, Assam

---

Penang, Sumatra[1, 2] Java

tabacicola Senna Bull. S. Ent. Ital. XXV, 1893, p. 294, t. II, f. 1, 1b[1]; Kln., Arch. Nat. 87, 1921, A. 6, p. 314, f. 6[2]; Beeson, Ind. For. Rec. XI, 1925, p. 177 (biol.). — Sibsagar, Assam Sunderbans, Bengal, Nilgiris N. Shan States Burma, Andamans[2]

Penang, Sumatra[1, 2], Java, Borneo[2]

westwoodi Parry, Tr. Ent. S. Lond. V, 1849, p. 182, t. 18, f. 9; Senna, Ann. Civ. Nat. Gen. XII (XXXII) 1892, p. 443[2]; Kln., Arch. Nat. 87, 1921, A. 6, p. 310, f. 5, 16[3]; Beeson, Ind. For. Rec. XI, 1925, p. 174 (biol.), Pl. III, f. 2. — Sibsagar, Assam Sunderbans, Bengal Kanara, Bombay Cochin State Nilgiris Katha, Burma[2, 3] Ceylon[1, 2]

Cyrtotrachelus J. Thoms, Arch. Ent. I, 1857, p. 119[1]; Pasc., Journ. Ent. I, 1862, p. 394; Lacord., Gen. Col. VII, 1866, p. 410.	Indo-China Perak[2, 3], Sumatra[3] Java[3], Borneo[2]

3. Parusambius Kleine (1924).

Type : *fraudulentus* Kleine.

Kleine, Ind. For. Rec. XI, 1925, p. 129.

fraudulentus Kln., *t. c.* .... 1925, p. 129.	Garo Hills, Assam, 3,500-3,900 ft.

4. Callipareius Senna (1892).

Type : *feœ* Senna.

Senna., Ann. Civ. Nat. Gen. XII (XXXII), 1892, p. 444; v. Schœnf., Gen. Ins. (Wytsman) 1908, p. 12; Cat. Col. (Junk), 7, 1910, p. 7.

feæ Senna, Ann. Civ. Nat. Gen. XII (XXXII) 1892, p. 445.[1]	Gopaldhara, Darjeeling, Coorg, East India Burma[1]
foveatus Senna, Ann. S. E., Belg. XXXVIII, 1894, p. 384[1].	Barway, Bengal[1] Sikkim, Burma[1]

5. Opisthenoxys Kleine (1921).

Type : *ochraceus* Kln.

Kln., Arch. Nat. 87, 1921, A. 10, p. 26.

arrogans Kln., Arch. Nat. 89, 1923, p. 133[1].	Nilgiris[1]
famulus Kln., Ind. For. Rec. XI, 1925, p. 128.	Pashok, Darjeeling (2,000 ft.)
ochraceus Kln., Arch. Nat. 87, 1921, A. 10, p. 28 (23)[1]; Kln., Jl. F. M. S. Mus. XIII, 1925.	Pahang, F. M. S. Penang, Perak Sumatra, Java Borneo[1], Philippines[1]

6. Sparganophasma Kleine (1924).

Type : *cara* Kln.

Kln., Capita Zoologica, Deel II, Afl. IV, 1926.

cara Kln., Capita Zoologica, Deel II, Afl. IV, 1926, p. 12[1].	Burma[1]
	Indochina[1], Sumatra[1] Borneo[1]

7. Eusebus Kleine (1922).

Type : *adelphus* Kolbe.

Kln., Arch. Nat. 88, 1922, A. 7, p. 173.

Adelphus Kolbe, Stett. E. Z. LIII, 1892, p. 169[1]; Kln., Arch. Nat. 88, 1922, A. 7, p. 173; Beeson, Ind. For. Rec. XI, 1925, p. 177 (biol.).	Pathri, Saharanpur, U. P., Ceylon[1]

8. Mesoderes Senna (1898).

Type : *sexnotatus* Senna.

Senna, Leyd. Mus. XX, 1898, p. 65; v. Schœnf., Gen. Ins. (Wytsman) 1908, p. 8; Cat. Col. (Junk) 7, 1910, p. 6.

Dysmorphorhynchus Kln., Ent. Mit. V, 1916, 1-4, p. 51.

aberrans Kln., Arch. Nat. 88, 1922, A. 3, P. 201[1].	Singapore[1]
argutus Kln., Arch. Nat. 89, 1923, A. 8, . 132[1].	Nilgiris[1]
aridus Kln., Arch. Nat. 89, 1923, A. 8, p. 132.[1].	Nilgiris[1]
maculatus Senna, Leyd. Mus. XX, 1898, p. 67[1]; Kln. Jl. F. M. S. Mus. XIII, 1925.	Pahang, F. M. S. Perak, Malacca[1]
amabilis Kln., Ent. Mitt. V, 1916, p. 52, t. II, f. 11, 14[2].	Simalbur Island
sexnotatus Senna, Leyd. Mus. XX, 1898, p. 66[1].	Java, Borneo N. Queensland[2] Singapore, Sumatra New-Guinea[1]

9. Paraclidorrhinus Senna (1902).

Type : *modiglianii* Senna.

Senna, Bull. S. Ent. Ital. XXXIV, 1902, p. 159; v. Schœf., gen. Ins. (Wytsman) 1908, p. 9; Cat. Col. (Junk) 7, 1910, p. 5.

modiglianii Senna, Bull. S. Ent. Ital. XXXIV, 1902, p. 160[1].	Penang, Sumatra[1]- Borneo

10. Dictyotopterus Kleine (1916).

Type : *splendens* Kln.

Kln., Ent. Mitt. V, 1916, 1-4, p. 75.

splendens, Kln., Ent. Mitt. V, 1916, 1-4, p. 77[1].	Andamans

11. Eterozemus Senna (1902). Formosa[1]

Type : *pubens* Senna.

Senna, Bull. S. Ent. Ital. XXXIV, 1902, p. 160.

lætus Senna, Ann. Civ. Nat. Gen. (2) XII, (XXXII), 1892, p. 441[1]; Ann. S. E. Belg. XXXVIII, 1894, p. 361[2]. — Burma[1, 2], ———— Java, Philippines

pubens Senna, Ann. Civ. Nat. Gen. (2) XII (XXXII), 1892, p. 439[1]; Ann. S. E. Belg. XXXVIII. 1894, p. 361[2]. — Burma[1, 2] ———— Perak, Formosa Philippines

12. Pseudocyphagogus Desbrochers dés Loges (1890).

Type : *squamifer* Desbr.

Desbr., Journ. As. Soc. Beng. LIX, 2, 1890, p. 221; v. Schœnf., Gen. Ins. (Wytsman) 1908, p. 11; Cat. Col. (Junk) 7, 1910, p. 6; Beeson, Ind. For. Rec. XI, 1925, p. 177 (biol.).

squamifer Desbr., Journ As. Soc. Beng. LIX, 2, 1890, p. 222[1]. — Lakhimpur, Assam Andamans[1] ———— Penang, Siam, Borneo

13. Allæometrus Senna (1902).

Type : *breviceps* Senna.

Senna, Bull. S. Ent. Ital. XXXIV, 1902, p. 157; v. Schœnf., Gen. Ins. (Wytsman). 1908, p. 9; Cat. Col. (Junk) 7, 1910, p. 6.

breviceps Senna, Bull. S. Ent. Ital. XXXIV, 1902, p. 158; Beeson, Ind. For. Rec. XI, 1925, p. 178 (biol.). — Garo Hills Kanara Bombay Buxa Bengal Jalpaiguri, Bengal ———— Indochina Sumatra, Java Borneo, Formosa Queensland

deformis Kln., Ind. For. Rec. XI, 1925, p. 127; Beeson, *t. c.*, 1925, p. 178. — Kachugaon, Assam ————

Zemioses sp., Stebbing, Ind. For. Ins. Col. 1914, p. 391 (biol.).

14. Autometrus Kleine (1922).

Type : *punclulatus* Kln.

Kln., Dentsch. E. Z. 1922, I. p. 130.

punctulatus Kln., Deutsch. E. Z. 1922, I, p. 131[1]. — Singapore[1]

STEREODERMINAE

Stereodermini Kln., Ent. Bl. 18, 1922, p. 161.

15. Jonthocerus Lacordaire (1866).

Type : *crematus* Lacord.

Lacord., Gen. Col. VII. 1866, p. 415; v. Schœnf., Gen. (Wytsman), 1908, p. 16; Cat. Col. (Junk) 7, 1910, p. 9, Kln., Arch. Nat. 85, 1919, A. 8, p. 30.

carinensis Senna, Ann. S. E. Belg. XXXVIII, 1894, p. 362[1]. Kln., Arch. Nat. 85, 1919, A. 8, p. 42. — Gopaldhara, Darjeeling Rungbong Valley, Darjeeling, 3,400-4,700 ft., Burma[1] ——— Indochina

canaliculatus (Cerobates) Motsch., var. *carinensis* Senna, Ann. Civ. Nat. Gen. (2) XII (XXXII), 1892, p. 450.

crematus Lacord., Gen. Col. VII, 1866, p. 416, nota. 1[1]; Kln., Arch. Nat. 85, 1919, A. 8 p. 47[2]. — Ceylon[1, 2]

modiglianii Senna, Ann. Civ. Nat. Gen. (2) XIX (XXXIX) 1898, p. 228[1]; K.M. Hell, Deutsch. E. Z. 1916, p. 297[2]; Kln., Arch. Nat. 85, 1919, A. 8, p. 43. — Andamans ——— Mentawei[1], Sumatra Philippines[2]

16. Cerobates Schœnherr (1840).

Type : *tristriatus* F.

Schœnh., Gen. Curc. V, 1840, p. 487; Imhof, Einf. Kol. 1856, p. 167; Lacord; Gen. Col. VII, 1866, p. 418; Senna, Leyd. Mus. XVII, 1895, p. 209; v. Schœnf., Gen. Ins. (Wytsman) 1908, p. 20; Cat. Col. (Junk) 7, 1910, p. 11.

adustus Senna, Leyd. Mus. XVI, 1894, p. 184[1]; XVII, 1895, p. 52[2]; *l. c.* p. 220[3]. — Gopaldhara, Darjeeling, 4,720 ft. Colombo Balangoda, Ceylon ——— Perak, Nias, Sumatra, Java[1, 2, 3], Borneo, Philippines

æmulus Kln., Arch. Nat. 88, 1922, A. 3, p. 204[1]. — Calcutta, ——— Pulo Penang[1]

æqualis Kln., Arch. Nat. 88, 1922, A. 3. p. 203[1]; Kln., Jl. F. M. S. Mus. 1925, p.	Calcutta[1], Nilgiris Ceylon, Andamans —————— Kuala Lumpur Perak, Indochina Sumatra[1]
andamanicus Senna, Leyd. Mus. XVII, 1895, p. 223[1].	Andamans[1] ——————
articulosus Kln., Arch. Nat. 1923, A. 1, p. 134[1].	Nilgiris[1] ——————
birmanicus Senna, Ann. S. E. Belg. XXXVIII, 1894, p. 364[1]; Leyd. Mus. XVII, 1895, p. 222[2].	Burma[1, 2] ——————
canaliculatus Motsch., Et. Ent. VII, 1858, p. 96; Senna, Ann. Civ. Nat. Gen. (2) XII (XXXII) 1892, p. 450[1]; Leyd. Mus XVII, 1895, p. 222[2]; Ann. Civ. Nat. Gen. (2) XIX (XXXIX) 1898, p. 227[3].	Formosa, Darjeeling East India[2] Burma[1], Andamans —————— Perak[2], Mentawei[3] Sumatra, Java Borneo, Formosa
collectivus Kln., Ind. For Rec. XI, 1925, p. 130[1].	Burma[1] ——————
concisus Kln., Ind. For. Rec. XI, 1925, p. 130[1].	Garo Hills Nilgiris, Ceylon[1] ——————
fossulatus Motsch., Et. Ent. VII, 1858, p. 96; Senna, Leyd. Mus. XVII, 1895, p. 221[1].	Sibsagar, Assam Nilgiris Kanara, Bombay Katha, Burma[1] —————— Perak
sexsulcatus Motsch., Et. Ent. VII, 1858, p. 95; Senna, Leyd. Mus. XIV, 1892, p. 162[1]; Ann. Civ. Nat. Gen. (2) XII (XXXII) 1892, p. 450[2]; *l. c.* (2) XIII (XXXIII) 1893, p. 264[3]; Leyd. Mus. XVII, 1895, p. 220; Ann. Civ. Nat. Gen. (2) XIX (XXXIX), 1898, p. 380[4]; Beeson, Ind. For Rec. XI, 1925, p. 179 (biol.); Gardner; *t. c.* 1925, p. 193 (descr. larva); Kln., Jl. F. M. S. Mus. XIII, 1925, p.	Dehra Dun Sibsagar, Assam East India[1], Kanara Bombay, Shwegu, Burma[1, 2] Tenasserim[1] Ceylon[1], Andamans ——————, F. M. S. Nias[4], Bua-Bua[2] Java[1], Borneo[1] Indochina Philippines Celebes, Sumbawa Moluccas[1] New-Guinea[3]

sumatranus Senna, Bull S. Ent. Ital. XXV, III, 1893, p. 306, t. III, f. I[1]; Leyd. Mus. XVII, 1895, p. 223[2]; Kln., Jl. F. M. S. Mus. XIII, 1925, p.	Sibsagar, Assam Pashok, Darjeeling, 1,500 ft., East India Andamans, Ceylon ———, Penang F. M. S., Siam Indochina, Formosa Sumatra[1,2], Borneo Timor[2], Celebes
tristriatus F., Syst. El. II, 1801, p. 554; N. Lund, Skrift, Nat. Hist. V, 2, 1802, p. 66; Schœnh., Gen. Curc. I. 1833, p. 332[1]; Senna, Leyd. Mus. XIV, 1892, p. 162[2]; Ann. Civ. Nat. Gen. (2) XII (XXXII) 1892, p. 448[3]; *l c.* (2) XIII (XXXIII) 1893, p. 264[4]; Leyd. Mus. XVII, 1895, p. 52[5]; *l. c.*, p. 218[6]; Ann. Civ. Nat. Gen. (2) XIX (XXXIX), 1898, p. 227[7]; *l. c.*, p. 380[8]; Beeson, Ind. For Rec. XI, 1925, p. 178, Pl. III, f. 3, 6 (biol.); Gardner, *t. c.* 1925, p. 189, Pl. IV, f. 1-8 (descr. larva); Kln., Jl. F. M. S. Mus. XIII, 1925, p.	Assam Pathri, Saharanpur, U.P. Singhbhum, Puri, Bihar & Orissa East India[3] Kanara, Bombay Nilgiris, Burma[3] Andamans, Ceylon[6] —————— Perak Penang, Indochina Nias[8], Mentawei[3] Engano[4], sumatra[1,3] Java[1,2,3,5], Borneo Philippines, Celebes Moluccas New Guinea[6]

TRACHELIZINÆ

Trachelizini, Lacord., Gen. Col. VII, 1866, p. 417; v. Schœnf., Gen. Ins. (Wytsman) 1908, p. 18; Cat. Col. (Junk) 7, 1910, p. 10; Kln., Ent. Bl. 18, 1992, p. 161.

17. Anocamara Kleine (1920).

Type : *proportionalis* Kln. (Sumatra).

Kln., Ent. Mitt. IX, 1920, p. 122.

catenata Kln., Ind. For Rec. XI, 1925, p. 132[1].	Penang[1]

18. Metatrachelizus Kleine (1922).

Type : *objectus* Kln.

Kln., Arch. Nat. 88, 1922, A. 3, p. 207.

abjectus Kln., Arch. Nat. 88, 1922, A. 3, p. 208[1].	Penang, Sumatra Borneo[1]

congruens Kln., Ind. For. Rec. XI, 1925, p. 132[1].	Sadyia, Assam[1] Garo Hills, 3,500-3,900 ft. ————

19. Trachelizus Schœnherr (1840).

Type : *bisulcatus* F.

Schœnh., Gen. Cure. V, 1840, p. 489; Lacord., Gen. col. VII, 1866, p. 419; Imhof, Einf. Kol. 1856, p. 167; v. Schœnf., Gen. Ins. (Wytsman) 1908, p. 21; Cat. Col. (Junk) 7, 1910, p. 12.

Brenthus F., Mant. Ins. I, 1787, p. 95.

Arrhenodes Schœnh., (pars) Cure. Disp. Meth. 1826, p. 70.

bisulcatus F., Syst. El. II, 1801, p. 548; N. Lund, Skrift Nat. Hist. Selsk. 1802, V, 2, p. 67; Gyll., in Schœnh., Gen. curc. I,1833, P. 331; Schænh; l. c., V, 1840, p. 491; Beeson, Ind. For. Rec. XI, 1925, p. 179 (biol.); Kln., Jl. F. M. S. Mus. XIII, 1925, p.	Sibsagar, Assam Singhbhum, Bihar and Orissa Pathri, Saharanpur, U. P., Sivaliks, U.P. Raipur, C.P.
impressus Haan., Dej. Cat. ed. 3, 1837, p. 265. *bicanaliculatus* (*Miolispa*) L. W. Schauf., Hor. Soc. Ent. Ross. XIX, 1885, p. 206; Senna, Bull. S. Ent. Ital. XXXI, 1899, p. 301.	Akeswar, S. Garhwal Dhauli Ganga, W. Almora, 4,000 ft. Khandesh, Bombay, N.Malabar, Madras Andamans
exophthalmus Lea, Proc. Linn. Ent. Soc. N. S. Wal. XXIII, 1895 p. 636 (sub *Ceocephalus*).	Ceylon ———— Australia (Common)
semivelatus (*Miolispa*) L. W. Schauf., *l. c.*, p. 206; Senna, *l. c.*, p. 301.	
dividuus Kln., Ind. For. Rec. XI, 1925, p. 131[1].	Gopaldhara, Darjeeling, 4,720-6,100 ft.[1]
insularis Senna, Leyd. Mus. XIV, 1892, p. 163[1].	Penang, Sumatra Java[1]
lævigatus Senna, Leyd. Mus. XX, 1898, p. 58[1]; Kln., Jl. F. M. S. Mus. XIII, 1925, p.	Burma ———— Pahang, Perak Sumatra, Java[1] Borneo
politus Senna, Ann. Civ. Nat. Gen. (2) XII (XXXII), 1892, p. 462[1].	Nilgiris, Burma[1] ———— Cambodia

rufomaculatus Senna, Ann. Civ. Nat. Gen. (2) XII (XXXII), 1892, p. 460[1]. — Burma[1]

---

20. Hemisamblus Kleine (1924).

Type : *contemptus* Kln.

Kln., Ind. For Rec. XI, 1925, p. 132.

contemptus Kln., Ind. For. Rec. XI, 1925, p. 133[1]. — Laos[1]

21. Miolispa Pascoe (1862).

Type : *suturalis* Pasc.

Pasc., Jour. Ent. I, 1862, p. 393; Lacord., Gen. Col. VII, 1866, p. 421; v. Schœnf., (Wytsman) 1908, p. 25; Cat. Col. (Junk) 7, 1910, p. 16; Kln., Stett. E. Z. 80, 1919, p. 176.

Cacotrachelus Sharp. Zool. Res. IV, 1899, p. 386.

adversaria Kln., Arch. Nat. 88, 1922, A. 3, p. 210[1]. — Calcutta[1], Andamans

---

discors Senna, Ann. S. E. Belg. XXXIX 1895, p. 358[1]; Kln., Stett. E. Z. 80, 1919, p. 241[2]. — Penang[1], Borneo Formosa Philippines[2] Celebes

granulata Kln., Stett. E. Z. 80, 1919, p. 303[1]. — Siam[1], Malacca[1] Sumatra[1], Java Borneo

honorata Kln., Jl. F. M. S. Mus. XIII, 1925, p. , f. 2. — Perak

impunctata Kln., Stetf. E. Z. 80, 1919, p. 251[1]. — Tengah-Geb, Malacca[1]

jordani Senna, Ann. S. E. Belg. XXXIX, 1895, p. 360[1]. — Penang[1]

lineata Senna, Leyd. Mus. XX, 1898, p. 57[1]; Kln., Stett. E. Z. 80, 1919, p. 233[2]. — Perak, Java[1, 2]

mariæ Senna, Leyd. Mus. XIII, 1891, p. 165[1]. — Penang[1] Siam, Sumatra

occulta Kln., Jl. F. M. S. Mss. XIII, 1925, p. , f. 1. — Pahang, F. M. S.

poweri Senna, Ann. S. E. Belg. XXXIX, 1895, p. 364[1]; Kln., Stett. E. Z. 80, 1919, p. 293[2]. — Malacca[1], Johore Sumatra[2]

pygmæa Senna, Ann. S. E. Belg. XXXIX, 1895, p. 360[1]; Kln., Stett. E. Z., 80, 1919, p. 222[2]. — Penang[1, 2], Ceram[1, 2] Amboina, Gilolo Ternate Gazelle, Penins[2]

siporana Senna, Ann. Civ. Nat. Gen. (2) XIX (XXXIX), 1898, p. 233[1]; Kln., Stett. E. Z. 80, 1919, p. 238[2]. — Penang, Sumatra[2] Mentawei[1], Java Borneo, Philippines Amboina[2]

suturalis Pasc., Journ. Ent. I, 1862, p. 393; Senna, Leyd. Mus. XIV, 1892, p. 166[1]; Kln., Stett. E. Z. 80, 1919, p. 218[2]. — Andamans

———

Penang[1], Sumatra[2] Java[2], Halmaheira[1] Amboina[1], Batchian[1]

22. Tulotus Senna (1894).

Type : *moculipennis* Senna.

Senna, Ann. S. E. Belg. XXXVIII, 1894, p. 370; v. Schœnf., Gen. Ins. (Wytsman) 1908, p. 23; Cat. Col. (Junk) 7, 1919, p. 15.

maculipennis Senna, Ann. S. E. Belg. XXXVIII, 1894, p. 371[1]; Kln., Ind. For. Rec. XI, 1925, p. 139[2]. — Manipur[2] Upper Burma[1] Ruby Mines, Burma[2]

C———

23. Miolispoides Senna (1894).

Type : *birmanicus* Senna.

Senna, Ann. S. E. Belg. XXXVIII, 1894, p. 368; v. Schœnf., Gen. Ins. (Wytsman) 1908, p. 26; Cat. Col. (Junk) 7, 1910, p. 17.

birmanicus Senna, Ann. S. E. Belg. XXXVIII, 1894, p. 369[1]. — Upper-Burma[1]

———

sculpturatus Calabr., Faun. Ent. Indo-Chin, (2) 1921, p. 5[1]. — Indochina[1]

24. Hypomiolispa Kleine (1918).

Type : *ceylonica* Desbr.

Kln., Ent. Bl. 14, 1918, p. 163. (*Miolispa* auct. pars.).

ceylonica Desbr., Journ. As. Soc. Beng. II, Nat. Sc. III, 1890, p. 222[1]; Kln., Ent. Bl. 14, 1918, p. 315.[2] — East India[2], Ceylon[1]

———

clavata Kln., Ent. Bl. 14, 1918, p. 327[1]. — Malacca, Sumatra Borneo

colorata Kln., Ind. For. Rec. XI, 1925, p. 136[1]. — Dikoya, Ceylon[1]

———

compressa Kln., Ent. Bl. 14, 1918, p. 341.[1] — Nilgiris

———

Perak, Sumatra Java, Borneo[1] Moluccas

conjugalis Kln., Ind. For. Rec. XI, 1925, p. 134.[1] — Andamans[1], ______

conjuncta Kln., Ind. For Rec. XI, 1925, p. 136[1]. — Tenasserim[1], ______

cruda Kln., Ind. For. Rec. XI, 1925, p. 135[1]. — Bogalantalawa[1]
Patkai Mts., Assam[1]
____
Selangor

demissa Kln., 1924, p. [1]. — Bukit Kutu,
Malacca[1] 3,457 ft.

dentigena Kln., Ent. Bl. 14, 1918, p. 331[1]. — Katha, Burma
____
Siam, Sumatra[1]

fausti Senna, Leyd. Mus. XVI, 1894, p. 188[1]; Ann. Civ. Nat. gen. (2), XIX (XXXIX), 1898, p. 233[2]; Kln., Ent. Bl. 14, 1918, p. 313[3]; Jl. F. M. S. Mus. XIII, 1925, p. — Pahang. Perak
Siam, Sumatra[1]
Mentawei[2], Borneo[3]

nitida Kln., Ent. Bl. 14, 1918, p. 311[1]. — Upper Assam[1]
____
Sumatra

nupta Senna, Leyd. Mus. XIV, 1892, p. 171[1]; Ann. Civ. Nat. Gen. (2), XIX (XXXIX), 1898, p. 232[2]; Kln., Ent. Bi. 14, 1918, p. 320[3]. — Upper Assam[2]
____
Sumatra
Siam[3], Mentawei[2]
Sumatra[3], Java[1, 3]
Borneo[3], Philippines[3]

trachelizoides Senna, Leyd. Mus. XVI, 1894, p. 193[1]; Kln., Ent. Bl. 14, 1918, p. 344[2]; Jl. F. M. S. XIII, 1925, p. [2]. — Pahang[2], Penang
Perak, Siam[2]
Sumatra[1], Java[2]
Borneo[2], Philippines[2]
Celebes

25. Higonius Lewis (1883).

Type : *cilo* Lewis.

kewis, Journ. Linn. Soc. Lond. XVII, 1883, p. 299; v. Schœnf., Gen. Ins. (Wytsman) 1908, p. 26; Cat. Col. (Junk) 7, 1910, p. 17.

cilo Lewis, Journ. Linn. Soc. Lond. XVII, 1883, p. 300, t. 12, f. 9, 10[1]. — Sibsagar, Assam
Lakhimpur, Assam
Gopaldhara,
Darjeeling
Burma
____
Formosa
Philippines, Japan[1]

crux Olliff, Journ. Linn. Soc. Lond. XVII, 1883, p. 300 note[1]; Senna, Bull, S. Ent. Ital. XXXIV, 1902, p. 167[2]; Beeson, Ind. For Rec. XI, 1925, p. 179 (biol.); Kln., Jl. F. M. S. Mus. XIII, 1925, p.	Buxa, N. Toungoo, Burma, Andamans[1] —————— Tonkin Penang, Pahang Sumatra[2], Borneo
hirsutus Senna, Bull. S. Ent. Ital. XXV, 1893, p. 117, t. I, f. I[1].	Perak[1]
poweri Lewis, Journ. Linn. Soc. Lond. XVII, 1883, p. 299[1]; Senna, Ann. Civ. Nat. Gen. (2) XII (XXXII), 1892, p. 447[2]; *l. c.* XIII (XXXIII), 1893, p. 260[3]; *l. c.* XIX (XXXIX), 1898, p. 380[4]; Leyd. Mus. XX, 1898, p. 56[5]; Bull. S. Ent. Ital. XXXIV, 1902, p. 167[6].	Burma[2], —————— Penang[1, 2] Sumatra[2, 6] Engano[3], Nias[4] Java[5], Borneo[2]

26. Araiorrhinus Senna (1893).

Type : *longirostris* Senna.

Senna, Bull. S. Ent. Ital. XXV, 1893, p. 325; v. Schœnf., Gen. Ins. (Wytsman) 1908, p. 24; Cat. Col. (Junk) 7, 1910, p. 15.

beesoni, Kln., Ind. For. Rec. XI, 1925, p. 138[1]; Beeson, *t. c.* 1925, p. 180 (biol.).	Katha, Burma[1] ——————
conquisitus Kln., Ind. For. Rec. XI, 1925, p. 138[1].	Garo Hills, Assam, 3,500-3,900 ft. —————— Perak[1], Malacca
exportatus Senna, Bull. S. Ent. Ital. XXV, 1893, p. 327[1].	Sumatra[1]
longirostris Senna, Bull. S. Ent. Ital. XXV, 1893, p. 326[1].	Patkai Mts., Assam —————— Perak, Sumatra[1]

27. Microtrachelizus Senna (1893).

Type : *lyratus* Perr. (New Caledonia).

Senna, Bull. S. Ent. Ital. XXV, 1893, p. 315; v. Schœnf., Gen. Ins. (Wytsman) 1908, p. 22; Cat. Col. (Junk) 7, 1910, p. 15.

accomodatus Kln., Arch. Nat. 88, 1922, A. 3, p. 208[1]; Beeson, Ind. For. rec. XI, 1925, p. 180 (biol.).	Margherita, Assam N. Malabar, Madras —————— Perak, Borneo[1]
apertus Kln., Ind. For. Rec. XI, 1925, p. 140[1]; Beeson, *t. c.,* 1925, p. 180 (biol.).	Lakhimpur, Assam[1] Sibsagar, Assam[1] Gopaldhara, Assam[1] ——————

benefleus Kln., Ind. For. Rec. XI, 1925, p. 139[1]; Beeson, *t. c.* 1925, p. 180-182, Pl. III, f. 4, 5 (biol.). — Margherita, Lakhipur, Assam

bhamoensis Senna, Ann. Civ. Nat. Gen. (2) XII (XXXII), 1892, p. 456[1]; *l. c.,* XIV (XXXIV), 1894, p. 557[2]. — Burma[1], New-Guinea[2]

brevitibia Senna, Ann. Civ. Nat. Gen. (2) XII (XXXII), 1892, p. 454[1]. — Burma[1]

cylindricornis Pow., Leyd. Mus. II, 1880, p. 187; Senna, Ann. Civ. Nat. Gen. (2) XII (XXXII), 1892, p. 456[1]; Bull. S. Ent. Ital. XXXIV, 1902, p. 170[2]. — Burma[1] / Sumatra[1, 2]

ghekuanus Senna, Ann. Civ. Nat. Gen. (2) XII (XXXII) 1892, p. 457[1]. — Burma[1]

pubescens Senna, Bull. S. Ent. Ital. XXV, 1893, p. 320, t. 3, f. 6[1]. — Perak Sumatra[1]

siamensis Kln., Jl. F. M. S. Mus. XIII, 1925, p. , f. 3. — Siam

tabaci Senna, Bull. S. Ent. Ital. XXV, 1893, p. 323, t. 4, f. 4[1]. — Burma / Sumatra[1], Borneo

28. Aneorhachis Kleine (1924).

Type : *astrictus* Kln.

Kln., Arch. Nat. 1923, 89, A. 8, p. 135,

astrictus Kln., Arch. Nat. 1923, 89, A. 8, p. 136[1]. — Nilgiris[1]

29. Hoplopisthius Senna (1892).

Type : *trichimerus* Senna.

Senna, Ann. Civ. Nat. Gen. (2) XII (XXXII), 1892, p. 451; Kolbe, Stett. E. Z. LIII, 1892, p. 173; v. Schœnt., Gen. Ins. (Wytsman) 1908, p. 23; Cat. Col. (Junk) 7, 1910, p. 15; Calabr., Bull. S. Ent. Ital. L. 1918, p. 66.

trichimerus Senna, Ann. Civ. Nat. Gen. (2) XII (XXXII) 1892, p. 452[1]; *l. c.* XIX (XXXIX) 1898, 224[2]; *l. c.,* p. 379[3]; Bull. S. Ent. Ital. XXXIV, 1902, p. 163[4]; Calabr., Bull. S. Ent. Ital. L, 1918, p. 67[5]; Beeson, Ind, for. Rec. XI 1925, p. 182 (biol.); Kln., Jl. F. M. S. Mus. XIII, 1925, p. . — Lakhimpur, Assam Burma[1, 5] / Perak, Selangor Sumatra[4, 5] Mentawei[2, 3] Nias[3, 5], Java[5] Borneo[5] Philippines Formosa

30. Carcinopisthius Kolbe (1892).

Type : *fruhstorferi* Kolbe.

Kolbe, Stett. E. Z. LIII, 1892, p. 174; v. Schœnf., Gen. Ins. (Wytsman) 1908, p. 27; Cat. Col. (Junk) 7, 1910, p. 17; Calabr., Bull. S. Ent. Ital. L. 1918, p. 68.

Pseudotaphroderes Bolkay, Ann. Hist. Nat. Mus. Hung. IX, I, 1911, p. 263.

fruhstorferi Kolbe, Stett. E. Z. LIII, 1892, p. 174[1].	Perak, Java[1] Moluccas
interrupticosta Senna, Ann. Civ. Nat. Gen. (2) XIX (XXXIX) 1898, p. 224.[1]	Burma —————— Perak, Mentawei[1] Java, Borneo
maculatus Senna, Bull. S. Ent. Ital. XXXIV, III, 1902, p. 163[1]; Beeson, Ind. For. Rec. XI, 1925, p. 182 (biol.).	Nakachari, Sibsagar Assam, Garo Hills Assam, 3,500-3,900 ft. Burma —————— Sumatra[1]
oberthüri Senna, Bull. S. Ent. Ital. XXIV, 1892, p. 255[1]; Ann. S. E. Belg. XXXVIII, 1894, p. 359[2]; Beeson, Ind. For. Rec. XI, 1925, p. 182 (biol.).	Nakachari, Sibsagar, Assam Burma[1, 2, 3] —————— Formosa[4]
felschei Kolbe, Stett, E. Z. LIII, 1892, p. 175[3];	
Pseudoth. formosanus Bolkay, Ann. Hist. Nat. Hung. IX, I, 1911, p. 265, fig. 4[4].	

AMORPHOCEPHALINAE.

Amorphocephalini Kln., Ent. Bl. 18, 1921, p. 162.

31. Symmorphocerus Schœnherr (1847).

Type : *frontalis* Ol. (South Africa).

Schœnh., Mant. Sec. Fam. Curc, 1847, p. 8; Lacord., Gen. Col. VII, 1866, p. 424; Pow., Ann. S. E. Fr. (5) VIII, 1878, p. 418; v. Schœnf., Gen. Ins. (Wytsman) 1908, p. 30; Cat. Col. (Junk) 7, 1910, p. 1910, p. 19.

cardoni Senna, Ann. S. E. Belg. XXXVIII, 1894, p. 385[1].	Fyzabad, Karachi Khandesh, Bombay Barway, Bengal[1] Bengal ——————

nagporensis Kln., Arch. Nat. 88, 1922, A. 3, p. 211[1].

Nagpur[1]
Coimbatore, Madras

32. Cordus Schœnherr (1847).

Type : *hospes* Germ. (Australia).

Schœnh., Mant, Sec. Fam. Curc. 1847, p. 10; Lacord., Gen. Col. VII, 1866, p. 422; Pow., Ann. S. E. Fr. (5) VIII, 1878, p. 479; v. Schœnf., Gen. Ins. (Wytsman) 1908, p. 27; Cat. Col. (Junk) 7, 1910, p. 18.

peguanus Senna, Ann. Civ. Nat. Gen. (2) XII (XXXII), 1892, p. 463[1]; Kln., Jl. F. M. S. Mus. XIII, 1925, p.

Palon, Insein, Burma[1], Tenasserim
F. M. S.
Sumatra

33. Paussobrenthus Gestro (1919).

Type : *backeri* Gestro.

Gestro, Ann. Civ. Nat. Gen. (3) VIII, (XLIII), 1919, p. 271.

bakeri Gestro, Ann. Civ. Nat. Gen. (3) VIII (XLVIII) 1919, p. 273.[1]

Singapore[1]

concitatus Kln., Ind. For. Rec. XI, 1925, p. 141[1].

Andamans[1]

conterminus Kln., Ind. For. Rec. XI, 1925, p. 141[1].

Cambodia[1]

34. Leptamorphocephalus Kleine (1918)

Type : *laborator* Kln.

Kln., Arch. Nat. 82, 1916 (1918), A. 12. p. 132.

cuneatus Kln., Ind. For Rec. XI, 1925, p. 142[1].

Patkai Mts., Assam.[1]

cupidus Kln., Ind. For Rec. XI, 1925, p. 143[1].

East India[1]

laborator Kln., Arch. Nat. 82, 1916 (1918), A. 12, p. 133[1].

Singapore, Formosa
Sumatra[1]

lævis Pow., Ann. S. E. Fr. (5) VIII, 1878, p. 486[1]; Kln., Arch. Nat. 82, 1916 (1918) A. 12. p. 136[2].

Darjeeling[2]
East India[1]
Nilgiris, Madras
Burma

sumatraus Senna, Leyd. Mus. XVI, 1894, p. 195[1]; Kln., Arch. Nat. 82, 1916 (1918), A, 12, p. 139[2].

Andamans
Malacca[2], Nias
Sumatra[1], Borneo

variolosus Pow., Ann. S. E. Fr. (5) VIII, 1878, p. 485[1]. — Malacca[1]

35. Paramorphocephalus Kleine (1920).

Type : *diabolus* Kln., (Sunda Isl).

Kln., Zool. Med. Leyd. 1920, V, p. 236.

binotatus Calabr., Bull. S. Ent. Ital. LIV, 1922, p. 109[1]. — Indochina[1]

36. Amorphocephalus Schœnherr (1840).

Type : *coronatus* Germ. (Mediterranean).

Schœnh., Gen Curc, V. 1840, p. 485; Lacord., Gen. Col. VII, 1866; p. 422, p. 422; Pow., Ann. S. E. Fr. (5) VIII, 1878, p. 480; v. Schœnf., Gen. Ins. (Wytsman) 1908, p. 29; Cat. Col. (Junk) 7, 1910, p. 18; Kln., Arch. Nat. 82, 1916 (1918), A 12, p. 52.

delicatulus Kln., Ind. For Rec. XI, 1925, p. 144[1]. — Karachi[1], ________

ARRHENODINAE

Arrhenodini, Lacord., Gen. Col. VII, 1866, p. 425; v. Schœnf., Gen. Ins. (Wytsman), 1908, p. 31; Cat. col. (Junk) 7, 1910; p. 20; Kln., Ent. Bl. 18, 1922, p. 162.

37. Agriorrhynchus Power (1878).

Type : *borrei* Pow.

Pow., Pet. Nouv. Ent. II, 1878, p. 241; v. Schœnf., Gen. Ins. (Wytsman) 1908, p. 34; Cat. Col. (Junk) 7, 1910, p. 21; Kln., Arch. Nat. 82, 1916 (1918), A. 11, p. 118.

borrei Pow., Pet Nouv. Ent. II, 1878, p. 241[1]; Kln., Arch. Nat. 82, 1916 (1918), A. II, p. 144[2]; Jl. F. M. S. Mus. XIII, 1925, p. . — Darjeeling, Burma ________ Siam, Pahang, Perak Penang, Singapore Sumatra[2], Java[1] Borneo

quadrituberculatus Senna, Ann. Civ. Nat. Gen. (2) XII (XXXII) 1892, p. 470[1]; Kln., Arch. Nat. 82, 1916 (1918), A. 11, p. 135[2]. — Darjeeling[2] Allahabad Tenasserim[1] ________ Indochina

undulatus Pow., Pet. Nouv. Ent. II, 1878, p. 241[1]; Kln., Arch. Nat. 82, 1916 (1918), A. 11, p. 140[2]; Jl. F. M. S. Mus. XIII, 1925, p. . — Assam, Burma ________, Pahang Penang, Malacca[1] Sumatra[2], Borneo[2], Java

38. Eupeithes Senna (1898).

Type : *dux* Senna.

Senna, Ann. Civ. Nat. Gen. (2) XIX (XXXIX) 1898, p. 381; v. Schœnf., Gen. Ins. (Wytsman) 1908, p. 37; Cat. Col. (Junk) 7, 1910, p. 24.

bardus Kln., Ent. Bl. 17, 1921, 7-9, p. 125[1].	Penang, Sumatra[1] Borneo
dux Senna, Ann. Civ. Nat. Gen. (2) XIX (XXXIX) 1898, p. 382[1].	Penang, Sumatra Nias[1], Borneo

39. Prophthalmus Lacordaire (1866).

Type : *tridentatus* F.

Lacord., Gen. Col. VII, 1866, p. 427; v. Schœnf., Gen. Ins. (Wytsman), 1908, p. 32; Cat. Col. (Junk) 7, 1910, p. 20; Kln., Stett. E. Z., 77, 1916, p. 217.

Brenthus F., Mant. Ins. I, 1787, p. 95.

Arrhenodes Schœnh., (pars.) Curc. Disp. Meth. 1826, p. 70.

apparatus Kln., Shornik. Entom. N. Mus. Praze, I, 1923, p. 48[1].	East India[1]
assimilis Kln., Stett. E. Z. 77, 1916, p. 248[1].	Indochina[1]
bourgeoisi Pow., Ann. S.E. Fr. (5) VIII, 1878, p. 38[1]; Kln., Stett. E. Z. 77, 1916, p. 259.[2]	East India[2] Ceylon[1,2]
brevis Pow., Ann. S. E. Fr. (5) VIII, 1878, p. 45[1].	Malacca[1], Sumatra
delesserti Pow., Ann. S. E. Fr. (5) VIII, 1878, p. 37[1]; Kln., Stett. E. Z. 77, 1916, p. 242.[2]	East India[1,2] Madura[2], Nilgiris Madras Trichinopoly[2]
heikertingeri Kln., Stett. E. Z. 77, 1916, p. 255[1].	Calcutta, Nilgiris Tenasserim Andamans[1]
longirostris Gyll., in Schœnh., Gen. Curc. I, 1833, p. 323[1]; Lacord., Gen. Col. VII, 1866, p. 428; Senna, Leyd. Mus. XIV, 1892, p. 176[2]; *l. c.* XVII, 1895, p. 53[3]; Bull. S. Ent. Ital. XXXI, 1899, p. 10; Kln., Stett. E. Z. 77, 1916, p. 245[4].	Malacca[1] Perak[4], Siam S. China, Hainan Sumatra[4] Java[1,2,3,4] Philippines Celebes[4]

obscurus Pow., Ann. S. E. Fr. (5) VIII, 1878, p. 45[1]. — East India[1]

potens Lacord., Gen. Col. VII, 1866, p. 428 note 2[1]; Senna, Ann. S. E. Belg. XXXVIII, 1894, p. 380[2]; Kln., Stett. E. Z. 77, 1916, p. 238[3]; Jl. F. M. S. Mus. XIII, 1925, p. — Nepal[3], Manipur Assam[1, 3] Cachar, Assam Sikkim[3], Madras[3] Dawna Hills, Burma, Tavoy Burma[2] — Kuala Lumpur Cambodia

sanguinalis Pasc., Ann. Mag. Nat. Hist. (4) X, 1872, p. 322, t. 15, f. 6.

pugnator Power, Bull. Soc. Ent. Fr. (5) VIII, 1878, p. 44; Kln., Jl. F. M.S. Mus. XIII, 1925, p. — Andamans — Selangor, Sunda Isl.

tridentatus F., Syst. El. II, 1801, p. 554; N. Lund., Skrift. Nat. Hist. V, 2, 1802, p. 91; Gyll., in Schœnh., Gen. Curc. I, 1833, p. 322[1]; Senna, Leyd. Mus. XIV, 1892, p. 176[2]; Bull. S. Ent. Ital. XXXIV, 1902, p. 174; Kln., Stett., F. Z. 77, 1916, p. 252[3]; Stebb., Ind. For Ins. 1914, p. 391, fig. 263 (Biol.)[4]; Kln., Jl. F. M. S. Mus. XIII, 1925, p. — Darrang, Assam[4] Andamans — Perak[2, 3] Sumatra[1, 3] Java[2, 3], Borneo[3] Celebes[2]

macrocephalus J. Sturm., Catal. 1826, p. 103.

versicolor Senna, Ann. Civ. Nat. Gen. (2) XII (XXXII), 1902, p. 466[1]. 1902, p. 466[1]. — Burma[1]

wichmanni Kln., Stett. E. Z. 77, 1916, p. 263[1]; Calabr., Faun. Ent. Indoch. 2, 1921[2]. — Gopaldhara, Assam Darjeeling — Indochina[1, 2] Formosa

40. Baryrrhynchus Lacordaire (1866).

Type : *latirostris* Gyll. (Sunda Isl.)

Lacard., Gen. Col. VII, 1866, p. 428; v. Schœnf., Gen. Ins. (Wytsman) 1908, p. 33; Cat. Col. (Junk) 7, 1910, p. 20; Kln., Ent. Bl. 12, 1916, p. 121.

Arrhenodes Schœnh., (pars.) Curc. Disp. Méth . 1826, p. 70.

Subg. *Fupsalomimus* Kln., Ent. Bl. 12, 1916, p. 155.

andamanicus Pow., Pet. Nouv. Ent. II, 1879, p. 298[1]; Kln., Ent. Bl. 12, 1916, p. 173[2]. — Cherrapunji, Assam, Andamans[1] Nicobars — Indochina[2]

dehiscens Gyll., in Schœnh., Gen. Curc. I, 1833, p. 324[1]; Senna, Leyd. Mus. XIV, 1892, p. 177[2]; Ann. Civ. Nat. Gen. (2) XIX (XXXIX), 1898, p. 382[3]; Bull. S. Ent. Ital. XXXIV, 1902, p. 175[4]; Kln., Ent. Bl. 12, 1916, p. 165[5]; Jl. F. M. S. Mus. XIII, 1925, p.	Sibsagar, Assam Margherita, Lakhipur, Assam. Burma ——— Perak[5], Sumatra[4, 5] Nias[3], Java[1, 2] Bonreo[2, 5], Celebes
latirostris Gyll., in Schœnh., Gen. Curc. I, 1833, p. 323; Senna, Not. Leyd. Mus. XVII, 1895, p. 53; *l. c.* XIV, p. 177; Kln., Ent. Bl. 1916, p. 161[1].	[Dikrang Valley, Dafla Hills, Bua, Sibsagar=**Assam** Sikkim, Tavoy, Andamans ——— Nias[1], Sumatra[1] Java[1], Borneo[1]
merocephalus Kln., Ent. Bl. 12, 1916, p. 178[1].	Darjeeling[1] Khasi Hills[1] ——— Indochina
miles Boh., in Schœnh., Gen. Curc. VIII, 1845, p. 372; Lacord, Gen. Col. VII, 1866, p. 429; Senna, Ann. Civ. Nat. Gen. (2) XII (XXXII) 1892, p. 469[1]; Ann. S. E. Belg. XXXVIII, 1894, p. 373[2]; Kln., Ent. Bl. 12, 1916, p. 170[3]; Beeson, Ind. For Rec. XI, 1925, p. 183 (biol.). *approximatus* F. Wlkr., Ann. Mag. Nat. Hist. (3) III, 1859, p. 262.	Himalaya[1, 3] Nepal, Darjeeling[3] Mangpu, Darjeeling 5,000 ft., Kalimpong, Darjeeling, Almora Garo Hills, Assam, 1,200-1,500 ft. Sibsagar, Assam Kurseong, Assam Dejoo, Upper Assam, Khasi Hills Ukhuri, Manipur Bengal[1], Buxa Pangamali, Chittagong Hill Tracts, Travancore Trichinopoly[3] Madras[3], Nilgiris Burma[1, 2, 3] Ceylon[2] ——— Salanga Isl.[3] Indochina[3], Shanghai[3] Hainan Island Java[3], Borneo[3]

poweri Rœlf., Ann. S. E. Belg. XXII, Bull. 1879, p. 54[1]; Lewis, Journ. Linn. Soc. Lond. XVII, 1883, p. 300, t. 12, f. 11[2]; v. Schœnf., D. E. Nat. Bibl. II, 1911 nr. 24, p. 190[3], Kiln., Ent., Bl. 12, 1916 p. 175.[4]	Indochina Formosa[3, 4] Hainan Isl.[4] Japan[1, 2, 4]
rudis Senna, Ann. Civ. Nat. Gen. (2) XIII (XXXIII) 1893, p. 280[1]; Kln. Ent. Bl. 1916, p. 159[2].	Nilgiris _________ Engano[1], Sumatra[2] Java
rugoscicollis Pow., Pet, Nouv. Ent. II, 1879, p. 298[1].	Malacca[1], Penang Indochina
schrœderi Kln., (*Eupsalomimus*) Stett. E. Z. 75, 1914, p. 172[1]; Ent. B1. 12, 1916, p. 182[2].	Siam[1], Philippines[1] Moluccas[2], Celebes Salawatti Isl.
speciosissimus Kln., Ent. Bl. 12, 1916, p. 180[1].	Madura[1] _________ Indochina[1]
umbraticus Kln., Ent. Bl. 12, 1916, p. 168[1].	Borneo[1], Indochina

41. Eupsalis Lacordaire (1866).

Type : *truncata* Boh.

Lacord., Gen. Col. VII, 1866, p. 430; Pow., Ann. S. E. Fr. (5) VIII, 1878, p. 492; v Schœnf., Gen. Ins. (Wytsman), 1908, p. 36; Cat. Col. (Junk) 7, 1910, p. 23; Kln., Arch. Nat. 82, 1916 (1917), A. 4, p. 55.

truncata Boh., Nouv. Mém. Soc. Nat. Mosc. I, 1829, p. 103; schœnh., Gen. Curc, I, 1833, p. 326[1]; *l. c.* V. 1840, p. 479; Kln., Arch. Nat. 82, 1916 (1917), A. 4, p. 118.[2]	Barway, Bengal[2] Bengal[1, 2] Kanara, Madras Colombo[2] _________
tuberculata Senna, Ann. S. E. Belg. XXXVIII, 1894, p. 373[1].	Upper Burma[1] _________

42. Stratiorrhina Pascoe (1872).

Type : *xiphias* Westw.

Pasc., Ann. Mag. Nat. Hist. (4) X, 1872, p. 322; v. Schœnf., Gen. Ins. (Wytsman)1908, p. 39; Cat. Col. (Junk) 7, 1910, p. 24; Calabr., Bull. S. ent. Ital. LIII, 1921, p. 42.

concors Kln., Journ. Malay. Br. R. As Soc. 1923, p. 272[1]; Jl. F. M. S. Mus. XIII, 1925, p. .[2]	Perak[1] Pahang[2]
major Calabr., Bull, S. Ent. Ital. LIII, 1921, p. 47[1]; Kln., Jl. F. M. S. Mus. XIII, 1925, p. [2].	Siam[1], Selangor[2] Penang, Sumatra[1] Borneo[1]

pascœi Kirsch., Mitt. zool. Mus. Dresd. I, 1875, p. 48[1]; Calabr., Bull. S. Ent. Ital. LIII, 1921, p. 45[2].	Malacca[1, 2], Perak[1] Penang, Borneo
xiphias Westw., Cab. Or. Ent. 1848 t. 15, f. I[1]; Calabr., Bull, S. Ent. Ital. LIII, 1921, p. 44[2]; Kln., Jl. F. M. S. Mus. XIII, 1925, p.[3].	Selangor[3] Penang[1] Borneo[2],

43. Cænorychodes Kleine (1920)

Type : *serrirostris* F.

Kln., Arch. Nat. 86, 1920, A. 9, p. 87. *Orychodes* Auct.

fasciatus Kln., Arch. Nat. 86, 1920, A. 9, p. 97[1].	Calcutta Andamans[1] ————
planicollis F. Wlk., Ann. Mag. Nat. Hist. II, 1859, p. 262, note 3.	Darjeeling[4] Sikkim4
indus Kirsch, Mitt. Zool. Mus. Dresd. I, 1875, p. 52[1]; Senna, Ann. Civ. Nat. Gen. (2) XII (XXXII), 1892, p. 472[2]; Ann. S. E. Belg. XXXVIII, 1894, p. 375[3]; Kln., Arch. Nat. 86, 1920, A. 9, p. 99[4]; Beeson, Ind. For. Rec. XI, 1925, p. 183 (biol.).	North Cachar[4] Kurseong, Assam[4] Dejoo, Assam S. Raipur, C. F. Bengal, Barway Bengal, Coorg Kanara, Nilgiris, Palnis, Madras, Katha, Burma[2, 3,4] Andamans, Ceylon[4] ———— Siam[2], Indochina Formosa[4]
serrirostris F., Syst. El. II, 1801, p. 553; Kln., Arch. Nat. 36, 1920, p. 88[1]; Jl. F.M. S. Mus. XIII, 1925, p. .	Burma ———— Penang
serrircstris N. Lund, Act. Hist. Nat. Hafn. 2, 86, 19, 1802.	Singapore Sunda Isl.[1] (common)
cinnamomi Senna, Bull. S. Ent. Ital. XXXI, 1899, p. 304[2].	Indochina, Formosa Philippines[1] Celebes[2, 1]

44. Hemiorhchodes Kleine (1920).

Type : *cambodjensis* Kln.

Kln., Arch. Nat. 86, 1920, A. 9, p. 83.

cambodjensis Kln., Arch. Nat. 86, 1920, A. 9, p. 84.[1]	Indochina[1]

contextus Kln., Ind. For Rec. XI, 1925, p. 148[1].	Cambodia[1]
continens Kln., Ind. For Rec. XI, 1925, p. 149[1].	Penang[1]
curvus Kln., Ind. For Rec. XI, 1925, p. 147[1].	Manipur[1], Burma[1].
dentipennis Senna, Leyd. Mus. XX 1998, p. 60[1].	Penang, Sumatra[1] Java[1]
dissonus Kln., Ind. For. Rec. XI, 1925, p. 146[1]; Jl. F. M. S. Mus. XIII, 1925, p. [2].	Perak[1] Pahang[2]
modestus Kln., Arch. Nat. 86, 1920, A. 9, p. 85[1].	Indochina[1] Sumatra[1]

45. Pseudorychodes Senna (1894)

Type : *lineolatus* Kirsch.

Senna, Ann. S. E. Belg. XXXVIII, 1894, p. 375; v. Schœnf., Gen. Ins. (Wytsman) 1908, p. 42; Cat. Col. (Junk), 7, 1910, p. 27.

crassus Senna, Ann. S. E. Belg. XXXVIII, 1894, p. 378[1].	Garo Hills, Assam Sikkim, Burma[1]
cruentatus Senna, Leyd. Mus. XX, 1898, p. 71[1]; Kln., Jl. F. M. S. Mus. XIII, 1925, p. .[2]	Malacca[1], Pahang[2] Sumatra
damnosus Kln., Ind. For Rec. XI, 1925, p. 149[1].	Perak[1]
honorus Kln., Jl. F. M.S. Mus. XIII, 1925, p. , figs. 4-6.	Pahang, Siam
lineolatus Kirsch., Mitt. Zool, Mus. Dresd. I, 1875, p. 49[1]; Senna, Bull. S. Ent. Ital. XXXIV, 1902, p. 176[2]; Kln., Jl. F. M. S. Mus. XIII, 1925, p. [3].	Malacca[1], Pahang[3] Perak, Siam Sumatra[2], Borneo
ritsemæ Senna, Leyd. Mus. XIII, 1891, p. 161[1].	Burma, Malacca[1], Sumatra
tenuirostris Senna, Ann. S. E. Belg. XXXVIII, 1894, p. 376[1].	Upper Burma[1]

46. Synorychodes Kleine (1921).

Type : *opacus* Kln.

Kln., Notul. Ent. I, 1, 1921, p. 9.

opacus Kln., Notul. Ent. I, 1, 1921, p. 10[1].	Amarapura, Burma[1]

47. Suborychodes Kleine (1917).

Type : *intermedius* Kln.

Kln., Ent. Mitt. VI, 1917, p. 328.

delectabilis Kln., Ind. For. Rec. XI, 1925, p. 150[1]. — East India[1]

intermedius Kln., Ent. Mitt. VI, 1917, p. 330[1]; Beeson, Ind. For. Rec. XI, 1925, p. 183 (biol.). — Dehra Dun, U.P. Patkai Mts., Assam Margherita, Lakhimpur, Assam Buxa; Perak, Sumatra[1] Borneo

48. Calorychodes Kleine (1924).

Type : *decens* Kln.

Kln., Ind. For. Rec. XI, 1925, p. 152.

decens, Kln., Ind. For. Rec. XI, 1925, p. 153[1]. — Ruby Mines, Burma[1]

49. Perorychodes Kleine (1924).

Type : *arrowi* Kln.

Kln. Ind. For. Rec. XI, 1925, p. 145.

arrowi Kln., Ind. For. Rec. XI, 1925, p. 146[1]. — Allahabad[1]

50. Parorychodes Kleine (1920).

Type : *degener* Senna.

Kln., Arch. Nat. 86, 1920, A. 9, p. 80.

Parabletus Calabr., Faun. Ent. Indochin. 2, 1921, p. 8.

cereus Kln., Ind. For. Rec. XI, 1925, p. 151[1]; Beeson, *t. c.* 1925, p. 184 (biol.). — Sibsagar, Assam[1]

comes Kln., Arch. Nat. 86, 1920, A. 9, p. 82[1]. — Indochina[1]

degener Senna, Ann. Civ. Nag. Gen. (2), XII (XXXII), 1892, p. 474[1] p. 474[1]; Kln., Arch. Nat. 86, 1920, A. 9, p. 81[2]; Calabr., Faun. Indochin. 2, 1921[3]. — Patkai Mts., Assam Burma[1], Andamans[2]; Indochina[3]

51. Orychodes Pascoe (1862).

Type : *pictus* Pasc. (Molluccas).

Pasc., Journ. Ent. I, 1862, p. 389; Lacord., Gen. Col. VII, 1866, p. 432; v. Schœnf., Gen. Ins. (Wytsman), 1908, p. 41; Cat. Col. (Junk) 7, 1910, p. 26; Kln., Arch. Nat. 86, 1920, A. 9, p. 77.

breviceps Senna, Ann. S. E. Belg. XXXVIII, 1894, p. 380[1]. — Burma[1]

---

52. Pseudomiolispa Kleine (1925).

Type : *pendleburyi* Kleine.

Kln., Jl. F. M. S. Mus. XIII, 1925, p. .

pendleburyi Kln., Jl. F. M. S. Mus. XIII, 1925, p. fig. 7-9. — Selangor Kuala Lumpur

BELOPHERINÆ

Belopherini, Lacord., Gen. Col. VII, 1866, p. 433; v. Schœnf., Gen. Ins. (Wytsman), 1908, p. 42; Cat. Col. (Junk) 7, 1910, p. 27; Kln., Ent. Bl. 18, 1922, p. 162.

53. Epicœnoneus Senna (1892).

Type : *femoralis* Senna.

Senna, Ann. Civ. Nat. Gen. (2) XII (XXXII) 1892, p. 476; v. Shœnf., Gen. Ins. (Wytsman) 1908, p. 47; Cat. Col. (Junk) 7, 1910, p. 31.

femoralis Senna, Ann. Civ. Nat. Gen. (2) XII (XXXII), 1892, p. 478[1]; Leyd. Mus. XX, 1898-99, p. 62.[2] — Burma[1], ——— Java[2], Sumatra Borneo

54. Euphenges Calaresi (1920).

Type : *ceylonicus* Calabr.

Calabr., Bull. S. Ent. Ital. LI, 1919 (1920), p. 62.

aciculatus F. Wlk., Ann. Mag. Nat. Hist. (3) III, 1859, p. 262[1]. — Ceylon[1]

---

ceylonicus Calabr., Bull. S. Ent. Ital. LI, 1919 (1920), p. 63[1]. — Ceylon[1]

---

delibratus Kln., Ind. For. Rec. XI, 1925, p. 154[1]. — Kanara, Bombay[1]

---

distributus Kln., Ind. For. Rec. XI, 1925, p. 154[1]. — Perak[1]

iridescens Calabr., Bull. S. Ent. Ital. LI, 1919 (1920), P. 65[1]. — Penang[1]

55. Heteroblysmia Kleine (1917).

Type : *borncensis* Kln. (Borneo).

Kln., Ent. Bl. 13, 1917, p. 285.

cava Kln., Ind. For. Rec. XI, 1925, p. 155[1]. — Penang[1]

56. Apocemus Calabresi (1921).

Type : *conciliator* Kirsch.

Calabr., Bull S. Ent. Ital. LIII, 1921, p. 58.

conciliator Kirsch., Mitt. Zool. Mus. Dresd. I, 1875, p. 48[1]; v. Schœnf., Cat. Col. (Junk) 7, 1910, p. 27, (sub *Eotoosmus*). — Malacca[1]

57. Megateras Kleine (1921).

Type : *horribilis* Kleine

Kln., Senkenbergiana, 1921, III, p. 62.

horribilis Kln., *l. c.* p. 64[1]; Jl. F. M. S. Mus. XIII, 1925, p.[2] — Pahang[2], Borneo[1]

58. Hopliterrhynchus Senna (1892).

Type : *emmæ* Senna.

Senna. Bull. S. Ent. Ital. XXIV, 1892, p. 26; v. Schœnf., Gen. Ins. (Wytsman) 1908, p. 44; Cat. Col. (Junk) 7, 1910, p. 28.

emmæ Senna, Bull. S. En. Ital. XXIV, 1892, p. 27[1]. — Malesia[1]

59. Ectocemus Pascoe (1862).

Type : *decemmaculatus* Montrz. (Moluccas, N. Guinea).

Pasc., Journ Ent. I, 1862, p. 388; Lacord., Gen. Col. VII, 1866, p. 434; v. Schœnf., Gen. Ins. (Wytsman) 1908, p. 43; Cat. Col. (Junk) 7, 1910, p. 27; Kln., Arch. Nat. 89, 1923, A. 4, p. 157.

Megacerus Motraz., Ann. S. Agr. Lyon, VII, 1855, p. 36.

cinnamomi Herbst., Fuessl. Arch. Ent. IV, 1783, p. 76, t. 24, f. 20; Käfer VII, 1797, p. 174, t. 107, f. 1; Kln., Jl. F. M. S. Mus. XIII, 1925, p. . — Andamans[3] Ceylon[1] ——— Malacca[2]

spinirostris F. Wlk., Ann. Mag Nat. Hist. (3) III, 1895, p. 262 (sub *Nemocephalus* et *Ithystenus*)[1]. — F. M. S., Indochina Siam,

spathulirostris Chevr. Lacord., Gen. Col. VII, t. 77, f. 3. — Sunda Isl. (common).

pubens Kirsch., Mitt. Zool. Mus. Dresd. I, 1875, p. 47[2]. — Moluccas

quatuordentatus Desbr., Journ. As S. Beng. LIX, (2) 1890, p. 219[3].

contractus Kln., Ind. For. Rec. XI, 1925, p. 158[1]. Cambodia.

60. Teraticorhynchus Kleine (1924).

Type : *defectus* Kln.

Kln., Ind. For Rec. XI, 1925, p. 155.

defectus Kln., Ind. For. Rec. XI, 1925, p. 157[1]. Tenasserim[1]

61. Desgodinsia Senna (1894).

Type : *spinirostris* Senna.

Senna, Ann. S. E. Belg. XXXVIII, 1894, p. 381; v. Schœnf., Gen. Ins. (Wytsman) 1908, p. 45; Cat. Col. (Junk) 7, 1910, p. 29.

spinirostris Senna, Ann. S. E. Belg. XXXVIII, 1894, p. 382[1]. Burma[1]

62. Pseudobelopherus Calabresi (1920).

Type : *orientalis* Calabr.

Calabr., Bull. S. Ent. Ital. LI, 1919 (1920), p. 59.

deductus Kln., Ind. For. Rec. XI, 1925, p. 157[1]. Perak[1]

orientalis Calabr., Bull. S. Ent. Ital. LI, 1919 (1920), p. 61[1]. Malacca[1]

63. Heterorhynchus Calabresi (1921)

Type : *ornatus* Calabr.

Calabr., Bull. S. Ent. Ital. LIII, 1921, p. 62[1].

ornatus Calabr., Bull. S. Ent. Ital. LIII, 1921, p. 63[1]. Tonkin, Borneo[1]

64. Anepsiotes Kleine

Type : *schenklingi* Kln.

Kln., Ent. Mitt. VI, 1917, p. 318.

bellus Kln., Ind For. Rec. XI, 1925, p. 159[1]. Katgal[1]

commendabilis Kln., Ind. For. Rec. XI, 1925, p. 159[1]. Travancore[1]

kleinei Calabr., Bull. S. Ent. Ital. LI, 1919, p. 68[1]. Penang[1]

schenklingi Kln., Ent. Mitt. VI, 1917, p. 321[1]. Ceylon

EUTRACHELINAE.

Lacord., Gen. Col. VII, 1866, p. 438; v. Schœnf., Gen. Ins. (Wytsman), 1908, p. 49; Cat. Col. (Junk) 7, 1910, p. 31; Kln., Ent. Bl. 18, 1922, p. 162.

65. Eutrachelus Latr. (1825).

Type : *temmincki* Latr.

Later., Fam. Nat. Regne An. 1825, p. 389; Imhoff, Stud. Kol. 1856, p. 167; Lacord., Gen. Col. VII, 1866, p. 439; v. Schœnf., Gen. Ins. (Wytsman) 1908, p. 49; Cat. Col. (Junk), 7, 1910, p. 32.

temmincki Latr., *l. c.,* p. 389; Gyll., in Schœnh., Gen. Curc, I, 1833, p. 338; Imhoff, *l. c.,* t. 15, f. 21; v. Schœnf., *l. c.,* t. I, f. 4, 4*a*; Kln., Jl. F.M.S. Mus. XIII, 1925, p.[1] — Selangor, Pahang; Sunda Isl.

sumatrensis Watrh., Trans. Ent. Soc. Lond. 1881, p. 489.

achilles Rothsch. et. Jord., Ann. Mag. Nat. Hist. (6) XII, 1893, p. 454.

borneensis Rothsch. et Jord., *l. c.,* p. 454.

ITHYSTENINÆ

Ithystenini Lacord., Gen. Col. VII, 1866, p. 464; v. Schœnf., Cat. Col. (Junk) 7, 1910, p. 42; Kln., Ent. Bl. 18, 1922, p. 162.

Leptorrhynchidœ v. Schœnf., Gen. Ins. (Wytsman) 1908. p. 69.

66. Cediocera Pascoe (1887).

Type : *longicornis* Pasc.

Pasc., Ann. Mag. Nat. Hist. (5) XX, 1887, p. 20; v. Schœnf., Gen. Ins. (Wytsman) 1908, p. 71; Cat. Col. (Junk) 7, 1910, p. 42.

longicornis Pasc., Ann. Mag. Nat. Hist. (5) XX, 1887, p. 20, t. 1, f. 5[1]; Senna, Ann. Civ. Nat. Gen. (2) XIX (XXXIX), 1898, p. 239[2]. — Calcutta, Andamans[1]

---

tristis Senna, Leyd. Mus. XIV, 1891, p. 181[1]; Ann. Civ. Nat. Gen. (2) XIII (XXXIII), 1893, p. 284[2]; Bull. S. Selangor[4] Ent. Ital. XXXIV, III, 1902, p. 177[3]; Kln., Jl. F. M. S. Mus. XIII, 1925, p. [4]. — Malacca; Sumatra, Mentawei[2]; Perak[1], Penang; Selangor[4]; Sumatra[1, 2], Engano[2]; Java[1], Borneo[1]; Philippines

67. Plesiophocylides Kleine (1924).

Type : *conditus* Kln.

Kln., In. For. Rec. XI, 1925, p. 160.

conditus Kln., Ind. For. Rec. XI, 1925, p. 161[1]. — Sylhet, Assam[1]

68. Achrionota Pascoe (1872).

Type : *bilineata* Pasc.

Pasc., Ann. Mag. Nat. Hist. 1872, (4) X, p. 325; v. Schœnf., Gen. Ins. (Wytsman) 1908, p. 75; Cat. Col. (Junk) 7, 1910, p. 44; Kln., Arch. Nat. 83, 1917 (1919) A. 7, p. 129.

billineata Pasc., Ann. Mag. Nat. Hist. 1872, (4) X, p. 325, t. 15, f. 4[1]. — Perak, Sumatra[1] Borneo, Philippines

69. Heterodiurus Senna (1909).

Type : *singularis* Senna.

Senna, Bull. S. Ent. Ital. XLI, 1909, p. 51.

singularis Senna, Bull. S. Ent. Ital. XLI, 1909, p. 52[1]. — India[1]

70. Diurus Pascoe (1862).

Type : *furcillatus* Gyll. (Sunda Isl).

Pasc., Journ. Ent. I, 1862, p. 392; Lacord., Gen. Col. VII, 1866, p. 471; v. Schœnf., Gen. Ins. (Wytsman) 1908, p. 77; Cat. Col. (Junk) 7, 1910, p. 45; Senna, Bull. S. Ent. Ital. XLI, 1909, p. 28.

albopilosus Kln., Jl. F. M. S. Mus. XIII, 1920, p. , f. 10. — Pahang

celsus Kln., Journ. Malay. R. As. Soc. 88, 1923, p. 274, f. 4[1]. — Gunong Tahan Pahang Penang, 2,500-3,000 ft.[1], Gunong Angsi Negri Sembilan, 2000-2,700 ft.

compendarius Kln., Ind. For. Rec. XI, 1920 p. 162[1]. — Perak[1]

completus Kln., Journ. Malay. R. As. Soc. 1923, p. 276, f. 5[1]; Jl. F. M. S. Mus. XIII, 1925, p.[2] — Malacca[1], Perak[2] Borneo[1]

forcipatus Westw., Cab. Or. Ent. 1848, p. 31, t. 15, f. 3[1]; Senna, Bull. S. Ent. Ital. XLI, 1909, p. 42[2]; Kln., Jl. F. M. S. Mus. XIII, 1925, p. .[3] — Sibsagar, Assam; Selangor[3], Penang[2] Sumatra[2], Java[1], Borneo[2]

furcillatus Gyll., in Schœnh., Gen. Curc. I, 1863, p. 359; Kln., Jl. F. M. S. Mus. XIII, 1925, p. — Selangor; Siam, Sumatra

ominosus Senna, Ann. Civ. Nat. Gen. (2) XII (XXXII), 1892, p. 492[1]. — Burma[1]; _________

sphacellatus Pasc., Ann. Mag. Nat. Hist. (5) XX, 1887, p. 19[1]; Kln., Jl. F. M. S. Mus. XIII, 1925, p. .[2] — Calcutta; N. Toungoo, Burma, Andamans[1]; Nicobars; ________, Siam[2]

PSEUDOCEOCEPHALINÆ.

Pseudoceocephalini, Kln., Ent. Bl. 18, 1922, p. 163.

Ceocephalini, Lacord., Gen. Col. VII, 1866, p. 444; v. Schœnf., Gen. Ins. (Wytsman) 1908, p. 55; Cat. Col. (Junk) 7, 1910, p. 35.

71. Opisthenoplus Kleine (1822)

Type : *cavus* F. Wlk.

Kln., Deutsch. E. Z. 1922, p. 139.

cavus F. Wlk., Ann. Mag. Nat. Hist. (3) III, 1859, p. 262[1]; Beeson, Ind. For. Rec. XI, 1925, p. 184 (biol.). — Garo Hills, Assam 1,200-1,500 ft.; Gopaldhara, Assam; Lakhimpur, Assam; Sikkim, Dehra Dun; Gonda, U. P.; N. Kanara, Bombay; Malabar, Madras; Nilgiris; N. Toungoo, Burma, Andamans; Ceylon[1]; _________; Indochina, Penang; Sumatra, Borneo; Java[1],

commodus Kln., Journ. Malay. R. As. Soc. 1923, p. [1]. — Maxwell Hills, Perak[1]

fascinatus Kln., Deutsch. E. Z. 1922, p. 140[1]; Beeson, Ind. For. Rec. XI, p. 184, (biol.). — Siwaliks; Saharanpur, U.P.; _______,; Sumatra; Formosa[1]; Philippines[1]

72. Hormocerus Schœnherr (1826)

Type : *reticulatus* F.

Schœnh., Curc. Disp. 1826, p. 70; Lacord., Gen. Col. VII, 1866, p. 447; v. Schœnf., Gen. Ins. (Wytsman), 1908, p. 56; Cat. Col. (Junk) 7, 1910, p. 35; Kln., Arch. Nat. 88, 1922, A. 7, p. 164.

Ceocephalus (pars.) Schœnh., Gen. Curc. I, 1833, p. 360; V, 1840, p. 510.

reticulatus F., Syst. El. II, 1801, p. 552; N. Lund., Skrift Nat. Hist. V, 1802, p. 81; Gyll., in Schœnh., Gen. Curc. I, 1833, p. 360; Senna, Leyd. Mus. XVII, 1895, p. 55; Stebbing, Ind. For. Ins. Col., 1914, p. 389, (biol.); Beeson, Ind. For. Rec. XI, 1925, p. 184 (Biol.); Kln., Jl. F. M. S. Mus. XIII, 1925, p. .	India Ceylon ———— Indo-china to Australia

73. Schizotrachelus Lacordaire (1866)

Type : *modens* Lacord.

Lacord., Gen. Col. VII, 1866, p. 454; v. Schœnf., Gen. Ins. (Wytsman) 1908, p. 62; Cat. Col. (Junk) 7, 1910, p. 39.

birmanicus Senna, Ann. Civ. Nat. Gen. (2) XII (XXXII), 1892, p. 485[1].	Burma[1] ————
cameratus Lacord., Gen. Col. VII, 1866, p. 456[1]; Kln., Jl. F. M. S. Mus. XIII, 1925. p.	Andamans ———— Malacca[1], Perak, Selangor, Siam Sumatra Borneo
carinensis Senna, Ann. Civ. Nat. Gen. (2) XII (XXXII) 1892, p. 483[1].	Burma[1] ————
carinirostris Senna, Ann. Civ. Nat. Gen. (2) XII (XXXII) 1892, p. 480[1].	Burma[1] ————
consanguineus Senna, Ann. Civ. Nat. Gen. (2) XII (XXXII) 1892, p. 485[1].	Burma[1] ————
consobrinus Lacord., Gen. Col. VII, 1866, p. 456[1]; Senna, Leyd. Mus. XIV, 1892, p. 180[2]; Ann. Civ. Nat. Gen. (2) XIII, (XXXIII), 1893, p. 283[3]; Ann. S. E. Belg. XXXVIII, 1894, p. 386.[4]	Barway, Bengal[4] Burma, Andamans ———— Engano[3] Java[1, 2], Borneo

intermedius Senna, Ann. Civ. Nat. Gen. (2) XII (XXXII) 1892, p. 481[1]; Ann. S. E. Belg. XXXVIII, 1894, p. 386[2]; Leyd. Mus. XVII, 1895, p. 56.[3] — Barway, Bengal[2] Nilgiris, Burma[1] ——— Indochina Sumatra, Java[2, 3] Borneo

madens Lacord., Gen. Col. VII, 1866, p. 455[1]; Senna, Ann. Civ. Nat. Gen. (2) XIX (XXXIX), 1898, p. 239[2]; Kln., Jl. F.M.S. Mus. XIII, 1925, p. (*Opisthenoplus*). — Perak, Selangor Indochina Malacca[1], Mentawei[2] Sumatra Philippines

marginatus Senna, Leyd. Mus. XX, 1898, p. 75[1]. — Perak, Labuan[1]

sculptiventris Senna, Ann. Civ. Nat. Gen. (2) XII (XXXII), 1892, p. 488[1]. — Burma[1] ——— Indochina[1]

vitalisi Calabr., Faun. Ent. Indoch. 1921, p. 12[1].

74. Holotrachelus Kleine (1923)

Type : *comparabilis* Kln.

Kln., Journ Malay R. As. Soc. 1923, p. 276.

comparabilis Kln., Journ. Malay R. As. Soc. 1923, p. 277[1]; Jl. F. M. S. Mus. XIII, 1925, p. . — Pahang, Selangor Malacca[1], Borneo[1]

75. Metatrachelus Kleine (1924)

Type : *comparativus* Kln.

Kln., Ind. For. Rec. XI, 1925, p. 162.

comparativus Kln., Ind. For. Rec. XI, 1925, p. 163[1]. — Penang[1]

differens Kln., Capita, Zoolog. 1924, Jl. F. M. S. Mus. XIII, 1925, p. . — Pahang, Selangor Sumatra

hospes Kln., Jl. F. M. S. Mus. XIII, 1925, p. — Selangor

76. Peraprophthalmus Kleine (1923)

Type : *applicatus* Kln.

Kln., Sbornik Ent. Nat. Mus. Praze, 1923, I, 4, p. 49.

applicatus Kln., Sbornik Ent. Nat. Mus. Praze, 1923, I, 4, p. 51.[1] — Dehra Dun Tenasserim[1]

PART XII – TABANIDAE

INTRODUCTION

A systematic catalogue may be one of three kinds. It may summarize chaos, may mark a stage in the development of our knowledge of a group, or it may be the logical outcome of a finished and "final" revision. Of those in the first category, the immortal *"Catalogue of the Diptera of South Asia,"* by van der Wulp, is a classical example, whilst the writer's *"Bombyliidæ"* in the present series is a type of the third, although in few groups of plants or animals can it be said as yet that our systematic knowledge is approximately complete. The present part of the *"Catalogue of Indian Insects"* falls into the second category, summarizing the second stage in our knowledge of the family Tabanidae in India. The first revision of the family was published by Miss Ricardo fourteen years ago. Odd species have been described since by various authors, and stray pieces of additional synonomy made out. As a basis for further, intensive, work on the family the present time is a convenient one to "take stock". The result shows how fragmentary our knowledge is. The very uncritical list of species in the three principal genera of the family throughout the World, recently published in *"Genera Insectorum"* by de Surcouf, shows 170 species of *Hæmatopota,* 1,140 of *Tabanus* and 200 of *Chrysops* as known. Allowing for these figures being somewhat too large owing to yet un-effected synonymising, the Indian list of 31 *Hæmatopota,* 75 *Tabanus* and 11 *Chrysops* appears remarkably poor. Seven other genera, three of them peculiar to the Region, with one or two species each, bring the Indian total in the family up to 126 species, but only the genus *Hæmatopota,* the Indian species of which are 18 per cent of the World total, indicates that here the present-day knowledge of the genus has kept pace with work on other areas. And yet this genus is probably the worst known of the major three in India, in the sense that more forms of it remain to be discovered than in the other two. However, if I am correct in this belief, then the Indian fauna of *Tabanus* and *Chrysops* is poor in comparison with the rest of the World, and the position not so bad as the above figures would at first sight indicate.

Still more defective than our systematic is our distributional knowledge. Tables I to III indicate our present knowledge of the latter in respect of the three principal genera, though it has not been possible to sub-divide the area to the same extent as the writer was able to do in dealing with *Sarcophaga* (Rec. Ind. Mus., xxvi, 260-273). The reason for this is simply paucity of information.*

---

* These statements require a certain amount of modification owing to the extra localities from the Pusa Collection added whilst in press.

Taking *Hæmatopota* (Table I), first, one is immediately struck by the fact that less than half the known Indian species are recorded from more than one zone. The rest would seem to be confined to very narrow limits for each species. Those which extend more widely are seen to be Malayan species entering India in various zones which have faunal affinities with those of further East, but a study of their distribution indicates that most of them must occur in intervening zones from which they are not yet known. The common *H. javana* Wd., for instance must occur in the Plains of Assam *en route* from further Asia and Burma to the Indo-Gangetic Plain East. Equally *H. lata* Ric. must occur in the latter and the Assam Plains on its way from the Assam Hills to the West. Zone by zone, the Assam Hills and Burma contain more known species of the genus than any others, but the writer (*Mem. Dept. Agric. Indian,* vii, 107), has seen an as yet undescribed species from the latter. North Ceylon, the Malabar Coast and the Indo-Gangetic Plain East, thanks to various collectors in the two first, and the location of Pusa in the third, are not far behind Burma, but Baluchistan, Rajputana, Central India, and most astonishing of all, the Eastern Himalaya, are completely unknown in respect of their *Hæmatopota*-fauna! From the Darjiling district I have seen several specimens in too bad order to describe, that probably represented at least three species, but the state of affairs for this zone is extraordinary in view of the fact that the dipterous fauna of Darjiling is probably better known than that of any other part of India, and is by far the richest in most families.

In regard to other Zones, the writer's determinations of Leese's Mohand collection (Fletcher and Senior White, *Proc. ivth. Entom. Meetg. Pusa,* p. 226), show that the Indo-Gangetic Plain West contains at least three undescribed species in this genus, whilst examination of a collection of Tabanidae sent by Capt. Crosse, of the Civil Veterinary Department, to Major Patton, the results of which are incorporated in the present catalogue, indicate that the Western Himalaya (Dharamsala and Pathankot), contain two undescribed species in Groups IV and V of the genus.

From the foregoing, then, it is clear that an enormous amount of more work has still to be done before the *Hæmatopota*-fauna of India is properly known, and the necessary additions to our knowledge can be made by anyone who will take the trouble to collect specimens and send them in for determination.

Turning now to *Tabanus,* we see a very similar state of ignorance. The Indo-Gangetic Plain East and Burma each contain twenty known species, whilst Rajputana yields not a single record, and Central India only one, being the locality given for a type by Walker three-quarters of a Century ago! Leese collected at least three undescribed species at Mohand, and the writer has recorded (*Mem. Dept. Agric. India,* vii, 106), an unidentified species from Chakrata in the Western Himalaya.

From the table it will be seen that Group I in the genus is mainly confined to intrudes from the Palearctic Region, to which Seistan really belongs, and it is possible that some of these species do not enter India proper at all. None the less, as Surra in that zone is of great importance, and it control an Indian problem, these species have been included in the key and the body of the Catalogue.

The peculiarly coloured species of Group II seem to be inhabitants of hill regions and the country at their feet.

In the small Group III the only true Indian species, *ditaeniatus,* is more widely distributed in Africa than it is in Asia, but covers a larger area of the Old-World Tropics than any other species of the family, extending to China, where its only Eastern relative is found. *T. sufis,* in this Group, is Palearctic, and just enters the Region in the Western Punjab.

Group IV are mainly Plains species.

The species of Group V and VI, with strikingly marked wings, are Malayan, only entering India in zones with strong affinities with that sub-Region.

Group VII contains most of the commonly and widely distributed species, as well as some localized rarities. Most of the species extend into Malaysia.

Group VIII is very poorly represented in India. It is Malayan, and its few Indian members are found in zones with strong affinities with Malaya.

The Indian species of Group IX have little affinity with further East, and are centered in India. *T. rufiventris* and *T. orientis* are widely distributed in North India. The known Indian distribution of *T. khasiensis,* the Khasia Hills and Meerut, is an excellent illustration of our lack of knowledge. The two localities have little in common, save that they are both Oriental! There is a species, not shown in the distribution table, peculiar, apparently, to the Nicobar Islands.

About half the species in Group X are purely Malayan. Those found in India are, as regards the zones in which they occur, distinctly of Malayan affinity, and some of them extend beyond the confines of the Empire eastward. One of the only two Tabanids known from the Andamans belongs to this Group, the species occurring also in Ceylon, Burma and Sumatra.

The sub-genus *Therioplectes,* Group XI, is purely Western Himalayan, and an intruder from the Palearctic.

The genus *Chrysops,* table III, is probably better known than the two previous, owing to the striking appearance of these insects. Only one species, *dispar* Fb., is commonly and widely distributed, and it probably occurs throughout the sub-Region, except to the North-West. *C. flavocinctus, fasciatus* and *fixissimus* are Malayan species only found in India in localities strongly associated faunistically with further East. *C. flavipes* is a Palearctic species not entering India proper, and *C. mloksiewiczi* is a Chinese species found in Yunnan, to which the last remark also applies. The remaining species are of localized distribution, or are found in similar stations, such as hilltops. No species are known from Rajputana and the Eastern Himalaya.

In spite of their importance, not only as probably carriers of Surra, but as worriers of domestic animals in general, it will thus be seen that our knowledge, systematic and geographical, of the Tabanidae is fragmentary in the extreme. If the present volume stimulates Veterinary Officers to accumulate material, which is easily done, as the flies do not have to be searched for, but are to be found on their patients for the looking, it is hoped that within a few years a very different Catalogue, especially zoo-geographically, may become not only possible but necessary.

In conclusion, the writer has to express his great obligation to Miss G. Ricardo, who has read through the first draft of the manuscript, and favoured him with additional references and localities, as well as a series of notes on doubtful of synonomy, which have been incorporated in the final manuscript for press.

TABLE 1.

	S. Ceylon Hills.	N. Ceylon.	S. Indian Hills.	Malabar Coast.	Carnatic.	Rajputana.	C. India.	Baluchistan and Seistan.	I. G. Plains, West.	I. G. Plains, East.	W. Himalaya.	E. Himalaya.	Assam Plains.	Assam Hills.	Burma.	Yunnan.	Siam.	Malay Peninsula.	Malayan Archip.	China.
H. roralis Fb.	..	+	..	+	+	..	+	..	..	+	..	..	..	..	..	..	..	..	+	..
H. limbata Big.	..	..	..	..	..	..	..	..	..	..	..	..	..	+	..	..	..	..	..	..
H. striatipennis Brun.	..	..	..	..	..	..	..	..	..	..	+	..	..	..	..	..	..	..	..	..
H. indiana Big.	..	..	..	..	..	..	..	..	..	..	..	..	+	+	..	..	..	..	..	..
H. cana Wlk.	..	..	..	..	..	..	..	..	..	+	..	..	..	..	..	..	..	..	..	..
H. dissimilis Ric.	..	..	..	+	..	..	..	..	..	..	..	..	..	..	..	..	..	..	..	..
H. unizonata Ric.	+	..	..	..	..	..	..	..	..	..	..	..	..	..	..	..	..	+	+	..
H. beesoni S.W.	..	..	..	..	..	..	..	..	..	..	..	..	..	..	+	..	..	..	..	..
H. annandalei Ric.	..	..	..	+	..	..	..	..	..	..	..	..	..	+	..	..	..	..	..	..
H. litoralis Ric.	..	..	..	..	+	..	..	..	..	..	..	..	..	..	..	..	..	..	..	..
H. rhizophoroe S.W.	..	+	..	..	..	..	..	..	..	..	..	..	..	..	..	..	..	..	..	..
H. rubida Ric.	..	..	..	..	..	..	..	..	..	..	..	..	..	..	+	..	..	+	..	..
H. javana Wd.	..	..	+	+	..	..	..	..	..	+	+	..	..	+	+	..	..	+	+	..
H. demellonis S.W.	..	..	..	+	..	..	..	..	..	..	..	..	..	..	..	..	..	..	..	..
H. albofasciatipennis Brun.	..	..	..	..	..	..	..	..	..	..	+	..	..	..	..	..	..	..	..	..
H. lata Ric.	..	..	..	..	..	..	..	..	+	..	..	..	..	+	+	+	..	..	+	..
H. assamensis Ric.	..	..	..	..	..	..	..	..	+	..	..	..	..	+	..	..	..	..	..	..
H. marginats Ric.	..	..	..	..	..	..	..	..	..	+	..	..	+	..	..	..	..	..	..	..
H. brevis Ric.	..	+	..	..	+	..	..	..	..	..	..	..	..	..	..	..	..	..	..	..
H. ricardonis S.W.	+	..	..	..	..	..	..	..	..	..	..	..	..	..	..	..	..	..	..	..

(contd.)

	S. Ceylon Hills.	N. Ceylon.	S. Indian Hills.	Malabar Coast.	Carnatic.	Rajputana.	C. India.	Baluchistan and Seistan.	I. G. Plains, West.	I. G. Plains, East.	W. Himalaya.	E. Himalaya.	Assam Plains.	Assam Hills.	Burma.	Yunnan.	Siam.	Malay Peninsula.	Malayan Archip.	China.
H. tessellata. Ric.	..	+	..	..	..	..	..	..	..	..	..	..	..	..	..	..	..	..	..	..
H. bilineata Ric.	..	..	..	+	..	..	..	..	..	..	..	..	..	..	..	..	..	..	..	..
H. cingalensis Ric.	..	+	..	..	..	..	..	..	..	..	..	..	..	..	..	..	..	..	..	..
H. latifascia Ric.	..	..	..	..	..	..	..	..	..	..	..	..	..	+	..	..	..	..	..	..
H. montana Ric.	..	..	+	..	..	..	..	..	..	..	..	..	..	+	..	..	..	..	..	..
H. fuscifrons Austen	..	..	..	..	..	..	..	..	..	+	..	..	..	..	+	..	..	..	..	..
H. fasciata Ric.	..	..	..	..	..	..	..	..	..	..	..	+	..	+	..	..	..	..	..	..
H. burmanica S.W.	..	..	..	..	..	..	..	..	..	..	..	..	..	..	+	..	..	..	..	..
H. inconspicua Ric.	..	..	..	+	..	..	..	..	..	..	..	..	..	..	..	..	..	..	..	..
H. hindostani Ric.	..	..	+	..	..	..	..	..	..	..	..	..	..	..	..	..	..	..	..	..
H. immaculata Ric.	..	..	..	+	..	..	..	..	..	..	..	..	..	..	..	..	..	..	..	..
	2	5	3	8	3	..	1	..	1	5	3	1	2	9	6	1	..	..	..	..

TABLE II.

	S. Ceylon Hills.	N. Ceylon.	S. Indian Hills.	Malabar Coast.	Carnatic.	Rajputana.	C. India.	Baluchistan and Seistan.	I. G. Plains, West.	I. G. Plains, East.	W. Himalaya.	E. Himalaya.	Assam Plains.	Assam Hills.	Burma.	Yunnan.	Siam.	Malay Peninsula.	Malayan Archip.	China.
T. leleani Aust.	..	..	..	..	..	..	..	+	..	..	+	..	..	..	..	..	..	..	..	..
T. hirtipalpis Ric.	..	..	..	..	..	..	..	..	..	..	..	+	..	..	..	..	..	..	..	..
T. eggeri Sch.	..	..	..	..	..	..	..	+	..	..	..	..	..	..	..	..	..	..	..	..
T. glaber Big.	..	..	..	..	..	..	..	+	..	..	..	..	..	..	..	..	..	..	..	..
T. excelsus Ric.	..	..	..	..	..	..	..	..	+	..	+	..	..	..	..	..	..	..	..	..
T. macer Big.	..	..	..	+	+	..	+	..	+	+	+	..	..	..	..	..	..	..	..	..
T. pulvifer Wlk.	..	..	..	..	..	..	..	+	..	..	..	..	..	..	..	..	..	..	..	..
T. pulcheelus Lw.	..	..	..	..	..	..	..	+	..	..	..	..	..	..	..	..	..	..	..	..
T. adjacens ric.	..	..	..	..	..	..	..	+	..	..	..	..	..	..	..	..	..	..	..	..
T. flavicinctus Ric.	..	..	+	?	..	..	..	..	+	..	..	..	..	+	..	..	..	..	+	..
T. leucocnematus Big.	..	..	+	..	..	..	..	..	..	..	..	+	..	+	+	..	..	..	..	..
T. bicinctus Ric.	..	..	+	+	..	..	..	..	+	..	..	..	..	..	..	..	..	+	..	+
T. sexcinctus Ric.	..	..	..	..	..	..	..	..	..	..	+	..	..	..	+	..	..	..	..	+
T. ditoeniatus Macq.	..	..	..	+	+	..	..	+	+	+	..	..	..	..	..	..	..	..	..	+
T. sufis Jænn.	..	..	..	..	..	..	..	..	..	+	..	..	..	..	..	..	..	..	..	..
T. virgo Wd.	..	..	..	..	+	..	..	..	+	+	+	..	..	..	..	..	..	..	..	..
T. nemocallosus Ric.	..	..	..	..	..	..	..	..	+	+	..	..	..	..	..	..	..	..	..	..
T. optatus Wlk.	..	..	..	..	..	..	..	..	+	+	..	..	..	..	..	..	..	+	+	..
T. melanognathus Big.	..	..	..	..	..	..	..	..	..	+	..	..	..	..	..	..	..	..	+	..
T. kakhyenensis S.W.	..	..	..	..	..	..	..	..	..	..	..	..	..	..	+	..	..	..	..	..
T. nephodes Big.	..	..	..	..	..	..	..	..	..	..	..	..	+	+	..	..	..	..	..	..
T. albofasciatus Ric.	..	..	..	..	..	..	..	..	..	..	..	..	..	+	..	..	..	..	..	..

(contd.)

	S. Ceylon Hills.	N. Ceylon.	S. Indian Hills.	Malabar Coast.	Carnatic.	Rajputana.	C. India.	Baluchistan and Seistan.	I. G. Plains, West.	I. G. Plains, East.	W. Himalaya.	E. Himalaya.	Assam Plains.	Assam Hills.	Burma.	Yunnan.	Siam.	Malay Peninsula.	Malayan Archip.	China.
T. auriflamma Wlk.	..	..	..	..	..	..	..	..	..	..	..	..	+	+	..	..	..	..	..	..
T. striatus Fb.	+	+	..	+	+	..	..	+	+	+	+	+	+	+	+	..	+	+	+	+
T. tenens Wlk.	+	..	..	+	+	..	..	..	+	+	+	..	+	..	..	..	..	..	+	+
T. conicus Big.	..	..	..	..	..	..	..	..	..	+	..	..	..	..	..	..	..	+	+	+
T. indiscriminatus Ric.	..	+	..	..	..	..	..	..	..	..	..	..	..	..	..	..	..	..	..	..
T. rubidus Wd.	..	..	..	+	+	..	..	..	+	+	+	..	+	+	+	..	+	+	+	+
T. speciosus Ric.	..	+	+	+	+	..	..	..	..	+	..	..	..	..	..	..	..	..	..	..
T. abscondens Wlk.	..	..	..	..	..	..	..	..	..	..	..	..	..	..	+	..	..	..	..	+
T. hirtistriatus Ric.	..	..	..	..	..	..	..	..	..	..	..	..	..	..	+	..	..	+	..	..
T. monotoeniatus Big.	..	..	..	..	..	..	..	..	..	..	..	..	+	+	+	+	..	..	..	..
T. brunnipennis Ric.	..	..	+	+	+	..	..	..	..	+	..	..	..	..	+	..	+	..	..	..
T. auristriatus Ric.	..	..	+	+	..	..	..	..	..	..	..	..	..	..	..	..	..	..	..	..
T. hybridus Wd.	..	..	..	..	..	..	..	..	..	..	..	..	+	..	+	..	..	+	+	+
T. jucundus Wlk..	+	..	..	+	..	..	..	..	+	..	+	..	..	..	..	..	..	..	..	+
T. puteus Ric.	+	+	..	..	..	..	..	..	..	..	..	..	..	..	..	..	..	..	..	..
T. rubicurdus Macq.	..	..	..	..	..	..	..	..	..	..	..	..	+	+	+	..	..	..	+	..
T. indianus Ric.	..	..	+	+	..	..	..	..	..	..	..	..	..	..	..	..	..	..	..	+
T. fuscomaculatus Ric.	..	..	..	..	..	..	..	..	..	..	+	+	..	+	+	..	..	..	..	..
T. leucopogon Big.	..	..	..	..	..	..	..	..	..	..	..	+	..	..	..	..	..	..	..	..
T. discrepans Ric.	+	..	..	..	..	..	..	..	..	..	..	..	..	..	..	..	..	..	..	..
T. rufiventris Fabr.	..	+	..	+	..	..	..	..	..	+	..	..	+	+	+	..	..	+	+	+
T. explicatus Wlk.	..	..	+	..	..	..	..	..	..	..	..	+	..	+	..	..	..	..	..	..
T. manipurensis Ric.	..	..	..	..	..	..	..	..	..	..	..	..	..	+	..	..	..	..	..	..
T. angustilmbatus S.W.	+	..	..	..	..	..	..	..	..	..	..	..	..	..	..	..	..	..	..	..

(*contd.*)

	S. Ceylon Hills.	N. Ceylon.	S. Indian Hills.	Malabar Coast.	Carnatic.	Rajputana.	C. India.	Baluchistan and Seistan.	I. G. Plains, West.	I. G. Plains, East.	W. Himalaya.	E. Himalaya.	Assam Plains.	Assam Hills.	Burma.	Yunnan.	Siam.	Malay Peninsula.	Malayan Archip.	China.
T. khasiensis Ric.	..	..	..	..	..	..	..	..	..	+	..	..	..	+	..	..	..	..	..	..
T. oxyceratus Big.	..	..	..	..	..	..	..	..	..	..	+	..	..	..	..	..	..	..	..	..
T. orientis Wlk.	..	..	..	..	..	..	..	..	+	+	+	+	..	..	..	+	..	..	..	..
T. provincialis Ric.	..	..	..	+	..	..	..	..	..	..	..	+	..	..	..	..	..	..	..	..
T. fulvimedius Wlk.	..	..	..	..	..	..	..	..	..	..	?	..	..	..	+	..	..	..	..	+
T. consanguineus Macq.	..	..	..	+	..	..	..	..	..	..	..	..	..	..	..	..	..	..	..	..
T. birmanicus Big.	..	..	..	..	..	..	..	..	..	..	..	..	..	+	+	..	..	+	..	+
T. tenebrosus Wlk.	..	..	..	+	..	..	..	..	..	..	..	..	..	..	..	..	..	..	..	..
T. leucohirtus Ric.	..	..	..	+	..	..	..	..	..	..	..	..	..	..	..	..	..	..	..	..
T. atrohirtus Ric.	+	..	..	..	..	..	..	..	..	..	..	..	..	..	..	..	..	..	+	..
T. flavissimus Ric.	+	..	..	..	..	..	..	..	..	..	..	..	..	..	..	..	..	..	..	..
T. joidus Big.	..	..	..	..	..	..	..	..	..	..	..	..	+	..	..	..	..	..	..	..
T. flaviventris Big.	+	..	..	..	..	..	..	..	..	+	..	..	+	..	?	..	..	..	..	..
T. diversifrons Ric.	..	..	..	..	..	..	..	..	..	+	..	..	+	+	..	..	..	..	..	..
T. demellonis S.W.	..	..	..	+	..	..	..	..	..	..	..	..	..	..	..	..	..	..	..	..
T. subcinerascens Ric.	..	..	..	..	..	..	..	..	..	..	..	..	..	..	+	..	..	..	..	..
T. obconicus Wlk.	..	..	..	+	..	..	+	..	..	+	..	..	..	..	..	..	..	..	..	..
T. tuberculatus Ric.	..	..	..	..	+	..	..	..	..	+	..	..	+	..	..	..	..	..	..	..
T. ceylonicus Sch.	..	+	..	..	..	..	..	..	..	..	..	..	..	..	..	..	+	+	+	..
T. fuscicauda Big.	+	..	..	..	..	..	..	..	..	..	..	..	..	..	+	..	..	..	+	..
T. subhirtus Ric.	..	..	..	+	..	..	..	..	..	+	..	..	..	..	..	..	..	..	+	..
T. subcallosus Ric.	..	..	..	..	..	..	..	..	..	+	+	+	..	..	+	..	..	..	..	..
T. hirtus Wlk.	..	..	..	..	..	..	..	..	..	..	+	..	..	..	..	..	..	..	..	..
T. wyvillei Ric.	..	..	..	..	..	..	..	..	..	..	+	..	..	..	..	..	..	..	..	..
	10	6	8	19	9	..	2	8	12	22	15	9	13	16	19	2	..	..	..	..

TABLE III.

	S. Ceylon Hills.	N. Ceylon.	S. Indian Hills.	Malabar Coast.	Carnatic.	Rajputana.	C. India.	Baluchistan and Seistan.	I. G. Plains, West.	I. G. Plains, East.	W. Himalaya.	E. Himalaya.	Assam Plains.	Assam Hills.	Burma.	Yunnan.	Siam.	Malay Peninsula.	Malayan Archip.	China.
Chry. flavocinctus Ric.	..	+	..	+	..	..	..	..	..	..	..	+	+	+	+	..	+	+	+	..
C. mlokosiewiczi Big.	..	..	..	..	..	..	..	..	..	..	..	..	..	..	..	+	..	..	..	+
C. dispar Fb.	+	+	+	+	+	..	+	..	+	+	+	..	+	+	+	..	+	+	+	+
C. designatus Ric.	..	..	..	..	..	..	..	..	+	..	+	..	+	..	..	+	..	..	..	..
C. pellucidus Fb.	..	..	+	+	+	..	..	..	..	+	..	..	..	..	..	..	..	..	..	..
C. pettigrewi Ric.	..	..	..	..	..	..	..	..	..	..	..	..	..	+	..	..	..	..	..	..
C. stimulans Wlk.	..	..	..	..	..	..	..	..	..	+	..	..	..	..	..	..	..	..	..	..
C. indianus Ric.	..	..	+	..	..	..	..	..	..	..	..	..	..	+	..	..	..	..	..	..
C. fasciatus Wd.	..	..	+	..	..	..	..	..	..	..	..	..	..	+	..	..	..	..	+	+
C. fixissimus Wlk.	..	+	..	..	..	..	..	..	..	..	..	..	+	..	..	..	..	..	+	+
C. flavipes Meig.	..	..	..	..	..	..	..	+	..	..	..	..	..	..	..	..	..	..	..	..
	1	3	3	2	2	..	1	1	2	3	2	1	4	4	2	2	..	..	..	..

KEYS TO SUB-FAMILIES AND TO GENERA, ADAPTED FROM DE SURCOUF.

Hind tibiae without spines. No ocelli, at most a vestigial tubercle	TABANINÆ
Hind tibiae always spines. Usually three well-developed ocelli	PANGONINÆ

TABANINAE

1. Apical portion of third antennal segment of three divisions	*Hæmatopota* Meig.
Apical portion of third antennal segment of four divisions	2
2. Third antennal segment simple, no basal, tooth or process.	3
Third antennal segment with a basal tooth or process.	4
3. Antennae arising from a frontal tubercle, third segment cylindrical. Palpi thickened basally, bent, apically pointed	*Udenocera* Ric.
Antennae not arising from a frontal tubercle, third segment conical. Palpi elongate, subulate	*Diachlorus* O.S.
4. Antennae not arising from a prominent tubercle. Frontal calli usually present.	*Tabanus* Linn.
Antennae arising from a prominent tubercle. No frontal calli.	*Neotabanus* Ric.

PANGONINAE

1. Third antennal segment of 8 divisions. Proboscis greatly prolonged	*Corizoncura* Rond.
Third antennal segment of 5 divisions Proboscis normal.	2
2. Third antennal segment with a basal tooth or process.	*Gastroxides* Saund.

Third antennal segment simple.	3
3. First and second antennal segments long, subequal.	*Chrysops* Meig.
First antennal segment long, second short.	*Silvius* Meig.
4. Tibiae normal	*Melissomorpha* Ric.
Tibiae large and flattened. Whole aspect of insect bee-like.	

TABANIDÆ.

Howlett, Ind. Ins. Life, 592-595, tab. lxii, [general acct.]; *id., loc. cit.*, tab. lxii, 1-1*c*, [life history of unamed sp.]; Lefroy, Pusa Bull. 7, (1907), 19-30, [preliminary acct., bitting flies of India]; Hine, U.S. Dept. Agric. Bull., Tech. ser. 12, pt. 11, (1906), E habits and bionomics, N. Amer. spp.]; King, 4th. Rept. Wellcome Lab., (B), 114-125, tab., [life histories of some Tabanid larvae]; Hill, Bull. Entom. Res., xii, 41-62, tabs. and t. figs., [bionomics of some Australian spp.]; Hine, Bull. U. S. Dept. Agric., xliv, 57-60, xlvi, 23-25, (1904), [bionomics of N. Amer. spp.]; March, Rockefeller Inst. Monog., xiii, [early stages, abst. in Rev. App. Ent., (B), ix, 122]; Jones, Entomologist, lv, 40-42, [habits of ♂ ♂, British]; Patton, Ind. Jo. Med. Res., vii, 735-750, tab. lxvii and t. fig., [Mesopotamian spp.]; Gaiger, Jo. Trop. Vet. Sci., v, 65-71, [check list of parasites of Indian domestic animals]; Shiraki, Taihoku Expt. Stn. Bull- (1918), [acct. of Formosan spp.]; McFie, Bull. Entom. Res., iii, 223-224, [attracted by scale insects, sweeping movements of tarsi]; Neave, *loc. cit.*, iii, 281,] if ♂♂ numerous, ♀♀ not biting]; Patt. and Cragg, T.B. Med. Ent., 271-276, [general acct.], 294-305, [bionomics and lab. breeding]; Surc., Gen. Ins. Fasc. 175, 17, [copulation], 22, [bionomics], 183-194 [do.]; Isaac., Ind. Agric. Ent. Mem. viii, 53, [method of rearing], 57, [number of moults].

Anatomy. Patt. and Cragg, T. B. Med. Ent., 22-28, tab. iv, 4, tab. v, 1-4, [mouth parts], 35, [Prestomum], tab. xiv, 3-4, [ovipositor], 69-74, [thorax], tab. xv, 3, [tarsus], tab. xx, 2-3, p. 98, [thorax and respiratory system], 102-108, tab. xxi, 1, 2, 5, 6, [alimentary tract], 124-5, tab xxvi, [digestion], tab. xxvii, 1, [nervous system], tab. xxviii, 1, [genital organs] 421, [dissection]; Lloyd, Bull. Ent. Res., iii, 96, [dissection of salivary glands]; Vogel, Zool. Jahrb., xlii (2), 259-282, tab. x, [mouth parts]; Cragg, Ind. Jo. Med. Res. vii, 648-663, tabs. lxi, lxii, lxiii, t. fig., [secretion and epithelial regeneration in mid gut]; Surc. Gen. Ins. Fasc. 175, 1-2, tab. i, 1-3, [head], 3, [mouth parts], 6, [thorax], 7, [wing], 11 tab. i, 4-5, [legs], 11, [abdomen], 12, tab. i, 6b-f, 7a-b, 8a-b, 9, 10, 11a-b, 12a-b, 13a-b, [♂ genitalia], 15, tab. i, 14-15, [♀ genitalia], 19, [larva], 21, [pupa]; Graber, Arch. f. mik. Anat., xvi, (1879), [Graber's organ].

Classification and Distribution. Surc.' Gen. Ins. Fasc. 175, 23, [classification] 201-204, [distribution]; Aust., Bull. Ent. Res., xii, 431-455, [Siamese]; Pratt, Jo. Trop. Vet. Sci., iv, 390-4, [principal F. M. S. spp.]; Villen., Ann. Soc., Ent. Fr., (1905), 306-310, [Meigen's types at Paris]; Ric., R. I. M., iv, 111-258, tab. xiii, 321-399. tabs. xv-xviii, [revision of Oriental spp.].

Enemies. d'Abreu, Proc. iiird. Ent. Meetg. Pusa iii, 866, [birds]; Shiraki, Taihoku Expt. Stn. Bull., (1918), 5, [American and Australian *Phanurus* spp.]; Bull. Ent. Res., i, 156-7, *loc. cit.*, ii, 182, [*Bembex* spp.]; Patel, Pusa Sci. Repts., (1919-20), 96, [Indian egg parasites]; Fletcher, *loc. cit.*, (1920-21), 47-48, [do.]; Surc., Gen. Ins. Fasc. 175, 22-23, [general notes]; Port., Petrograd, 1899, (Rossicé), (Summary in Zool. Centralbl., vii, 807, (1900), [methods of destruction].

Disease Transmission. Pease, Jo. Trop. Vet. Sci., i, 70, 91, [2 African spp. carrying Surra]; Lingard, *loc. cit.*, i, 92-112, [carrying over of *Tryp. evansi* from season to season]; Holmes, *loc. cit.*, i, 119, [biting flies in Trypanosomiasis]; Sergent, Ed. & Et., *loc. cit.*, i, 214-241, (abs. from Ann. Inst. Pasteur, [el Debab]; Birt, Jo. R. A. M. C., vi, 653-58, (abs. Jo. Trop. Vet. Sci. i, 449-451), [Herpetomonads of 2 European spp.]; Laveran & Mesnil, Ann. Inst. Pasteur, xx, 296-303, (abs. Jo. Trop. Vet. Sci., i, 467), [trypanosomiasis of horses in Annam compared with Surra, Tabanidae carry]; Holmes Jo. Trop. Vet. Sci., iii, 157-172, tabs. xiv, xv, [investigation of surra out-break, acct. of Naini Tal road]; Marra, Rev. Gen. Méd. Vét. xi, 49-57, (abs. Jo. Trop. Vet. Sci., iii, 409, [spreading epizootic aphtha]; Leese, Jo. Trop. Vet. Sci., iv, 107-132, [expts. on natural trans. of surra]; Fraser

Symonds & Pratt, Pub. ix, F. M. S. Inst. Med Res., (abs. Ind. Med. Gaz., xliii, (11), 427-8, Jo. Trop. Vet. Sci., vi, 203-4; reprinted, *loc. cit.*, iv, 345-389), [surra in F. M. S., 4 spp. of Tabanidae carry]; Old, Jo. Trop Med. Hyg., xii, (2), (1909), (reprinted. Jo. Trop. Vet. Sci., iv, 395-416, [Trypanosomiasis]; Baldrey, Jo. Trop. Vet. Sci., vi, 271-282, tabs. xiii, xiv, [developmental cycle of surra in Tabanidae]; Leese, *loc. cit.*, vii, 19-32, (abs. Trop. Vet. Bull. i, 15-17), [biting flies and surra]; Patton, Ind. Jo. Med. Res., vii, 735-750, [summary of knowledge to 1920 on Tabanidae and trypanosomiasis]; *id.*, Arch. f. Protist., xv, 333-362, tab. xxx, and t. figs., [*Crithidia tabani*]; Fletcher & S. White, Proc. ivth. Ent. Meetg. Pusa, 222-235, tab. xl, [review of knowledge of vectors of surra]; Cross & Patel, Punjab Vet. Bull., (6), 1921, (abs. Rev. Appl. Ent., (B)), x, 63, [surra, ticks carry, Tabanidae only vectors]; Sergent, C. R. Soc. Biol. Paris, lxxxv, 408 (abs. Rev. App. Ent. (B)), ix, 181, [Tabanidae transmit Sarcocystis].

TABANINÆ.

Shiraki, Taihoku Expt. Stn. Bull., (1918), 87-88; SurC., Gen. Ins. Fasc. 175, 25, [key to genera of World].

KEY TO THE KNOWN INDIAN SPECIES OF *HAEMATOPOTA*. BASED ON RICARDO'S GROUPING.

[Generally applicable to ♀♀ only].

1. Legs uniform in colour	Group I.
Legs not uniform in colour	2
2. Tibiae without pale rings, but base of front pair, at least, pale	Group II.
At least mid tibiae with median and sub-apical pale rings	3

GROUP IV.

1. Apical band of wing double	2
Apical band of wing single	*annandalei* Ric.

2. Abdomen with lateral grey spots	*litoralis* Ric.
Abdomen with only hind margins of segments greyish	*rhizophorae* S. W.

GROUP V.

1. Wing with a pale, oblique transverse streak only	*rubida* Ric.
Wing with the usual rosette markings	2
2. Apical band of wing double	3
Apical band of wing single	4
3. Hind tibiae strongly fringed	*javana* Wd.
Hind tibiae not strongly fringed	*demellonis* S. W.
4. Apical band very broad, other pale markings very narrow.	*albofasciatipennis* Brun.
All pale markings of wing approximately equal in breadth, as usual	5
5. First antennal segment incrassate, as long as first division of third segment	*lata* Ric.
First antennal segment narrower, not as long as first division of third segment	*assamensis* Ric.

GROUP VI.

1. Apical band of wing double	2
Apical band of wing single	7
2. Apical bands very broad, almost filling up spex of wing.	*marginata* Ric.
Apical bands not almost filling up apex of wing.	3
3. Both branches of apical band joined at their base, then diverging.	4
Branches of apical band separate, commencing on opposite borders of wing, and usually overlapping in the middle	6
4. First segment of antenna shorter than basal annulation of third	5
First segment of antenna longer than basal annulation of third	*tesselata* Ric.

5. Frontal callus yellow-brown, paired sports touching it — *brevis* Ric.

   Frontal callus shining black-brown, paired spots well separated — *ricardonis* S. W.

6. First antennal segment incrassate, as long as first division of third segment — *bilineata* Ric.

   First antennal segment narrower, but nearly as long as whole third segment — *cingalensis* Ric.

7. Apical band very broad, reaching from outer to inner border of wing — 8

   Apical band of wing narrower — 9

8. A broad black band on face — *latifascia* Ric.

   No broad black facial band — *montana* Ric.

9. Apical band reaching from outer to inner border of wing — 10

   Apical band not reaching from outer to inner border — 11

10. Frons very narrow. A large cordiform dark spot in place of usual paired spots — *fuscifrons* Aust.

    Usual paired spots present on frons — *fasciata* Ric.

11. Hind femora and apical two-thirds of hind tibiae black fringed — *burmanica* S. W.

    Hind legs without noticeable fringes — 12

12. First antennal segment more than half as longs as third — 13

    First antennal segment shorter than this — *inconspicua* Ric.

13. A black band on upper part of face — *hindostani* Ric.

    No black facial band, but usually a spot on each side below — *immaculata* Ric.

1. Hæmatopota, Meigen

Hæmatopota, Meig., Illig. Mag., ii, 267, 40, (1803); Wulp, Cat. Dipt. S. Asia, 66; Howlett, Ind. Ins. Life, tab. lxii, 5, [unnamed sp. figd.]; Ric., A. M. N. H. (7) xviii, 113; *id.,* R. I. M., iv, 321, [papers on Oriental spp. with keys]; Shiraki, Taihoku Expt. Stn. Bull. (1918), 88-92, [Formosan spp.]; Patt. & Cragg, T.

B. Med. Ent., 291, 292, 295, [feeding point], 297, [oviposition]; Cragg, Sci. Mem. M. O. Ind., iv, [structure of *H. pluvialis*].

Chrysozona, Meig., Nouv. Class., 23, 34, (1800); Kert., Cat. Dipt., iii, 201; Verrall, "British Flies", v, 772, [reasons against adoption of name]; Aust., African Blood-Sucking Flies, 121, [do.]; Surc., Gen. Ins. Fasc. 175, 29, [key to sub-genera], 31, [sub-genera defined].

Austenia, Surc., Bull. Mus. H. N. Paris, 455, (1909); *id.* Gen. Ins. Fasc. 175, 29, [sub-genus].

Parhaematopota, Grünb., Zool. Anz., xxx, 360, (1906); Surc., Gen. Ins. Fasc. 175, 36, [sub-genus].

Potisa, Surc., Bull. Mus. H. N. Paris, 455, (1909); *id.* & Gonz., Dipt. Vuln. Venezuela, ii, 29; *id.,* Gen. Ins. Fasc. 175, 30. [sub-genus].

albofasciatipennis, Brunetti.

Hæmatopota albofasciatipennis, Brun., R. I. M., vii, 458-60, tab. xxxvii, 1,[1] (1912). — Bhowali (Kumaon)[1]

Chrysozona albofasciatipennis, Surc. Gen. Ins. Fasc. 175, 32.

annandalei, Ricardo.

Hæmatopota annandalei, Ric., R. I. M., iv, 335-6, tab. xvii, 21,[1] (1911); Brun., *loc. cit.*, xxvi, 100, [♂ descr.]. — Shillong[1]; Umling (Assam)[1]

Chrysozona annandalei, Surc., Gen. Ins. Fasc. 175, 32. — [Baroma (Assam) Igatpuri (Bombay)]*

assamensis, Ricardo.

Hæmatopota assamensis, Ric., R. I. M., iv, 343-4, tab. xvii, 29, (1911)[1]. — Nongpoh (Khasia)[1]; Shillong[1]

Chrysozona assamensis, Surc., Gen. Ins. Fasc. 175, 32. — [Dogaon]

beesoni, Senior White.

Hæmatopota beesoni, S. W. Mem. Dept. Agric. Ind., vii, 145, tab. xi, 4, 5, (1922), 107, [note][1]. — Bondaung (Burma)[1]

bilineata, Ricardo.

Hæmatopota bilineata, Ric., R. I. M., iv, 350-1, tab. xviii, 34, (1911)[1]. — Igatpuri (W. Ghats)[1]

Chrysozona bilineata, Surc., Gen. Ins. Fasc. 175, 32.

---

* Records of localities enclosed within square brackets have been added from material in the Pusa collection.—T. B. F.

brevis, Ricardo.

Hæmatopota brevis, Ric., A. M. N. H., (7) xviii, 122-3, tab. vi, 19,[1] (1906); *id.,* R. I. M., iv, 349-50, tab. xvi, 16[2]. — Trinco Dis. (Ceylon)[1,2] Bangalore[1, 2] Punjab Madras Pres. (B. M. Colln.)

burmanica, Senior White.

Hæmatopota burmanica, S. W., Mem. Dept. Agric. Ind., vii, 147, tab. xi, 6, 7,[1] (1922); 107, [note]. — Bondaung (Burma)[1]

cana, Walker.

Hæmatopota cana, Wlk., List Dipt. B. M., i, 207,[1] (1843); Big., Bull. Soc. Zool. Fr., xvi, 79[2]; *id.* T. A. S. B., lx, 264[3]; Wulp, Cat. Dipt. S. Asia, 66[4]; Ric., A. M. N. H., (7) xviii, 116-17[5]; *id.* R. I. M., iv, 330[6], [redescribed].

Chrysozona cana, Kert., Cat. Dipt., ill, 202[7]; Surc., Gen. Ins. Fasc. 175, 328.

— [1, 2, 3, 4, 5, 6, 7, 8] N. Bengal. (Probably from Terai). The Collector, Dr. Campbell, was here stationed (*vide* Hooker's Himalayan Journals') Poona (B. M. Colln.)

cingalensis, Ricardo.

Hæmatopota cingalensis, Ric., A. M. N. H., (7) xviii, 119, tab. v, 16[1], (1906); *id.,* R. I. M., iv, 351-2, tab. xvi, 15[2]; Henry, Trop. Agric., xlvii, 100[3]. — Trinco Dis. (Ceylon)'[1, 2] Anuradhapura (Ceylon)[1, 2]

Chrysozona cingalensis, Surc., Gen., Ins. Fasc. 175, 32. — Habarane (N. C. P. Ceylon)[2] Matale (Ceylon)[3]

demellonis, Senior White.

Hæmatopota demellonis, S. W., Mem., Dept. Agric, Ind., vii, 146, tab. xi, 1-3[1], (1922) — Goa (Port. Ind.)[1]

dissimilis, Ricardo.

Hæmatopota dissimilis, Ric., R. I. M., iv, 331-2, tab. xvii, 22[1], (1911). — N. Kanara[1]

Chrysozona dissimilis, Surc., Gen. Ins. Fasc. 175, 33.

fasciata, Ricardo.

Hæmatopota fasciata, Ric., R. I. M., iv, 358[1], (1911). — Shillong[1] Darjiling District (B. M., Colln.)

Chrysozona fasciata, Surc. Gen. Ins. Fasc. 175, 33.

fuscifrons, Austen.

Hæmatopota cordigera, Big., Bull. Soc. Zool. Fr., xvi, 76[1], (1891); *id.,* Mem. Soc. Zool. Fr., v, 626[2]; Wulp, Cat. Dipt. S. Asia, 67[3]; Ric., A. M. N. H., (7) xviii, 125, tab. vi, 21[4].

Bengal[1, 2, 3, 4, 5]
Bondaung (Burma)[1]
Mohnyin (Burma)[6]

Chrysozona cordigera, Kert., Cat. Dipt., iii, 202; Surc., Gen. Ins. Fasc. 175, 33.

Hæmatopota fuscifrons, Aust., A. M. N. H., (8) i, 411, (1908), [renamed]; Ric., R. I. M., iv, 357-8, tab. xvii, 17[5]; S. W., Mem. Dept. Agric. Ind., vii, 107[6].

[Preoccupied by another *H. cordigera,* Bigot, from Abyssinia].

hindostani, Ricardo.

Hæmatopota hindostani, Ric., A. M. N. H., (8) xix, 226[1], (1917).

Bababuddin Hills (Mysore)[1]
[Dodabetta (Nilgiris)]

Chrysozona hindostanti, (*laps.*), Surc. Gen. Ins. Fasc. 175, 33.

immaculata, Ricardo.

Hæmatopota immaculata, Ric., R. I. M., iv, 359-60, tab. xviii, 33[1], (1911).

Kanara (Bombay)[1]
Bombay[1]

Chrysozona immaculata, Surc., Gen. Ins. Fasc. 175, 33.

inconspicua, Ricardo.

Hæmatopota inconspicua, Ric., R. I. M., iv, 358-9, tab. xviii, 31[1], (1911).

Igatpuri (W. Ghats)[1]

Chrysozona inconspicua, Surc., Gen. Ins. Fasc. 175, 33.

indiana, Bigot.

Hæmatopota indiana, Big., Bull. Soc. Zool. Fr., xvi, 78[1], (1891); *id.,* Mem. Soc. Zool. Fr., v, 626; Wulp, Cat. Dipt. S. Asia, 67; Ric., A. M. N. H., (7) xviii, 124, tab. vi, 20; *id.,* R. I. M., iv, 328-9, tab. xvii, 19[2], [redescribed].

Margherita (Assam)[1, 2]
Tezpore (Assam)[2]
[Shillong (Khasis)]

Chrysozona indiana, Kert., Cat. Dipt., iii, 203; Surc., Gen. Ins. Fasc. 175, 33.

javana, Wiedemann.

Hæmatopota javana, Wd., Dipt. Exot., 100, (1821); *id.,* Auss. Zweifl. Ins. i, 218[1]; Wulp., Dipt. Sumat. Exp., 19, tab. i, 12[2]; *id.,* Cat. Dipt. S. Asia, 66[3]; Wlk., List

Pusa[4], Calcutta[4]
Bengal (generally)[4]
Lushai Hills[4]

Dipt. B. M., i, 208; *id., loc. cit.*, Suppl. i, 297, 13; Big., Bull. Soc. Zool. Fr., xvi, 77; Howlett, Ind. Ins. Life, tab. lxii, 5; Ric., R. I. M., iv, 340-3, tab. xvii, 25[4]; Fletcher, Pusa Sci. Reprs., 1920-21, 46, tab. iii, [bionomics, early stages]; S. W., Mem. Dept. Agric. Ind., vii, 107[5].

Banhar (Bihar)[5]
Pyinmana (Burma)[5]
Bondaung (Burma)[5]
[Combatore (S. India) Bihar (generally Murree hills (Punjab) Manikganj (Bengal) Taru, near Peshwar Nongpoh (Assam) Khasi hills (Assam) Kathgodam (United Provinces) Katteri (Nilgiris, S. India)]

Chrysozona javana, Kert., Cat. Dipt., iii, 204; Surc., Gen. Ins. Fasc. 175, 34.

? *Hæmatopota asiatica,* Rond., Ann. Mus. Civ. Genoa, vii, 461, (1875), [*laps.*]; Kert., Cat. Dipt., iii, 204.

? *Hæmatopota nigra,* Wd., Dipt. Exot, 101, 7[6], (1821); Kert., Cat. Dipt., iii, 204-5.

Java[1, 3, 6,]
Sumatra[2, 3]
Borneo[3]
F. M. S.[4]

lata, Ricardo.

Hæmatopota lata, Ric., A. M. N. H., (7) xviii, 121-2[1], (1906); *id.*, R. I. M., iv, 344-5 tab. xvii, 26[2]; *id., loc., cit.*, iv, 401[3]; Edw, Jo. F. M. S. Mus., viii, 27[4], [?]; C. Vet Dept. Mem., ii, tab. v, B, [?]; Fletcher and S. W., Proc ivth. Ent. Meetg., Pusa, 226[5].

Khasia Hills[1, 2]
N. Chin Hills[1, 2]
Yunnan[3]
Mohand (U. P.)[5]

Sumatra, Montane[4]

Chrysozona, lata, Surc., Gen. Ins. Fasc. 175, 34.

latifascia, Ricardo.

Hæmatopota latifascia, Ric., R. I. M., iv, 356, tab. xvi, 10[1], (1911).

Shillong[1]

Chrysozona, latifascia, Surc., Gen. Ins. Fasc. 175, 34.

limbata, Bigot.

Hæmatopota limbata, Big., Bull. Soc. Zool. Fr., xvi, 78[1], (1891); *id.*, Mem. Soc. Zool. Fr., v, 626[2]; Wulp, Cat. Dipt. S. Asia, 66[3]; Ric., A. M. N. H., (7) xviii, 115[4]; *id.,* R. I. M., iv, 325-6[5], [redescribed].

Shillong[1, 2, 3, 4, 5]
Khasia Hills[4, 5]

Chrysozona limbata, Kert., Cat. Dipt., iii, 205; Surc., Gen. Ins. Fasc. 175, 34.

litoralis, Ricardo.

Hæmatopota litoralis, Ric., A. M. N. H., (8), xi, 546[1], (1913). — Puri (E. Coast)[1]

Chrysozona litoralis, Surc., Gen. Ins. Fasc. 175, 34.

marginata, Ricardo.

Hæmatopota marginata, Ric., R. I. M., iv, 347-8, tab. xviii, 32[1], (1911). — Pusa[1] Goalbathan (E. Bengal)[1] Tezpore (Assam)[1] Bihar (B.M. Col. [Saran (Bihar)]

Chrysozona marginata, Surc., Gen. Ins. Fasc. 175, 34.

montana, Ricardo.

Hæmatopota montanus, Ric., A. M. N. H., (8) xix, 225[1], (1917). — Bababuddin Hills[1] Coonoor[2]

Hæmatopota montana, Patton, Ind., J. Med. Res., x, 67[2]. — Mahableshwar (B.M. Col.) [Santikoppa, Sidapur and pol betta (Coorg) Shillong, (Assam)]

Chrysozona montana, Surc., Gen. Ins. Fasc. 175, 34.

rhizophorae, Senior White.

Hæmatopota rhizophorae, S. W., Spol. Zeyl., xi, 387, tab. i, I, (1921). — Trincomali (Ceylon)[1]

ricardonis, Senior White.

Hæmatopota ricardonis, S. W., Spol. Zeyl., xiii, 209, tab. vii, 1-3[1], (1925). — Ratnapura (Ceylon)[1]

roralis, Fabricius.

Hæmatopota roralis, Fb., Syst. Antl., 107, 2[1], (1805); Wd., Dipt. Exot., 97, 1[2]; *id.,* Auss. Zweifl. Ins., i, 215[3]; Big., Bull. Soc. Zool. Fr., xvi, 78; *id.,* T. A. S. B., lx, 264[4]; Wlk., List Dipt. B. M., i, 208; Wulp, Cat. Dipt. S. Asia, 66[5]; Ric., A. M. N. H. (7) xviii, 116; *id.,* R. I. M., iv, 328-5, tab. xvii, 24[6], [redescribed]. — [1, 2, 3, 4, 5, 6]Tranquebar (E. Coast) Trinco District (Ceylon)[6] Calcutta[6] Goalbathan (E. Bengal)[6] Gorakhpur (U.P.)[6] Bihar[2] [Manaparai (Madras Pres.) Nagpur Bhandara (C.P.) Surat (Bombay)

Chrysozona roralis, Kert., Cat. Dipt., iii, 208, Surc., Gen. Ins. Fasc. 175, 35.

Balugaon (Orissa)
Bihar (generally)]
Nalanda (Ceylon)
S.W.

———

Java[5]
Borneo[5]

rubida, Ricardo.

Hæmatopota rubida, Ric., A. M. N. H., (7) xviii, 120, tab. vi, 18[1], (1908); *id.,* R. I. M., iv, 338, tab. xvii, 18[2]; Pratt, Jo. Trop. Vet. Sci., 390-4[2], [F. M. S. distribution].

Burma[1,2]

———

F. M. S.[3]

Potisa rubida, Surc., Gen. Ins. Fasc. 175, 31.

striatipennis, Brunetti.

Hæmatopota striatipennis, Brun., R. I. M., vii, 460-1, tab. xxxvii, 2[2], (1912); S. W., Mem. Dept. Agric. Ind., vii, 107[2].

Dehra Dun[1, 2]

Chrysozona striatipennis, Surc., Gen. Ins. Fasc. 175, 35.

tessellata, Ricardo.

Hæmatopota tessellata, Ric., A. M. N. H., (7) xviii, 120, tab. v, 7[1], (1906); *id.,* R. I. M., iv, 348-9, tab. xvi, 14[2].

[1, 2]Trinco District (Ceylon)

Chrysozona tessellata, Surc., Gen. Ins. Fasc. 175, 35.

unizonata, Ricardo.

Hæmatopota unizonata, Ric., A. M. N. H., (7) xviii, 118, tab. v, 15, 15A[1], (1906); *id.,* R. I. M., iv, 332-3, tab. xvi, 12, 13[2]; Pratt, Jo. Trop. Vet. Sci., iv, 390-4[3], [F. M. S. distribution]; Longstaff, 'Butterfly Hunting,' 384[4]; Edw., Jo. F. M. S. Mus., vii, 27[5], [?]

Hakgala[1, 2]
Newera Eliya[2]
Pundaluoya[2]
Hatton[4]
Galagedera [all Ceylon Hills]

———

F. M. S.[2]
Sumatra (mountain)[3]

Chrysozona unizonata, Surc., Gen. Ins. Fasc. 175, 36.

2. Udenocera, Ricardo.

Udenocera, Ric., A. M. N. H., (7) xiv, 352, (1904); Kert., Cat. Dipt., iii, 210; Ric., R. I. M., 361-2; Surc., Gen. Ins. Fasc. 175, 48.

brunnea, Ricardo.

Udenocera brunnea, Ric., A. M. N. H. (7), xiv, 355[1], (1904); *id.,* R. I. M., iv, 362-3, tab xvi, 8[2]; Kert; Cat. Dipt.,

Kandy (Ceylon)[1, 2]

———

iii, 210; Surc., Gen. Ins. Fasc. 175, 49, [description requoted], tab. ii, 12a.

? *Chrysops parallelus,* Wlk., Proc. Linn. Soc., Lond., v, 276[3], (1861). — Batjan Ids.[3, 4]

? *Diachlorus parallelus,* Ric., A. M. N. H., (7) ix, 372[4].

3. Diachlorus, Osten-Sacken.

Diachlorus, O.-S., Mem. Boston N. H. S., ii, 475, (1876); Kert., Cat. Dipt., iii, 211; Surc., Gen. Ins. Fasc. 175, 49.

Diabasis, Macq., Suit. à Buff., i, 207, (1834), [*nec* Hoffmsg., 1819].

fulvescens, Brunetti.

Diachlorus fulvescens, Brun., R. I. M., vii, 456-8[1], (1912); Surc., Gen. Ins. Fasc., 175, 51, [wrong date quoted]. — Dha Kuri (Kumaon)[1]

KEY TO THE KNOWN INDIAN SPECIES OF *TABANUS*.

[Adapted from Ricardo. Generally applicable to females only.]

1. Eyes bare	2
Eyes hairy (sub-genus *Therioplectes*)	Group XI,
2. Frons with varying arrangement of calli	3
Frons with no calli	Group IV
3. Frons with two separate calli.	4
Frons with one square, narrow or oblong callus, which is always prolonged towards the vertex by a more or less narrow line	6
4. First callus usually transverse, second irregular in shape, smaller. Frons 3-6 times as long as broad. Small spp., 9-14 mm. long.	5
Both calli small, round, spot-like, one above the other	Group III
5. Thorax and scutellum concolorous. Abdomen unbanded	Group I
Scutellum pale haired. Abdomen with pale bands	Group II
6. Wings with spots or bands	Group V

	Wings distinctly dark marked, but not banded. Large spp. 18-23 mm. long	Group VI
	Wing not noticeably marked	7
7.	Abdomen with stripes and or bands	8
	Abdomen without stripes or bands	Group X
8.	Species with one or more usually continuous stripes on abdomen.	Group VII
	Species with median and or lateral spots on abdomen, not usually forming a continuous stripe.	Group VIII
	Species with pale bands and sometimes spots on a abdomen.	Group IX

GROUP I.

1.	Femora black or grey	2
	Femora yellow	6
2.	Femora black, front tibiae pale in whole or part.	3
	Femora grey, front tibiae black	l*eleani* Aust
3.	Thorax and abdomen shining black, latter with narrow red segmentations	*Hirtipalpis* Ric.
	Thorax with grey tomentum, usually more or less forming stripes	4
4.	Frons 7-8 times as long as wide. 1st posterior cell, (in ♀ only) narrowed or closed.	*eggeri* Schin.
	Frons broader, 5 times as long as wide. 1st posterior cell normal.	5
5.	Antennae black	*glaber* Big.
	Antennae reddish	*excelsus* Ric.
6.	Abdomen black, with three grey stripes	*macer* Big.
	Abdomen mainly black, unstriped, some reddish laterally	7
7.	Abdomen black, frons less than 3 times as long as wide	*pulverifer* Wlk.
	Antennae yellow. Frons 3½-4½ times as long as wide	8
8.	Four round black dots on fore broder of all abdominal segments except last	*pulchellus* Lw.

	Abdomen with black median stripe, extended to sides along hind margins, and occupying nearly all of last 3 segments dorsally	*adjacmmcens* Ric.

GROUP II.

1.	Brown species, scutellum and abdominal bands yellow haired	2
	Black species, scutellum and abdominal bands white-haired	3
2.	Abdomen with 6 narrow bands. Front tibiae pale on basal two-thirds only	*flavicinctus* Ric.
	Abdomen with 2 broad bands. Front tibiae with only extreme tips black.	*leucocnematus.* Big.
3.	Abdomen with 2 bands. Front tibiae pale on basal two-thirds only	*bicinctus* Ric.
	Abdomen with 6 bands. Front tibiae with only extreme tips black	*sexcinctus* Ric.

GROUP III.

1.	Abdomen yellowish, with median and lateral brown or black stripes	*ditaenatus* Macq.
	Abdomen blackish grey with 3 series pale spots	*Sufis Joenn.*

GROUP IV.

1.	Abdomen brown, reddish basally, covered with grey tomentum, with two small greyish spots on each segment. Wings with no appendix. 7-11 mm	*virgo* Wd.
	Abdomen blackish brown, with median and lateral grey tomentose spots. Wings with an appendix. Long 13 mm.	*nemocallosus* Ric.

GROUP V.

1.	First posterior cell open	2
	First posterior cell closed	*kakhyenensis* S.W.
2.	Frons narrowest at vertax. Abdomen reddish, with small white median spots.	*optatus* Wlk.
	Frons narrowest anteriorly Abdomen yellowish, with larger white median spots	*melanognathus* Big.

GROUP VI.

1.	First posterior cell closed or greatly narrowed apically	2
	First posterior cell open as usual	*auriflamma* Wlk.
2.	Abdomen reddish brown, with grey tomentose bands and median spots. First posterior cell closed. Palpi slender.	*nephodes* Big.
	Abdomen brown, with grey tomentose bands but no median spots. First posterior cell narrowed in the margin. Palpi large, broad.	*albofasciatus* Ric.

GROUP VII.

1.	Frontal callus oblong or narrow with linear extension. Frons at least 5 times as long as broad.	2
	Frontal callus large, somewhat protuberant, quadrate, with short linear extension. Frons 3-4 times as long as broad	12
2.	Abdomen with lateral stripes and spots in addition to the median stripe	3
	Abdomen with no lateral spots, or only isolated ones on second segment	8
3.	Frontal callus with a spindle shaped linear extension	4
	Frontal callus with a narrower linear extension.	5
4.	Grey abdominal median and lateral stripes all of same length.	*tenens* Wlk.
	Median stripe beginning only on 3rd segment, lateral stripes ending on 3rd or 4th segment	*striatus* Fb.
5.	Lateral spots large and distinct	6
	Lateral spots small and indistinct. Thorax blackish. Abdomen reddish yellow	*conicus* Big.
	Lateral spots small but distinct. Fore-head narrower anteriorly than in above species	*indiscriminatus* Ric.
6.	Median stripe composed of sports of equal size, not very large	7
	Median stripe composed of unequal sized spots, those on 3rd and 4th segments very large and conspicuous. Femora blackish	*speciosus* Ric.

7. Thorax and abdomen lilac or blackish brown. Femora blackish. — *rubidus* Wd.

   Thorax, abdomen and femora reddish brown — **priscus* Wlk.

8. Median stripe grey tomentose, or white or yellow haired — 9

   Median abdominal tripe golden haired — 11

   Median abdominal stripe black, with a narrow grey stripe often apparent on it — *abscondens* Wlk.

9. Frons barely half as wide anteriorly as at vertex. Small brown species — *hirtistriatus* Ric.

   Frons parallel sided or nearly so — 10

10. Large black or reddish brown species, median abdominal stripe narrow. Femora black. Wings tinged brown — *monotoeniatus* Big.

    Brown species, median abdominal stripe broad, and with 2 round spots on 2nd segment. Femora reddish. Wings brown on fore border — *brunninpennis* Ric.

11. Frons parallel sided. Abdomen yellowish-brown, median stripe dull, indistinct — *auristriatus* Ric.

    Frons narrowed anteriorly. Abdomen yellowish-brown, median stripe broad. Front femora yellowish. Wings brown, especially on fore border — *hybridus* Wlk.

12. Small black species. Median abdominal grey stripe beginning on third segment, lateral stripes ending on second segment. — *jucundus* Wlk.

    Small brown species. Abdomen with complete median grey stripe — *puteus* Ric.

GROUP VIII.

1. Frons narrow, 6-10 times as long as wide anteriorly, with long, narrow frontal callus — 2

   Frons broader, 4-5 times as long as wide, with large, square or pyriform callus — 5

2. Brown, reddish or yellowish, median sized or large species, median abdominal spots triangular, greyish or yellowish — 3

---

*Recently sunk by Austen in *rubidns* Wd.

Reddish species with black spots on 2nd and 3rd abdominal segments	4
3. Abdomen reddish yellow with small median white spots	*rubicundus* Macq.
Abdomen reddish brown, with long triangular median white spots, usually joining one another	*indianus* Ric.
4. Tibiae and tarsi blackish	*fuscomaculatus* Ric.
Tibiae and tarsi reddish	*fuscomaculatus* var. *altemaculatus* Ric.
5. Frontal callus square. Small black species with white-haired median abdominal spots. Legs blackish Tibiae yellowish	*leucopogon* Big.
Frontal callus pyriform. Large reddish brown species with median grey triangular abdominal spots. Legs reddish-brown	*discrepans* Ric.

GROUP IX.

1. Abdomen blackish-brown, the first 2 segments covered with grey tomentum	*rufiventris* Fb.(♂)
Abdomen of usual pattern and appearance	2
2. Frons narrow, 7-8 times as long as broad. Fronal callus narrow, with a linear extension	3
Frons broader, 4-5 times as long as broad. Frontal callus large, with a stout, broad or spindle-shaped extension	8
3. Large species, abdomen with very narrow pale bands	4
Smaller species, abdomen with whitish bands	6
4. Brown species, scutellum and sides of thorax with pale tomentum	*albocostatus* Big.
Black species, no such pale tomentum	5
5. Abdomen fulvous segmented	*explicatus* Wlk.
Abdomen not fulvous segmented	*manipurensis* Ric.
6. Frons narrower anteriorly	7
Frons parallel sided. Frontal callus narrow, not reaching eyes. Antennae, palpi and legs black. Wings faintly brownish	*khasiensis* Ric.

7. Larger species. Antennae, palpi and legs wholly black — *nicobarensis* Schin.

   Smaller species. Antennae black, palpi grey. Basal half of front and basal 4-5ths of posterior tibiae yellowish — *angustilimbatus* S.W.

8. Brown or black species — 9

   Reddish-yellow species — 10

9. Abdomen with broad white bands and large triangular median spots — *rufiventris* Fb. (♀)

   Bands and spots on abdomen narrower. Smaller species — *oxyceratus* Big.

10. Abdomen reddish-yellow, darkened apically — 11

    Abdomen basally reddish-yellow, but last four segment black — *fulvimedius* Wlk.

11. Femora black — 12

    Femora yellow — *consanguineus* Macq

12. Sub-callus dull. Palpi thin — *orientis* Wlk.

    sub-callus shining. Palpi stout — *provincialis* Ric.

GROUP X.

1. Black or brown species, frons narrowed anteriorly. 16-19 mm. long — 2

   Dark reddish-brown species. 18-22 mm. long — 3

   Reddish-yellow species. 16-24 mm. long. — 4

   Small species of all colours from black to yellow, under 16 mm. long — 6

2. Thorax paler than abdomen, with yellowish tomentum and pubescence — *birmanicus* Big.

   Thorax not paler than abdomen, reddish-brown with grey-blue tomentum — *tenebrosus* Wlk.

3. Frons 7 times as long as broad. Front coxae with white hair — *leucohitus* Ric.

   Frons 10 times as long as broad. Front coxae with black hair — *atrohirtus* Ric.

4. Large species, 20-24 mm. Thorax, antennae, palpi and legs yellow — *flavissimus* Ric.

   Smaller species, 16-19 mm. long — 5

5. Frons narrowest at vertex. Legs wholly red.	*nigropictus* Macq
Frons narrowest anteriorly. Legs black, tibiae white at base	*joidus* Big.
6. Frons 5-6 times as long as broad	7
Frons 7-10 times as long as broad, narrowest anteriorly	12
7. Frons narrowest anteriorly. Small reddish-brown species	*trinominata* S.W.
Frons parallel sided, or nearly so	8
8. Reddish-brown species. Frontal callus always distinct	9
Yellow species. Frontal callus often indistinct	*flaviventris* Big.
9. Thorax and scutellum blackish. Antennae black at tip	10
Thorax and scutellum reddish. Antennae not black at tip	11
10. Femora black. Wings slightly tinged brown	*diversifrons* Ric.
Femora yellowish Wings grey, with hyaline streaks in cells	*demellonis* S. W.
11. Abdomen reddish-brown, unicolorous. Legs reddish-yellow, femora often darker.	*Subcinerascens* Ric.
Abdomen with indistinct zig-zag bands of grey tomentum. Legs blackish, femora yellowish	*obconicus* Wlk.
12. Sub-callus bare, shining and tuberculous Sub-callus not bare, shining, tuberculous	13 14
13. Legs reddish-yellow, front tarsi darker. Abdomen yellowish.	*tuberculatus* Ric.
Legs black, tibiae white. Abdomen reddish-brown or black.	*ceylonicus* Schin.
14. Femora black, tibiae yellowish	*fuscicauda* Big.
Femora yellowish	*subhirtus* Ric.

GROUP XI.

1. Frons with 2 transverse calli, subcallus bare and shining. Abdomen black with white-haired bands	*subcallosus* Ric.
Frons with one callus. Subcallus not bare and shining	2

2. Frontal callus black, claviform, with spindle-shaped extension. Abdomen black, reddish at sides of basal segments. Segmentations yellow haired, often produced medianly as basal spots — *hirtus* Ric.

   Frontal callus small, indistinct. Adbomen black, only reddish on sides of 2nd segment, the whole covered with golden-yellow pubescence — *wyvillei* Ric.

[Not included in key: *T. fuscicrura* Big.]

The following species, originally described from "East India" or "India Or.", are not included in the catalogue of species in this genus, as they have been recognized, if at all, subsequently from *extra*-Indian localities only, and were probably originally described from such.

Species.	Synonyms.	Described from	Since recognized from
basalis Macq.	..	India Or.	Sumatra. Siam.
brunneus Macq.	..	India Or.	Java, F. M. S. Timor.
orientalis Wd.	..	East India	. . .
servillei Macq.	*Pyrausta* Wlk.	East India	Java.

4. Tabanus, Linnaeus.

Tabanus, Linn., F. Suec., 462, (1761); Kert., Cat. Dipt., iii, 217; Ric., R. I. M., iv, 111-258, [revision of Oriental spp.]; Shiraki, Taihoku Expt. Stn. Bull., (1918), 119-123, 188-192, [Formosan spp.]; Patt. & Cragg, T. B. Med. Ent., 277-288, [key to Oriental spp.]; Szilady, Ent. Mitteil., iv, 93; Surc., Gen. Ins. Fasc. 175, 54-55, [subdivisions]; Steinheil, Arb. Lab. Warschen, 1896, [abs. in Zool. Centralbl., iv, 873, (1897)], [mouth and method of suction]; Cornwall & Patt., I. Jo. Med. Res., ii, 569-93, [salivary secretion]; Hill, Bull. Ent. Res., xii, 41-59, figs., [bionomics of 3 Australian spp.]; Cragg, I. Jo. Med. Res., vii, 648-63, tabs. lxi-lxiii, [secretion and epithelial regeneration in mid-intestine]; Aust., Bull. Ent. Res., xii, 433-55, [Siamese spp. with key]; *id., loc. cit.*, x, 292-321, [Palestine spp. with key].

Atylotus, O.-S., Mem. Boston N. H. S., ii, 426, (1876); Shiraki, Taihoku Expt. Stn. Bull., (1981) 162, [Formosan spp.].

[A valid sub-genus, used quite incorrectly by Bigot.]

Therioplectes, Zeller, Isis, xi, 819, (1842).

A valid sub-genus.

Bellardia, Rond., Arch. Zool. Modena, iii, 79, (1863).

Invalid.

abscondens, Walker.

Tabanus abscondens, Wlk., Trans. Ent. Soc., v, 275[1], (1860); Wulp. Cat. Dipt. S. Asia, 60[2]; Big., T. A. S. B., lx, 264[3]; Ric., R. I. M., iv, 161[4]; Kert., Cat. Dipt., iii, 217; Surc., Gen. Ins. Fasc. 175, 58. — Burma[1, 2, 3] N. Chin Hills[4] Dawnat Range (Tenasserim)[4]

China[4]

adjacens, Ricardo.

Tabanus adjacens, Ric., R. I. M., iv, 248[1], (1911); Surc., Gen. Ins. Fasc. 175, 58. — Perso-Baluchistan Frontier[1]

albocostatus, Bigot.

Atylotus albocostatus, Big., Mem. Soc. Zool. Fr., v, 649[1], (1892). — "India"[1, 2]

Tabanus albocostatus, Wulp, Cat. Dipt. S. Asia, 63[2]; Kert. Cat. Dipt., iii, 219 ; Ric., R. I. M., iv, 190[3], [*alboscutatus, (laps.),* in key on p. 122]; Surc., Gen. Ins. Fasc. 175, 59.

Not recognized since original description. Type redescribed by Ricardo. — Type redescribed by

albofasciatus, Ricardo.

Tabanus albofasciatus, Ric., R. I. M., iv, 146-7[1], (1911); Surc., Gen. Ins. Fasc. 175, 59. — Shillong[1]

angustilimbatus, Senior White.

Tabanus angustilimbatus, S. W., Spol. Zeyl., xii, 389[1], (1924) — Marlbe Valley (Ceylon)[1]

atrohirtus, Ricardo.

Tabanus atrohirtus, Ric., R. I. M., iv, 209[1], (1911); Edw., F. M. S. Mus., Jo., viii, 28[2]; Surc., Gen. Fasc. 175, 61. — Ceylon[1]

Sumatra (Mount)[2]

auriflamma, Walker.

Tabanus auriflamma, Wlk., List. Dipt. B. M., i, 155[1], (1848); Wulp, Cat. Dipt. S. Asia, 59[2]; Big., T. A. S. B., lx, 264[3]; Kert., Cat. Dipt., iii, 225; Ric, R. I. M., iv, 148- — Sylhet[1, 2, 3, 4] Sibsagar (Assam)[4, 5] Naga Hills[4]

9[4] ; –S. W. Mem. Dept. Agric. Ind., vii, 104[5]; Surc., Gen. Ins. Fasc. 175, 61.

Cachar[4]

auristriatus, Ricardo.

Tabanus auristriatus, Ric., R. I. M., iv, 161-2[1], (1911); Patt., I. Jo. Med. Res., x, 67[2], Surc., Gen. Ins. Fasc. 175, 61.

[The Burmese record, (S. W., Mem. Dept. Agric. Ind., vii, 105), is an error in determination].

Gersoppa (N. Kanara)[1]
Coonoor[2]

bicinctus, Ricardo.

Tabanus bicinctus, Ric., R. I. M., iv, 132-3[1], (1911); Shiraki, Taihoku Expt. Stn. Bull. (1918), 234-40, tab. v, 3[2]; C. Vet. Dept. Mem., ii, tab. iii, 9[3]; Fletcher & S.W., Proc., iv, Ent. Meetg. Pusa, 226; Patton, I. Jo. Med. Res., x, 67[4]; Surc., Gen. Ins. Fasc. 175, 62.

N. Kanara[1]
Mohand (U.P.)[3]
Coonoor[4]
[Pollibetti (S. Coorg)
Sampaji Ghat (Coorg)]
———————
F.M.S.[1], Formosa[2]

birmanicus, Bigot

Atylotus birmanicus, Big., Mem. Soc. Zool. Fr., v, 653[1], (1892)

Tabanus birmanicus, Wulp, Cat. Dipt. S. Asia, 63[2]; Kert., Cat. Dipt., iii, 228; Ric. R. I. M., iv, 200-1[3]; Shiraki, Taihoku Expt. Stn. Bull., (1918), 420-4, tab. viii, 8, xi, 13-14[4]; Surc., Gen. Ins. Fasc. 175, 62.

Burma[1, 2, 3]
Lushai Hills[3]
Nongpoh (Khasia)[3]
———————
F.M.S.[3]
Formosa[4]

brunnipennis, Ricardo.

Tabanus brunnipennis, Ric., R. I. M., iv, 160-1, tab. xiii, 9[1], (1911); Patel, Pusa Sci. Repts. (1919-20), 97[2]; Aust. Bull. Ent Res., xii, 444[3]; S.W. Mem. Dept. Agric. Ind., vii, 105[4]; Surc. Gen. Ins. Fasc. 175, 63.

Basi (B. Kanara)[1]
Gorakhpur (U.P.)[1]
Pusa[2,] S. Malabar[3]
Mohnyin (U. Burma)[4]
[Bistupur (Bankura District)
Puri
Sidapur (Coorg)
Khurda
Santikoppa (N. Coorg)
polibetta (S. Coorg)
Ranchi
Samastipur (Bihar)
Kasaragode (S. Kanara)]
———————
Siam[1, 2]

ceylonicus, Schiner.

Tabanus ceylonicus, Schin., Novara Reise, Dipt., 93[1]; (1868); Wulp, Cat. Dipt. S. Asia, 61[2]; Kert., Cat. Dipt. iii, 233; Ric., R. I. M., iv, 223-4[3]; Surc., Gen. Ins. Fasc. 175, 64.	Ceylon[1, 2] Trincomali, (Ceylon)[3] Trincomali (Ceylone)[2] --- Malay Penin.[3] Java[2, 4, 5]
Tabanus nitidulus, Big., Mem. Soc. Zool. Fr., v, 679[4]; (1892); Wulp, Cat. Dipt. S. Asia, 65; Kert., Cat. Dipt. iii, 265.	Sumatra[3]

conicus, Bigot.

Atylotus conicus, Big., Mem. Soc. Zool. Fr., v, 650[1], (1892).	"Indian"[1, 2] Calcutta[6]
Tabanus conicus, Wulp, Cat. Dipt. S. Asia, 63[2]; Kert., Cat. Dipt., iii, 236.	--- Java[3, 4, 6]
Atylotus abbreviatus, Big., Mem. Soc. Zool. Fr., v, 670[3], (1892).	Formosa[5, 6,] F.M.S.[6]
Tabanus abbreviatus, Wulp, Cat. Dipt. S. Asia, 63[4]; Kert., Cat. Dipt., iii, 217; Shiraki, Taihoku Expt. Stn. Bull., (1918), tab. vii, fig. 1[5]; Ric., R. I. M., iv, 153-4[6]; Surc., Gen. Ins. Fasc. 175, 58.	

consanguineus Macquart.

Tabanus consanguineus, Macq., Dipt. Exot., i, 50[1], (1850); Big., T. A. S. B., lx, 264[2]; Wulp, Cat. Dipt. S. Asia, 59[3]; Ric., R. I. M., iv, 197-8[4]; Surc., Gen. Ins. Fasc. 175, 65; Kert., Cat. Dipt., iii, 236.	Malabar Comast[1, 2, 3, 4]
[Not recognized since original description from type]	Redescribed by Ricardo[1]

demellonis, Senior White

Tabanus demellonis, S. W., Spol. Zeyl., xii, 391[1], (1924).	Combarjua (Port. Ind.)[1]

discrepans, Ricardo.

Tabanus discrepans, Ric., R. I. M., iv, 185, tab. xiv, 17[1], (1911); Surc., Gen. Ins. Fasc. 175, 66.	Ceylon[1]

ditaeniatus, Macquart.

Tabanus ditaeniatus, Macq. Dipt. Exot., i, 126[1], (1838); Wlk., List Dipt. B. M., v, 236, 355; Bezzi, Ann. Mus. Civ. Genoa, (2) xii (32) 184; Aust., Rept. Wellc. Lab. (1906) 57, f. 22; *id.*, 'African B. S. Flies,' 118, tab. xi, 82[3]; Kert., Cat. Dipt., iii, 238; Brun.,	E. India[20] Jhelum District[17] Perso-Baluchistan Frontier[5] Pusa[5], Bengal[5] Madras Coast[5]

Spol. Zeyl., vi, 170[4]; Ric., R. I. M., iv, 134[5]; Old, Jo. Trop. Vet. Sci., iv, 409; Gaiger, *loc. cit.* v, 69; Lefroy, Pusa Bull. 7, p. 21; D. Brock., Bull. Ent. Rers., i, 56,[7] [habits]; King., *loc. it.*, i, 265-74,[8][bionomics, larva descrd.] Sanderson, *loc., cit.* ii, 111, [habits]; Simpson, loc. cit., ii, 215-321[10]; *id., loc. cit.*, iii, 185[11]; Neave, *loc. cit.*, iii, 298,[12] [eye colours]; King., *loc. cit.*, v, 247-8, tab.xxvi, 1, 2,[13][bionomics]; Ric., *loc. cit.*, vi, 407[14]; Patt. & Cragg, T.B. Ent., 295, tab. xlii, 7, [feeding point]; de Mello, Proc. ivth. Ento. Meetg. Pusa, 44,[15] [?] C. Vet. Dept. Mem. ii, tab. i, 1[16]; Fletcher & S. W., Proc., ivth. Ento. Meetg. Pusa, 226; Aust., Trans. R. S. T. M. & H., xv., 264, abs, R. A. E. (B), x, 143[17]; Surc., Gen. Ins. Fasc. 175, 66.

Tabanus bipunctatus, Wulp, Leyd. Mus. Notes, vii, 75, tab. v, 5, [18](1885).

Tabanus bivittatus, Mats., Thous. Ins. Japan, addit. ii, 384, tab. xxii, 12,[22] (1916).

Tabanus fuscinervis, Macq., Dipt., Exot., i, 184, (1839).

Atylotus nigromaculatus, Ric., A. M. N. H., (7) vi, 165,[19] (1900); Newst., Ann. Trop. Med., (1907), 44.

Tabanus pyrrhus, Wlk., Dit. Saunds., i, 47, tab. ii, 45,[20] (1850); Big., T. A. S. B., lx, 264; Wulp, Cat. Dipt. S. Asia, 60; Coq., Proc. U.S. Mus., xxi, 311[21]; Kert., Cat. Dipt., iii, 272.

Umballa[5]
Balighai (Orissa)[5]
Port India[15]
Mohand (U.P.)[16]
Tuticorin[4]
Pathankot(Crosse)
D. Ghazipur (Crosse)[1]
[Dera Ismail Khan
Mianwali (Punjab)
Seistan
Tank (N. W. F.)
Coimbatore (s. India)]
Japan[21, 22], China[5]
Hong Kong.[14]
Africa (throughout)[3, 7, 8, 9, 10, 11, 12, 13, 18, 19.]

Mauritius[1]

Réunion[1]

diversifrons, Ricardo.

Tabanus diversifrons, Ric., R. I. M., iv, 214-5,[1] (1911); Surc., Gen. Ins. Fasc. 175, 66.

Shillong[1]
Khasis, Lower Ranges[1], Buxar Duars[1] Sylhet[1]
Darjiling District.[1]
[Calcutta]

eggeri, Schiner.

Tabanus eggeri, Schin., Novara Reise Dipt., 81,[1] (1868); Kert., Cat. Dipt., iii, 239[2]; Ric., R. I. M., iv, 249[3]; Aust., Bull. Ento. Res. x, 294, 301.[5]

Tabanus intermedius, Egg., Verh. zool.- bot. Ges. Wien. ix, 389, (1859), (*nec.* Wlk. 1850)]; Brauer, Denk. Akad. Wien, xlii, 182, 42, tab. iii, vi, 42; Gob., Mem. Soc. Linn. N. Fr., (1881), 21, 6; Pand., Rev. Entom., ii, 202, 4; Surc., Gen. Ins. Fasc. 175, 72.

Pers. Seistan[3]

South Europe[3]
Asia Minor[2]
Egypt[1,2] Palestine[5]
Mashobra[1]

excelsus, Ricardo.	
Tabanus excelsus, Ric., A. M. N. H., (8) xi, 543,[1] (1913); Leese, Ind. Vet. Dept. coll. Notes, 2, tab. iv, 13[2]; S. W., Mem. Dept. Agric. Ind., vii, 104[3]; Surc., Gen. Ins. Fase. 175, 67.	Mohand (U.P.)[2] Dehra Dun[3] Kangra Dist. (Crosse)
explicatus, Walker.	Sikkim[1, 2] Khasia Hills[2] Coonoor[2]
Tabanus explicatus, Wlk., List Dipt. B. M., v, 328,[1] (1854); Big., T. A. S. B., lx, 264; Wulp, Cat. Dipt. S. Asia, 60; Kert., Cat. Dipt., iii, 240; Ric. R. I. M., iv, 191[2]; Surc., Gen. Ins. Fasc. 175, 67.	
flavicinctus, Ricardo.	Khasia Hills[1] Nilgiris[1] ?N. Kanara[1] Mohand (U.P.)[2] ——————— Java[2]
Tabanus flavicinctus, Ric., R. I. M., iv, 130-1,[1] (1911); C. Vet. Dept. Mem. ii, tab. ii, 6[2]; Fletcher & S.W. Proc. ivth. Ento. Meetg., Pusa 226; Surc., Gen. Ins Fasc. 175, 68[3].	
flavissimus, Ricardo.	Pundaluoya (Ceylon)[1]
Tabanus flavissimus, Ric., R. I. M., iv, 207[1], (1911); Surc., Gen. Ins. Fasc. 175, 68.	
flaviventris, Bigot.	Sibsagar (Assam)[1, 2]
Atylotus flaviventris, Big., Mem. Soc. Zool. Fr., v, 657,[1] (1892).	Ceylon[2], ?Calcutta[2] ?Tenasserim[2] Bihar[2]
Tabanus flaviventris, Wulp., Cat. Dipt. S. Asia, 63; Kert., Cat. Dipt., iii, 242; Ric., R. I. M., iv, 213-4;[2] Surc., Gen. Ins. Fasc. 175, 68.	
fulvimedius, Walker	?Nepal[1] N. China Hills[2] 'India"[4, 5]
Tabanus fulvimedius, Wlk., List Dipt. B. M., i, 152,[1] (1848); Kert., Cat. Dipt., iii, 243; Ric., R. I. M., iv, 197[2]; Shiraki, Taihoku Expt. Stn. Bull., (1918), 399-407, tab. x, 14-15,[3]; Aust., Bull. Ent. Res. xii, 451, [Ric. *nec.* Wlk.]; Surc., Gen. Ins. Fasc. 175, 69.	Formosa[2, 3]
Atylotus melanopygatus, Big., Mem. Soc. Zool. Fr., v, 651,[4] (1892).	
Tabanus melanopygatus, Wulp, Cat. Dipt. S. Asia, 63; Kert., Cat. Dipt., iii, 260.	
Atylotus pagodinus, Big., Mem. Soc. Zool. Fr., v, 654,[5] (1892).	
Tabanus pagodinus, Wulp, Cat. Dipt. S. S. Asia, 63; Kert., Cat. Dipt., iii, 267.	

fuscicauda, Bigot.

Atylotus fuscicauda Big., Mem. Soc. Zool. Fr., V, 658,[1] (1892).

Ceylon[1, 2]
Pundaluoya (Ceylone)[2]

Tabanus fuscicauda, Wulp, Cat. Dipt. S. Asia, 63[2]; Kert., Cat. Dipt. iii, 244; Ric., R. I. M., iv, 217-8[3]; S. W., Mem. Dept. Agric. Ind., vii, 106[4]; Surc., Gen. Ins. Fase. 175, 69.

Shewgu (Burma)[4]
Peradeniya (Ceylone)[3]
Andaman Islands[5, 6]

? *Atylotus andamanicus,* Big., Mem. Soc. Zool. Fr., v, 668,[5] (1892).

Tabanus andamanicus, Wulp, Cat. Dipt. S. Asia, 63;[6] Kert., Cat. Dipt., iii, 221; Surc., Gen. In.s Fasc. 175, 60.

Miss Ricardo (m.s.) has cancelled her Sumatra record.

fuscicrura, Bigot.

Tabanus fuscicrura, Big., Mem. Soc. Zool. Fr., v, 662,[1] (1892); Surc., Gen. Ins. Fasc. 175, 69[2].

Ceylon[1, 2]

[Not recognized since its original description.]

fuscomaculatus, Ricardo.

Tabanus fuscomaculatus, Ric., R. I. M., iv, 183-4,[1] (1911); S. W., Mem. Dept. Agric. Ind., vii, 106[2]; Surc., Gen. Ins. Fasc. 175, 69, subsp. *altermaculatus,* Ric., A.M.N.H., (8) xi, 544,[1] (1913).

Myitkina Dist. (U. Burma)[1]
Sikkim[1], Masuri[2]
Kumaon[2]
Manipur[3]

glaber, Bigot.

Tabanus glaber, Big., Mem. Soc. Zool. Fr., v, 678[1] (1892); Wulp, Cat. Dipt. S. Asia, 63; Kert., Cat. Dipt., iii, 246; Ric., R. I. M., iv, 249-50, tab. xiv, 23;[2] Patton, I, Jo. Med. Res. vii, 738-9, tab. lxvii, 2, t. f. 1 (2)[3]; Surc., Gen. Ins. Fasc. 175, 70.

Helmand River (Afghanistan)[1, 2]
Kashgar[2]
Mesopotamia[3]
Persian Seistan[3]

hirtipalpis, Ricardo.

Tabanus hirtipalpis, Ric. R. I. M., iv, 129,[1] (1911); Surc., Gen. Ins. Fasc. 175, 70.

Bichiakoh (Nepal)[1]

hirtistriatus, Ricardo.

Tabanus hirtistriatus, Ric., R. I. M., iv, 158-9,[1] (1911); Surc., Gen. Ins. Fasc. 175, 70.

Dawnat Range (Tenasserim)[1]

The Upper Burma record (S. W., Mem. Dept; Agric. Ind., vii, 105); is an error in determination.

F. M. S[1]

hirtus, Walker.

Tabanus hirtus, Wlk., Ins. Saunds., i, 52,[1] (1850); Wulp, Cat. Dipt. S. Asia, 60; Kert., Cat. Dipt. iii, 249; Big., T. A. S. B., lx 264; Gaiger, Jo, Trop. Vet. Sci., v, 69[2]; Surc., Gen. Ins. Fasc. 175, 71.

Therioplectes hirtus, Ric., R.I.M., iv, 228-9,[3] (1911).

India[1]
Phagu (Simla D.)[3]
Naini Tal D.[3]
United Provinees[2, 3]
[Muktesar (Kumaon)]

hybridus, Wiedemann.

Tabanus hybridus, Wd., Auss. Zweifl. Ins., i 557,[1] (1828); Wlk., Proc. Linn. Soc. Lond., i, 110[2]; Wulp, Cat. Dipt. S. Asia, 59,[3] Ric, R. I. M., iv, 162[4]; *id.,* Bull. Ento. Res. vi. 407;[5] Surc., Gen. Ins. Fasc. 175. 71.

Burma[1], Sylhet[4]
F. M. S[4]
Borneo[3]
S. China[1, 2, 3, 5]

indianus, Ricardo.

Tabanus indianus, Ric., R. I. M., iv, 175-6,[1] (1911); Shiraki, Taihoku Expt. Stn. Bull., (1918), 284-94, tab. vi, 1[2]; Ric. Bull. Ento. Res., vi, 406[3]; de Mello, Proc. ivth. Ento. Meetg. Pusa, 44[4]; Surc., Gen. Ins. Fasc. 175, 71.

? *Tabanus mentitus,* Wlk., List Dipt. B.M., i, 162,[5] (1848); Wulp, Cat. Dipt. S. Asia, 59.

Kadra (N. Kanara)[1]
Goa (Port. India)[4]
[Santikoppa (Coorg)
Polibetta (Coorg)[1]
Makut (Coorg)
Mercara (Coorg)
Poona]
Foochow (China)[5]
Formosa[1, 2]
Hong Kong[3]

indiscriminatus, Ricardo.

Tabanus indiscriminatus, Ric., Arch. Naturgesch. lxxx, (A8) 123,[1] (1915).

Ceylon[1]

joidus, Bigot.

Atylotus joidus, Big., Mem. Soc. Zool. Fr., v, 657,[1] (1892).

Tabanus joidus, Wulp, Cat. Dipt. S. Asia, 63; Kert., Cat. Dipt., iii, 252; Ric., R. I. M., iv, 211-2[2]; Surc., Gen. Ins. Fasc. 175, 72.

Sibsagar (Assam)[2]

jucundus, Walker.

Tabanus jucundus, Wlk., List Dipt. B.M., i, 187,[1] (1848); Wulp., Cat. Dipt. S. Asia 60; Ric., R. I. M., iv, 164-5, tab. xiii, 10[3]; *id.,* Bull. Ento. Res., vi, 407[4]; C. Vet. Dept.Mem. ii, tab. iv, 12[5]; Fletcher & S.W., Proc. ivth. Ento. Meetg. Pusa, 226; S. W. Mem. Dept. Agric. Ind., vii, 105[6]; Surc., Gen. Ins. Fasc. 175, 72.

Pundaluoya (Ceylon)[3]
Kohat, N. W. F. P.[3]
Masuri[3, 6], Igatpuri[3] (W. Ghats)
Mohand (U.P.)[5]

Hong Kong[1, 4]

kakhyenensis, Senior White.

Tabanus kakhyenensis, S. W., Mem. Dept. Agric. Ind., vii, 143, tab. xii,[1] (1922), 104, [note]. — Mohnyin[1] N. Toungoo[1] Pyinmana[1] (All U. Burma)

khasiensis, Ricardo.

Tabanus khasiensis, Ric., A. M. N. H., (8) iii, 487,[1] (1909); *id.,* R. I. M., iv, 193-4[2]; Surc., Gen. Ins. Fasc. 175, 72. — Khasia Hills[1, 2] Meerut[2]

———

Sinabang (Siam) Penang (B. M. Colln.)

leleani, Austen.

Tabanus leleani, Aust. Bull. Ento. Res., x, 312-6, figs. 13, 14,[1] (1920); Surc., Gen. Ins. Fasc. 175, 73; *id.,* Ann. Soc. Ent. Fr., xci, 238, [bionomics]; Aust., Bull. Ento. Res., xiii, 287[2]; *id., loc. cit.,* xvi, 17. — Kangra Valley[1, 2] [Seistan, Quetta Attock Near Srinagar (Kashmir)]

———

Algeria to Mesopotamia[1, 2]

leucocnematus, Bigot.

Atylotus leucocnematus, Big., Mem. Soc. Zool. Fr., v, 656,[1] (1892). — India[1, 2]

Tabanus leucocnematus, Wulp., Cat. Dipt. S. Asia, 63,[2] [*leucocrematus,* laps.] Kert., Cat. Dipt., iii, 254; Ric., R. I. M., iv, 131-2[3]; Patton, I. Jo. Med. Res., x, 67[4]; S. W., Mem. Dept. Agric. Ind., vii, 104[5]; Surc., Gen, Ins. Fasc. 175, 73. — Lushai Hills[3] Khasia Hills[3] Coonaor[4] Katha (U. Burma) [Darjiling][1]

leucohirtus, Recardo.

Tabanus leuchirtus, Ric., A. M. N. H., (8) Kanara (Bombay)[1,2] iii, 488,[1] (1900); id., R. I. M., iv, 208-9[2]; Surc., Gen. Ins. Fasc. 175, 73. — Kanara (Bombay)[1, 2]

leucopogon, Bigot.

Atylotus, leucopgon, Big., Mem. Soc. Zool. Fr., v, 651,[1] (1892). — Sikkim[1, 3]

Tabanus leucopgon, Wulp, Cat. Dipt. S. Asia 63; Kert Cat. Dipt., iii, 254; Ric., R. I. M., iv, 184, tab. xiv, 16[2]; Surc., Gen. Ins. Fasc. 175, 73.

Macer, Bigot.

Atyltus macer, Big., Mem. Soc. Zool. Fr., v, 649[1] (1892), [only]. — India[1], Jhelum Dist.[2]

Tabanus macer, Wulp, Cat. S. Asia, 62; Ric., R. I. M., iv, 223; Aust., Trans. R. S. T. M. & H., xv, 264,[2] (abs. R. A. E. (B) x, 143]; Isaac, Mem. Dept. Agric. Ind., ix, (2) 21, tab. v, 1, 6, vi, 1, 1a, vii, 1, viii, la-d, [biomonics].

Tabanus bicallosus, Ric., A. M. N. H., (8) iii, 489-90,[3] (1909); *id.,* R. I. M., iv, 129-30,[4] (1911); Patt. & Cragg. T. B. Med. Ent., 295, [feeding point], 297, [oviposition], tab. xlii, 1-6; Fletcher, Pusa Sci. Repts., (1916-17), 96, [oviposition]; Patel, *loc. cit.,* (1919-20), 97, [do.]; Fletcher, *loc. cit.,* (1920-21), 47, 48, (parasites]; de Mello, Proc. ivth, Ento. Meetg. Pusa, 44,[5] [?]; C. Vet. Dept. Mem. ii, 4, tab. i, 2[6]; Fletcher & S. W., Proc. ivth. Ent. Meetg. Pusa, 225; Surc., Gen. Ins. Fasc. 175, 62.

Tabanus trichinoplis, Ric., A. M. N. H., (8) xiv, 359-60,[7] (1914).

Pusa[3, 4]
Goa (Port. India)[5]
Mohand (U. P.)[6]
Trichinopoly[7]
Pathankot (Crosse)
[Murree Kathgodam
Dogaon
Rawalpindi
Chamba State (Punjab)
Nagpur, Dharwar
Coimbatore
Taliparamba (N. Malabar)
In train between Rameswaram and Madras]

manipurensis, Bicords.

Tabanus manipurensis, Ric., A. M. N. H., (8) xi, 544,[1] (1913); Surc., Gen. Ins. Fasc. 175, 75.

Manipur[1]

melanognathus, Bigot.

Atylotus melanognathus, Big., Nouv. Arch. Mus. H. N. Paris, (3) ii, 204,[1] (1890).

Tabanus melanognathus, Wulp, Cat. Dipt. S. Asia, 62; Ric., R. I. M., iv, 226[2].

Tabanus nonoptatus, Ric., R. I. M., iv, 140-1,[3] (1911); Surc., Gen. Ins. Fasc. 175, 77.

Purneah (N. Bengal)[3]

Laos (Siam)[1, 2]

monotaeniatus, Bigot.

Atylotus monotaeniatus, Big., Mem. Soc. Zool. Fr., v, 655,[1] (1892).

Tabanus monotaeniatus, Wulp, Cat. Dipt. S. Asia, 63[2]; Kert., Cat. Dipt., iii, 262; Ric., R. I. M., iv, 159-60[3]; *id., loc. cit.,* iv. 401[4]; S. W., Mem. Dept. Agric. Indi., vii, 105[5]; Surc., Gen. Ins. Fasc. 175, 76.

India[1, 2]

N. Khasia Hills[3]
Nongpoh (Khasia)[3]
Sibsagar (Assam)[2]
Yunnan[4]
Mohnyin (U. Burma)[5]

nemocallosus, Ricardo.

Tabanus nemocallosus, Ric., A.M.N.H., (8) iii, 490,[1] (1909); *id.,* R.I.M., iv, 137[2]; Fletcher, Pusa Sci. Repts., (1915-

Pusa[1, 2]
Jhelum Dist.[3]

-16), 79, [bionomics] ; *id., loc cit.,* 1916-17), 97, [do]; Patel, *loc. cit., (*1919-20) 97, [notes]; Aust., Trans. R.S.T.M. & H., xv, 264, [abs. R.A.E. (B), x, 143]; Surc., Gen. Ins. Fasc. 175, 76; Issac, Mem. Dept. Agric. Ind., ix (2), 24, tab. v, 5, vi, 6, 6a, vii, 6, x, 6a-e, [bionomics].

nephodes, Bigot.

Atylotus nephodes, Big., Mem. Soc. Zool. Fr., v, 656,[1] (1892).

Tabanus nephodes, Wulp, Cat. Dipt. S. Asia, 63; Kert., Cat. Dipt. iii, 264; Ric., R. I. M., iv, 145-6; Surc., Gen. Ins. Fasc. 175, 77.

Tabanus salvazai, Surc. Bull. Soc. Ent. Fr. (1921) (19) 286, (Dec. 1921 or Jany. 1922), [var].

Naga Hills[1]
Sibsagar (Assam)[1]

nicobarensis, Schiner.

Tabanus nicobarensis, Schin., Novara Reise Dipt., 81,[1] (1866); Wulp, Cat. Dipt. S. Asia, 61; Kert., Cat. Dipt., iii, 264; Ric., R. I. M. iv, 191-2[2]; Surc., Gen. Ins. Fasc. 175, 77.

Nicobar Islands[1, 2]

nigropictus, Macquart

Tabanus nigropictus, Macq., Dipt. Exot., suppl. v, 24,[1] (1855); Wulp, Cat. Dipt S. Asia, 60; Kert. Cat. Dipt. iii, 265; Big., T.A.S.B., lx, 264; Ric., R.I.M., iv, 210-11[2]; Sur., Gen. Ins. Fasc. 175, 77.

India[1, 2]

[Not recognized since original description. Type redescribed by Recardo]

obconicus, Walker.

Tabanus obconicus, Wlk., Dipt. Saunds., i, 54[1] (1850); Wulp, Cat. Dipt. S. Asia, 60; Kert., Cat. Dipt. iii, 266; Big., T.A.S.B., lx, 264; Ric., R.I.M., iv, 215-6[2]; Surc., Gen. Ins. Fasc. 175, 77.

Central India]
Bombay[2]
Belgatchia (Bengal)[2]

optatus, Walker.

Tabanus optatus, Wlk., Proc. Linn. Soc. Lond., i, 111, 33,[1] (1856); Wulp, Cat. Dipt. S. Asia, 61[2]; Ric., R.I.M. iv, 139[3]; Pratt, Jo. Trop. Vet. Sci., 390-4,[4] [F. M. S. distribution] ; Edw., Jo. F. M. S. Mus., viii, 27[5]; Surc., Gen. Ins. Fasc. 175, 78.

Tabanus alboscutatus, Rond., Ann. Mus. Civ. Genoa, vii, 456,[6] (1875)

Belgatchia (Bengal)[3]
[Bangaon (Bhagalpur Dist.)]

Sumatra[3, 5, 7]
Borneo[1, 2, 6]

Tabanus equestris, Wulp, Leyden Mus. Notes, vii, 77, 31, tab. v, 6, (1885).

orientis, Walker.

Tabanus orientis, Wlk., List Dipt. B. M., i, 152,[1] (1848); Wulp, Cat. Dipt. S. Asia, 59; Kert. Cat. Dipt., iii, 267; Big., T.A.S.B., lx, 264[6]; Ric, R.I.M., iv, 195-7[2]; Baldrey, Jo. Trop. Vet. Sci., vi, 271-282[4] [surra transmission expets.]; C. Vet. Dept. Memo. ii, tab. iii, 18[5]; Fletcher & S. W. Proc. ivth. Ento. Meetg. Pusa, 226; Surc., Gen. Ins. Fase. 175, 78; Ric., R. I.M., iv, 401.[3]

Nepal[1, 2] United Provinees[2] Naini Tal[2] Thandiani(Himalaya)[2] W. Bhutan[2]Masuri[2] Dongagalli(Murree)[2] Yunnan[3] Muktesar[4] Mohand[5]

Tabanus consocius, Wlk., Dipt. Saunds., i, 56, (1850); Big., T.A.S.B., lx 264; Wulp, Cat. Dipt. S. Asia, 60; Kert., Cat. Dipt., iii, 236; Gaiger, Jo. Trop. Vert. Sci., v, 69; Aust., Trans. R.S.T.M. & H., v, 264,[7] [abs. R.A.E. (B), x, 143].

Sikkim[6] Jhelum Dist.[7] Kangra Dist. (Crosse) [Dalhousie]

Tabanus pertinea, Wlk., Dipt. Saunds., i, 56; (1850); Big., T.A.S.B., lx, 264; Wulp, Cat. Dipt. S. Asia, 60; Kert., Cat. Dipt. iii, 268.

The Burmese locality; (S. W. Mem. Dept. Agric. Ind., vii, 106), is an error in determination.

oxyceratus, Bigot.

Atylotus oyceratus, Big., Mem. Soc. Zool. Fr., v, 652,[1] (1892).

India[1]

Tabanus oxyceratus, Wulp., Cat. Dipt. S. Asia, 63; Kert., Cat. Dipt., iii, 267; Ric., R.I.M., iv, 195[2]; S. W., Mem. Dept. Agric. Ind., vii, 106[3]; Surc., Gen. Ins. Fasc. 175, 78.

Cheena Forest (Himalayas)[2] N. Chin Hills[2] Masuri[2, 3]

provincialis, Ricardo.

Tabanus provincialis, Ric., A.M.N.H., (8) xi, 545,[1] (1913); Surc., Gen. Ins. Fasc. 175, 80.

Arian kava (Travancore)[1] Kurseong[1]

pulchellus, Loew.

Tabanus pulchellus, Lw., Verh. Zool-bot. Ges. Wien, viii, 597, 26,[1] (1858); Brauer, Denk. Akad. Wien, xlii, 206. 60; Kert., Cat. Dipt., iii, 271[2]; Ric., A. M. N. H. (7) xvi, 200; Patton, I. Jo. Med. Res., vii, 736-8, tab. lxvii, 1, and fig. 1 (1),[2] [probably local race of *ditaeniatus* Macq.] ; Surc., Gen. Ins. Fasc. 175, 80.

Tabanus cyprianus, Ric., R. I. M., iv, 248-9,[4] (1911).

Seistan[4]

Mesopotamia[1, 3] Asia Minor[2] N. Africa[2] Perso-Afghan Frontier[2]

Tabanus pulverifer, Wlk., List Dipt. B. M., v, supp. i, 236,[1] (1854); Surc., Gen. Ins. Fasc. 175, 80.

Tabanus persis, Ric. R. I. M., iv, 251-2, tab. xiv, 24,[2] (1911); Patton, I. Jo. Med. Res. vii, 740, fig. 1 (4).[3]

puteus, Ricardo.

Tabanus puteus, Ric., R. I. M., iv, 165-6, tab. xiii, 11,[1] (1911); Surc., Ben. Ins. Fasc. 175, 81.

Trincomali Dist.[1]
Colombo[1]
Habarane N. C. P. (S. W.)
(All Ceylon).

rubicundus, Macquart.

Tabanus rubicundus, Macq., Dipt. Exot., supp. i, 32,[1] (1846); *id., loc.cit.,* supp. iii, 170; *id., loc. cit.,* v, 27[2]; Big., T.A.S.B. lx, 264; Wulp, Cat. Dipt. S. Asia, 59; Kert., Cat. Dipt., iii, 274; Ric., R.I.M., iv, 172-4, tab. xiii, 12; S. W., Mem. Dept. Agric. Ind., vii; 105,[4] [? *monilifer* a good sp.]; Surc., Gen. Ins. Fasc. 175, 81.

India[1]
Chargola Valley (Sylhet)[2]
Sylhet[3, 5]
Khasia Hills[3, 6]
Petsut, Katha (U. Burma)[4]
Sibsagar (Assam)[6]

Tabanus internus, Wlk., List Dipt. B.M., i, 164, (1848)[5]; Wulp, Cat. Dipt. S. Asia, 60; Kert., Cat. Dipt., iii, 252.

Java[2]

Atylotus monilifer, Big., Mem. Soc. Zool. Fr., v, 654,[6] (1892).

Tabanus monilifer, Wulp., Cat. Dipt. S. Asia, 63; Kert., Cat. Dipt., iii, 262.

rubidus, Wiedemann.

Tabanus rubidus, Wied., Dipt. Exot., 69,[1] (1821); *id.,* Ausser. Zwelifl. Ins., i, 127-8; Big., T.A.S.B., lx, 264; Wulp, Leyden Mus. Notes, vii, 71. 23[2]; *id.,* Cat. Dipt. S. Asia, 59[3]; Kert., Cat. Dipt., iii, 274; Ric., R. I. M., iv, 155-6[4]; *id.,* Bull. Ento. Res., vi, 405[5]; Gaiger, Jo. Trop. Vet. Sci., v, 69, [? sp.]; Lefroy, Pusa Bull. 7, 21,[6] [?sp.]; d'Abreu, Proc. iiird, Ento. Meetg. Pusa, iii, 866, [eaten by birds] ; Aust., Bull. Entr. Res. xii, 448,[7] [*priscus* Wlk. and *virgatulus* Aust. possibly only forms]; Surc., Gen. Ins. Fasc. 175, 82; Issac, Mem. Dept. Agric. Ind., ix (2) 22, tab. v, 2, 2a, 3, vi, 2, 2a, vii, 2, viii, 2a-e, [bionomics]; *id., loc. cit.,* viii, 93, tab. xi, [larva].

Bengal[1, 3, 4]
Moulmein[4]
Bombay[4, 6 ,7], Pusa[4]
Sibsagar (Assam)[4]
Khasia Hills[4]
Partabgahr (U.P.)[4]
Nepal[4, 9]
'East India[8, 19]
Trinco Dist (Ceylone) [9, 10]
Umballa[9], Calcutta[9]
Muktesar[11], Matheran (W. Ghats)[13]
Goa (Port Ind.)[14]
Jhelum Dist[15]
Masuri[16]

Tabanus albimedius, Wlk., Dipt. Saunds., i, 48,[8] (1850) ; Big., T. A. S. B., lx, 264; Wulp, Cat. Dipt. S. Asia, 60; Kert., Cat. Dipt., iii, 219; Ric., R. I. M., iv, 156-7, tab. xiii, 8[9]; Burn., Spol. Zeyl. vi 170;[10] Baldrey, Jo. Trop. Vet. Sci., vi, 271-82,[11] [transmission expts.,? sp] ; Cragg, I. Jo. Med. Res., vii, 648-63, tabs. lxi-lxiii, [intestine]; Cornwall & Patton, *loc. cit.,* ii, 569-93, [salivary secretion] ; Ric., Bull. Ent. Res., vi. 405[12]; Howlett, Ind. Ins. Life, tab. lxii, 1-1c, [life history [; *id.,* Pusa Sci. Repts., (1909-10), 71; Fletcher, *loc. cit.,* (1915-16), 80, [bionomics], 82, [surra]; *id., loc. cit.,* (1916-17), 96, [oviposition]; Patel, *loc. cit.,* (1919-20), [parasites]; Fletcher, *loc. cit.,* (1920-21), 47, 48, [do]; Patt. & Cragg. T. B. Med. Ent., tab. ii, 7, [antenna], xiv, 2, [thorax], xv, 4, [medifurca], xviii, 5, [genitalia], xxvii, 9, [salivary gland]; 288-9. 294, [feeding point] 297, [oviposition]; Longstaff 'Butterfly Hunting,' 393[13]; de Mello, Proc. ivth. Ento. Meetg. Pusa, 44[14]; Aust., Trans. R.S.T.M., & H., xv, 264,[15] [abs. R.A.E. (B), x, 143]; S. W., Mem. Dept. Agric. Ind., vii 105[16]; Surc., Gen. Ins. Fasc. 175, 59.

Dharamsala (Crosse)
Kangra (Crosse)
D. Ghazipore (Crosse)
[Hopin (Upper Burma)
Chapra (Bihar)
Coimbatore
Rawalpindi
Taru, near Peshawar, Giridhi
Khumbala, Khurda
Khasi Hills
Bareilly, Midnapur
Kashmir
Nongpoh (Assam)
Manhum, Delhi
Cuttack
Dera Ismail Khan
Dumraon (Bihar)
Kasargode (S. Kanara)]

? Asia'[17], Java[3, 18, 20]
Sumatra[2, 3]
F. M. S.[4, 7, 21]
Siam[4, 7]
Hong Kong[5, 7, 12]
Cochin China[1]
Annam[7]

? *Tabanus calidus,* Wlk., Dipt., Saunds., i, 57,[17], (1850); Wulp, Cat. Dipt. S. Asia, 60.

Atylotus lachrymans, Big., Mem. Soc. Zool. Fr., v, 669,[18] (1892).

Tabanus lachrymans, Wulp, Cat. Dipt. S. Asia, 63.

Tabanus priscus, Wlk., List Dipt. B. M., i, 176, (1848); Aust., Bull. Ent. Res., xii, 448, [Synonomy].

Tabanus umbrosus, Wlk., Dipt. Saunds., i, 52,[19] (1850); Wulp, Cat. Dipt. S. Asia, 60; Kert., Cat. Dipt., iii, 290.

Tabanus vagus, Wlk., Dipt. Saunds., i, 50,[20] (1850); Big., T.A.S.B., lx, 264; Wulp, Cat. Dipt. S. Asia, 60; Kert.,

Cat Dipt., iii, 291; Fraser, Jo. Trop Vet. Sci., iv, 373, 375 381-2, [infection expts., sp.]; Pratt., *loc. cit.*, iv, 390-4,[21] [F. M. S. distribution.].

rufiventris, Fabricius.

Tabanus rufiventris, Fb., Syst. Antl., 96, 13,[1] (1805); Wied. Dipt. Exot., i, 66, 9; *id.*, Ausser. Zweifl. Ins., i, 118; Wlk., List Dipt. B. M., v, 245, 398; Ric., R. I. M., iv 185.[10]

Atylotus assamensis, Big., Mem. Soc. Zool. Fr., v, 653,[2] (1892); Kert., Cat. Dipt., iii, 222.

Tabanus crassus, Wlk., Dipt. Saunds., i, 50,[3] (1850) ; Big., T. A. S. B., lx, 264; Wulp, Cat. Dipt. S. Asia, 60[4]; Kert., Cat. Dipt., iii, 237; Ric., R. I. M., iv, 189-90[5]; Shiraki Taihoku Expt. Stn. Bull. (1918) 330-5, tab. vii, 5[6]; Ric., Bull. Ento. Res., vi, 407[7]; Patel, Pusa Sci. Repts., (1919-20) 97,[8] [parasites]; Surc., Gen. Ins. Fasc. 175, 66; Isaac, Mem. Dept. Agric. Ind., viii, 62, tab, viii, [recognized].

Tabanus sanguineus, Wlk., Dipt. Saunds., i, 54,[11] (1850); Wulp. Cat. Dipt. S. Asia, 60[12]; Kert., Cat. Dipt., iii, 277; Ric., R. I. M. iv, 194-5[13]; Shiraki, Taihoku Expt. Stn. Bull. (1918), 343-50, tab. vii, 6[14]; Ric., Bull. Ent. Res., vi, 405[15]; Fletcher, Pusa Sci. Repts., (1915-16), 80; Patel, *loc. cit.* (1919-20), 97, [parasites]; Surc., Gen. Ins. Fasc. 175, 83.

E. India[1, 3, 4, 10]
Assam[2, 13]
Karwar (N. Kanara)[5]
Pusa[8, 9, 13]
Lr. Khasia Hil[13]
Lushai Hills[13]

Sumatra[13]
F. M. S.[13]
Hongkong[11, 12, 13, 15]
Formosa[5, 6, 13, 14]

sexcinctus, Ricardo.

Tabanus sexcinctus, Ric., R. I. M., iv, 133,[1] (1911); Shiraki, Taihoku Expt. Stn. Bull., (1918), 234-40, tab. v, 2[2]; S. W., Mem. Dept. Agric. Ind., vii, 104[3]; Surc., Gen. Ins. Fasc. 175, 83.

Lushai Hills[1]
Masuri[3] [?sub-form]

Formosa[2]

speciosus, Ricard.

Tabanus speciosus, Ric., R. I. M., iv, 157-8,[1] (1911); Patton & Cragg, T. B. Med. Ent., 289, tab. xl. 294, [feeding point], 297, [oviposition]; Surc., Gen. Ins., Fasc. 175, 84.

Travancore[1]
Trinco Dist.[1]
(Ceylone)
Madras (B.M.Coln.)
Angul, Bombay
Cuddapah (Madras Pres.)
Chapra, Khurda
Santikoppa (N. Coorg)

Coimbatore.
Dumraon
Buxar
Bhagalpur.

striatus, Fabricius.

Tabanus striatus, Fb., Syst. Ent., iv, 371,[1] (1794); *id.,* 'Syst. Antl., 103; Aust., Bull. Ent., Res., xii, 445[2]; *id.,* Trans. R.S.T.M. & H., xv, 264; Isaac, Mem. Dept. Agric. Ind., viii, 97-109, tabs. xii-xv, [bionomics, larval structure, adult genitalia].

Tabanus hilaris, Wlk., Dipt. Saunds., i, 49, ta. ii, 3,[4] (1850); Big., T.A.S.B., lx, 264; Wulp, Cat. Dipt. S. Asia, 60[5]; Kert., Cat. Dipt., iii, 249; Ric., R. I. M. iv, 153[6]; *id.,* Bull. Ent. Res., vi, 407[7];; Patt. & Cragg, T. B. Med. Ent. 294, [feeding point], 297, [oviposition]; Patt., Arch. f. Protist., xv, 333-62, [*Crithidia tabani*]; Fletcher, Pusa Sci. Repts, (1915-16), 80; Patel, *loc. cit.,* (1919-20), 97; Surc., Gen Ins. Fasc. 175, 70.

Tabanus partitus, Wlk., Proc. Linn. Soc. Lond., i, 9,[8] (1856); Wulp, Cat. Dipt. S. Asia, 60; Lefroy, pussa Bull. 7, 21[9]; Fraser, Jo. Trop. Vet. Sci., iv, 373, 375, 381-2, [infection expts., ? sp.]; Pratt. *loc. cit.,* v, 390-4,[10] [F.M.S. distribution]; Gaiger, *loc. cit.,* v, 69.

Tabanus rufocallosus, Big., Mem. Soc. Zool. Fr., v, 679,[11] (1892); Wulp, Cat. Dipt. S. Asia, 63.

Kohat[2], Jhelum D.[3]
E. India[4, 5]
Pusa[6], Punjab[6]
Bombay[9]
[Hissar (Punjab)
Belgachia (Bengal)
Midnapur, Angul
Monghyr
Rawalpindi
Lahore, Bhagalpur
Suri, Cuttack
daltonganj (Bihar)[1]
Darjiling, Burdwan
Puruha
Ludhiana (Punjab)
Kathgodam
Coimbatore]

subcallosus, Ricardo.

Therioplectes subcallosus, Ric., R. I. M., iv, 227-8,[1] (1911).

Tabanus subcallosus, Baldrey, Jo. Trop. Vet. Sci., vi, 271-282,[2] [transmission expts.]; Surc., Gen. Ins. Fasc. 175, 85.

Masuri[1]
Dalhousie[1]
Muktesar[2]

[Dogaon
Meerut
Muzafarnagar
Dera Dun (U.P.)
Pokokku (Burma)
Dalhousie
Dunera hills
(Gurdaaspur
Distrcit)
Kangra (Punjab)
Amparad (near
Muktesar).]

subcinerascens, Ricardo.

Tabanus subcinerascens, Ric., R. I. M., iv, 215,[1] (1911); Surc., Gen. Ins. Fasc. 175, 85.	N. Chin Hills[1]

subhirtus, Ricardo.

Tabanus subhirtus, Ric., R. I. M., iv 219,[1] (1911); Surc., Gen. Ins. Fasc. 175, 85.	Bengal[1], Bombay[1]
Atylotus cinerascens, Big., Mem. Soc. Zool. Fr., v, 669,[2] (1892), [*nec* king 1827].	Java[2]
Tabanus cinerascens, Wulp, Cat. Dipt. S. Asia, 68; Kert., Cat. Dipt., iii, 234.	

sufis, Jaennicke.

Tabanus sufis, Jaenn., Abhandl. Senck. Naturf. Ges., vi, 332, 24,[1] (1867); Verral, Br. Flies, v, 9 syst. list app.) 11.[2]	Punjab (B. M. Colln.)
Tabanus alboveitralis, News.	N. Africa[1, 2] E. Africa[2]

tenebrosus, Walker.

Tabanus tenebrosus, Wlk. List Dipt. B. M., v, supp. i, 242, 389,[1] (1854); Big., T. A. S. B., lx, 264; Wulp, Cat. Dipt. S. Asia, 60; Kert., Cat. Dipt., iii, 285; Ric., R. I. M., iv, 203-4;[2] Surc., Gen. Ins. Fasc. 175, 86.	Kanara[1, 2]
[Not recognized since original description. Ricardo.]	Type redescribed by

tenens, Walker.

Tabanus tenens, Wlk., Dipt. Saunds., i, 49,[1] (1850); Wulp, Cat. Dipt. S. Asia, 60; Gaiger, Jo. Trop. Vet. Sci., v, 69; Aust., Bull. Ent. Res., xii, 445,[2] [synonomy] Big., T. A. S. B., lx, 264; Kert., Cat. Dipt., iii, 285; Lefroy Pusa Bull. 7, 21[3]; Issac, Mem. Dept. Agric. Ind., viii, 97, tabs. xii-xiv, [larval structure]; *id., loc., cit.*, ix, 23, tab. v, 4, vi, 4, 4a, vii, 4, ix, 4a-e, [bionomics]	E. India[1] Ceylon-general[2, 8] Bassein[3, 8, 9] (Bombay), Madras[2] Pusa[2], Bengal-general[8, 13], Burma[8] Gonda D. (U. P.)[8] E. Coast[8, 12] Sylhet[8] Manipur[8]
? *Tabanus chinensis,* Thun., N. Acta R. S. Upsala, ix, 53, 62, (1827); Bezzi, Wien. Ent. Zeit., xxvii, vols. ii & iii, (1908).	Mohmand Pass (N. W. F. P.)[8] S. India-general[10] S. India-general[10]
? *Tabanus costalis,* Licht., Cat. 213, Hamburg, (1796); Aust., A. M. N. H., (8), i, 346.	Goa (Port. Ind.)[11] Dehra Dun[13]

Tabanus dorsilinea, Wd., Anal. Ent., 22, 10, (1824); *id.,* Ausser. Zweifl. Ins., i, 184, 10; Big., T. A. S. B., lx, 264, [*dorsoliniae,* laps.]. — [Kasargoda (S. Kanara) Combatore Berhampur

? *Tabanus manilensis,* Schin. Novara Reise Dipt., 84 (1868)[15]; Wulp, Cat. Dipt. S. Asia, 61.

———————

Java[4, 6]
Sumatra[5, 6]

Tabanus megalops, Wlk., List Dipt. B. M. v, supp. i, 247, (1854); Wulp, Cat. Dipt. S. Asia, 60. — F. M. S[8] Philippines[8, 15] China[7, 14]

Tabanus striatus, Wd. (*nec* Fb.), Dipt. Exot., 79; *id.,* Ausser. Zweifl. Ins., i, 155; Wulp, Dipt. Sumat. Exp., 16[5]; *id.,* Leyden Mus. Notes, vii, 61[6]; *id.,* Cat. Dipt. S. Asia, 58[7]; Kert., Cat. Dipt., iii, 281; Ric., R. I. M., iv, 149- 153[8]; Gaiger, Jo. Trop. Vet. Sci., v, 69; Lefroy, Pusa Bull. 7, 21[9]; Howlett, Ind. Ins. Life, tab. lxii, 2; Mitzmain, Bull. 28, Phil. Ag. Dept., [abs. T. Vet. Bull., i, 285-6], [transmission of surra]; *id.,* Phil. Jo. Sci., (B) viii, 23-9; *id, loc. cit.,* viii, 197-221, tabs. i-vii, [life history]; Fletcher, S. Ind. Ins., 358-60, figs. 218-219[10]; Patton & Cragg, T. B. Med. Ent., 288, 294, [feeding point], 297, [oviposition]; Fletcher, Pusa Sci. Repts., (1915-16), 79, [no. of broods], 82, [surra]; de Mello, Proc. ivth. Ento. Meetg. Pusa, 44[11]; Dover, R. I. M., xxii, 394[12]; S. W., Mem. Dept. Agric. Ind., vii, 105[13]; Surc., Gen. Ins. Fasc. 175, 84.

The above represents Major Austen's views on the synonomy of the species generally known as *striatus.* Their correctness turns on whether the specimen at Kiel or at Copenhagen is the actual type. Major Patton holds diametrically opposite views on this point to Major Austen, and if the former is proved correct, the whole synonomy will need reversing to that given by Ricardo in R. I. M., iv, 149-153.

trinominata, nom. nov.

Atylotus angustus, Big., Mem. Soc. Zool. Fr., v, 652[1], (1892), [*nec* Macq. 1838]. — India[1, 2]

Tabanus angustus, Wulp, Cat. Dipt. S. Asia, 63; Kert., Cat. Dipt., iii, 221; Surc., Gen. Ins. Fasc. 175, 60.

Tabanus palpalis, Ric. R. I. M., iv, 212-3, tab. xiv, 19, 20[1]; (1911), [*nec* Brethès, 1909].

[Not recognized since original description. Type redescribed by Ricardo.]

tropicus, auctt. Ind. *nec* Panzer (1794).

Tabanus tropicus, Lingard, Jo. Trop. Vet. Sci., i, 92-112[1]; Baldrey, *loc. cit.*, vi, 271-82[1], [transmission expts.]; Schat, S. S. Bur. Bull., 15, vol. ii[3], [transmits surra]; Waterho. in Aitch., Zool Afghan Delimit. Comn., 131[4].

Kathgodam[1]
Muktesar[2]
(both Himalaya)
Padda Sultan
(Helmand)[4]

Java[3]

These records probably reer to more than one species. Unless the specimens with which these authors worked can be found, the reference will have to remain doubtful.

tuberculatus, Ricardo.

Tabanus tuberculatus, Ric., R. I. M., iv, 220-1[1], (1911); Surc., Gen. Ins. Fasc., 175, 87.

Companiganj
(Sylhet)[1]
Belgatchia (Bengal)[1]
Pusa[1]
[Villupuram (S. Arcot)
stehanda (Kistna)]

virgo, Wiedemann.

Tabanus virgo, Wd., Anal. Ent., 22[1]; (1824); *id.*, Ausser. Zweifl. Ins., i, 184; Big., T. A. S. B., lx, 264; Wulp, Cat. Dipt. S. Asia, 59[2]; Kert., Cat. Dipt., iii, 292; Ric., R. I. M., iv, 136[3]; Patt. & Cragg, T. B. Med. Ent., 297, [ovi-position], tab. xlii, 13-16; Fletcher, Pusa Sci. Repts., (1916-17), 96, [ovi-position]; Patel, *loc. cit*, (1919-20), 100; C. Vet. Dept. Mem. 2, tab. ii, 7[4]; Fletcher & S. W., Proc., ivth. Ento. Meetg. Pusa, 226; Aust., Trans. R. S. T. M. & H., xv, 264[7], [abs. R. A. E. (B), x, 143]; Surc., Gen. Ins. Fasc. 175, 83; Isaac, Mem. Dept. Agric. Ind., ix (2) 24, tab. vi, 5, vii, 5, x, 5a-e, [bionomics].

India[1, 2, 5, 6]
Madras[1], Mysore[3]
Calcutta[3], Pusa[3]
Mohand (U. P.)[4]
Jhelum Dist.[7]
Vizianagram (S.W.)
D. Ghazipore
(Crosse)
Kangra (Crosse)
[Chapra (Bihar)
Chicacole Road
(Madras
Presidency)
Suri (Bihar)
Sohawa (Punjab)
Hazaribagh

Tabanus albulus, Wlk, Dipt. Saunds., i, 46[5], (1850); Big., T. A. S. B., lx, 264; Wulp, Cat. Dipt. S. Asia, 60; Kert., Cat. Dipt., iii, 220.

Rawalpindi
Khurda
Cuttack
Bareilly

Tabanus puella, Wlk., Dipt. Saunds., i, 53[6], (1850); Big., T. A. S. B., lx, 264; Wulp, Cat. Dipt. S. Asia, 60; Kert., Cat. Dipt., iii, 271. — Purulia (Bihar) Angul, Bhowrah (Champaran)]

wyvillei, Ricardo.

Therioplectes wyvillei, Ric., R. I. M., iv, 229-30[1], (1911). — Narkunda[1] Kasauli[1] (both W. Himalaya)

Tabanus wyvillei, Surc., Gen. Ins. Fasc. 175, 88.

5. Neotabanus, Ricardo.

Neotabanus, Ric., R. I. M., iv, 363, (1911); Surc., Gen. Ins. Fasc. 175, 93.

There is an invalid sub-genus of *Tabanus* of the same name, erected by Lutz.

ceylonicus, Ricardo.

Neotabanus ceylonicus, Ric., R. I. M., iv, 363-4[1], (1911); Surc., Gen. Ins. Fasc. 175, 93. — Pundaluoya (Ceylmon Mts.)[1]

PANGONINAE.

Ric., A. M. N. H., (7) v, 105-121, 167-181, [revision, tables of genera]; Shiraki, Taihoku Expt. Stn. Bull., (1918), 9, [notes on sub-family]; Surc., Gen. Ins. Fasc. 175, 94-5, [key to genera of World].

6. Corizoneura, Rondani.

Corizoneura, Rond., Arch. Zool. Anat. Fisiol., iii, 85, (1864); Ric., R. I. M., iv, 365; Shiraki, Taihoku Expt. Stn. Bull., (1918), 17-18; D-Brock., Bull. Ento. Res., i, 55-6, [an African sp. a bad animal biter]; Aust., *loc. cit.,* xi, 139, [the sub-divisions of *Pangonia* Latr.]; Patton & Cragg, T. B. Med. Ent., 294; Surc., Gen. Ins. Fasc. 175, 125.

KEY TO THE INDIAN SPECIES OF *CORIZONEURA.*

1.	Proboscis much longer than the whole body	2
	Proboscis scarcely longer than the body.	*taprobanes* Wlk.
2.	First two joints of front tarsi modified, widened in both sexes.	*longirostris* Hardw.
	First two joints of front tarsi simple	*longirostirs* var. *varipes* Ric.

longirostris, Hardwicke.

Pangonia longirostris, Hard., Trans. Linn. Soc. Lond., xiv, 135, tab. vi, 5, 6[1], (1823); Wd., Ausser Zweifl. Ins., ii, 621, 14[2]; Gay in Griff., Cuvier's Règ. Anim., xv, tab. cxiv, 3; Macq., Ann. Soc., Ent. Fr., vi, 429, tab. xv, 20-23; Wlk., List Dipt. B. M., i, 131; *id, loc. cit.*, v, supp. i, 139. 101; Röder, Stettin. Ent. Zeit., xlii, 384[3], [redescr]; Wulp, Cat. Dipt. S. Asia, 64; Kert., Cat. Dipt., iii, 173; Ric., A. M. N. H., (7) v, 167-8; Tetle, Bull. Ento. Res., viii, 253-267, tab. vi, 11t. figs., [structure of mouth parts]; Howlett, Ind. Ins. Life, tab. lxii, 3.

Nepal[1, 2]
Allahabad[3]
Muktesar (Himalaya)[4]
Tibet[4], Sikkim[5, 6, 7]
Trivandrum (Travancore)[5]
Sadiya Dist. (Assam)[5]
Shillong, Simla.
Masuri.

Corizoneura longirostris, Ric., R. I. M., iv, 365-6[4]; Mitter, I. Jo. Med. Res., v, 522-8, tab. [method of feeding]; Big., T. A. S. B., lx, 264; Aust., Bull. Ento. Res., xi, 139-152; Surc., Gen. Ins. Fasc. 175, 134.

subsp. **varipes,** Ric., R. I. M., iv, 367[5], (1911).

Pangonia amboinensis, aucct. *nec.* Fab., Big., T. A. S. B., lx, 264; Wulp. Cat. Dipt. S. Asia, 64[7]; Ric., R. I. M., iv, 369.

taprobanes, Walker.

Pangonia, taprobanes, Wlk., List Dipt. B. M. v, suppl. i, 134, (1854)[1] Wulp, Cat. Dipt. S. Asia 64[2]; Kert., Cat. Dipt., iii, 74; Ric., A. M. N. H., (7) v, 167-8.

Corizoneura taprobanes, Ric., R. I. M., iv, 367-8[3]; Aust., Bull. Ento. Res., xi, 139-152; de Mello, Proc. ivth. Ento. Meetg. Pusa, 45[5]; Surc., Gen. Ins. Fasc. 175, 134.

Ceylone (general)[1, 2, 3]
Karwar (N. Kanara)[3]
Bangalore[3], Nilgiris[3]
Arnadi[3] (Berar)
Bombay[4]
Goa (Port. Ind.)[5]

? Pangonia, rufa, Macq., Dipt. Exot. suppl. iv, 18[4], (1850), [nec *id.*, 1838]; Kert., Cat. Dipt., iii, 160; Big., T. A. S. B., lx, 264 ; Wulp, Cat. Dipt. S. Asia, 64; Surc., Gen. Ins. Fasc. 175, 129.

[Taliparamba (N. Malabar
Coimbatore (Hills near)
Nagpur
Shevaroy Hills (S. India)
gundalur (Nilgiris)
Mercara (Coorg)
Kallar (Nilgiris]

7. Gastroxides, Saunders.

Gastroxides Saunds., Trans. Ent. Soc. Lond., iii, 59, (1841); Kert., Cat. Dipt., iii, 181; Ric., R. I. M., iv, 370; Surc., Gen. Ins. Fasc. 175, 137.

Ditylomyia, Big., Rev. et Mag. Zool., (2) xi, 305, (1859).

KEY TO THE INDIAN SPECIES OF *GASTROXIDES.*

1. Thorax and abdomen black, latter in with basal segments bright rufous	*ater* Saund.
Thorax brown with 2 yellowish stripes. Abdomen in ♀ black, 1st. and 3rd. segments with yellowish, 2nd and 4th with grey hind borders, remaining segments with some red at sides. In abdomen reddish, shining, a median balck spot on each side of first 3 segments, and 2 extra lateral spots on last of these. 2nd segment narrowly black laterally, 4th black, posteriorly dull yellow; remainder black.	*ornatus* Big.

ater, Saunders.

Gastroxides ater, Saund., Trans. Ent. Soc. Lond., iii, 59, tab. v, 4[1], (1841); *id., loc. cit.,* iv, 223, tab. xiv, 3; Wlk., List Dipt. B. M., i, 209; *id., loc. cit.,* v, suppl. i 293, 1; Lw., Dipt. Fauna Süd Afrik., i, 15; Wulp., Cat. Dipt. S. Asia, 64[2]; Ric., A. M. N. H., (7) viii, 298; *id., loc. cit.,* (7) xiv, 372; Kert., Cat. Dipt., iii, 181; Ric., R. I. M., iv, 370-1[4], tab. xvi, 9; Howlett, Ind. Ins. Life, tab lxii, 4; Fletcher, Pusa Sci. Repts., (1915-16), 81, [bionomics]; *id., loc. cit.,* (1920-21), 21, 49, [tree-hole breeder]; Surc., Gen. Ins. Fasc. 175, 137, tab. v, 6a-b; Isaac, Mem. Dept. Agric. Ind., ix, (2), 25, tab. vi, 3, vii, 3, ix, 3a-e, [bionomics].

Central India[1, 2]
Calcutta[4]
Pusa[6]
Coimbatore (B. M. Colln.)
[Chapra
Fyzabad (U. P.)
Surat]

ornatus, Bigot.

Ditylomyia ornatus, Big., Rev. et Mag. Zool., (2) xi, 306[1], (1859); Wulp, Cat. Dipt. S. Asia, 63.

Ceylone[1]
Colombo[2]

Gastroxidus ornates, Ric., A. M. N. H., (7) xiv, 372; Kert., Cat. Dipt., iii, 181; Ric., R. I. M., iv, 371-3[2]; Surc., Gen. Ins. Fasc. 175, 137.

Heneratgoda
(Ceylon plains[2])

8. Silvius, Meigen.

Silvius, Meig., Syst. Besch., ii, 27, (1820); Wd., Ausser. Zweifl. Ins., i, 93; Macq., Dipt. Exot., I, i, tab. vi, 1; Schin., F. Auster., i, 43; Ric., A. M. N. H., (7) viii, 290; Surc., Bull. Mus. H. N. Paris, ii, 46; Kert., Cat Dipt. iii, 177; Ric., R. I. M., iv, 373; Shiraki, Taihoku

Expt. Stn. Bull., (1918), 10; Patton & Cragg, T. B. Med. Ent., 292; Hill, Bull. Ent. Res., xii, 59-62, figs., [bionomis of an Australian sp.] ; Surc., Gen. Ins. Fasc. 175, 141.

Erodiorhynchus, Macq., Dipt. Exot., I, 111, tab. vi, 1 (1338).

? Philoliche, Wd., Ausser. Zweifl Ins., I, 95, (1828).

indianus, Ricardo

Silvius indianus, Ric. R. I. M., iv, 374[1], (1911); Surc., Gen. Ins. Fasc. 175, 143. — Bhura (Naini Tal D.)[1]

9. **Melissomorpha,** Ricardo.

Melissomorpha, Ric., Proc. Zool. Soc. Lond., (1906), 97; Surc., Gen. Ins. Fasc. 175, 145. — Darjiling[1]

indiana, Ricardo

Melissomorpha indiana, Ric., Proc. Zool. Soc. Lond., (1906), 98,[1] (1906); Surc., Gen. Ins. Fasc. 175, 146, tab. iii, 12a.

10. **Chrysops,** Meigen.

KEY TO THE INDIAN SPECIES OF *CHRYSOPS*

[adapted from Ricardo].

1. Wings with a dark transverse band and an apical spot	2
Wings with a dark transverse band but no apical spot.	*flavicinctus* Ric.
2. Wings with a hyaline sinus in the band in 5th posterior cell	3
Wings with no hyaline sinus in band in 5th cell	8
3. Abdomen yellowish, with black stripes.	4
Abdomen black, whitish or yellowish at base.	7
4. Discal cell hyaline	5
Discal cell not hyaline	6
5. Abdomen with four long black stripes	*mlokosiewiczi* Big.
Abdomen with a black bifid stripe on 2nd. segment	*flavipes* Big.

6. Abdomen with short, black, bifid stripe on 2nd. sgement, often exending to 3rd., or 4th	*dispar* Fb.
Abdomen with bifid stripe on 2nd segment, and a network of black markings on 3rd, leaving only three yellow spots free	*designatus* Ric.
7. Abdomen black, 2nd. segment yellowish on fore border. Face blackish. Apical border of wing band straight	*pellucidus* Fb.
Ist. abdominal segment yellow, with median black spot. 2nd. yellow, 3rd. blackish, yellow at sides and with 3 yellow spots. 4th-6th. segments black. Face black. Transverse band of wing irregular on both borders	*pettigrewi* Ric.
8. A clear spot in discal cell. Abdomen blackish with pale spots.	*stimulans* Wlk.
No clear spot in discal cell	9
9. Abdomen yellow, 2nd. segment with a black bifid stripe	*indianus* Ric.
Abdomen yellow, with a black band in the middle	10
10. Legs and face blackish	*fasciatus* Wd.
Legs and face yellowish brown	*fixissimus* Wlk.

Chrysops, Meig., Nouv. Class., 23, (1800); *id.,* Illig. Mag., 267, (1803); Ric., A. M. N. H., (7) ix, 366-381, 424-431, [notes on Oriental spp.]; *id.,* R. I. M., iv, 374-7, [key to Oriental spp.]; Kert., Cat. Dipt., iii, 181; Shiraki,, Taihoku Expt. Stn. Bull., (1918), 22-25; Patton & Cragg, T. B. Med. Ent., 290-1, [key to Oriental spp.]; Kröb., Zool. Jahrb., Syst., xliii, 41-160, 12 figs., tabs. i, ii; Hoffm., Arch. Schiffs. Trop. Hyg., xxvi, 244-8, fig., [medical importance]; Surc., Gen. Ins. Fasc. 175, 148-9, [bionomics.]

designatus, Ricardo.

Chrysops designata, Ric., R. I. M., iv, 383-4, tab. xv, 1[1], (1911); *id.,* loc. cit., iv, 401[2]; Kröb., Zool. Jahrb. Syst., xliii, 94-96, tab. i. 21[3]; C. Vet. Dept. Mem. 2, tab. v, E[4]; Fletcher & S. W., Proc., ivth. Ento. Meetg. Pusa, 226; Surc., Gen. Ins. Fasc. 175, 151.

Sarah (Nepal)[1, 3]
Jaulasal (Naini Tal Dist.)[1]
Yunnan[1, 2]
Mohand (U.P.)[4]
Gandarbal (Kashmir)
Saikhoa (Assam)

dispar, Fabricius.

Tabanus dispar, Fb., Syst. Ent., supp., 567[1]. (1798).

Chrysops dispar, Fb., Syst. Antl., 112. 5; Wd., Dipt. Exot., 102.1; *id.*, Ausser. Zweifl. Ins., i, 196. 3; Masq., Dipt. Exot., I, i, 159. 5[2]; *id.*, *loc. cit*, supp. 3, 14[3]; Wlk., List Dipt. B.M., i, 195; *vid.*, *loc. cit.*, v, supp. i, 229. 67; Wulp, Dipt. Sumat Exp., 19, 1[4]; O-S., Ann. Mus. Civ. Genoa, xvi, 418[5]; *id.*, Berl. Ent. Zeit., xxvi, 97; Röder, Ent. Nachtr., xix, 234. 3[7]; Wulp, Cat. Dipt. S. Asia, 65[8]; Ric, A. M. N. H., (7), ix, 374; Kert., Cat. Dipt., iii, 186; Howlett, Ind. Ins. Life, tab. lxii, 6; Ric., R. I. M., iv, 380-2, tab. xv, 3[9]; Pratt, Jo. Trop. Vet. Sci., iv, 390-4[10], [F. M. S. distribution]; Lefroy, Pusa Bull. 7, 24[22]; Big., T. A. S. B., lx, 264; Shiraki, Taihoku Expt. Stn. Bull., (1918), 35-43,tab. 1, 4[11]; Ric., Bull. Ento. Res., vi, 407[12]; Patton & Cragg, T. B. Med. Ent., 291, tab. xv, 1, [thorax]; 2, [halter], tab. xlii, 12, [egg mass]; Edw., Jo. F. M. S. Mus., viii, 27[31]; d'Abreu, Proc. iiird Ento. Meetg. Pusa, iii, 866[16], [eaten by birds]; Kröb., Zool. Jahrb. Syst., xliii, 82-85, fig., B. tab. i, 14, 15; de Mello, Proc. ivth. Ento. Meetg. Pusa, 45[15]; S. W., Mem. Dept. Agric. Ind., vii, 107[14]; Surc., Gen. Ins. Fasc. 175, 151.

'East India[1, 14, 20, 21]
Ceylon[7, 8, 9]
Nepal[9]
Bangalore[9]
Moulmein[9]
(L. Burma)
Rangoon[9]
Orissa Coast[9]
Kulattupuzha (W. Ghats)[9]
Travancore[9]
Morapai[9]
(S. Bengal)
Shillong[22]
Nongpoh[22] (Khasia)
Lakhimpur D. (Assam)[14]
Goa (Port. Ind.)[15]
Central Provinces[16]
'Bengal'[17, 18, 23]
[Hazaribagh Daltongani (Bihar)][2]
Siripur (Bihar)
Suri (Bihar)
Angul, Peradeniya (Ceylone)
Erode, Chapra
Bitanla (Balaghat, C.P.)
Peshawar
Maymyo (Burma)
Tilip (Puri Dist.)
Prome (Burma)
Purulia
Bashra (Midnapur Dist.)
Nagpur
Manantoddy (Wynned)
Belgachia (Bengal)
Bankura
Margheriata (Assam)
Sidapur (Coorg)

Virajpet (Coorg)
Dalhousie
Pollibetta (Coorg)
Palur (Madras Pres.)
Coimbatore
Pyinmana (U. Burma)
Taliparamba (N. Malabar)
Ernad (S. Malabar)
Kasargode (S.Kanara)

Chrysops bifasciata, Macq., Dipt. Exot., I, i, 161, (1838)[17].

Chrysops bifasciatus, Wulp, Cat. Dipt. S. Asia, 65 : Kert., Cat. Dipt., iii, 182.

Chrysops impar, Rond., Ann. Mus. Civ. Genoa, vii, 460, (1875).

Chrysops ligatus, Wlk., List Dipt. B. M., i, 195[18], (1848); Big., T. A. S. B., lx, 264; Wulp, Cat. Dipt. S. Asia, 65.

Hæmatopota lunatus, Gray in Griff., Cuvier Règ. Anim., xv, 696, tab. cxiv, 4, (1832).

Chrysops semicirculus, Wlk., List Dipt. B. M., i, 196[19], (1848); Big., T. A. S. B., lx, 264; Wulp, Cat. Dipt. S. Asia, 65; Kert. Cat. Dipt., iii, 196; Ric., A. M., N. H., (7) ix, 375.

Chrysops terminalis, Wlk., List Dipt. B. M., i, 195[20], (1848); Big., T. A. S. B., lx, 264; Wulp, Cat. Dipt. S. Asia, 65.

? *Chrysops flaviventris,* Macq, Dipt. Exot. supp. i, 44[21], (1845); Wlk., List Dipt. B. M., v, supp. i, 292; Big., T. A. S. B., lx, 264; Wulp, Cat. Dipt. S. Asia, 65; Kert., Cat. Dipt., iii, 187; Ric., A. M. N. H. (7) ix, 373; *id.,* R. I. M., iv., 396; Surc., Gen. Ins. Fasc. 175, 151.

Siam[9]
Annam[9]
Sumatra[4, 8, 9, 13]
Java[2, 3, 5, 8, 9]
Borneo[8, 17]
Philippines[8]
Hong Kong[9, 12]
Formosa[9, 11]

fasciatus, Wiedemann.

Chrysops fasciatus, Wd., Dipt. Exot., 103. 3[1], (1821); *id,* Ausser. Zweifl. Ins., i, 198. 5; Wlk., List Dipt. B.

Sibsagar (Assam)[4]
Andaman Ids.[4]

M., i, 195; *id., loc. cit.*, v, supp. i, 291. 66; Dol., Naturkund. Tijd. Ned. Ind., xvii, 84[2]; Wlk., Proc. Linn. Soc. Lond., iv, 104. 38; Wulp, Cat. Dipt. S. Asia, 65[3]; Ric., A. M. N. H., (7) ix, 375; Kert., Cat. Dipt., iii, 186; Surc., Gen. Ins. Fasc. 175, 151.

F. M. S.[4]
Amboina[2, 3]
Java[1, 3]
Borneo[3, 4]

Chrysops fasciata, Ric., R. I. M., iv, 391-2, tab. xv, 7[4].

fixissimus, Walker.

Chrysops fixissimus, Wlk., Proc. Linn. Soc. Lond., i, 112, 37[1], (1856); Wulp, Cat. Dipt. S. Asia, 65; Kert., Cat. Dipt., iii, 187; Ric., A. M. N. H., (7), ix, 376; Pratt, Jo. Trop. Vet. Sci., iv, 390-4[3], [F. M. S. distribution]; Surc., Gen. Ins. Fasc. 175, 151.

Trincomali[2]
(Ceylon)

F. M. S.[2, 3]
Borneo[1, 2, 4]

Chrysops fixissima, Ric., R. I. M., iv, 392-3[2].

Chrysops testaceicallosus, Meij.

Chrysops unizonatus, Rond., Ann. Mus. Civ. Genoa, vii, 459[6], (1873); O.-S., Berl. Ent. Zeit., xxvi, 97.

flavipes, Meigen.

Chrysops flavipes, Meig., Klass, i, 159, 3, tab. ix, 13[1], (1804); Villen., Ann. Soc. Ent. Fr., lxxiv, 310. 6; Kert., Cat. Dipt., iii, 187[2]; Ric., R. I. M., iv, 397[3]; Surc., Gen Ins. Fasc. 175, 151.

Baluchistan[3]

Asia Minor[3, 4, 6]
C. & S. Europe[1, 5, 6]

Chrysops flavipes var. *askahabadensis,* Szil. Arch. f. Naturg., 111[4], (1919).

Chrysops perspicillaris, Lw., Neue Beitrag., iv, 25. 5[5], (1856); *id.,* Verh. Zool. bot. Ges. Wien, viii, 663; Schin., F. Austr., i, 41; Gob., Mem. Soc. Linn. Fr., 1881, 48; Pand., Rev. Ent., ii, 226; Griff., Boll. Mus. Zool. & An. comp. Tornio, x, no. 218. 6; Ric., A. M. N. H., (7) ix, 426, 430; Aust., Bull. Ento. Res., xvi, 8[6].

Coonor
Amlapuram
(Coimbatore (Dist.)
Angul]

flavocinctus, Ricardo.

Chrysops flavocinctus, Ric., A.M. N.H., (7), ix, 380[1], (1902); Kert., Cat. Dipt., iii, 187; Praft, Jo. Trop. Vet. Sci., iv, 390-4[2], [F. M. S. distribution]; Lefroy, Pusa Bull. 7, 24[3]; Surc., Gen. Ins. Fasc. 175, 151.

Chrysops flavocincta, Aust., Bull. Ento. Res., xii, 431[4]; Ric., R. I. M., iv, 395-6, tab. xv, 4[5]; S. W., Mem. Dept. Agric. Ind., vii, 107[5].

Khasia Hills
(Assam)[1, 3]
Trincomali (Ceylon)[1]
Travancore
Mohnyin (U. Burma)[5]
[Jebong (Darjiling)
Gauhati
Pakokku (Burma)
Malwun (Burma)]

Borneo[1]

F. M. S.
Siam[4]

indianus, Ricardo.

Chrysops indianus, Ric., A. M. N. H., (7) ix, 379[1], (1902); Kert Cat. Dipt., iii, 189; Surc., Gen. Ins. Fasc. 175, 152.

Chrysops indiana, Ric., R. I. M., iv, 390-1[1].

Nilgiri Hills[1, 2]
Khasia Hills[1, 2]

mlokosiewiczi, Bigot.

Chrysops mlokosiewiczi, Big., Ann. Soc. Ent. Fr., (5) x, 146[1], (1880); Ric., A. M. N. H., (7) ix, 378; Kert., Cat. Dipt., iii, 191[2]; Ric., R. I. M., iv, 378-9[3] 401[4]; Shiraki Taihoku Expt. Son. Bull. (1918), 25-35, tab. i, 3[5]; Ric., Bull. Ento. Res., vi, 407[6]; a Kröb., Zool. Jahrb. Syst., xliii, 128-9, tab. ii, 55[7]; Surc., Gen. Ins. Fasc. 175, 153.

Chrysops iranensis, Big., Mem. Soc. Zool. Fr., v, 602[8], (1891).

Chrysops striatus, Wulp, Leyden Mus. Notes, vii, 79,[9] (1885); *id.,* Cat. Dipt. S. Asia, 66; Ric., A. M. N. H., (7) ix, 379; Kert., Cat. Dipt., iii, 197.

Yunnan[4]

? Caucasus[1, 2]
? Persia[1, 2, 8]
Chusan Archin[3]
Amoy (China)[3, 9]
Formosa[5]
Hong Kong[6]
S. Ussuri[2]

pellucidus, Fabricius.

Chrysops pellucidus, Fb., Syst. Antl.,113. 11[1], (1805); Wd., Dipt. Exot., 107. 8; *id.,* Ausser. Zweifl. Ins., i, 206; Wlk., List Dipt. B. M., i 200; Kert., Cat. Dipt., iii, 194; Big., T. A. S. lx, 264; Ric., A. M. N. H., (7) ix, 373; Surc., Gen. Ins. Fasc. 175, 154.

Chrysops pellucida, Ric., R. I. M., iv, 386-7, tab. xv, 2[2].

Tranquebar[1, 2]
Trichinopoly[2]
[Coimbatore[2]
Coonoor
Amlapuram
(Coimbatore Dist.)
Agnul]

pettigrewi, Ricardo.

Chrysops pettigrewi, Ric., A. M. N. H., (8) xi, 452[1], (1913); [Surc., Gen. Ins. Fasc.] 175, 154.

Manipur[1]

stimulans, Walker.

Chrysops stimulans, Wlk., Saunds., i, 73[1], (1850); *id.,* List Dipt. B. M., v, supp. i, 265; Big., T. A. S. B., lx, 264; Wulp, Cat. Dipt. S. Asia, 65; Kert., Cat. Dipt., iii, 197; Lefroy, Pusa Bull. 7, 24[3]; Ric., A. M. N. H., (7) ix, 376; *id.,* R. I. M., iv, 389-90, tab. xv, 5[2]; Fletcher, Pusa Sci. Repts., (1915-16), 80, [bionomics]! *id., loc. cit.,* (1916-17), 96, [oviposition]; Patel, *loc. cit.,* (1919-20), 97, [parasites]; Surc., Gen. Ins. Fasc. 175, 154.

'India'[1, 2]
pusa[2, 3]

PART XIII – CICINDELIDAE.

INTRODUCTION

BIONOMICS. The bionomics of the Cicindelidæ have been little studied in India and Malaya, but the facts that are known indicate that they are of considerable interest. The late Mr. Shelford has discussed the mimicry of *Collyris* by the remarkable Locustid *Condylodera tricondyloides* and a species of *Sarpedon* (Diptera) in *P. Z. S.*, II, 1902, and a recent contribution to the same question has been made by Mr. Uvarov in *T. E. S.*, 1922. Koningsberger and Shelford have discussed the life-history of *Neocollyris emarginata* in Java, and the early stages of *N. bonelli, N. tuberculata,* and *Tricondyla cyanea* have been described by van Leeuwen. Outside our limits very extensive observations have been made on the life-history and ecology of certain species of *Cicindela,* particularly by Mr. Vernon Shelford in America, but in India no species has been thoroughly studied from these aspects. Professor Houlbert of Rennes has, however, discussed the external morphology of the larva of *C. biramosa* in great detail. On the island of Barkuda in the Chilka Lake the late Dr. Annandale and one of the present authors have taken the larvæ of *C. hœmorrhoidalis* from termite-mounds, with which they are constantly associated, as Horn has already stated, (*D. E. Z.*, pp. 237, 395; 1899). The curiously local habits of various species, which are a feature of the Indian Cicindelidæ, are discussed in the various papers listed elsewhere in this catalogue, particularly by Annandale and Gravely. Much of this information is summarized by Fowler.

ECONOMIC IMPORTANCE. The Cicindelidæ are not important as pests, and are probably beneficial in destroying other pests, but as they are usually not very numerous, this is not of much consequence. The eggs of *Neocollyris,* however, are laid mainly in borings made by the adults in tea and coffee and in saplings of forest trees, and this may rank them among the insects of minor economic importance. As Fowler has suggested, these borings, if numerous, must have a destructive effect, but *per contra* the voracious larvæ may consume large numbers of injurious insects. Within our limits, *Neocollyris,* probably, *N. crassicornis,* has only been reported from tea and coffee from South India. As much has been written about the importance of *N. emarginata* in Malaya, it may be useful to point out that this species does not occur in India. The species of *Tricondyla* are said to have

Table showing distribution of Indian Cicindelidæ (1923).

Under the genera the column on the left hand side contains the total number of forms in a particular region, the column on the right hand side the number of endemic forms. The number of Indian species extending outside our limits indicated, and where the genus, but no Indian extends, is marked with an X. The table is based on records available till 1923 only. The word" has no precise zoogeographical meaning here, nor do the divisions necessarily correspond with recognised zoogeographical divisions.

	Faunistic Regions.	Collyris.		Necocolrlyrs.		Tricondyla.		Derocrania.		Therates.		Prothyma.		Heptodonta.		Cicindela.		Apteroessa.		Megacephela.		Tox	
Distribution in B. India	Assam-Burmese region	2	-	40	12	5	1	-	-	9	4	7	2	6	1	64	11	-	-	-	-	133	
	E. Himalaya region	-	-	12	1	1	-	-	-	4	1	1	-	3	-	37	-	-	-	-	-	56	2
	W. Himalayan region	-	-	4	-	-	-	-	-	-	-	2	1	1	-	31	-	-	-	1	-	39	-
	Gangetic Plain	1	-	11	1	1	-	-	-	-	-	5	1	-	-	55	2	-	-	-	-	72	4
	Southern Indian region	3	2	15	5	2	2	3	1	-	-	2	-	-	-	51	11	1	1	-	-	87	22
	Western Indian region	1	1	5	1	-	-	2	-	-	-	3	1	-	-	48	10	-	-	1	-	60	13
	Ceylon	1	1	9	7	4	4	10	10	-	-	1	-	-	-	30	13	-	-	-	-	55	35
	Andamans and Nicobars	-	-	4	3	-	-	-	-	-	-	-	-	-	-	4	-	-	-	-	-	8	3
Extra-Indian Distribution	Malayan region	2		18		-		-		2		1		1		22		-		-			
	Indo-Chinese & Chinese region	1		16		4		-		1		2		4		39		-		-			
	Austro-Malayan region (Celetes, etc.)	-		7		-		-		-		-		-		13		-		-			
	Australlan region	-		-		-		-		-		-		-		1		-		-			
	African region	-		-		-		-		-		-		-		3		-		-			
	Palaearctic region	-		-		-		-		-		-		-		9		-		-			
	New World	-		-		-		-		-		-		-		-		-		-			
	Total No. of Indian Forms	6	4	65	42	11	7	13	13	10	7	12	10	6	2	161	112	1	1	2	-	287	198

similar habits as larvæ. *Cicindela sexpunctata* preys on the rice bug, *Leptocorisa varicornis,* and on the rice Hispid. *Hispa armigera.*

GEOGRAPHICAL DISTRIBUTION

From the analysis on the opposite page it will be seen that the Cicindelid fauna of our region is very compact. More than two-thirds of the species are endemic, and the only outside elements worthy of recognition are the distinct Malayan element, and the relationship of the Assam-Burmese fauna to that of Indo- and Western China. The North African and Southern European element so characteristic in other groups, such as the Aculeate Hymenoptera among insects, is practically absent, except in a few species of *Cicindela,* while only one species penetrates to Australia. The only cosmopolitan genera are *Cicindela* and *Megacephala,* the remainder being almost confined to the Oriental region.

The Assam-Burmese region contains the largest number of forms, one genus (*Therates*) being almost peculiar to it and the Eastern Himalayas. In the Western Himalayas the predominant forms belong to the genus *Cicindela,* but the comparative paucity of species in this region is not so striking when it is considered that a rich vegetation is not an essential to the preponderance of adephagous species, as it is in the Phytophaga.

The Gangetic Plain (chiefly Bengal) contains a fair proportion of species, belonging chiefly to *Cicindela* and *Neocollyris,* but only about four out of more than seventy species are apparently endemic. Next to the Assam-Burmese region the largest number of forms (mainly endemic) is found in the South, only three small genera being absent in its fauna; it also contains the curious *Apteroessa grossa.* The Western Indian fauna is well represented and contains a species of the genus *Megacephala* which, apart from its occurrence in the Palæarctic region, is chiefly South American. The Palæarctic element is only noticeable here, eleven forms, nine of *Cicindela,* being common to both regions. Though its South Indian affinities are quite evident, Ceylon is remarkable for the high proportion of forms peculiar to it. *Derocrania* is mainly a Ceylonese genus. The paucity of the Andamanese Cicindelid fauna is not surprising, especially as these islands have not been very thoroughly investigated. Broadly speaking the tiger-beetles of the Ceylonese sub-region, a defined by Wallace, except for a strong Western element, are very distinct from those which inhabit the area between Bengal and Burma.

The apparent discontinuity of distribution in certain species such as *C. aurulenta* and *C. angulata* var. *plumigera,* has been discussed by Dr. Horn in a paper entitled "Die Weddabrucke", *but we refrain from subscribing to any theory of former land-connections until more extensive collections have been made.

ACKNOWLEDGMENTS

It remains to be said that in the preparation of this catalogue we have had the advantage of consulting, jointly or singly, the extensive collections belonging to the Zoological Survey of India, to the British Museum (including the Nevinson collection deposited there), and to Mr. H. E. Andrewes. These examinations have enabled us to add many localities hitherto unrecorded.†

The main part of the work was done in the excellent library of the Entomological Society of London, and we must express our thanks to the Assistant Librarian of that Institution for his unfailing courtesy. We are also indebted to Dr. C. J. Gahan, Keeper of the Entomological Department, British Museum, for access to the collections and library in his care, to Mr. H. E. Andrewes for his hospitality and interest, and to Dr. R. Gestro for a list of the types in the Genoa Museum. Mr. T. Bainbrigge-Fletcher, Imperial Entomologist, has placed us under a special obligation by checking the manuscript and filling in certain details which circumstances compelled us to omit, while Dr. C. F. C. Beeson, Imperial Forest Zoologist, has also kindly looked over the M. S.Finally we must record our gratitude to Dr. Walther Horn of the Deutsches Entomologisches Institut at Berlin-Dahlem, who has on many occasions placed his special knowledge at our disposal, for reading through our manuscript and making certain additions and corrections. He is also responsible for most of the information regarding the types. The support of his distinguished authority will be as much appreciated by those who use this work, as it is by the authors.

London : 1924

MERCIA EHYNES-WOOD
CEDRIC DOVER

* Mem. Ist Congress Int. Ent., p. 314 : Brussels, 1911.

† Since this was written we have also seen the collections of the Raffles Museum in Singapore, and the F. M. S. Museum at Kuala Lumpur.

PRINCIPAL ABBREVIATIONS

A. M. N. H.	– Annals and Magazine of Natural History.
Ann. Gen.	– Annali del Museo di Storia Natura'e di Genova.
A. S. E. B.	– Annales de la Societe Entomologique de Belge.
A. S. E. F.	– Annales de la Societe Entomologique de France.
B. E. Z.	– Berliner Entomologisches Zeitschrift.
B. S. E. F.	– Bulletin de la Societe Entomologique de France.
B. S. N. M.	– Bulletin de la Societe Naturale de Moscou. Comp. Rend.
S. E. B.	– Compte Rendus de la Societe Entomologique de Belge.
D. E. Z.	– Deutsches Entomologisches Zeitschrift.
Fowler, F. B. I.	– The Fauna of British India including Ceylon and Burma. Coleoptera, General Introduction and Cicindelidæ and Paussidæ. By W. W. Fowler, London : 1912.
J. B. N. H. S.	– Journal of the Bomay Natural History Society.
R. I. M.	– Records of the Indian Museum.
S. E. Z.	– Stettizer Entomologisches Zeitschrift.
T. E. S.	– Transactions of the Entomological Society of London.

PART 13 – CICINDELIDAE.

[*Explanatory Note.*—The species are arranged alphabetically under the genera, and the authorities responsible for the records are indicated by the use of numbers. Where a number has not been given against a reference usually no exact record has been mentioned therein, but in a very small number of cases this is due to the fact that we have not had access to the reference concerned.

The original record of the species or variety is stated first, the other Indian localities being listed alphabetically under it : extra-Indian localities are given below the line. The records against which no numbers have been given are new records, the letters in brackets after them indicating the collection in which the specimens may be found. Localities in Italics are now known to be erroneous records.

With regard to the localities, the system of indicating the authorities responsible for them has necessitated a certain amount of repetition and obscurity. To avoid confusion we have not tried to exercise any selective influence, as in a catalogue of this type it would be difficult to know where to stop. Thus under the same species one may find the locality "Penang" given in its ordinary English rendering and also in the Malay rendering "Pulo Pinang" simply because two authors have recorded them in two different ways, or one may find "Talewadi" and "Castle Rock", which are both practically the same. Again one may find two localities such as "Deli" and "Sumatra" given under one species. As "Delhi" is in Sumatra it is obvious that "Deli" is the only definite record in Sumatra for the species, and that "Sumatra" is an older and more general record, but the system followed here requires the recording of both names in different places. Moreover, where localities such as "Bombay", "Tenasserim" or "Ceylon" are given it does not necessarily follow that the species is found all over the area recorded, but that the authority responsible did not provide more accurate data.

The majority of the localities will be familiar to students of the Oriental fauna, but a few are obscure and practically unknown. Their exact position, however, may be found in the *Times Atlas and Gazetteer,* or in the various Indo-Malayan Gazetteers and Maps published by local Governments. Most of the commoner Indian localities will be found in Brunetti's "Catalogue of the Oriental and S. Asiatic Nemocera" in *Rec. Ind. Mus.* XVII, 1920, and in the map in Maulik's volume on Chrysomelinæ and Halticinæ in the *Fauna of British India* Series, 1926. For easy reference a general index of localities has also been added at the end of this work.

A word may also be added on the location of the types. It has been the custom of many authors (including Dr. Horn) to designate all the specimens before them at the time of description as types, with the result that a "type" of a particular species is now to be found in more than one collection. In this catalogue the location of the type has the sanction of Dr. Horn's authority, so that should types be found in collections other than those mentioned here, it would be convenient to regard the "type" recorded by us as the actual type and the others as paralectotypes. The selection of lectotypes in the Cicindelidae deserves the attention of workers on this group.

A=Andrewes collection; N=Nevinson collection in British Museum ; B. M. = British Museum collection; H=localities supplied by Dr. Horn from his collection; F=localities supplied by Mr. Fletcher or Dr. Beeson from the collections at Pusa or Dehra Dun.

GENERAL ACCOUNTS

Lefroy, Indian Insect Life, pp. 259-262, 1909; Fowler, F. B. I., 1912; Horn, Wyts. Genera Insectorum, Cicindelinæ, 1908-15; Westwood, Modern Classification of Insects, pp. 47-57, London, 1839; Sharp, The Cambridge Natural History, Insects, II, p. 201, 1899.

CATALOGUES AND LISTS

Annandale & Horn : 1909 : Annotated List of Asiatic Beetles in the collection of the Indian Museum, Part I, Family Carabidæ, subfamily Cicindelinæ : Calcutta. Referred to as "Annand. & Horn, List."

Atkinson : 1889 : Catalogue of the Insecta of the Oriental Region. No. I. Order Coleoptera, Family Cicindelidæ—in *Jour. As. Soc, Bengal,* 1819 pp. 1-24, Ref. to as "Atkinson, Cat."

Chaudoir : 1865 : Catalogue de la Collection de Cicindelites de M. le Baron de Chaudoir : Brussels. Ref. to as "Chaud., Cat. Coll."

Codina : 1917 : Cataleg de la Col. leccie de Cicindelinæ dipositada en el Museu da Catalunya—in *Junta Cienc. Nat. Barcelona,* II, 2, pp. 327-529. Ref. to as "Codina, Barcel, Cat."

Dover & Ribeiro : 1923 : A List of the Indian Cicindelidæ with Localities—in *Rec. Ind. Mus.,* xxv, pp. 345-363 (with special reference to the collection of the Indian Museum).

Fleutiaux : 1892 : Catalogus Systematique des Cicindelidæ apres Linne : Liege. Ref. to as "Fleut., Cat."

Gemminger & Harold : 1868 : Catalogus Coleopterorum. I. Cicindelidæ Carabidæ : Munich. Ref. to as "Gemm. & Har., Cat."

Horn : 1905 : Systematische Index der Cicindeliden-in *Deuts. Entom. Zeitschrift* (Beilage), pp. 1-56. (This list has not been referred to throughout this catalogue, as it only a list of species with the region in which they occur indicated by a letter, such as "I" for "India.")

Horn : 1908-1915 : Wytsman's Genera Insectorum, Cicindelinæ, fascs. 82, 84 : Brussels.

Horn : 1926 : in Junk and Schenkling's Coledpterorum Catalogus, Pars 86, Carabidæ—Berlin. Ref. to as "Col. Cat., Cic." (This catalogue was received when our M. S. was already in print. It has not, therefore, been possible for us to refer to it throughout the text, but we have referred to it under the subfamilies and genera; in a few cases, where we have thought it would be useful, we have cited it under species. Changes in arrangement have been indicated by means of footnotes.)

Schaum : 1863 : Contributions to our knowledge of the Cicindelidæ of Tropical Asia—in *Journ. Entom.*, II, pp. 57-74. (A list of the Oriental species with synonymy.)

BIONOMIC AND ECONOMIC

Westermann, Rev. Ent. Silberm., I, p. 106, 1833; Saunders, T. E. S., p. 61, 1834; Barlow, Ind. Mus. Notes, IV, p. 46, 1896; Peal, Ind. Mus. Notes, V, pp. 40, 48, 1900; Robinson, Fasc. Malay Zool., p. 179, 1903; Koningsberger, Mededeclingen uit's Lands Plantentuin, XLIV, p. 113, 1905; Shelford, Journ. Str. Br. R. As. Soc., p. 283, 1906; P. E. S., p. LXXII, 1905; T. E. S., p. 83, 1907; A Naturalist in Borneo, p. 156; London, ed. 1918; P. Z. S., p. 26, 1902; van Leeuwen, Meded. Alg. Proefstat. Java, Salatiga (2), No. 15, p. 137, 1909; Tijds. V, Ent., p. 19, 1910; Fletcher, Spol. Zeyl. V, p. 62, 1907; Lefroy, Indian Insect Life, pp. 259-262, Calcutta, 1909; J. B. N. H. S., XIX, p. 1008, 1910; Annandale and Horn, List, 1909; Horn, D. E. Z., p. 223, 1899; Fowler, F. B. I., 1912; Gravely, R. I. M., VII, p. 207, 1912; Houlbert, Insecta, Rennes, p. 169, 1912; Poyser, Planters Chronicle, XI, p. 622, 1916; Fletcher, Planters Chronicle, XII, p. 14, 1917; Annandale and Dover, R. I. M., XXII, p. 335, 1921; Dover and Ribeiro, R. I. M.; XXII, p. 721, 1921; Uvarov, T. E. S., p. 209, 1922; Dover, Planters Chronicle, p. 473, July 19, 1924. For extra Oriental references for comparison see Horn, Gen. Ins., 1908-1915, and Col. Catalogue, Pars 86, Cicindelinae, 1926.

Division : *ALOCOSTERNALIÆ,* Horn.

Alocosternaliae, Horn, Ent. Nachr., p. 214, 1900; Ent. Zeitung, Berl., ii, p. 5, 1905; Gen. Ins. pp. 83 and 84, 1908; Col. Cat., Cic., p. 11, 1926; Fowler, F. B. I., p. 222, 1912.

Sub-family: COLLYRINÆ, Chaudoir (1860).

Collyrinœ, Fowler, F. B. I., p. 223, 1912 (Key to genera).

Collyrini, Horn, Gen. Ins., p. 93, 1908; Col. Cat., Cic., 86, p. 21, 1926.

Collyridœ, Chaudoir, B. S. N. M., p. 270, 283, 1860; Horn, D. E. Z., p. 33, 1901, p. 7, 1905.

Genus : **Collyris,** Fabricius (1801).

Collyris, Fabr., Syst. Eleuth., i, p. 226, 1801; Fischer, Ent. Ross., i, p. 97, 1821; Chaud., (ex parte) B. S. N. M., p. 284, 1860; A. S. E. F., pp. 483-536, 1864 (monograph); Cat. Col., p. 7, 1865; Horn, D. E. Z., p. 353, 1892, p. 49, 1897, p. 36, 1901, p. 421, 1907; Ann. Gen., p. 371, 1893; Ann. Gen., p. 674, 1895; Gen. Ins., p. 98, 1908; Col. Cat., Cic., p. 30, 1926 (as subgenus); Maindron, A. S. E. F., p. 6, 1905; Lac., Gen. Col., p. 29, 1854; Fleut., Cat., p. 138, 1892; Fowler, F. B. I., p. 223; 1912.

Colliurus, Latreille (ex parte), Gen. Crust. Ins., i, p. 174, 1806; Castelnau, Hist. Nat. Ins., i, p. 24, 1851.

Archicollyris, Horn, D. E. Z., p. 43, 1901, p. 421, 1907; Gen. Ins., p. 99, 1908.

Collyrides ingenuœ, Chaud., A. S. E. F., p. 489, 1864.

Genotype : *Collyris longicollis,* Fabricius (1787).

Distribution : India and Malaya.

1. **brevipennis,** Horn, D. E. Z., p. 44, 1905[1], p. 59, 1905; Gen. Ins., p. 100, 1908; Fowler, F. B. I., p. 226, 1912[2]; Dover & Rib., R. I. M., xxii, p. 721, 1921[3] (habits), xxv, p. 346, 1923[4].

   Type : Coll. W. Horn.

   Bombay[1, 2, 4]
   Castle Rock[3, 4]
   Gudhali (A)
   Talewadi[3, 4]

2. **brevipennis** var. **elongato-subtilis,** Horn, Arch. Nat., lxxix, A. heft ii, p. 2, 1913[1]; Gen. Ins., p. 426, 1915; Dover & Rib., R. I. M., xxv, p. 346, 1923[2].

   Type : Coll. W. Horn.

   Anamalai hills[1, 2]

3. **brevipennis** var. **subtilesculpta,** Horn, D. E. Z., p. 44, 1901[1]; p. 422, 1907; Gen. Ins., p. 100, 1908; Maindron & Fleut., A. S. E. F., p. 6, 1905[2]; Fowler, F. B. I., p. 226, 1912[3]; Dover & Rib., R. I. M., xxv, p. 346, 1923[4].

   Type : Coll. W. Horn.

   S. India
   Coonoor[2, 3, 4]
   Coorg[3, 4]
   Dindigul[2, 3, 4]
   Kanara (H)
   Ouchterlony valley, 3-3,500 ft.[4]
   Shembaganur (H)

4. **dohrni,** Chaudoir, B. S. N. M., ii, p. 286, 1860[1]; A. S. E. F., p. 490, 1864[2]; Cat., p. 7, 1865; Atkinson, Cat., p. 20, 1889[3]; Gemm. & Har., Cat., p. 37, 1868;

   Colombo[1, 2, 3, 5, 6]
   E. Province of Ceylon[4]

Fleut., Cat. p. 138, 1892; Horn., Spol, Zeyl., p. 30, 1904[4]; D. E. Z., p. 422, 1907; Gen. Ins., p. 100, 1908; Fowler, F. B. I., p. 225, 1912[5]; Dover & Rib., R. I. M., xxv, p. 346, 1923[6].

Kandy (H)

Type : Mus. Paris.

5. **longicollis,** Fabricius, Mant. Ins., p. 185, 1787[1]; Ent. Syst., p. 168, 1792[2]; Syst. El., i, p. 226, 1801[3]; Hcrbst, Kaf., x, p. 215, pl. 173, f. 9, 1806[4]; Latr., Gen. Crust. Ins., i, p. 174, 1806[5]; Hist. Nat. Crust. Ins., viii, p. 190, pl. 11, f. 3, 1804; Castelnau, Hist. Nat. Ins., i, p. 24, 1851[6]; Mots., Etud. Ent., p. 34, 1855[7]; Schaum, B. E. Z., p. 79, 1861 (synonymy); Chaud., A. S. E. F., p. 490, pl. 7, f. 1, 1864[8]; Cat. Coll., p. 7, 1865[8a]; Gemm. & Har. Cat., p. 38, 1868[9]; Atkinson, Cat., p. 21, 1889[10]; Fleut., Comp. Rend., S. E. B., p. 168, 1890; Cat., p. 139, 1892; Horn, D. E. Z., p. 354, 1892; p. 422, 1907; Gen. Ins., p. 100, 1908; Col. Cat. Cic., p. 30, 1926[b]; Annand. & Horn, List, p. 2, pl. 1, f. 2, 1909[11]; Fowler, F. B. I., p. 225, f. 100, 1912[12]; Dover & Rib., R. I. M., xxv, p. 346, 1923[13].

Indes Or. [1, 5, 6, 7, 15]
Assam[12]
Barway (N)
Bengal[b]
Chota Nagpur[12]
Kumaon[b]
Mysore[10, 12, 1]
Mungphu[12, b]
Nowatoli (H)
North Malabar[b]
Sahibgang[11, 12, 13]
Sikkim[10]
Surada[b]
Borneo[9, 10]
Java[9, 10]
Siam[8a, 8, 10, 12]

C. caviceps, Klug, Jahrb. Insekt., i, p. 45, 1834; Chaud. (ex parte), A. S. E. F., p. 491, 1864[14]; Gemm. & Har., Cat., p. 37, 1868; Atkinson, Cat. p. 19, 1889; Fleut., Cat., p. 139, 1892.

C. herbsti, Gistl. Syst. Ins. I, p. 85, 1837.

C. lafertei, Chaudoir, B. S. N. M., p. 284, 296, 1860[15].

C. cylindricollis, Dejean, in litt., Cat. Col., ed. 3, p. 7, 1837.

Type : Brit. Mus. (Banks Cab.)

6. **mniszechi,** Chaudoir, Rev. Mag. Zool., p. 75, 1864[1]; A. S. E. F., p. 492, 1864[2]; Cat. Coll., p. 7, 1865[3]; Atkinson, Cat., p. 21, 1889[4]; Gemm. & Har., Cat., p. 38, 1868[5]; Fleut., Cat., p. 139, 1892[6]; Horn., D. E. Z., p. 354, 1892; p. 60, 1897[7]; p. 82, 1904; Ann. Gen., p. 372, 1893[8]; Gen. Ins., p. 100, 1908; Fowler, F. B. I., p. 227, f. 101, 1912[9]; Dover & Rib., R. I. M., xxv, p. 346, 1923[10].

Siam[1, 2, 3, 6, 7]
Bhamo (B. M.)
Bhamo hills 4,000 ft. (B. M.).
Carin Cheba (Karan-Ni)[8]
Indes Or.[12, 13]
Karen hills[7, 10]
Momeit[9, 10]

C. grandis, Hope, in litt., Chaud., in litt., Thomson in litt.; Horn, D. E. Z., p. 354, 1892; p. 422, 1907[11].

C. caviceps, Chaudoir (ex parte), A. S. E. F., p. 491, 1864[12].

C. dormeri, Horn, D. E. Z., p. 196, 1898[13], p. 422, 1907[14].

C. robusta, Dohrn., S. E. Z., p. 292, 1891; Horn, S. E. Z., p. 176, 1896; D. E. Z., p. 422, 1907.

C. gigas, Lesne, B. S. E. F., p. 361, 1901[15]; Mission Pavie, Indo-Chine, 1879-95, etud. div. iii, p. 15, 1904; Horn, D. E. Z., p. 82, 1904.

Mysore[11]
Naga hills[9, 10]
Borneo[7, 10, 14]
? Java[10, 14]
Laos (between Luang-Prabang and Muong-Theng)[4, 8, 9, 10, 14, 15]
Malacca [4, 5, 7, 8, 9, 10]
Penang (H)

Type : Mus. Paris.

Genus : **Neocollyris,** Horn (1901).

Neocollyris, Horn, D. E. Z., p. 45, 1901; Spol. Zeyl. ii, p. 38, 1904; Gen. Ins., p. 100, 1908; Fowler, F. B. I., pp. 229-238, 1912 (structure, life-history, key to spp.).

Collyris, Fabr., ex parte, see p. 3.

Colliurus, Latr., ex parte, see p. 4.

Collyrides spuriæ, Chaudoir, A. S. E. F., p. 493, 1864.

Genotype : *Collyris bonelli,* Guerin (1834).

Distribution : India and Malaya.

1. **acute-apicalis,** Horn, Arch. Nat., lxxix, A, ii, p. 4, 1913[1]; Gen., Ins., p. 427, 1915[2]; Dover & Rib., R. I. M., xxv, p. 346, 1923[3].

   Chandkhira[1–3]
   Sylhet[2, 3]

   Type : Coll. W. Horn.

2. **ampulicollis,** Horn, Arch. Nat. lxxix, A, ii, p. 5, 1913[1]; Gen. Ins., p. 430, 1915; Dover & Rib., R. I. M., xxv, p. 346, 1923[2].

   Wallardi[1, 2]

   Type : Coll. W. Horn.

3. **andrewesi,** Horn, D. E. Z., p. 170, 1894[1]; p. 59, 1905[2]; Spol. Zeyl., ii, p. 35, 1904[3]; Gen. Ins., p. 106, 1910; Maindron & Fleut., A. S. E. F., p. 8, 1905[4]; Fowler, F. B. I., p. 270, 1912, f. 120[5]; Dover & Rib., R. I. M., p. 346, 1923[6].

   N. Kanara[1, 4, 5, 6]
   Anamailai hills, 3000-4000 ft.[2, 4, 5, 6]
   Ceylon[3, 4, 5, 6]
   Mahe[4, 6]

Type : Coll. W. Horn.

Malabar (B. M.)
Nilgiri hills[2, 4, 5, 6]
Ouchterlony valley (A.) Permaad (Travancore Tea Co.) (A.)
Trivandrum[5, 6]
Wallardi[4, 6]

4. **apicalis,** Chaudoir, Rev. Mag. Zool., p. 105, 1864[1]; Cat. Coll., p. 9, 1865[2]; A. S. E. F., p. 517, 1864[3]; Gemm. & Har., Cat., p. 37, 1868[4]; Atkinson, Cat., p. 18, 1889[5]; Fleut., Cat., p. 143, 1892[6]; Horn, Ann. Gen., p. 676, 1894[7]; Gen. Ins., p. 105, 1910; A. S. E. B., lvii, p. 363, 1913[8]; Col. Cat., Cie., p. 41, 1926[a]; Annand. & Rob., Fasc. Malay. Zool., i, p. 182, 1903 (habits)[9]; Fowler, F. B. I., p. 267, f. 119, 1912[10]; Codina, Barcel, Cat., p. 348, 1917[11]; Dover & Rib., R. I. M. xxv, p. 346, 1923[12].

Type : Mus. Paris.

Malacca[1, 2, 3, 4, 5, 6, 10, 11, 12]
Karen Mts. (B. M.)
Mergui (B. M.)
Patkai hills[10, 12]
Tavoy (B.M.)
Tenasserim (H.)

Borneo[11, a]
Gunong Tahan
Lao-Kay[8]
Perak (H)
Selangor (B.M.)
Singapore[10, 12]
Sumatra[10, 11, 12]
Tonkin [a]

5. **aptera,** Lund, Skrivt. Nat. Selsk., i, p. 65, pl. 6, f. 1, 1790[1]; Schriften Naturf. Gesellsch. Kopenhagen, i, (1), pp. 60-72, pl. 6, f. a, d, 1793[2]; Fabr., Ent. Syst., i, p. 169, 1792[3]; Syst. Eleuth., i, p. 226, 1801[4]; Brulle, Hist. Nat. Ins., iv, Col. 1, p. 99, pl. 3, f. 1, 1834[5]; Schaum, S. E. Z., p. 50, 1847; Chaud., B. S. N. M., p. 286, 1860[6]; A. S. E. F., p. 518, pl. 8, f. 15, 1864[7]; Cat. Coll., p. 9, 1865[8]; Gemm. & Har. Cat. p. 37, 1868[9]; Fleut. Cat., p. 143, 1892[10]; Horn, D. E. Z., p. 224, 1892[11]; p. 58, 1901[12]; Gen. Ins., p. 105, 1910, p. 429, 1915[13]; Fowler, F. B. I., p. 266, 1912[14]; Dover & Rib., R. I. M., xxv, p. 347, 1923[15].

Collyris lundi, Crotch, Col. Hefte, p. 106, 1869; Atkinson, Cat., p. 21, 1889.[16]

Type : Mus. Copenhagen.

Indes Or.[1, 2, 3, 4, 8, 10]
Bengal[6, 9, 11, 16]
N. India[7, 12]
N. Manipur[14]
Naga hills[13, 14, 15]
Punjab (H.)
Sylhet[14, 15]

Tenasserim[14, 15]
Java[5]

6. **apteroides,** Horn, D. E. Z., p. 59, 1901[1]; Gen. Ins., p. 105, 1910[2]; Fowler, F. B. I., p. 266, 1912[3]; Dover & Rib., R. I. M., xxv, p. 347, 1923.[4]

Type : Coll. W. Horn.

Manipur[1–4]
Chandkhira (H.)
Nongpoh (A.)

7. **attenuata,** Redtenbacher, in Hügel's Kaschmir, iv, p. 498, 1848[1]; Chaudoir, Cat. Coll., p. 9, 1865[2]; Gemm. & Har., Cat., p. 37, 1868[3]; Atkinson, Cat., p. 19, 1889[4]; Fleut., Cat., p. 143, 1892[5]; Horn, Gen. Ins., p. 100, 1908[6]; Fowler, F. B. I., p. 240, 1912[7]; Codina, Barcel. Cat., p. 343, 1917; Dover & Rib., R. I. M., xxii, p. 721, 1921; xxv, p. 347, 1923[8].

C. maculicornis Chaudoir, B. S. N. M., 1, p, 19, 1850[9]; Cat. Coll., p. 9, 1865; A. S. E. F., p. 524, 1864; Gemm. & Har., Cat., p. 38, 1868; Atkinson, Cat., p. 2, 1889[10]; Fleut., Cat., p. 144, 1892.

Type : Mus. Vienna.

Kashmir[1, 3, 4]
Assam[7], Bhimtal (H.)
Calcutta[7, 8]
Kangra Valley (B. M.)
Khasi Hills[6, 8]
Mungphu[8]
N. Manipur (N.)
Nepal[2, 5, 6, 8]
5,000-8,500 ft.
Sikkim[7, 4], Simla[4, 6, 10]
Singla, 1,500 ft.[8]
Soom, 4,000-5,000 ft.[8]

8. **auripennis,** Horn. D. E. Z., p. 67, 1902[1]; p. 7, 1905; A. S. E. B., lvii, p. 362, 1913[2]; Gen. Ins., p. 427, 1915; Fowler, F. B. I., p. 247, f. 107, 1912[3]; Dover & Rib., R. I. M., xxv, p. 347, 1923[4].

Type : Coll. W. Horn.

Mt. Mauson[1, 2]
Manipur[3, 4]

Annam[4]

9.* **bonelli,** Guérin, in Belanger's Voy. Ind. Or. Zool., p. 481, pl. 2, f. 1, 1834 (*Colliuris*)[1]; Castelnau, Hist. Nat. Ins. Col., i. pl. 2, f. 7, 1840[2]; Schmidt-Gœbel, Faun. Col., Birm., p. 17, 1846[3]; Chaudoir, A. S. E. F., p. 503, pl. 7, f. 7, 1864[4]; Cat., p. 8, 1865[5]; Gemm. & Har., Cat., p. 37, 1868[6]; Atkinson, Cat., p. 19, 1889[7]; Lesne, B. S. E. F., p. 55, 1891[8]; Horn, Ann. Gen., p. 674, 1895[9]; D. E. Z., p. 356, 1892, p. 49, 1897[10]; Gen. Ins., p. 102, 1908; A. S. E. B., lvii, p. 363, 1913; Col. Cat., Cic., p. 34, 1926; Fleut, Cat., p. 140, 1892[11]; B. S. E. F., p. 49, 1917[12]; Schilder, Ent. Blat. Berl., p. 175 1911[13]; Fowler, F. B. I., p. 248, 1912[14]; Codina, Barcel, Cat., p. 345, 1917[15]; Annand. & Dover, R. I. M., xxii, p. 335, 1921[16]; Dove & Rib., R. I. M., xxii, p. 721, 1921[17]; xxv, p. 347, 1923[18].

Bengal[1, 23, 24]
Barkuda Id.[16, 18]
Calcutta[14, 16, 18]
Central Provinces[18]
Deccan [c]
Dehra Dun (F.)
Dhansiri Valley[7]
Gokteir (A.)
Kharagpur[16, 17, 18]
Khasi Hills[16, 17, 18]
Maymyo[14, 18]
Naga Hills[14, 18]
Patkai hills[14, 18]
Pegu[14, 18]
Rangamati [17, 18]
Saran[18]
Sibsagar[16, 17, 18]
Sikkim[14, 16, 18]

---

* The type of this species is either from Bengal or Java, but without reference to it we cannot give a definite opinion.

Collyris postica, Brullé, Arch. Mus. Paris, i, p. 138, pl. 9, f. 8, 1839[19]; Chaud., A. S. E. F., p. 504, 1864[20]; Cat., p. 8, 1865[21]; Atkinson, Cat., p. 22, 1889[a]; Fleut., Cat., p. 141, 1892[22].

C. ruficornis, Brullé, Arch. Mus. Paris, i, p. 139, 1839[23]; Gemm. & Har., Cat., p. 37, 1868 (var.); Atkinson, Cat., p. 19, 1889 (var.); Fleut., Cat., p. 140, 1892 (var.)[24].

flavitarsis, Brullé, Arch. Mus. Paris, i, p. 141, 1839[25]; Gemm. & Har., Cat., p. 37, 1868 (var. of *cribrosa*); Atkinson, Notes Zool. N. W. P., p. 117, 1882; Cat. p. 19, 1889 (var.).

C. filiformis, Chaudoir, B. S. N. M., p. 697, 1843[26]; A. S. E. F., p. 507, pl. 8, f. 9, 1864; Cat., p. 8, 1865[27]; Gemm. & Har., Cat., p. 38, 1868[b]; Horn, D. E. Z., p. 92, 357, 1892[28]; Ann. Gen., p. 674, 1895[29]; Fleut., Cat. p. 141, 1892[c]; Atkinson, Cat., p. 20, 1889.

C. melanopoda, Schmidt-Gœbel, Faun. Col. Birm., p. 13, 1846; Chaudoir, A. S. E. F., p. 508, 1864; Horn, D. E. Z., p. 131, 1899 (var.).

C. cribellata, Chaudoir, B. S. N. M., p. 290, 1860[30].

C. cribrosa, Chaudoir, A. S. E. F., p. 507, 1864; Cat., p. 8, 1865[31]; Gemm. & Har., Cat., p. 37, 1868; Fleut., Cat., p. 141, 1892[32]; A. S. E. F., p. 57, 1902[33]; Horn, D. E. Z., p. 92, 1892[34]; Ann. Gen., p. 674, 1895[35]; Lesne, B. S. E. F., p. 292, f. 3, 1895; Atkinson, Cat., p. 19, 1889[36].

C. puncticollis, Chaudoir, B. S. N. M., p. 291, 1860; Gemm. & Har., Cat., p. 39, 1868 (var.); Atkinson, Cat., p. 22, 1889 (var. of *ortygia*).

C. terminalis, Chaudoir, A. S. E. F., p. 509, 1864[37]; Cat. Coll., p. 8, 1865[38]; Gmm. & Har., Cat., p. 23, 1889[d]; Atkinson, Cat., p. 23, 1889[93]; Fleut., Cat., p. 141, 1892[40]; Horn, D. E. Z., p. 357, 1892[41]; p. 50, 1897.

C. modesta, Motschulsky, B. S. N. M., pt. 3, p. 177, 1864[42].

Siripur[16, 17, 18]
Tenasserim[14, 18]
Tharawaddy[14, 18]
Trichur[18], Tura[17, 18]

Balighe[35]
Banguey[15, 18]
Benculen[9]
Borneo[34, 41]
Caner[29]
Java[1, 3, 4, 5, 6, 10, 11, 14, 18, 19, 22, 25, 27, 28, 30, 32, 36, 42, 43, 44, 45, 46, 48, a, b]
Kompong-Thom[12]
Laokay[33]
Lass (Luang-Prabang to Theng)[8]
Malacca[31, 32, 34, 38, 40, 41, d]
Nais[41], Padang[35]
Pangherang-Pisang[9, 35]
S. China[13, 18], Siboga[35]
Sidjoendjoeng[29]
Singapore[34]
Si-Rambe[47]
Sumatra[20, 21, 41, 22, a]
Sumba[15, 18]
Sumbawa[15, 18]
Taiping (hills) [17, 18]
Tourakhom[12]
Tringano[37, 39]
Vientian[12]

C. fuscicornis, Motschulsky, B. S. N. M., pt. 3, p. 177, 1864; Gemm. & Har., Cat., p. 38, 1868; Atkinson, Cat., p. 20, 1889; Fleut., Cat., p. 140, 1892.

C. nigrocornis, Motschulsky, B. S. N. M., pt. 3, p. 177, 1864; Gemm. & Har., Cat., p. 38, 1868; Atkinson, Cat., p. 21, 1889; Fleut., Cat., p. 140, 1892.

C. thoracica, Horn, D. E. Z., p. 356, 1892, p. 50, 1897.

C. obscura, Castelnau, Etud. Ent., p. 40, pl. 1, f. 7, 1835[43]; Brullé, Arch. Mus. Paris, I, p. 140, 1839[44]; Chaudoir, A. S. E. F., p. 503, 1864[45]; Gemm. & Har., Cat., p. 37, 1868 (var.); Atkinson, Cat., p. 19, 1889 (var.); Fleut., Cat., p. 140, 1892[46]; Horn, Ann. Gen., p. 674, 1895 (var.)[47]; D. E. Z., p. 49, 1897 (var.)[48]; Maindron, A. S. E. F., p. 7, 1905.

Type : Mus. Paris.

10. *__bonelli__ var. __batesi,__ Horn, D. E. Z., p. 355, 1892[1]; Lesne, Mission Pavie, Indo-Chine, iii, p. 59, 1904[2]; Fowler F. B. I., p. 249, 1912[3]; Dover & Rib., R. I. M., xxv, p. 347, 1923.[4]

Type : Coll. W. Horn.

Indes Or.[1], Assam[2]
Dehra Dun (F.)
Khasi hills[3, 4]
Momeit (B. M.)

Cambodia[2]
Cochin China[2, 3, 4]
Siam[2, 3, 4]

11. †**bonelli** var. **diversipes**, Fowler, F. B. I., p. 250, 1912[1]; Dover & Rib., R. I. M., xxii, p. 722, 1921[2]; xxv, p. 347, 1923[3].

Collyris cruentata, Horn (nec. Schmidt-Gœbel, 1846), D. E. Z., p. 224, 1894[4].

Type : Coll. W. Horn.

Java[1, 3, 4]
Burma[1, 3]
Pashok, 3, 500 ft.[2, 3]
Tura, 1,200-1,500 ft.[2, 3]

Borneo[1, 3]

12. **bonelli** var. **distincta,** Chaudoir, B. S. N. M., p. 290, 1860[1]; A. S. E. F., p. 501, 1864[2]; Cat. Coll., p. 8, 1865[3]; Gemm. & Har., Cat., p. 37, 1868[4]; Atkinson, Cat., p. 20, 1889[5]; Fleut., Cat., p. 140, 1892[6]; Horn, D. E. Z., p. 54, 1901[7]; p. 59, 1905[8]; Gen. Ins., p. 103, 1908; Lefroy, Ind. Ins. Life, pl. 16, f. 11, 1909; Annand. & Horn, List, p. 3, 1909[9]; Fowler, F. B. I., p. 250, 1912[10];

Indes Or.[1, 6, 4, 19–23]
Anamailais[8]
Balighai[11, 13]
Bandra(A.)
Bombay[7, 14, 15, 16, 17, 18]
Calcutta[9, 10, 12]

---

* In *Col. Cat., Cic,* p. 35, 1926, Horn gives this form as a synonym of *N. bonelli,* but it is generally regarded as a good variety.

† This form is also relegated to the synonymy of *N. bonelli* by Horn *(loc. cit.)*

Codina, Barcel. Cat., p. 345, 1917[11]; Dover & Rib., R. I. M., xxii, p. 722, 1921[12]; xxv, p. 347, 1923[13].

Collyris procera, Chaudoir, A. S. E. F., p. 501, 1864[14]; Cat. p. 8, 1965[15]; Gemm. & Har., Cat., p. 39, 1868[16]; Atkinson, Cat., p. 22, 1889[17]; Fleut., Cat., p. 140, 1892[18]; Horn, D. E. Z., p. 355, 1892[19]; p. 193, 1898 (va.).

C. conicollis, Motschulsky, B. S. N. M., pt. 3, p. 176, 1864[20]; Gemm. & Har., Cat., p. 37, 1868[21]; Atkinson, Cat., p. 19, 1889[22]; Fleut., Cat., p. 142, 1892[23].

Castle Rock[12]
Chatrapur[9, 10, 12]
Darjiling[10, 12]
Gudhali (A.)
Igatpuri (A.)
Kanara [7, 8, 10]
Khasi hills[10, 12]
Mahe[11]
Pashok, 3,000 ft. [12]
Teppukadu, 2,500 ft. (A.)
Walayar (A.)
China[19], Java[19]
Malacca (H.)
Manilla[19]
Sumatra[19]

Type : Mus. Paris.

13.* **bonelli** var. **ortygia,** Buquet, A. S. E. F., p. 604, 1835 (*Colliuris*)[1]; Chaud., A. S. E. F., p. 502, pl. 7, f. 6, 1864[2]; Cat., p. 8, 1865[3]; Gemm. & Har., Cat., p. 39, 1868[4]; Atkinson, Cat., p. 21, 1889[5]; Bates, Sci. Rep. 2nd Yarkand Mission, p. 2, 1891[6]; Fleut., Cat., p. 140, 1892[7]; A. S. E. F., p. 570, 1902[8]; Horn, D. E. Z., p. 356, 1892[9]; p. 357, 1896, p. 49, 1897[10]; p. 52, 1901,[11] p. 65, 1902[12]. Ann. Gen., p. 377, 1893[13]; p. 674, 1895 (var.)[14]; Spol. Zeyl., ii, p. 35, 1904[15]; Maindron, A. S. E. F., p. 7, 1905[16]; Annand. & Horn, List, p. 3, 1909[17]; Robinson, Fasc. Malay, Zool., I, pt. 1, p. 179, 1903 (habits)[18]; Lesne, Mission Pavie, Indo-Chine, iii, p. 59, 1904[19]; Fowler, F. B. I., p. 249, f. 108, p. 250, 1912[20]; Codina, Bar. Cat., p. 345, 1917[21]; Dover & Rib., xxii, p. 722, 1921[22]; xxv, p. 347, 1923[23].

Type : Unknown to Dr. Horn and the authors.

Java[1, 3, 4, 5, 7, 11, 19, 23]
Almora (B. M.)
Bhamo [13, 20, 23]
Calcutta[6, 11, 17, 20, 23]
? Ceylon [11, 15]
Deccan[45]
Dehra Dun (F.)
Jhelum valley[5, 6]
Kashmir[20, 23]
Mahe[16, 20, 21, 23]
Nepal[11, 19, 20, 23]
N. Cachar[20, 23]
Pusa (A.)
Pyinmana (F.)
Rajmahal[20, 23]
Rangoon[13], Saran[23]
Sibsagar (F.)
Singla 1,500 ft.[22, 23]
Sirpur[22, 23]
Tharawaddy [20, 23]

* This form is also regarded by Horn (*Col. Cat., Cic.*, p. 34, 1926) as a synonym of *N. bonelli.* In his Catalogue (p. 20) Atkinson records *N. emarginata,* Dejean, from Sibsagar, and there is a specimen in the Nevinson collection from Rangoon, which is included under this species. We are convinced, however, that these are misidentifications, as the species is very like *bonelli* var. *ortygia*.

Thayetmyo (B. M.)
Tura, 1,500 ft.[22, 23]
Teinzo[13]
Karin hills[13]

———————

Annam[23]
Bali Id.[11, 20,23]
Balighe[14]
Banguey Id.[11, 17, 20, 23]
Borneo[11, 17, 19]
Bukit Besara, 2,500 ft.[18]
Cochin China[11, 20]
Cambodia[29]
Hainan Id. (B.M.)
Logo Toba[14], Laokay[8]
Luang Prabang[19]
Manson-Gebirge[12]
Malacca[9, 11, 20, 23]
Nias Id.[11, 20]
Palabuan[10]
Pangherang-Pisang[14]
Penang (B.M.)
Philippine Is.[13]
Puloweh Is. (B.M.)
Saigon[19]
S. China[11, 13, 20, 23]
Siam[11, 20, 23]

13. **bonelli** var. **ortygia.**

Siboga[14]
Singapore[11, 20, 23]
Son Tay[18]
Sukabumi, 1,500-2,000 ft.[10]
Sandaran Agong, Korinchi Lake (B.M.)
Scekaranda[21]
Sumba[11, 17, 20, 23]
Sumbawa[11, 17, 20, 23]
Sunda Is.[5, 2, 13]
Theng[19]

14. **bipartita,** Fleutiaux, B. S. E. F., p. 24, 1897[1]; Horn, D. E. Z., p. 57, 1901[2]; Gen. Ins., p. 104, 1908[3]; Fowler, F. B. I., p. 264, 1912[4]; Dover & Rib., R. I. M., xxv, p. 347, 1923[5].
India[1–5]
Karen Mts.[4, 5]

Type : Coll. E. Fleutiaux.

15. **brevilabris,** Horn, Ann. Gen., p. 381, 1893[1]; Gen. Ins., p. 100, 1908[2]; Fowler, F. B. I., p. 239, 1912[3]; Dover & Rib., R. I. M., xxv, p. 347, 1923[4].
Karen hills[1, 3, 4]
Assam (Naga hills)[2, 3, 4]

Borneo[3]
Martapura (N.)
Sarawak (H.)

Type : Mus. Genova.

16. **ceylonica,** Chaudoir, A. S. E. F., p. 529, 1864[a]; Gemm. & Har., Cat., p. 37, 1868[1]; Atkinson, Cat., p. 19, 1889[2]; Fleut., Cat., p. 144, 1892[3]; Bates, A. M. N. H. (5), xvii, p. 71, 1886[4]; Horn, D. E. Z., p. 63, 1901[5]; Spol. Zeyl., ii, p. 31, 1904[6]; Gen. Ins., p. 106, 1910[7]; Fowler, F. B. I., p. 272, 1912[8]; Dover & Rib., R. I. M., xxv, p. 347, 1923[9].
Ceylon,[1, 3, 5, 7, 8, 9]
Bogawantalawa[2, 4, 6]

Type : Mus. Paris.

17. **compressicollis,** Horn, A. S. E. B., p. 445, 1909[1]; Gen. Ins., p. 106, 1910[2]; Dover & Rib., R. I. M., xxv, p. 347, 1923[3].
Assam[1–2]

Type : Coll. W. Horn.

18. ***crassicollis,** Chaudoir, A. S. E. F., p. 497, 1864[1]; Gemm. & Har., Cat., p. 37, 1868[2]; Atkinson, Cat., p. 19, 1889[3]; Fleut., Cat., p. 140, 1892[4]; Horn, D. E. Z., p. 56, 1901[5]; Gen. Ins., p. 104, 1908; Fowler, F. B. I., p. 259, 1912[6]; Dover & Rib., R. I. M., xxv, p. 347, 1923[7].
Indes Or.[1–4]
Sikkim[5, 6, 7]

Type : This ought to be in the Paris Museum, but is apparently lost.

19. **crassicornis,** Dejean, Spec. Col. I, p. 166, 1825[1]; Schmidt-Gœbel, Faun. Col. Birm., p. 12, 1846[2]; Chaud., B. S. N. M., p. 297, 1860 (note on synonymy); A. S. E. F., p. 494, pl. 7, fig. 2, 1864[3]; Cat. Col., p. 8, 1865[4]; Gemm. & Har., Cat. p. 37,
Indes Or. [1–4, 23, a]
Anamalai hills, 3, 500-4,000 ft.[14, 16, 20, 22]
Andaman Is.[5]
Asansol[27]
Bangalore (B.M.)

---

* Horn *(Col. Cat., Cic.,* p. 38, 1926) doubtfully regards this form as a synonym of *N. insignis,* Chaud.

1868[a]; Atkinson, Cat., p. 19, 1889[5]; Fleut., A. S. E. F., p. 137, 1889[6]; A. S. E. B., p. 168, 1890[7]; Cat. p. 139, 1892[8]; A. S. E. F,., p. 570, 1902[9]; Horn, D. E. Z., p. 354, 1892[10]; p. 49, 1897[11], p. 56, 1901[12], p. 59, 1905[13]; Ann. Gen., p. 373, 1893[14]; p. 674, 1894[15]; Spol. Zeyl., ii, p. 31, 1904[16]; Gen. Ins., p. 104, 1908; A. S. E. B., p. 363, 1913; Annand. & Horn, List, p. 4, 1909 (habits)[16]; Robinson, Fasc. Malay, Zool., i, p. 182, 1903 (habits)[17]; Lesne, Mission Pavie, Indo-Chine, iii, p. 60, 1904[18]; Schilder, Ent. Blatter, Berl., p. 176, 1911[19]; Fowler, F. B. I., p. 261, fig. 115, p. 515 (*Collyris* sp. boring in tea) 1912[20]; Fleut., B. S. E. F., p. 48, 1917[21]; Codina, Barcel. Cat., p. 347, 1917[6]; Dover & Rib., R. I. M., xxv, p. 347, 1923[22]; Dover, Planters chronicle, p. 473, 1924[b].

Bhamo[14] (B.M.)
Calcutta (and environs)[16, 20, 22]
Chandipur[22]
Chota-Nagpur[12, 20, 22]
Ceylon[12, 16, 20, 22a]
Gopkuda Id.[16, 20, 22]
Kumaon[7]
Kunber Nowatoli (F.)
Maldah[20, 22]
Nicobar Is. (B.M.)
N. Shan States[22]
Ouchterlony valley, 3,000-3,500 ft. (boring in tea plants,[b]
Pondicherry[22]
Ramnad[12], Ranchi[20, 22]
Sibsagar[20, 22]
Travancore[22], Tetara[7]

Collyris longicollis, Dejean nec. Fabr. Spec. Col., i, p. 166, 1825[23].

C. diardi, Mac. Leay, Annul. Javan., ed. 1, i, p. 10, 1825[24]; ed. 2, p. 104, 1833; Chaud., B. S. N. M., p. 297, 1860 (syn.).

C. macleayi, Brullé, Rev. Ent. Silberm., ii, p. 101, 1834; Hist. Nat. Ins., i, p. 102, 1834[25]; Gistl. Syst. Ins., i, 86, 1837; Naturg. Tierreichs, ed. i, p. 111, 1848; ed. 2, p. 111, 1851; Gistl. & Bromme, Handbuch. Naturg Reiche, p. 364, 1850.

C. pleuritica, Schmidt-Gœbel, Faun. Col. Birm., p. 13, 1846; Castelnau Hist. Nat. Ins., i, p. 24, 1851[26].

Colliuris clavicornis, Mannerheim, in litt. Motschulsky, Etudes Ent., p. 22, 1856.

Collyris gibbicollis, Motschulsky Etudes Ent., p. 110, f. 4, 185; B. S. N. M., xxxvii, pt. 3, p. 178, 1864; Gemm. & Har., Cat., p. 37 (var.), p. 38, 1868 (var.) ; Atkinson, Cat., p. 19 (var.), p. 20, 1889; Fleut., Cat., p. 139, 1892.

Balighe[15]
Batavia (B.M.)
Battambang[18]
China[20, 30]
Celebes[10, 12, 22, 29]
Cochin-China[9, 8]
Combodia[9, 18]
Hongkong[3–5, 12, 14, 19]
Kompong-Thong[21]
Lago Toba[15]
Lahat[10], Laokay[9]
Mabek[17]
Martaban[2]
Mons Tjikorai, 4,000 ft., Java[5, 25, 28]
Moluccas[3, 5, 12, 14, 22]
Mytho, [6, 12]
Nias [a]
Pangherang-Pisang[15]
Saigon[18]
Siboga[15] Si-Rambe[15]
Sunda Is.[3, 5, 14]

C. purpurata, Klug, Jahrb., Insek., i, p. 46, 1834; Gemm. & Har., Cat., p. 37, 1868 (var.); Fleut., A. S. E. B. p. 168, 1890[27]; Cat., p. 139, 1892 Horn, D. E. Z., p. 49, 1897[28].

Sumatra[12, 20, 22]
Sockaranda [a]
Siam [4, 20, 22]
Sukabumi, 2,000 ft.[11, 28]

C. vollenhovenii, Chaudoir, A. S. E. F., p. 495, 1864[29].

C. dejeani, Horn, D. E. Z., p. 82, 1895.[30]

Type : Mus. Paris.

20. **cyaneipalpis,** Horn, Zool. Jahrb. Syst., p. 47, 315, 1923[1] *(Collyris).* Type : Coll. W. Horn.

Nilgiri hills[1]

21. **cruentata,** Schmidt-Gœbel, Faun., Col. Birm., p. 14, 1846[1]; Chaud., A. S. E. F., p. 505, 1864[2]; Cat. Coll., p. 8, 1865[3]; Gemm. & Har., Cat., p. 37, 1868[4]; Atkinson, Cat., p. 19, 1889[5]; Fleut., Cat. p. 141, 1892[6]; Horn, D. E. Z., p. 357, 1892[7], p. 54, 1901[8]; Ann. Gen., p. 376, 1893[9]; Gen. Ins., p. 103, 1908[10]; Lesne, Bull. Mus. Hist. Nat. Paris, p. 238, 1896[11]; Fowler, F. B. I., p. 252, 1912[12]; Codina Barcel, Cat., p. 345, 1917, Dover & Rib., R. I. M., xxv, p. 347, 1923[14].

Burma[1, 4, 5, 8, 10]
Karen hills[9, 12, 14]
Pegu[12, 14], Palon[9, 13]
Rangoon[12, 14]
Road to Shan States (N.)
Sadiya (B.M.)
Sibsagar[12, 14]
Taung-ngu[12, 14]
Tharawaddy[12, 14]
Tenasserim[12, 14]
Tavoy (B.M.)

Collyris spuria, Horn, D. E. Z., p. 360, 1892[15]; Ann. Gen., p. 377, 1893; p. 675, 1894[16].

Type : Mus. Prague

Borneo[2, 5, 7, 8, 9, 10, 12, 14]
Laos[11]
Malacca[3, 6, 8, 12, 14]
Menado[16], Penang[16]
Prom-Penh (H.)
Siam[8, 10, 12, 14]
Sumatra[8, 10, 12, 14, 15]

22. ***cylindrica,** Schmidt-Gœbel, Faun. Col. Birm., p. 15, 1946[1]; Horn, D. E. Z., p. 131, 1899[2]; Gen. Ins., p. 104, 1908[3]; Fowler, F. B. I., p. 255, 1912[4]; Dover & Rib., R. I. M., xxv, p. 348, 1923[5].

Burma[1–8], Assam[7]

Malacca[8]
"Malaya"[7]

Collyris fuscitarsis, Schmidt-Gœbel, Chaud., A. S. E. F., p. 499, 1864; Gemm. & Har., Cat. p. 38, 1868[6]; Atkinson, Cat., p. 20, 1889[7]; Fleut., Cat., p. 190, 1892[8] (all *ex-parte*).

---

* This form is given as a synonym of *N. fuscitarsis,* Schmidt-Goebel in Horn's catalogue (Col. Cat., Cic., p. 37, 1926).

Type: Mus. Prague.

23. **cylindripennis,** Chaudoir, Rev. Mag. Zool., p. 106, 1864[1]; A. S. E. F., p. 514, pl. 8, t. 13, 1864[2]; Cat. Coll., p. 9, 1865[3]; Gemm. & Har., Cat., p. 37, 1868[1]; Atkinson, Cat., p. 19, 1889[5]; Fleut., Cat., p. 142, 1892[6]; Horn, Ann. Gen., p. 377, 1893[7]; D. E. Z., p. 53, 1901[8]; Gen. Ins., p. 103, 1908[9]; Fowler, F. B. I., p. 252, f. 109, 1912[10]; Dover, & Rib., R. I. M., xxv, p. 348, 1923[11].	Siam[1–6, 8–11] Karen hills[7, 10, 11]
Type : Mus. Paris.	
24. **feai,** Horn, Ann. Gen., p. 373, 1893[1]; D. E. Z., p. 58, 1901; Gen., Ins., p. 105, 1910; Fowler, F. B. I., p. 264, 1912[2]; Dover & Rib., R. I. M., xxii, p. 722, 1921[3]; xxv, p. 348, 1923[4].	Karen hills[1, 2, 4] Rangamati[3, 4]
Type : Mus. Genova.	
25. **foveifrons,** Horn, D. E. Z., p. 60, 1901[1], p. 65, 1902[2]; Gen. Ins., p. 105, pl. 7, f. 11, 1910; A. S. E. B., 1, vii, p. 363, 1913[3]; Fowler, F. B. I., p. 267, 1912[4]; Dover & Rib., R. I. M., xxv, p. 348, 1923[5].	Khasi hills[1, 4, 5] ——————— Mt. Mauson[2, 3]
Type : Coll. W. Horn.	
26. **fuscitarsis,** Schmidt-Gœbel, Faun. Col. Birm., p. 16, 1846[1]; Chaudoir, A. S. E. F., p. 499, 1864; Cat. Coll., p. 8, 1865; Motschulsky, B. S. N. M., pt. 3, p. 175, 1864[2]; Gemm. & Har., Cat. p. 38, 1868; Atkinson, Cat., p. 20, 1889[3]; Fleut., Cat., p. 140, 1892[4]; A. S. E. F., p. 570, 1902[5]; B. S. E. F., p. 49, 1917[6]; Horn, D. E. Z., p. 355, 1892, p. 55, 1901[7], p. 66, 1902[8]; Ann. Gen., p. 375, 1893[9]; Gen. Ins., p. 104, 1908[10]; A. S. E. F., p. 363, 1913[11]; Col. Cat., Cic., p. 37, 1926; Robinson, Fasc. Malay, Zool., I, p. 183, 1903[12]; Annand. & Horn., List, p. 3, 1909; Fowler, F. B. I., p. 256, f. 112, 1912[13]; Codina, Barcel Cat., p. 346, 1917[14]; Dover & Rib., R. I. M., xxii, p. 722, 1921[15]; xxv, p. 348, 1923[16].	Moulmein[1, 13, 16] Assam[3, 13, 16] Bengal[17, 18, 19, 20] Bhamo (B.M.) Buxa-duars (F.) ? Calcutta[21] Dacca (B. M.) Karen hills[9, 14] N. Shan States (N.) Rangoon[13, 16] Sikkim[7, 10, 13, 14, 16] Singla 1,500 ft. [15, 16] Tharawaddy[13, 16] Tikekee[9] Upper Maymyo, 3,000 ft. (N.) ———————
Collyris rufipes, Motschulsky, B. S. N. M., pt. 3, p. 175, 1864[17]; Gemm. & Har., Cat. p. 39, 1868[18]; Atkinson, Cat., p. 22, 1889[19]; Fleut., Cat., p. 140, 1892[20].	Annam[7] Bukit Besar, 2,500 ft.[12]

C. violacea, Motschulsky, B. S. N. M., pt. 3, p. 176, 1864; Gemm. & Har., Cat. p. 39, 1868; Atkinson, Cat., p. 23, 1889; Fleut., Cat., p. 142, 1892.

C. longicornis, Motschulsky, B. S. N. M., pt. 3, p. 176, 1864; Gemm. & Har., Cat., p. 38, 1868; Atkinson, Cat., p. 21, 1889; Fleut., Cat., p. 142, 1892.

C. diffracta, Schmidt-Gœbel, Faun., Col. Birm., p. 17, 1846[21]; Chaudoir, A. S. E. F., p. 499, 1864; Gemm. & Har., Cat. p. 37, 1868; Atkinson, Cat., p. 19, 1889; Fleut., Cat., p. 140, 1892; Horn, D. E. Z., p. 131, 1899 (var.).

Type : Mus. Prague.

Canton[7]
Cochin China[11, 13]
Java[7, 13, 16]
Martaban[2]
Kompong-Thom[6]
Laokay[5]
Malacca[4, 7, 10, 13, 14, 16]
Penang (B. M.)
Siam[7]
Sumatra[13, 16]
Than Moi[8]
Tuyen-Quan[5]
Vientian[6]

27. **insignia,** Chaudoir, Rev. Mag. Zool., p. 76, 1864[1]; A. S. E. F., p. 496, pl. 7, f. 4, 4*a*, 1864[2]; Cat. Coll., p. 8, 1865; Gemm. & Har., Cat. p. 38, 1868[3]; Atkinson, Cat., p. 20, 1889[4]; Fleut., Cat., p. 139, 1892; Horn, D. E. Z., p. 355, 1892[5]; p. 56, 1901[6]; Gen. Ins., p. 104, 1908[7]; Annand. & Horn, List, p. 3, 1909[8]; Fowler, F. B. I., p. 258, f. 11, 1912[9]; Codina, Barcel. Cat., p. 347, 1917[10]; Dover & Rib., R. I. M., xxii, p. 722, 1921[11]; xxv, p. 348, 1923[12].

Type : Mus. Paris.

Sylhet[1–4]
Bhutan[7–10, 12]
Burma[9, 12]
Above tura, 3,900 ft.[11, 12]
Darjiling[5, 7, 9, 10, 12]
Khasi hills [9, 12]
Madura (H.)
Mungphu[8, 9, 12]
Pamkabari, [9, 12]
Pashok[12]
Sikkim[4, 5, 6, 9, 12]
Singla, 1, 500 ft.[12]
Road to S. Shan States (N.)

28. **kollari,** Horn, D. E. Z., p. 17, 1901[1]; p. 59, 1905; Gen. Ins., p. 101, 1908[2]; Fowler, F. B. I., p. 245, 1912[3]; Dover & Rib., R. I. M., xxv, p. 348, 1923[4].

Type : Coll. W. Horn.

Central India[1–4]
Bengal (H.)
Ouchterlony valley 3,500 ft. (A.)
Teppukadu (A.)

29. ***lesnei,** Horn, Ann. Gen., p. 374, 1893[1]; D. E. Z., p. 54, 1901, p. 83, 1904; Gen. Ins., p. 103, 1908; Fowler, F. B. I., p. 253, 1912[2]; Dover & Rib., R. I. M., xxv, p. 348, 1923[3].

Type : Coll. W. Horn.

Karen hills[1, 2, 3]

Laokay (H.)

* Treated as a variety of *N. emarginata,* Dej. by Horn (*Col. Cat., Cic.,* p. 37 1926).

30. **linearis,** Submidt-Gœbel, Faun. Col. Birm., p. 15, 1846[1]; Chaud., A. S. E. F., p. 528, 1864[2]; Cat. Coll., p. 9, 1865[3]; Gemm. & Har., Cat., p. 38, 1868[4]; Atkinson, Cat., p. 21, 1889[5]; Fleut., Cat., p. 144, 1892[6]; B. S. E. F., p. 49, 1917[7]; Horn, Ann. Gen., p. 379, 1893[8]; D. E. Z., p. 45, 1901[9]; Gen. Ins., p. 100, 1908[10]; A. S. E. F., lvii, p. 362, 1913[11]; Fowler, F. B. I., p. 243, f. 104, 1912[12]; Codina, Barcel Cat., p. 343, 1917; Dover & Rib., R. I. M., xxv, p. 348, 1923[13].

Type : Mus. Prague.

Burma[1, 4, 5, 9, 10, 12]
Assam[12, 13]
Karen hills[8]
Momeit (B. M.)
Patkai hills (B. M.)
Tavoy (B. M.)
Tharawaddy (N.)
Maymyo[12, 13]
Pegu[12, 13], Sylhet[4]

Chapa[7]
Saigon [9, 10, 11, 12, 13]
Siam[2, 3, 5, 6, 10, 12, 13]
Lao-Kay[11]
Prom-Penh[11]
Tonkin[11, 13], Laos (A.)

31. **linearis** var. **srnkæ,** Horn, D. E. Z., p. 15, 1894[1]; Gen. Ins., p. 100, 1908[2]; Schilder, Ent. Blatt. Berl., p. 175, 1911[3]; Fowler, F. B. I., p. 243, 1912[4]; Codina, Barcel Cat., p. 344, 1917[5]; Dover & Rib., R. I. M., xxv, p. 348, 1923[6].

Type : Coll. W. Horn.

Ruby Mines[1, 4, 5, 6]
Pegu dist.[4, 6]

Cambodia[6], China[4]
Saigon[6], Tonkin[6]
Yunnan[2, 3, 5]

32. **mœsta,** Schmidt-Gœbel, Faun. Col. Birm., p. 14, 1846[1]; Chaud., A. S. E. F., p. 505, 1864[2]; Gemm. & Har., Cat. p. 39, 1868[3]; Atkinson, Cat., p. 39, 1889[4]; Fleut., Cat., p. 141, 1892[5]; B. S. E. F., p. 49, 1917[6]; Horn, Ann. Gen., p. 376, 1893[7]; D. E. Z., p. 54, 1901[8]; Gen. Ins., p. 103, 1908; A. S. E. B., lvii, p. 263, 1913[9]; Lesne, Mission Pavie, Indo-Chine, iii, p. 59, 1904[10]; Fowler, F. B. I., p. 251, 1912[11]; Codina, Barcel. Cat., p. 346, 1917[12]; Dover & Rib., R. I. M., xxv, p. 348, 1923.[13]

Coltyris porrecta, Chaudoir, in litt., A. S. E. F., p. 513, 1864[14].

C. flavicornis, Chaudoir, B. S. N. M., p. 292, 1860[15]; A. S. E. F., p. 512, pl. 8, f. 11, 1864[16]; Cat. Coll., p. 9, 1865[17]; Gemm. & Har., Cat. p. 38, 1868; Atkinson, Cat., p. 20, 1889[18]; Horn, D. E. Z., p. 360, 1892; Fleut., Cat. p. 142, 1892[19]; A. S. E. F., p. 570, 1903[20].

Burma[1, 5, 8, 11, 13, 18]
Karen hills[7]
N. India [4, 7, 8, 14, 15, 16, 18]

Cambodia[8, 11, 13]
Soprea Than[10]
Cochin China [4, 7, 8, 11, 14, 16, 18]
Malacca [8, 11, 12, 13, 17, 19]
Lao-Kay[20]
Kompong-Thom[6]
Kompong-Toul[9]
Perak[11, 13]
Siam[8, 11, 13]
Pangeon (A.)
Laigon (A.)
Haut Mekong (A.)
Tringano (B.M.)
Tonkin (H.)

Type : Mus. Prague.

Borneo (H.)
Vientian[6]

33. **orichalcina,** Horn, D. E. Z., p. 149, 1896[1]; P. 57, 1901[2]; Gen. Ins., p. 104, 1908[3]; A. S. E. B., lvii, p. 137, 1914[4]; Schilder, Ent. Blatter Berl., p. 176, 1911[5]; Fowler, F. B. I., p. 263, f. 117, 1912[6]; Codina, Barcel, Cat., p. 347, 1917[7]; Fleut., B. S. E. F., p. 49, 1917[8]; Dover & Rib., R. I. M., xxv, p. 348, 1923[9].

Type : Col. W. Horn.

Naga hills[1, 3, 6, 7, 9]
3,500–5,000 ft.
Nilgiris hills[2, 6, 9]
Bhamo hills, 4,000 ft. (B.M.)
Maymyo (A.)
N. Manipur[1, 2, 3, 6, 7]
Punjab[1, 2]
W. Manipur, 3,500–5,000 ft. (B. M.)

Lao-Kay[4]
Vientian[8]
Yunnan[3, 5, 7, 9]

34. **nilgirica,** Fowler, F. B. I., p. 260, 1912[1]; Horn, Gen. Ins., p. 429, 1915[2]; Dover & Rib., R. I. M., xxv, p. 348, 1923[3].

Type : Coll. H. E. Andrewes.

Nilgiri hills [1, 3]
(The unique type is from the Coorg jungle.)

35. **parvula,** Chaudoir, B. S. N. M., p. 17, 1848; p. 295, 1860[1]; A. S. E. F., p. 527, 1864[2]; Cat. Coll., p. 9, 1865[3]; Gemm. & Har., Cat. p. 39, 1868[4]; Atkinson, Cat., p. 22, 1889[5]; Fleut., Cat., p. 144, 1892[6]; Horn, D. E. Z., p. 169, 1894; p. 46, 1901[7]; p. 59, 1905; Gen. Ins., p. 101, 1908[8]; Maindron & Fleut., A. S. E. F., p. 8, 1905[9]; Fowler, F. B. I., p. 244, f. 105, 1912[10]; Dover & Rib., R. I. M., xxv, p. 348, 1923[11].

Bombay[1–6, 8, 9]
Anamalai hills (and neighbourhood)[7, 9, 11]
Bombay City (A.)
Bandra (A.)
Belgaum[9, 10, 11, 12, 13]
Castle Rock[11, 13]
Coonoor and environs[9, 11]
Mahe[9, 11]
Martapura (N.)
Sidapur[11]

Collyris amæna, Chaudoir, B. S. N. M., p. 295, 1860; A. S. E. F., p. 527, 1864; Cat. Coll., p. 9, 1865; Gemm. & Har., Cat., p. 37, 1868; Atkinson, Cat., p. 18, 1889; Fleut., Cat., p. 144, 1892; Horn, D. E. Z., p. 169, 1894[12]; as var. Fowler, F. B. I., p. 244, 1912; Dover & Rib., R. I. M., xxv, p. 348, 1923[13].

Type : Mus. Paris.

Teppukadu, 2,500 ft. (A.)
Yercaud[11]
Shevaroy hills, 3,000–5,000 ft.

36. **parvula** var. **maindroni,** Horn, D. E. Z., p. 294, 1905[1]; Gen. Ins., p. 101, pl. 7, f. 6, 1908; Zool. Med. Leiden, p. 91, 1922[a]; Fowler, F. B. I., p. 245, 1912[2]; Dover & Rib., R. I. M., xxv, p. 348, 1923[3]. — Wallardi[1, 2, 3, a], Anamalais (H.), Karkur Ghat, 1,500 ft., Shembaganur [a]

 Type : Coll. W. Horn.

37. **planifrons,** Horn, D. E. Z., p. 293, 1905[1]; Gen. Ins., p. 100, 1908[2]; Fowler, F. B. I., p. 239, 1912[3]; Dover & Rib., R. I. M., xxv, p. 348, 1923[4]; Horn, Ent. Rec., xxxvii, p. 166, 1925[5]. — Ceylon[1–4], Urugalla[5]

 Type : Coll. W. Horn.

38. **planifrontoides** Horn, Ent. Rec., xxxvii, p. 165, 1925[1]. — Anuradhapura[1], Wellawaya[1]

 Type : Brit. Mus.

39. **plicaticollis,** Chaudoir, A. S. E. F., p. 534, 1864[1]; Gemm. & Har., Cat. p. 39, 1868[2]; Atkinson, Cat., p. 22, 1889[3]; Fleut., Cat., p. 145, 1892[4]; Horn, D. E. Z., p. 63, 1901[5]; Spol. Zeyl., ii, p. 31, 1904[6]; Gen. Ins., p. 106, 1910[7]; Fowler, F. B. I., p. 270, 1912[8]; Dover & Rib., R. I. M., xxv, 348, 1923[9]; Horn, Ent. Rec., xxxvii, p. 166, 1925[10]. — Ceylon[1–9], Urugalla[10]

 Type : Brit. Mus. (♂)

40. **plicicollis,** Horn, D. E. Z., p. 63, 1901[1]; Gen. Ins., p. 106, 1910[2]; Fowler, F. B. I., p. 272, 1912[3]; Dover & Rib., R. I. M., xxv, p. 348, 1923[4]. — Nilgiri hills[1–4]

 Type : Coll. W. Horn.

41. **plicicollis** var. **subtile-flavescens,** Horn, Arch. Nat. 79-A, heft. ii, p. 4, 1913[1]; Gen. Ins. p. 429, 1915; Dover & Rib., R. I. M., xxv, p. 348, 1923[2]. — Shembaganur[1, 2]

 Type : Coll. W. Horn.

42. **punctatella,** Chaudoir, A. S. E. F., p. 525, 1864[1]; Gemm. & Har., Cat., p. 39, 1868[2]; Atkinson, Cat., p. 22, 1889[3]; Bates, A. M. N. H., (5) xvii, p. 70, 1886[4]; Fleut., Cat., p. 144, 1892[5]; Horn, D. E. Z., p. 51, 1901[6]; Spol. Zeyl., ii, p. 31, 1904[7]; Gen. Ins., p. 102, 1908[8]; Col. Cat. Cic., p. 34, 1926[a]; Fowler, F. B. I., p. 248, 1912[9]; Dover & Rib., R. I. M., xxv, P. 348, 1923[10]; Horn, Ent. Rec., xxxvii, p. 166, 1925[11]. — Ceylon[1, 2, 5, 6, 8, 10, a], Balangoda[3, 4, 7, 9, 11], Wellamaya (H.), Urugalla[11]; Nias [a].

Collyris nietneri, Horn, D. E. Z., p. 357, 1895[11].

 Type : Brit. Mus. (♀)

43. **redtenbacheri,** Horn, D. E. Z., p. 12, 1894, p. 48, 1901[1]; p. 67, 1902; Gen. Ins., p. 101, 1908[2]; Fowler, F. B. I., p. 239, f. 103, 1912[3]; Dover & Rib., R. I. M., xxii, p. 721, 1921[4]; xxv, p. 348, 1923[5].

Collyris attenuata Redt., Chaudoir, (nec. Redt.), A. S. E. F., p. 523, pl. 9, f. 19, 19*a*, 1864.

Type : A good specimen of Chaudoir's *C. attenuata* in the Paris Museum may conveniently be selected as the type.

Bhutan[2, 5]
Above Tura, 3,000 ft.[5, 4]
Arrakan[2]
Bhim Tal (H.)
Calcutta[1, 2]
Darjiling (H.)
Dehra Dun (F.)
Ghumti[5]
Gopaldhara (A.)
Khasi hills[3, 5]
Kousanie, 6,075 ft.[4, 5]
Kurseong (H.)
Mungphu[3, 5]
Naga hills[3, 5]
Nepal[3, 5]
Padong, Bhutan (H.)
Pashok, 1–3,500 ft.
Patkai hills[3, 5]
Pinjore[5], Simla[1, 2, 3, 5]
Singla[5], Sikkim[1, 3]
Tavoy (B. M.)
Tura, 200–1,500 ft.[4, 5]

44. **rœschkei,** Horn, D. E. Z., p. 365, 1892[1], p 48, 1901[2]; Gen. Ins., p. 101, 1908[3]; Fowler, F. B. I., p. 247, 1912[4]; Dover & Rib., R. I. M., xxv, p. 348, 1923[5].

Type : Coll. W. Horn.

Calcutta[1–5]

45. **rubens,** Bates, Cist, Ent., ii, p. 336, 1878[1]; Atkinson, Cat., p. 22, 1889[2]; Fleut., Cat., p. 145, 1892[3]; Horn, D. E. Z., p. 62, 1901[4]; Gen. Ins., p. 106, 1910[5]; Fowler, F. B. I., p. 269, 1912[6]; Dover & Rib., R. I. M., xxv, p. 348, 1923[7].

Type : Coll. R. Oberthur.

Assam (plains)[1–7]
Margherita (H.)

46. **rufipalpis,** Chaudoir, A. S. E. F., p. 504, 1864[1]; Gemm. & Har., Cat., p. 39, 1868; Atkinson, Cat., p. 22, 1889[2]; Fleut., Cat., p. 141, 1892; B. S. E. F., p. 369, 1917[3]; Horn, D. E. Z., p. 356, 1892, p. 50, 1897[4], p. 55, 1901[5], p. 66, 1902[6]; Ann. Gen., p. 374, 1893[7]; Gen. Ins., p. 103, 1908[8]; A. S. E. B., lvii, p. 363, 1913[9]; p. 140, 1914[10]; Annand. & Horn, List, p. 3, 1909[11]; Lesne, Mission Pavie, Indo-Chine, 1879-95, iii, pl. 8, f. 1*b*, 1904[12]; Fowler, F.

N. India[1, 2, 4, 7, 16]
Assam[8, 11, 12, 13, 14, 15]
Karen hills,[11, 15]
Sylhet (H.)

Cambodia[10, 15]
Cochin-China[4, 5, 12, 13, 16]

B. I., p. 254, f. 111, 1912[13]; Codina, Barcel, Cat., p. 346, 1917[14]; Dover & Rib., R. I. M., xxv, p. 349, 1923.[15]

Java[4, 5, 7, 8, 12, 13, 14, 15, 16]
Mons Gede (B. M.)
Prabang[12]
Sukabumi, 2,000 ft.[4, 14, 16]
Sumatra[5, 12, 13, 15]
Tonkin[6, 8, 9, 13, 14, 15]

Collyris obscura, Lesne, B. S. E. F., p. 55, 1891,p. 292, f. 1 1895; Horn, D. E. Z., p. 357, 1896, p. 51, 1897 (var.).[16]

Type : Mus. Paris.

Tranninh[3]
Theng[12]

47. **saphyrina,** Chaudoir, B. S. N. M., xxii, p. 18, 1850; A. S. E. F., p. 498, pl. 7, f. 5, 1864[1]; Cat. Coll., p. 8, 1865; Gemm. & Har., Cat., p. 39, 1868; Atkinson, Cat., p. 22, 1889[2]; Fleut., Cat., p. 140, 1892; Horn., D. E. Z., p. 355, 1892, p. 55, 1901[3]; Gen. Ins., p. 104, 1908[4]; Fowler, F. B. I., p. 257, f. 113, 1912[5]; Codina, Barcel. Cat., p. 346, 1917[6]; Dover & Rib., R. I. M., xxv, p. 349, 1923.[7]

N. India[1, 2, 7]
Assam[3, 4, 5, 6, 7]
Bengal[4, 6, 7]
Kangra valley, 4,500–6,500 ft.
Kumaon (F)
Mungphu[3, 5, 7]
Nepal valley[7]
Nepal[3, 4, 5, 6]
Pashok7, 2,500 ft.
Pedong [3, 5, 7]
Sikkim[3, 6]

Collyris boysii, Chaudoir, φ nec. ♀, B. S. N. M., xxxviii, p. 288, 1860.

Type : Mus. Paris.

Malacca (H.)
Siam (H.)
Sumatra (H.)

48. **saphyrina** var. **boysii,** Chaudoir, B. S. N. M., p. 288, 1860[1], ♂ nec. ♀; A. S. E. F., p. 498, 1864[2]; Lesne, B. S. E. F., p. 291, 1895; Horn, D. E. Z., p. 55, 1901; Gen. Ins., p. 104, 1908[3]; Col. Cat., Cic., p. 38, 1926[a]; Zool. Med. Leiden, p. 91, 1922; Dover & Rib., R. I. M., xxv, p. 347, 1923.[4]

Type : Mus. Paris.

N. India[1, 2]
Above Tura[4], 3,000 ft.
Bhutan [3, 4]
Gopaldhara (H.)
Mt. Pu. [a]
Rungbong valley (A.)
Yunnan[a].

49. ***sarawakensis,** Thomson, Arch. Ent., i, p. 133, 1957[1]; Chaud., A. S. E. F., p. 531, pl. 9, f. 22, 1864[2]; Cat. Coll., p. 10, 1865[3]; Gemm. & Har., Cat., p. 39, 1868[4]; Atkinson, Cat., p. 22, 1889[5]; Fleut., Cat., p. 144, 1892[6]; Horn, D. E. Z., p. 367, 1892[7]; p. 61, 1901[8]; Gen. Ins.,

Sarawak[1, 2, 3, 5, 6]
Burma[8, 9, 13, 14]
Sylhet[12, 14]

Batoe Is. (H.)

---

*Dr. Horn queries the Indian localities.
† This form is given specific rank in Horn's Col. Cat., Cic., p. 41; 1926.

p. 106, 1910[9]; Shelford, P. Z. S., p. 234, pl. 19, f. 5, 1902 (mimicry)[10]; Naturalist in Borneo, p. 210, 1916 (mimicry); Annand. & Rob., Fasc., Malay. Zool., 1, p. 182, 1903 (mimicry : habits)[11]; Fowler, F. B. I., p. 268, 1912[12]; Codina, Barcel. Cat., p. 344, 1917[13]; Dover & Rib., R. I. M., xxv, p. 349, 1923.[14]

Type : Not known to Dr. Horn or to the authors.

Borneo[1, 4, 5, 6, 7, 9, 12, 13, 14]
Bukit Besar,[11], 2,500 ft. Malacca [2, 7, 8, 9, 12, 13, 14]
Sandaron-Agong, Korinchi Lake (B. M.)
Pontianak[10]
Singapore (B. M.)
Sockaranda : N. [13]
Sumatra[8, 9, 12, 14]
Tebbing-Tinggi[13]

50. †**sarawakensis** var. **dohertyi,** Horn, D E. Z., p. 83, 1895[1], p. 61, 1901[2]; Gen. Ins., p. 106, 1910[3]; Fowler, F. B. I., p. 268, 1912[4]; Dover & Rib., R. I. M., xxv, p. 349, 1923.[5]

Type : Coll. W. Horn.

Malacca[1, 2, 4, 5]
Burma[2–5]

Marang[1]
Mentawei Is.[3]
Siam[2–5], Sumatra[3–4]

51. **saundersi,** Chaudoir, A. S. E. F., p. 496, 1864[1]; Cat. Coll., p. 8, 1865[2]; Gemm. & Har., Cat., p. 39, 1868[3]; Atkinson, Cat., p. 22, 1889[4]; Fleut., Cat., p. 139, 1892[5]; Bates, A. M. N. H., (5) xvii, p. 71, 1886[6]; Horn, D. E. Z., p. 392, 1899 (habits), p. 56, 1901[7]; Spol. Zeyl., ii, p. 31, 1904[8]; Gen. Ins., p. 104, 1908[9]; Maindron & Fleut., A. S. E. F., p. 8, 1905[10]; Fowler, F. B. I., p. 259, 1912[11]; Dover & Rib., R. I. M., xxv, p. 349, 1923.[12]

Type : Mus. Paris.

Ceylon[1, 2, 3, 5, 7, 9]
Bandarawela [8, 10, 11, 12]
Colombo[4, 6]
Kandy[10, 11, 12]

52. **saundersi** var. **continentalis,** Horn., D. E. Z., p. 295, 1905[1]; Gen. Ins., p. 104, 1908; Fowler, F. B. I., p. 269, 1912[2]; Dover & Rib., R. I. M., xxv, p. 349, 1923.[3]

Type : Coll. W. Horn.

Wallardi[1, 2, 3]
Nilgiri hills, 3,500 ft. (A.)

53. **saundersi** var. **lætior,** Horn., Spol. Zeyl., ii, pp. 31, 35, 1904[1]; Gen. Ins., p. 104, 1908; Bates, A. M. N. G. (5) xvii, p. 71, 1886 (*Collyris* sp.); Fowler, F. B. I., p. 259, 1912[2]; Codina, Barcel. Cat., p. 347, 1917[3]; Dover & Rib., R. I. M., xxv, p. 349, 1923[4]; Horn, Ent. Rec., xxxvii, p. 166, 1925.[5]

Type : Coll. W. Horn.

Morawaki[1, 2, 4]
Andaman Is.[3]
Colombo[2, 4]
Deniyaya[5]
Korale[1, 2, 4]
Kandy[2, 4]
Wellawaya (H.)

54. **schaumi,** Horn, D. E. Z., p. 366, 1892[1]; p. 48, 1901[2]; Gen. Ins., p. 101, 1908[3]; Zool. Med. Leiden,

Cornwallis1
Andaman Is. [2–4, a]

vii, p. 112, 1922[a]; Fowler, F. B. I., p. 242, 1912[4]; Dover & Rib., R. I. M., xxv, p. 349, 1923[5].

var. *chevrolati*, Horn, D. E. Z., p. 16, 1894[6]; Fowler, F. B. I., p. 242, 1912.

Type : Coll. W. Horn.

55. **similis,** Lesne, B. S. E. F., p. 55, 1891,[1] p. 292, f. 2, 1895; Bull. Mus. Hist. Nat. Paris, p. 238, 1896[2]; Mission Pavie, Indo-Chine, 1879-1895, iii, pl. 8, f. 1, 1*a*, 1904[3]; Horn, D. E. Z., p. 54, 1901,[4] p. 83, 1904, p. 59, 1905[5]; Gen. Ins., p. 103, 1908[6]; Fleut., Cat., p. 140, 1892[7]; Fowler, F. B. I, p. 254, 1912[8]; Dover & Rib., R. I. M., xxv, p. 349, 1923.[9]

Type : Mus. Paris.

Laos (Luang-Prabang to Theng), [1-4, 6, 7, 9]
Assam [6, 8, 9]
Khasi hills (H.)
Nilgiri hills[5, 6, 8, 9]
Ouchterlony valley, 3,500 ft. (A). (Labelled "black aberr.")

Celebes (N.)

56. **smaragdina,** Horn, D. E. Z., p. 220, 1894[1]; Gen. Ins., p. 104, 1908[2]; Annand. & Horn, List, p. 4, 1909[3]; Fowler, F. B. I., p. 258, 1912[4]; Dover & Rib., R. I. M., xxv, p. 349, 1923.[5]

Type : Coll. W. Horn.

Perak[6, 8, 9]
Tuyen Quangin (Tonkin)[3]

57. **smithii,** Chaudoir, A. S. E. F., p. 518, 1864[1]; Gemm. & Har., Cat., p. 39, 1868[2]; Atkinson, Cat., p. 22, 1889[3]; Fleut., Cat., p. 143, 1892 (gives wrong ref. and spelling)[4]; Horn, D. E. Z., p. 361, 1892[5]; p. 58, 1901[6]; Gen. Ins., p. 105, 1910[7]; Annand. & Horn, List, p. 4, 1909[8]; Fowler, F. B. I., p. 265, f. 118, 1912[9]; Codina, Barcel. Cat., p. 348, 1917[10]; Dover & Rib., R. I. M., xxv, p. 349, 1923.[11]

Indes Or.[1], Bhutan[2–5]
Kurseong[2–5]
Mungphu[4, 5]
Pedong (H.)
Soom,[5] 4,000–5,000 ft.
Indes Or. [1–4]
Assam[5, 6, 11]
Cachar[7, 10, 11]
Dacca [9, 11]
Margherita[10]
Naga hills[7, 8, 9, 10, 11]
Pegu (& Dist.)[8, 9, 10, 11]
Tibet[6, 8, 9, 11, 12]

Collyris fasciatoplicata, Chaudoir, in litt., Horn, D. E. Z., p. 193, 1898.

C. macleayi, Horn, D. E. Z., p. 81, 1895[12]; p. 58, 1901.[13]

Type : Brit. Mus.

58. **subtilis,** Chaudoir, Rev. Mag. Zool., xviii, p. 712, 1863[1]; A. S. E. F., p. 525, 1864[2]; Cat. Coll., p. 9, 1865[3]; Gemm. & Har., Cat., p. 39, 1868[4]; Atkinson, Cat., p. 23, 1889[5]; Fleut., Cat., p. 144, 1892[6]; Horn, Ann. Gen., p. 377, 1893[7]; D. E. Z., p. 45, 1901[8]; Gen. Ins., p. 100, 1908[9]; Fowler, F. B. I., p. 240, 1912[10]; Codina, Barcel. Cat., p. 343, 1917[11]; Dover & Rib., R. I. M., xxv, p. 349, 1923.[12]

Type : Mus. Paris.

Siam[1–12]
Karen hills[10, 12]
Kodaikanal[10, 12]
Palni hills[10, 12]

Java[8–12]
Malacca (H.)
Perak (B. M.)
Selangor (B.M.)Sumatra[8–12]

59. ***subtilis** var. **brachycephala,** Horn, Ann. Gen., p. 378, 1893[1]; D. E. Z., p. 51, 1897[2]; Fowler, F. B. I., p. 241, 1912[3]; Dover & Rib., R. I. M., xxv, p. 349, 1923.[4]

Type : Coll. W. Horn.

Karen hills [1, 3, 4]
Pengalengang[2], 4,000 ft.

60. **†subclavata,** Chaudoir, B. S. N. M., xxxiii, pt. 4, p. 289, 1860[1]; A. S. E. F., p. 495, pl. 7, f. 3, 1864[2]; Cat. Coll., p. 8, 1865; Gemm. & Har., Cat., p. 39, 1868[3]; Atkinson, Cat., p. 23, 1889[4]; Fleut., Cat., p. 139, 1892; Horn, D. E. Z., p. 56, 1901[5]; Gen. Ins., p. 104, 1908[6]; p. 438, 1915; A. S. E. B., lviii, p. 137, 1914[7]; Zool. Med. Leiden, vii, p. 112, 1922[a]; Fowler, F. B. I., p. 262, 1912[8]; Codina, Barcel. Cat., p. 347, 1917[9]; Dover & Rib., R. I. M., xxv, p. 349, 1923.[10]

Collyris subdilatata, Chaudoir, in litt., Schaum, Journ. Ent., p. 73, 1863.[11]

Type : Mus. Paris.

Deccan [1, 3, 4]
Andaman Is. (Port Blair)[5, 6 9]
Bengal[5, 6, 8, 9, 10]
Kandy (N.)
Kollar Dist.[10]
Malda[4], Mysore (F.)
Nilgiris hills [6, 8, 9, 10]
N. India[2]
Shembaganur (H.)

China[8]
Penang[6]
Tonkin (Daung-Trua)[7]

61. **subclavata** var. **andamana,** Bates, Cist. Ent., ii, p. 335, 1878[1]; Atkinson, Cat., p. 18, 1889[2]; Fleut., Cat., p. 140, 1892[3]; Horn, D. E. Z., p. 354, 1892[4], p. 60, 1897[5]; Gen. Ins., p. 104, 1908; Fowler, F. B. I., p. 262, f. 116, 1912[6]; Dover & Rib., R. I. M., xxv, p. 349, 1923[7].

Type : Coll. R. Oberthur.

Andaman Is.[1, 2, 3, 6, 7]

Java[4, 5]

62. **variicornis,** Chaudoir, A. S. E. F., p. 530, 1864[1]; Gemm. & Har., Cat., p. 39, 1868[2]; Atkinson, Cat.,

Sylhet[1–4, 6, 7]
Cochin (B. M.)

---

* This variety is relegated to the synonymy of *N. subtilis* by Horn (Col. Cat., Cic., P. 31, 1926).

† *N. subclavata* is considered a variety of *N. crassicornis* by Horn (Col. Cat., Cic., p. 39, 1926) with the form *andamana* as a synonyn.

p. 23, 1889[3]; Fleut., Cat., p. 144, 1892[4]; Robinson, Fasc. Malay. Zool., i, p. 183, 1903[5]; Horn, D. E. Z., p. 47, 1901; Gen. Ins., p. 101, 1908; Fowler, F. B. I., p. 245, f. 106, 1912[6]; Dover & Rib., R. I. M., xxv, p. 349, 1923.[7]

Type : Mus. Paris.

Darjiling (H.)
Gopaldhara (H.)
Khasi hills[6, 7]
Manipur (B. M.)
Momeit (B. M.)
Moulmein (B. M.)
Naga hills[6, 7]
Patkai hills[6, 7]
Rangoon (B. M.)
Singla[7]
Sitong[7], 3,800–4,000 ft.
Tharawaddy[6, 7]
Tura[7], Uppter Teesta valley, 6, 7 4000 ft.

Bukit Besar,[5, 6, 7] 2,500 ft.
Tonkin (B. M.)

63. **variicornis** var. **flavolabiata,** Horn, D. E. Z., p. 366, 1892[1]; p. 47, 1901[2]; Gen. Ins., p. 101, pl. 7, f. 8, 1908[3]; Zool. Med. Leiden, vii, p. 112, 1922[a]; Annand. & Horn, List, p. 2, 1909[4]; Fowler, F. B. I., p. 246, 1912[5]; Dover & Rib., R. I. M., xxv, p, 349, 1923.[6]

Type : Coll. W. Horn.

Cornwallis1, [a]
Andaman Is. [2–6]
Nicobar Is. [4, 5, 6]
Port Blair [a]

64. **variicornis** var. **gestroi,** Horn, Ann. Gen., p. 380, 1893[1]; D. E. Z., p. 47, 1901; Gen. Ins., p. 101, 1908; Fowler, F. B. I, p. 246, 1912[2]; Dover & Rib., R. I. M., xxv, p. 349, 1923.[3]

Type : Coll. W. Horn.

Karen hills[1, 2, 3]
Tenasserim[1, 2, 3]

65. **variitarsis,** Chaud., B. S. N. M., xxxiii, p. 295, 1860[1]; A. S. E. F., p. 523, 1864[2]; Cat. Coll., p. 9, 1865[3]; Gemm. & Har., Cat., p. 39, 1868[4]; Atkinson, Cat., p. 23, 1889[5]; Fleut., Cat., p. 144, 1892[6]; Horn, D. E. Z., p. 45, 1901[7]; Gen. Ins., p. 100, 1908[8]; A. S. E. B., lvii, p. 362, 1913[9] p. 140, 1914[10]; Lesne, Mission Pavie, Indo-Chine iii, p. 59, 1904[11]; Annand. & Horn, List, p. 2, 1909 (habits)[12]; Fowler, F. B. I., p. 241, 1912[13]; Dover & Rib., R. I. M., xxii, p. 721, 1921[14]; xxv, p. 349, 1923.[15]

N. India [1, 2, 4, 5]
Bengal [13 15]
Bhamo[18, 19, 20]
Buxa Duars (A.)
Chandkhira (H.)
Darjiling[7, 13, 15] (H.)
Dehra Dun (F.)
Dunsiri valley[12, 15]
Gokteik (A.)
Karen hills[13, 15, 16, 18, 19, 20]

Collyris schmidt-gœbeli, Horn, Ann. Gen., p. 378, 1893[16]; Fleut., B. S. E. F., p. 49, 1917.[17]

C. variitarsis var. *brachycephala,* Horn, Ann. Gen., p. 379, 1893[18]; D. E. Z., p. 45, 1901; Gen. Ins., p. 100, 1908; Fowler, F. B. I., p. 242, 1912 (habits)[19]; Dover, & Rib., R. I. M., xxv, p. 349, 1923.[20]

Type : Mus. Paris.

Margherita (H.)
Nepal[3, 6, 8, 11, 13, 15]
Rangoon [18, 10, 20]
Sibsagar (F.)
Sikkim[11, 13]
Singla [14, 15]
Sukna[12, 13, 15]
Teinzo[13, 15]
Tenasserim[13, 15, 16]
Tura[14, 15]

Luang Prabang[11]
Bao-Lae[10]
Mt. Mauson[9]
Penang[13, 15]
Then[11]
Vientian[17]

Genus : **TRICONDYLA**, Latreille (1822).

Tricondyla, Latreille, in Cuvier's Règne Anim., ed. I, III, p. 179, 1817; Latr. & Dejean, Hist. Nat. Icon. Col. Eur., i, p. 65, 1822; Dejean, Spec. Col., i, p. 160, 1825; Castelnau, Rev. Ent. Silber., ii, p. 38, 1834; Lacordaire, Gen. Col., i, p. 28, 1854; Chaudoir, B. S. N. M., p. 284, 299, 302 (list of species) 1860; Schaum, B. E. Z., p. 76, 1861, p. 182, 1862; Gemm. & Har., Cat., p. 35, 1868; Atkinson, Cat., p. 16, 1889; Horn, D. E. Z., p. 52, 1897, p. 17, 1906 (monograph); Gen. Ins., p. 94, 1908; Col. Cat., Cic., p. 22, 1926 (cites numerous bionomic references) Fleut., Cat., p. 135, 1892; A. S. E. F., p. 498, 1893; Fowler, F. B. I., pp. 273-276 (structure, life-history, key to species).

Collyris, Fabricius (*ex parte*) Syst. Eleuth., i. p. 226, 1801; Herbst, Nat. Ins., X, p. 215, 1806.

Colliuris, Latreille, (*ex parte*) in Cuvier's, Règne Anim., ed. 1, iii, p. 179, 1817; Lamarck, Hist. Nat. Anim. Sans. Vert., ed. 4, II, p. 677, 1835.

Genotype : *Tricondyla aptera* Olivier (1790).

Distribution : Mainly South India and Ceylon; also found from N. India to Malaya.

1. **coriacea,** Chevrolat, Rev. Zool., p. 221, 1841[1]; Chaudoir, B. S. N. M., xxxiv, p. 359, 1861[2]; Cat. Coll., p. 11, 1865; Schaum., B. E. Z., p. 78, 1861; Journ. Entom., II, p. 71, 1863; Gemm. & Har., Cat., p. 36, 1868; Atkinson, Cat., p. 16, 1889[2]; Fleut., Cat., p. 137, 1892; Horn, Spol. Zeyl., ii, p. 39, 1904[3], p. 24, 1906; Gen. Ins., p. 96, 1908; Fowler, F. B. I., p. 276, f. 121, 1912[4]; Codina, Barcel. Cat., p. 339, 1917; Dover & Rib., R. I. M., xxv, p. 349, 1923[5]; Horn, Ent. Rec., xxxvii, p. 166, 1925.[6]

   Type : Coll. E. Fleutiaux.

   Trincomali[1–5]
   Chilaw[4, 5]
   Colombo (B.M.)
   Hambantota (B.M.)
   Horawapotana[6]
   Kanthalai[3, 4, 5]
   Kekirawa[3, 4, 5]
   Olumadu (B. M.)
   Palatupana[3, 4, 5]

2. **cyanea** var. **annulicornis,** Schmidt-Gœbel, Faun. Col. Birm., p. 10, 1846[1]; Schaum, B. E. Z., p. 77, 1861,[2] p. 184, 1862[a]; Journ. Ent., ii, p. 71, 1863[3]; Chaud., Cat. Coll., p. 11, 1865[4]; Gemm. & Har., Cat., p. 35, 1868[5]; Atkinson, Cat., p. 16, 1889[6]; Fleut., A. S. E. F., pp. 498, 501, 1893[7]; B. S. E. F., p. 308, 1920 (key); Horn, D. E. Z., p. 53, 1897[8], p. 131, 1899; p. 27, 1906[9]; Gen. Ins., p. 97, 1908[10]; A. S. E. B., p. 362, 1913[11]; Annand. & Horn, List, p. 2, 1909[12]; Fowler, F. B. I., p. 279, 1912[13]; Codina, Barcel. Cat., p. 341, 1917[14]; Dover & Rib., R. I. M., xxv, p. 349, 1923[15]; Shelford, Naturalist in Borneo, p. 160, ed. 1918, (mimicry).

   Tricondyla gibba, Chaudoir, B. S. N. M., p. 358, 1861[16]; Cat. Coll., p. 11, 1865[17]; Schaum., Journ. Ent., ii, p. 71, 1863[18]; Gemm. & Har., Cat., p. 36, 1868[19]; Atkinson, Cat., p. 17, 1889[20]; Fleut., Cat., p. 136, 1892[21]; Gestro, Ann. Gen., p. 369, 1893.[22]

   Type : Mus. Prague.

   Burma [a, 1–8, 10–15]
   Karen Mts.[12]
   Mooleyit[12]
   Pegu (Palon forest)[12]
   Thagata[12]
   Tenasserim[13, 15]
   Tharawaddy (N.)
   Toungoo (F.)
   ——————
   Cambodia[7–21]
   Chantaboun (B.M.)
   Siam[8, 9, 10, 12, 13, 14, 15]

3. **gestroi,** Fleutiaux, A. S. E. F., p. 500, 1893[1]; B. S. E. F., p. 48, 1917,[2] p. 308, 1920 (footnote to key); Horn, D. E. Z., p. 25, 1906[3]; Gen. Ins., p. 96, 1908; A. S. E. B., p. 362, 1913[4]; Fowler, F. B. I., p. 281, f. 125, 1912[5]; Dover & Rib., R. I. M., xxv, p. 349, 1923.[6]

   Tricondyla mellyi, Gestro (nec. Chaudoir, 1850), Ann. Gen., p. 370, 1893.[7]

   Type : Coll. E. Fleutiaux.

   Karen hills [1, 5, 6, 7]
   Sylhet[5, 26]
   ——————
   Chapa[2]
   Cochin China[3, 4, 5]
   Lao-Kay[4]
   Laos (A.)
   Prom-Penh[4]
   Yunnan (H.)

4. **gounelli,** Horn, D. E. Z., p. 361, 1900[1]; p. 59, 1905[2], p. 24, 1906[3]; Gen. Ins., p. 96, 1908[4]; Annand. & Horn, List, p. 1, 1909[5]; Fowler, F. B. I., p. 278, 1912[6]; Dover & Rib., R. I. M., xxii, p. 722, 1921[7]; xxii, p. 349, 1923.[8]
Ramnad [1, 3, 4, 5, 6, 8] Anamalai hills [2, 4, 5, 6, 8] 3,500–4,000 ft. Trivandrum[7, 8]

Type : Coll. W. Horn.

4. **gounelli,** var. **horni,** Maindron, B. S. E. F., p. 263, 1904[1]; Horn, D. E. Z., p. 24, 1906; Gen. Ins., p. 96, 1908; Annand. & Horn, List, p. 1, pl. 1, f. 6, 1909; Fowler, F. B. I., p. 278, 1912[2]; Dover & Rib., R. I. M., xxv, p. 350, 1923[3].
Wallardi[1] Trivandrum[3] Travancore (mountains)[2, 3]

Type : Coll. G. Babault.

5. ***granulifera,** Motschulsky, Etud. Ent., p. 110, f. 3, 1857[1]; Schaum, B. E. Z., p. 78, 1861; Chaud., Cat. Coll., p. 11, 1865; Gemm. & Har., Cat., p. 36, 1868; Atkinson, Cat., p. 17, 1889[2]; Fleut., Cat., p. 138, 1892; Horn, D. E. Z., p. 362, 1900, p. 24, 1906; Spol. Zeyl., ii, pp. 32, 38, 1904[3]; Gen. Ins., p. 96, 1908; Fowler, F. B. I., p. 277, f. 122, 1912[4]; Codina, Barcel. Cat., p. 340, 1917; Dover & Rib., R. I. M., xxv, p. 350, 1923.[5]
Colombo[1, 2] Haragaru[3, 4, 5] Nalanda[3, 4, 5]

Tricondyla, femorata, Walker, A. M. N. H., (3), iii, p. 202, 1858; Tennent, Ceylon, etc., London, 1859, p. 274; Schaum, B. E. Z., p. 76, 1861; Gemm. & Har., Cat., p. 36, 1868; Atkinson, Cat., p. 17, 1889; Fleut., Cat., p. 138, 1892.

Type : ?Mus. Moscow.

6. **granulifera** var. **rugosa,** Chaudoir, A. S. E. F., p. 447, 1863[1]; Cat. Coll., p. 11, 1865[2]; Horn., D. E. Z., p. 209, 1892 (var.), p. 24, 1906[3]; Gen. Ins., p. 96, 1908[4]; Fowler, F. B. I., p. 277, 1912[5]; Codina, Barcel, Cat., p. 340, 1917[6]; Dover & Rib., R. I. M., xxv, p. 350, 1923.[7]
Ceylon[1–5, 7] Kalani valley near Colombo (B. M.) Kandy[6] Nalanda (H.)

Type : Mus. Paris.

7. ***macrodera,** Chaudoir, B. S. N. M., iv, p. 300, 1860[1]; Cat. Coll., p. 11, 1865; Schaum, Journ. Ent.,
N. India[1, 2, 4, 5,] Above Tura,[15, 16] 3,500–3,900 ft.

---

* We do not understand the position of this species in Horn's Catalogue (Col. Cat. Cic., p. 24, 1920). He gives Motschulsky's *granulifera*–an older name–as a synonym of his own *T. nigripalpis* with the remark "vera sec. type!" Walker's *femorata* is regarded as a variety of *T. nigripalpis,* with Indian references as synonyms. It would, therefor,e seem that Dr. Horn regards the species treated here as *T. granulifera* as *T. nigripalpis* var. *femorata* Walk., but we find it difficult to subscribe to this nomenclature. The form *rugosa* is regarded by Horn as a synonym of *T. nigripalpis.*

* Drs. Horn has a recent paper convergence in on this species and its allies in *Ent., Mitt.,* xiv, pp. 327-329, 1925.

ii, p. 71, 1863[2]; Gemm. & Har., Cat., p. 36, 1868[3]; Atkinson, Cat., p. 17, 1889[4]; Fleut., Cat., p. 137, 1892[5]; A. S. E. F., p. 499, 1893[6]; B. S. E. F., p. 48, 1917, p. 309, 1920 (key : footnote)[7]; Horn, D. E. Z., p. 66, 1902[8], p. 25, 1906[9]; Gen. Ins., p. 96, 1908[10]; A. S. E. B., p. 362, 1913[11]; Zool. Med. Leiden, vii, p. 90, 1922; Annand. & Horn, List, p. 1, 1909[12]; Fowler, F. B. I., p. 278, f. 123, 1912[13]; Codina, Barcel, Cat., p. 340, 1917[14]; Dover & Rib., R. I. M., xxii, p. 722, 1921,[15] xxv, p. 350, 1923.[16]

Type : Mus. Paris.

Bhutan (Bhutan Duars)[9, 12, 13, 16]
Cachar[4, 13, 16]
Darjiling[8, 13, 16]
Ghumti[16], 4,000 ft.
Gopaldhara[16]
Hindostan[3]
Kalimpong, [16] 600–4,500 ft.
Khasi hills (B. M.)
Mungphu[13, 16]
Naga hills[6]
Maria Barti, Bhutan (H.)
Pashok,[16] 2,000–3,000 ft.
Pasighat[16], Pegu[4]
Rangoon (B.M.)
Rungbong valley[16]
Sibsagar[4, 13, 16]
Singla,[16] 1,500 ft.
Sikkim[4, 9, 10, 13, 14, 16]
Tura[15, 16]

Chapa[7]
Mt. Sauson[8, 11, 14]
Yunnan (H.)

8. **macrodera** var. **tuberculata,** Chaudoir, B. S. N . M., xxxiii, (4) p. 301, 1860[1]; Cat. Coll., p. 111, 1865[2]; Schaum, Journ. Ent., ii, p. 71, 1863[3]; Gemm. & Har., Cat., p. 36, 1868; Atkinson, Cat., p. 18, 1889[4]; Fleut., Cat., p. 137, 1892; A. S. E. F., p. 500, 1893[5]; Horn, D. E. Z., p. 356, 1896, p. 25, 1906[6]; Gen. Ins., p. 96, 1908; Zool. Med. Leiden, vii, p. 90, 1922; Annand. & Horn, List, p. 1, 1909[7]; Fowler, F. B. I., p. 280, 1912[8]; Dover & Rib., R. I. M., xxv, p. 350, 1923.[9]

Type : Mus. Paris.

Sylhet[1, 6, 8, 9]
Siliguri[7, 8, 9]

China[1–4]

9. **nigripalpis,** Horn, D. E. Z., p. 224, 1894,[1] p. 24, 1906[2]; Spol. Zeyl., ii, pp. 32, 39, 1905[3]; Gen. Ins., p. 96, pl. 8, f. 5, 1908; Fowler, F. B. I., p. 276, 1912[4]; Dover & Rib., R. I. M., xxv, p. 350, 1923.[5]

Type : Coll. W. Horn.

Central Ceylon[1, 2, 4, 5]
Kongawella[3]
Kandy (H.)

10. **mellyi,** Chaudoir, B. S. N. M., i, p. 17, 1850[1]; Cat. Coll., p. 11, 1865; Schaum, B. E. Z., p. 78, 1861; Journ. Ent., ii, p. 71, 1863; Gemm. & Har., Cat., p. 36, 1868; Atkinson, Cat., p. 17, 1889[2]; Fleut., Cat., p. 137, 1892; A. S. E. F., p. 500, 1893[3]; B. S. E. F., p. 369, 1917,[4] p. 308, 1920 (key); Gestro, Ann. Gen., p. 370, 1893[5]; Horn., D. E. Z., p. 66, 1902,[6] p. 25, 1906; Gen. Ins., p. 97, 1908; A. S. E. B., p. 362, 1913,[7] p. 139, 1914[8]; Annand. & Horn, List, p. 2, 1909[9]; Fowler, F. B. I., p. 280, f. 124, 1912[10]; Codina, Barcel Cat., p. 340, 1917; Dover & Rib., R. I. M., xxii, p. 722, 1921,[11] xxv, p. 350, 1923.[12]

Tricondyla tumidula, Walker, A. M. N. H., (3), iii, p. 50, 1859[13]; Tennent, Ceylon, etc., London, p. 274, 1859[a]; Gemm. & Har., Cat., p. 36, 1868[14]; Atkinson, Cat., p. 18, 1889[15]; Horn, Spol. Zeyl., ii, p. 35, 1904 (footnote : note).[16]

Type : Mus. Paris.

Himalaya[1, 2, 3]
Alipur[3]
Bhamo (B. M.)
Ceylon[13, 14, 15, 16]
(plains[a] or coast country)
Karen hills[5]
Maini Mukh[12]
Siliguri[9, 10, 12]
Sylhet (B. M.)
Above Tura,[11, 12] 3,500–3,900 ft.

Chapa[4]
Dien-Bien-Phu[4]
Lao-Kay[4]
Mt. Mauson,[6, 7] 2,000–3,000 ft.
Phong-Tho[4]
Trinninh[4]
Vientianm[8]

Genus : **DEROCRANIA**, Chaudoir (1860).

Derocrania, Chaudoir, B. S. N. M., pp. 284, 297, 1860; Schaum, B. E. Z., p. xxxvi, 1861; Horn, Spol. Zeyl., ii, p. 39, 1904; D. E. Z., p. 21, 1906; Gen. Ins., p. 95, 1908 Col. Cat., Cic., p. 23, 1926 (as sub-gen. of *Tricondyla*); Fleut., Cat., p. 138, 1892; A. S. E. F., p. 502, 1893; Fowler, F. B. I., p. 282-283, 1912 (structure, key to spp.).

Tricondyla, Latreille, (*ex parte*) Gemm. & Har., Cat., p. 35, 1868; Atkinson, Cat., p. 16, 1889·

Genotype : *Derocrania dohrni,* Chaudoir, 1860 (?=*Tricondyla scitiscabra* Walker 1859).

Distribution : Mainly S. India and Ceylon.

1. **agnes,** Horn, D. E. Z., p. 64, 1905[1]; p. 22, 1906[2]; Gen. Ins., p. 96, pl. 8, f. 4, 1908[3]; Fowler, F. B. I., p. 286, 1912[4]; Dover & Rib., R. I. M., xxv, p. 350, 1923.[5]

Type : Coll. W. Horn.

Ceylon[1, 2, 4]
Central Ceylon[3, 5]

2. **brevicollis,** Horn., D. E. Z., p. 152, 1905[1]; p. 22, 1906[2]; Gen. Ins., p. 96, 1908; Fowler, F. B. I., p. 285, 1912[3]; Dover, & Rib., R. I. M., xxv, p. 350, 1923.[4]

Trichinopoly[1, 2, 3, 4]

Type : Coll. W. Horn.

3. **concinna,** Chaudoir, B. S. N. M., p. 298, 1860[1]; Cat. Coll., p. 10, 1865[2]; Schaum, Journn. Ent., ii, p. 71, 1863,[3] Gemm. & Har., Cat., p. 36, 1868[4]; Atkinson, Cat., p. 16, 1889[5]; Bates, A. M. N. H., (5), xvii, p. 70, 1886[6]; Fleut., Cat., p. 138, 1892[7]; Horn, D. E. Z., p. 22, 1906[8]; Spol. Zeyl., ii, p. 32, 1904[9]; Gen. Ins., p. 96, 1908[10]; Fowler, F. B. I., p. 287, f. 128, 1912[11]; Codina, Barcel. Cat., p. 339, 1917[12]; Dover & Rib., R. I. M., xxv, p. 350, 1923[13]; Horn, Ent. Rec., xxxvii, p. 166, 1925.[14]

   Ceylon[1, 2, 3, 4, 7, 8, 10, 12]
   Kandy[5, 6, 9, 11, 13]
   Nalanda[9]
   Puttalam[9]
   Belihul-Oya (H.)
   Urugalla[14]
   Lady Horton[14]

   Type : Mus. Paris.

4. **flavicornis,** Horn, D. E. Z., p. 92, 1892,[1] p. 346, 1893, p. 22, 1906[2]; Spol. Zeyl., ii, p. 39, pl. 7, f. 2, 1904[3]; Gen. Ins., p. 96, 1908[4]; Fleut., A. S. E. F., p. 502, 1893[5]; Fowler, F. B. I., p. 287, 1912[6]; Dover & Rib., R. I. M., xxv, p. 350, 1923.[7]

   Ceylon[1–7]

   Type : Coll. W. Horn.

5. **fusiformis,** Horn, Spol. Zeyl., ii, p. 35, pl. 7, f. 1, 1904[1]; D. E. Z., p. 22, 1906[2]; Gen. Ins., p. 96, 1908[3]; Fowler, F. B. I., p. 286, 1912[4]; Dover & Rib., R. I. M., p. 350, 1923.[5]

   Ceylon[1, 2, 4]
   Central Ceylon[3, 5]

   Type : Coll. W. Horn.

6. **gibbiceps,** Chaudoir, B. S. N. M., p. 298, 1860[1]; Cat. Coll., p. 10, 1865[2]; Schaum, Journ. Ent., p. 71, 1863[3]; Gemm. & Har., Cat., p. 17, 1889[4]; Fleut., Cat., p. 138, 1892[5]; Horn, Spol. Zeyl., ii, p. 31, 1904[6]; D. E. Z., p. 22, 1906[7]; Gen. Ins., p. 96, 1918[8]; Fowler, F. B. I., p. 287, 1912[9]; Dover & Rib., R. I. M., xxv, p. 350, 1923[10]; Horn, Ent. Rec., xxxvii, p. 166, 1925.[11]

   Ceylon[1–5]
   Central Ceylon[7–10]
   Kandy (B.M.)
   Nalanda[6]
   Urugalla[11]

   Type : Mus. Paris.

7. **halyi,** Horn, D. E. Z., p. 132, 1899, name only, p. 193, 1900,[1] p. 23, 1906[2]; Spol. Zeyl., ii, p. 39, pl. 7, f. 3, 1904[3]; Gen. Ins., p. 96, 1908[4]; Fowler, F. B. I., p. 292, f. 132, 1912[5]; Dover & Rib., R. I. M., xxv, p. 350, 1923.[6]

   Ceylon[1, 2, 4, 5, 6]
   Anuradhapura[3]
   Kanthalai[3]
   Pankulam[3]
   Puttalam[3]
   Trincomali[3]

   Type : Coll. W. Horn.

8. **honorei,** Fleutiaux, A. S. E. F., p. 502, 1893[1]; Dierckx, La Cellule, xviii, p. 289, 1901 (pygidial

   "Continent asiatique"[1]

glands); Horn, D. E. Z., p. 22, 1906[2]; Gen. Ins. p. 96, 1908[3]; Fowler, F. B. I., p. 283, f. 126, 1912[4]; Codina, Barcel. Cat., p. 339, 1917[5]; Dover & Rib., R. I. M, xxv, p. 350, 1923.[6]

Type : Coll. E. Fleutiaux.

Bombay[2, 3, 4, 6]
Cochin State (B. M.)
Kodaikanal[2, 4, 6]
Nilgiri hills (B. M.)
Ramnad[2, 4, 6]
Shembaganur[6]
Trichur[6]
Trichinopoly[2, 4, 6]

9. **longesulcata,** Horn, D. E. Z., p. 194, 1900[1]; p. 59, 1905,[2] p. 22, 1906[3]; Gen. Ins., p. 96, 1908; Maindron & Fleut., A. S. E. F., p. 8, pl. 1, f. 1, 1905[4]; Fowler, F. B. I., p. 284, 1912[5]; Dover & Rib., R. I. M., xxv, p. 350, 1923.[6]

Type : Coll. W. Horn.

Madikere[1, 3, 4]
Anamalai hills [3, 5, 6]
Castle Rock[6]
Coonoor [2, 3, 6]
Kodaikanal[4]
Nilgiri hills, [3, 5,6]
4,500–5,000 ft.
Wallardi[6]

10. **nematodes,** Schaum, Journ. Ent., ii, pp. 61, 71, pl. 4, f. 1, 1863[1]; Chaud., Cat., Coll., p. 10, 1865[2]; Gemm. & Har., Cat., p. 36, 1868[3]; Bates, A. M. N. H., (5), xvii, p. 70, 1886[4]; Atkinson, Cat., p. 17, 1889[5]; Fleut., Cat., p. 138, 1892[6]; Horn, Spol. Zeyl., ii, p. 31, 1904[7]; D. E. Z., p. 22, 1906[8]; Gen. Ins., p. 96, 1908[9]; Fowler, F. B. I., p. 289, f. 130, 1912[10]; Dover & Rib., R. I. M., xxv, p. 350, 1923.[11]

Type : Mus. Berlin.

Ceylon[1, 2, 3, 6]
CentralCeylon[8, 9, 10, 11]
Bogawantalawa[4, 5, 7]

11. **nietneri,** Motschulsky, Etud. Ent., viii, p. 25, 1859[1]; xi, p. 23, 1862; Chaud., Cat. Coll., p. 10, 1865; Gemm. & Har., Cat., p. 36, 1868; Bates, A. M. N. H., (5), xvii, p. 70, 1886[2]; Atkinson, Cat., p. 17, 1889[3]; Fleut., Cat., p. 138, 1892; Horn, Spol. Zeyl., ii, p. 31, 1904[4]; D. E. Z., p. 22, 1906; Gen. Ins., p. 96, 1908; Ent. Rec., xxxvii, p. 166, 1925[a]; Fowler, F. B. I., p. 285, f. 127, 1912[5]; Codina, Barcel, Cat., p. 339, 1917; Dover & Rib., R. I. M., xxv, p. 350, 1923.[5]

Nuwara Eliya[1–4]
Balangoda[3, 4]
Balangoda Ridge[2, 5, 6]
Bogawantalawa[3, 4, 7]
Kandy[5, 6, 8]
Nalanda[4]
Wadduwa (H.)
Urugalla[a]

D. lævigata, Chaudoir, B. S. N. M., p. 299, 1860; Gemm. & Har., Cat., p. 36, 1868.

Tricondyla rhaphidioides, Schaum, B. E. Z., p. 75, 1861; Journ. Ent., ii, p. 71, 1863; Gemm. & Har., Cat., p. 36, 1868.

D. obscuripes, Bates, A. M. N. H., (5), xvii, p. 70, 1886[7]; as. var. Atkinson, Cat., p. 17, 1889; Fleut., A.S.E.F.,

p. 270, 1893[8]; Cat., p. 138, 1892 (as var. of *concinna*).

Type : Mus. Moscow.

12. **schaumi,** Horn, D. E. Z., p. 67, 1892,[1] p. 23, 1906[2]; Spol. Zeyl., ii, p. 32, 1904[3]; Gen. Ins., p. 96, 1908[4]; Fleut., Cat., p. 138, 1892[5]; Fowler, F. B. I., p. 289, f. 129, 1912[6]; Codina, Barcel. Cat., p. 339, 1917[7]; Dover & Rib., R. I. M., xxv, p. 350, 1923.[8] — Ceylon[1, 2, 4–8]; Kekirawa[3]

Type : Coll. W. Horn.

13. ***scitiscabra,** Walker, A. M. N. H., (3), iii, p. 51, 1859[1]; Tennent, Ceylon, etc., London, 1889, p. 274[a]; Gemm. & Har., Cat., p. 36, 1868[2]; Atkinson, Cat., p. 18, 1889[2]; Fleut., Cat., p. 138, 1892[3]; 1892[4]; Horn, Spol. Zeyl., ii, p. 32, 1904[5]; D. E. Z., p. 23, 1906[6]; Gen. Ins., p. 96, 1908[7]; Ark. Zool. Stockholm, xii, No. 11, p. 21, 1920 ("Weddabrucke" theory); Annand. & Horn, List, p. 1, 1909 (habits); Codina, Barcel, Cat., p. 339, 1917[9]; Fowler, F. B. I., p. 291, f. 131, 1912[10]; Dover & Rib., R. I. M., xxv, p. 350, 1923.[11]

D. dohrni, Chaudoir, B. S. N. M., p. 297, 1860[12]; Cat., Coll., p. 10, 1865[13]; Schaum, Journ. Ent., ii, p. 71, 1863[14]; Putzeys, Comp. Rend., S. E. B., p. 45, 1875[15]; Gemm. & Har., Cat., p. 36, 1868[16]; Atkinson, Cat., p. 16, 1889.[17]

Ceylon[1–4, 6, 7, 12, 13, 14, 16, 17]; Belihul-Oyal (H.); Colombo[5] (coast country or plains)[a]; Galle[15]; Hambantata (B. M.); Kandy[8–11]; Madulsima (B. M.); Nalanda[5], Ombalota[6]; Paranthan (B.M.); Puttalam (H.); Sigiriya[5]

Type : See footnote.

Division : ***PLATYSTERNALIÆ,*** Horn.

Platysternaliae, Horn, Ent., Nach., p. 214, 1900; D. E. Z., p. 10, 1905; Gen. Ins., p. 107, 1910; Fowler, F. B. I., p. 293, 1912.

Sub-family : THERATINÆ, Latrielle (Fowler, 1912).

*The type of this species was not labelled by the author. A specimen in the British Museum, however, which is undoubtedly synonymous with *D. nematodes,* Schaum, is labelled "probably the true type of *scitiscabrar,* Walk." We have not been able to decide the point from Walker's description, as the colour of the legs, the most obvious character of Schaum's species, is not mentioned. If this specimen is the type the synonymy of the species concerned will be as follows : *scitiscabra* Wlk, *nematodes* Schaum, not *dohrni* Chaud., which would be distinct. We have followed Horn for the present, but it would be satisfactory to have this joint settled, as it involves the fixation of the type of the genus. Horn (Ent. Rev. XXXVII, p. 166, 1925) also records the species from Urugalla, Horawapotana, and Badulla in Ceylon.

Theratinœ, Fowler, F. B. I., p. 293, 1912.

Theratidœ, Horn, D. E. Z., p. 16, 1898; p. 466, 1907.

Theratites, Horn, D. E. Z., p. 466, 1907.

Theratina, Horn, Gen. Ins., p. 189, 1910; Col. Cat., Cic., p. 109, 1926.

Genus : THERATES, Latreille (1817).

Therates, Latreille, in Cuvier's Regn., Anim, ed. i, iii, p. 179, 1817; and Dejean, Hist. Nat. Icon. Col. Eur., p. 63, 1822; Dejean, Spec. Col., i, p. 157, 1825; Brulle, His. Nat. Ins., iv, p. 91, 1834; Guérin & Percheron, Gen. Ins., i. pl. 1, 1835; Castelnau, Hist. Nat. Col., i, p. 25, 1810; Lacordaire, Gen. Col., i, p. 28, 1854; Schaum, B. E. Z., p. 182, 1860, p. 78, 1861, p. 180, 1862, p. 67, 1863; Journ. Ent., p. 74, 1863; Thomson, Musée Scientif., pp. 41-55, 1860; Gemm. & Har., Cat., p. 34, 1868; Atkinson, Cat., p. 14, 1889; Horn, D. E. Z., p. 209, 1892, p. 345, 1893; Gen. Ins., p. 193, 1910; Col. Cat., Cic., p. 110, 1926; Fleut., Cat., p. 133, 1892; Fowler, F. B. I. p. 294, 1912; (key to spp.).

Eurychile, Bonelli, Mem. R. Acad. Sci., Torino, xxiii, p. 236, pl. 4, ff. 1-11, 1818.

Genotype : *Cicindela labiata,* Fabricius (1801).

Distribution : Eastern Himalayas, Assam and Burma to Celebes and N. Guinea.

1. **chenelli,** Bates, Cist. Ent., ii, p. 335, 1878[1]; Atkinson, Cat., p. 15, 1889[2]; Fleut., Cat., p. 134, 1892[3]; Gestro, Ann. Gen., p. 368, 1893[4]; Horn, D. E. Z., p. 196, 1900; Gen. Ins., p. 194, 1910[5]; Ent. Medd., xiv, p. 216, 1923; Fowler, F. B. I., p. 297, f. 136, 1912[6]; Codina, Barcel, Cat., p. 391, 1917[7]; Dover & Rib., R. I. M., xxii, p. 722, 1921[7]; xxv, p. 350, 1923.[8]

Naga hills[1–5, 8]
Karen hills[6–8]
N. Shan States[7, 8]
Upper Burma (H.)
———
Penang (B. M.)
Singapore (B. M.)

Type : Coll. R. Oberthur.

2. **chenalli** var. **concinnus,** Gestro, Ann. Gen., p. 105, 1888[1], p. 368, 1893[2]; Atkinson, Cat., p. 24, 1889[3]; Fleut., Cat., p. 134, 1892 Horn, D. E. Z., p. 224, 1892[4]; p. 53, 1889; Gen. Ins., p. 194, 1910; Fowler, F. B. I., p. 298, 1912[5]; Dover & Rib., R. I. M., xxv, p. 350, 1923.[6]

Ruby mines (Thagata)[1, 2, 5, 6]
Karen hills[2]
Tenasserim[3, 4]

Type : Mus. Genova.

3. **crebrepunctatus,** Horn, Ent. Medd., xiv, p. 215, 1923.[1] Type : Coll. W. Horn.	Upper Burma (Namtu)[1]
4. **dohertyi,** Horn., S. E. Z., p. 277, 1905[1]; Gen. Ins., p. 194, 1910; Ent. Medd., xiv, p. 215, 1923; Fowler, F. B. I., p. 296, f. 135, 1912[2]; Dover & Rib., R. I. M., xxii, p. 722, 1921[3]; xxv, p. 350, 1923.[4] Type : Coll. W. Horn.	Patkai hills[1, 2, 4] Margherita (H.) Pashok, 3, 4 5,000 ft. Pegu[2, 4] tenasserim[2, 4]
5. ***gestroi,** Horn, D. E. Z., p. 196, 1900[1]; Gen. Ins., p. 194, 1910[2]; Ent. Medd., xiv, p. 215, 1923; Fowler, F. B. I., p. 298, 1912[3]; Dover & Rib., R. I. M., xxii, p. 722, 1921[4]; xxv, p. 350, 1923.[5] Type : Coll. W. Horn.	Lakhon[1 2, 3, 5] Above Tura,[4, 5] 3,500–3,900 ft. Khasi hills[2, 3, 5]
6. **gestroi** var. **annandalei,** Horn, R. I. M., ii, p. 412, 1908[1]; Gen. Ins., p. 194 1910[2]; Annand. & Horn, List, p. 6, pl. i. f. 4, 1909 (habits)[3]; Fowler F. B. I., p. 299, 1912[4]; Dover & Rib., R. I. M., xxii, p. 722, 1921 (habits); xxv, p. 350, 1923.[5] Type : Indian Museum, Calcutta.	Kurseong[1–5] 47–5,000 ft. Ghumti,[5] 4,000 ft. Kalimpong,[5] 600–4,500 ft. Mahanadi,[5] 4,120 ft. Pashok,[5] 2,000–5,000 ft.
7. **hennigi,** Horn, Ent. Nach., p. 178, 1898[1]; Gen. Ins., p. 194, 1910[2]; Fowler F. B. I., p. 296, 1912[3]; Dover & Rib., xxii, p. 722, 1921 (habits)[4]; xxv, p. 351, 1923.[5] Type : Coll. W. Horn.	Khasi hills[1, 2, 3, 5] Above Tura,[4, 5,] 3,900 ft.
8. **hennigi** var. **dormeri,** Horn, D. E. Z., p. 197, 1898[1]; p. 296, 1905[2]; Gen. Ins., p. 194, pl. 12, f. 9, 1910[3]; Fowler, F. B. I., p. 295, f. 134, 1912[4]; Dover & Rib., R. I. M., xxv, p. 350, 1923[5]; Horn, Mem. Dept. Agric. Ind., viii, No. 9, p. 91, 1924.[6] Type : Coll. W. Horn.	Borneo[1, 2, 5] N. Manipur[3, 6] Patkai hills[4, 5, 6]
9. ***maindroni,** Horn, D. E. Z., p. 197, 1900[1]; Gen. Ins., p. 194, 1910[2]. Type ; Coll. W. Horn.	S. W. Borneo, [1, 2] Taun-Gyo (N.)=?Taungno. ——————— Pontianak (N.)

* In Col. Cat., Cic., p. 113, 1926 both *gestroi* and its variety *annuandalei* are given as separate varieties of T. Kreatzi Horn, a Malayan insect which has not been recorded from India.

* This is the first record of this species from our limits, represented by a single purpose specimen in the Navinaon collection.

10. **obliquus,** Fleut., A. S. E. F., p. 497, 1893[1]; Horn, D. E. Z., p. 197, 1900; Gen. Ins., p. 194, f. 70, (on p. 29), 1910; Fowler, F. B. I., p. 208, 1912[2]; Codina, Barcel. Cat., p. 391, 1917[3]; Dover & Rib., R. I. M., xxii, p. 722, 1921[4]; xxv, p. 351, 1923.[5] — Ruby mines[1, 2, 3, 5] Momeit[1, 2, 5] Pashok, [4, 5] 5,000 ft. —————— Yunnan[5] Formosa[3]

Type : Coll. E. Fleutiaux.

11. **waagenorum,** Horn, D. E. Z., p. 198, 1900[1]; Gen. Ins., p. 194, p. 12, f. 10, 1910[2]; Annand., & Horn, List, p. 7, 1909[3]; Fowler, F. B. I., p. 199, 1912[4]; Dover & Rib., R. I. M., xxv, p. 351, 1923.[5] — Darjiling[1, 2, 4, 5] Pegu[2–5] Tenasserim[2–5]

Type : Coll. W. Horn.

Genu : PROTHYMA, Hope (1838).

Prothyma, Hope, Col. Man., ii, p. 27, 1838; Horn, D. E. Z., p. 86, 1906; Gen. Ins., p. 170, 1910, p. 436, 1915; Col. Cat., Cic., p. 96, 1926; Fowler, F. B. I., p. 300, 1912 (key to species).

Euryoda, Lacordaire, (*ex parte*) Mem. Soc. Liege, p. 107, 1843; Gen. Col., i, p. 21, 1854; Fleut., Cat., p. 129, 1892; Horn, D. E. Z., p. 101, 1895; pp. 39-41, 1899; pp. 84-86, 1904.

Rhytidophœna, Bates, Entomologist, Suppl., p. 7, 1891.

Jansenia, Chaudoir, (*ex parte*), Cat., p. 55, 1865; Gemm. & Har., Cat., p. 33, 1868; Atkinson, Cat., p. 14, 1889; Fleut., Cat., p. 38, 1892.

Dromicidia, Chaud., B. S. N. M., i, p. 21, 1852; Gemm. & Har., Cat., p. 34, 1868; Atkinson, Cat., p. 14, 1889; Fleut., Cat., p. 38, 1892.

Genotype : *Cicindela quadripunctata,* Fabricius (1801).

Distribution : Mainly Indo-Malaya and Africa.

1.***belloides,** Horn, A. S. E. B., p. 311, 1907[1]; Gen. Ins., p. 175, 1910[2]; Fowler, F. B. I., p. 300, footnote, p. 309, f. 140, 1912[3]; Dover & Rib., R. I. M., xxv, p. 351, 1923. — N. Kanara[1–4]

Type : Coll. W. Horn.

2. †**bouvieri,** Horn, Bull, Mus. Hist. Nat. Paris, p. 328, 1896, p. 328, 1896[1]; Gen. Ins., p. 174, 1910[2]; — Lakhon[1–4] Maymyo[3, 4]

---

* This is the only Indian species in the genus with pubesence on the underside and Fowler (p. 300, footnote) thinks it may be referable to a separate genus.

† Included as a variety of *P. exormata* by Horn (Col. Cat., Cic., p. 99, 1926).

Fowler, F. B. I., p. 307, 1912[3]; Dover & Rib., R. I. M., xxv, p. 351, 1923[4].

Type : Coll. W. Horn.

3. **exornata,** Schmidt-Gœbel, Faun. Col. Birm., p. 1, pl. 1, f. 7, 1846 (*Cicindela*)[1]; Chaud., Cat. pp. 18, 57, 1865 (Cic.)[2]; Gemm & Har., Cat., p. 15, 1868 (*Cic.*)[3]; Atkinson, Cat., p. 7, 1889; (Civ.)[4]; p. 7, 1892. Gestro, Ann. Gen., p. 88, 1889 (*Cic.*)[5], p. 367, 1893 (*Euryoda*)[6]; Fleut., Cat., p. 38, 1892 (*Jansenia*)[7]; B. S. E. F., p. 49, 1917[8]; Horn, D. E. Z., p. 130, 1899[9]; p. 201, 1900 (*Eur.*); Gen. Ins., p. 174, 1910[10]; A. S. E. B., p. 363, 1913,[11] p. 140, 1914[12]; Annand. & Horn, List, p. 5, 1909,[13]; Fowler, F. B. I., pp. 305, 306, f. 139, 1912[14]; Dover & Rib., R. I. M., xxv, p. 351, 1923.[15]

Type : Mus. Prague.

Burma[1, 3, 4], Ceylon[9]
Karen hills [9, 14, 15]
Cheba[6], 900–1,100 metres
Ghecu.[6], 1,300–[1, 4] metres
N. Chin hills [14, 15]
Katha (F.)
Pegu[13, 14, 15]
Rangoon (N.)
Tharawaddy[14, 15]
Teinzo[5]

Annam[10, 11, 14, 15]
Cambodia[2, 7, 10, 11, 14, 15]
Cochin-China[12]
Laos[10, 15]
Luang-Prabang[12]
Vientian[8]

4. ***hennigi,** Horn, Ent. Nach., p. 177, 1898[1] (*Heptodonta* or ? *Euryoda*); Gen. Ins., p. 175, pl. 11, f. 6, 1910[2]; Fowler, F. B. I., p. 308, 1912[3]; Dover & Rib., R. I. M., xxv, p. 351, 1923.[4]

Type : Coll. W. Horn.

Khasi hills[1–4]

5. **limbata,** Wiedemann, Zool. Mag., ii, p. 64, 1823 (*Cicindela*)[1]; Gemm & Har., Cat., p. 20, 1865[2] (*Cic.*); Atkinson, Cat., p. 9, 1889 (Cic.)[3] ; Bates, Sci., Reps. 2nd Yarkand Miss., p. 2, 1891[4]; Fleut., Cat., p. 38, 1892 (*Jansenia*)[5] Horn, Gen. Ins., p. 174, 1910[6]; Annand. & Horn, List, p. 5, 1909[7]; Fowler, F. B. I., p. 304, f. 138, 1912[8]; Dover & Rib., R. I. M., xxv, p. 351, 1923.[9] *tetraspilota*, Chaudoir, B. S. N. M., xxv, p. 29, 1852.

Type : Unknown to the authors.

Bengal[1, 2, 3, 5, 6]
Calcutta[8, 9]
Jhelum valley [3, 4, 6, 7, 8, 9]
Simla[8, 9]

---

* The inclusion of this species in *Prothyme* is provisional. *See* refs above.

6. **limbata** var. **inornata**. Horn, D. E. Z., p. 368, 188[9]; (*Euryoda*)[1], p. 13, 1905; Gen. Ins., p. 174 1910[2]; Fowler, F. B. I., p. 305, 1912[3]; Codina, Barcel, Cat., p. 382, 1917[4]; Dover & Rib., R. I. M., xxv, p. 351, 1923.[5]

Kulu[1–6]
Assam[3, 5]

Cic. (*Rhytidophœna*) *limbata, Wied.*, Bates, (nec. Wied.) Entomologist, suppl., p.7, 1891.[6]

Type : Coll. W. Horn.

7. **limbata** var. **feæ,** Gestro, Ann. Gen., p. 88, 1889[1] (*Cicindela*), p. 368, 1893[2]; Fleut., Cat., p. 130, 1892; Horn, Gen. Ins., p. 174, 1910; Fowler, F. B. I., p. 304, 1912[3]; Dover & Rib., R. I. M., xxv, p. 351, 1923.[4]

Bhamo[1–4]
Teinzo[1–4]

Type : Genova Mus.

8. **paradoxa,** Horn, D. E. Z., p. 75, 1892 (*Cicindela*)[1]; Spol. Zeyl., ii, p. 35, pl. 7, f. 5, 1904 (*Euryoda*)[2]; Gen. Ins., p. 174, 1910[3]; Maindron & Fleutiaux, A. S. E. F., p. 9, pl. 1, f. 2, 1905 (habits)[4]; Annand. & Horn, List, p. 5, 1909 (habits)[5]; Fowler, F. B. I., pp. 303, 304, 1912 (habits)[6]; Codina, Barcel Cat., p. 382, 1917[7]; Dover & Rib., R. I. M., xxv, p. 351, 1923.[8]

Type : Coll. W. Horn.

Ceylon[1, 3, 4, 5, 7, 8]
Colombo[2]
Kanara[3, 6, 7, 8]
Karunagapalli[8]
Kumpta (H.)
Madras (B. M.)
Mahe[4, 6, 8]
Malabar (H.)
Matale[2]
Minneri (B. M.)
Negombo[2]
Ombolata[7]
Pondicherry[3, 6, 7, 8]
Puttalam[2]
S. Mysore (B.M.)
Trivandrum [6, 8]
Weligama[2]

Nias[7]

9. **proxima,** Chaudoir, B. S. N. M., xxxiii, (4), p. 325, 1860 (*Cicindela*)[1]; Cat. Coll., p. 118, 1965; Gemm. & Har., Cat., p. 24, 1868 (*Cic.*)[2]; Atkinson, Cat., p. 11, 1889, (*Cic.*)[3]; Fleut., Comp. Rend., S. E. B., p. 168, 1890 (*Euryoda*)[4]; Cat., p. 130, 1892; Annand. & Horn, List, p. 5, pl. 1, f. 1, 1909[5]; Horn, D. E. Z., p. 59, 1905[6]; Gen. Ins., p. 174, 1910[7]; Fowler, F. B. I., p. 302, 1912[8]; Dover &

N. India[1, 2, 3]
Balugaon [9, 10]
Birbhum[5, 8, 10]
Calcutta[10]
Coconada[4, 9, 10]
Coimbatore[9, 10]
Cuttack[10]
Dharwar[8, 10]
Katiawar[7, 10]

Rib., R. I. M, xxii, p. 722, 1921[9]; xxv, p. 351, 1923.[10]

Kurseong[4, 8, 10]
Kanara[6, 7, 8]
Nagpur[8, 10]
Ramnad[7, 8, 10]
Sikkim[7]

Prothyma quadripunctata, Schmidt-Gœbel, nec. Fabr., Faun. Co. Birm., p. 1, 1846 (*Cicindela*; Horn, D. E. Z., p. 131, 1899 (*Euryoda*; synonymy); Codina, Barcel, Cat., p. 381, 1917.[11]

Type : Mus. Paris.

Java (B.M.)
Sanal (H.)
Semarang[11]

10. **reconciliatrix,** Horn, D. E. Z., p. 200, 1900 (*Euryoda*)[1]; Gen. Ins., p. 175, 1910[2]; Col., Cat., Cic., p. 99, 1926[a]; Annand. & Horn, List, p. 5, 1909[3]; Fowler, F. B. I., p. 308, 1912[4]; Dover & Rib., R. I. M., xxii, p. 722, 1921[5]; xxv, p. 351, 1923.[6]

Type : Coll. W. Horn.

Tenasserim[1, 2, 4, 6, a]
Above Tura,[5, 6], 3,500 ft.
?Bengal[2, a]
Dacca [3, 4, 8]
Khasi hills [a]

Penang [a]

11. **schmidt-gœbeli,** Horn, D. E. Z., p. 87, 1895 (*Euryoda*)[1]; p. 201, 1900; Gen. Ins., p. 175, 1910[2]; Fowler, F. B. I., p. 306, 1912[3]; Dover & Rib., R. I. M., xxv, p. 351, 1923.[4]

Type : Coll. W. Horn.

Laos[1–4]
? Bengal [4]
Gokteik[4]

Cambodia[3, 4]

12. **scrobiculata,** Wiedemann, Zool. Mag. ii, p. 65, 1823 (*Cicindela*)[1]; Gemm. & Har., Cat., p. 34, 1868[2]; Atkinson, Cat., p. 14, 1889[3] (*Dromicidia*); Chaud., Cat., Coll., pp. 12, 54, 1865; Fleut., Cat., p. 38, 1892 (*Jansenia*)[4]; Horn, A. S. E. B., p. 537, 1892[5]; Gen. Ins., p. 173, pl. 20, f. 7, 1910[6]; Fowler, F. B. I., p. 302, f. 137, 1912[7]; Codina, Barcel, Cat., p. 380, 1917[8]; Dover & Rib., R. I. M., xxv, p. 351, 1923.[9]

Type : Not known to the authors.

Bengal[1–4, 8]
Barway[5, 8]
Balasore[6, 8, 9] (sea-coast)
Chota-Nagpur (H.)
Maldah[3, 7, 9]

Genus : HEPTODONTA, Hope (1838).

Heptodonta, Hope, Col. Man., ii, pp. 18, 22, 25, 1838; Lacordaire, Mem. Soc. Sc. Liége, i, p. 107, 1843; Gen. Col. i, p. 21, 1854; Schaum, B. E. Z., p. 174, 1862; Fleutiaux, Cat., p. 128, 1892; Fowler, F. B. I., p. 310, 1912 (key to species).

Odontochila, group *Heptodonta*, Horn, Gen. Ins. pp. 196, 203, 1910; Col. Cat., Cic., p. 123, 1926.

Pronyssa, Bates, Ent. Mon, Mag., x, p. 267, 1874; Atkinson, Cat., p. 13, 1889; Fleut., Cat., p. 126, 1892.

Aenictomorpha, Chaudoir, B. S. N. M., p. 11, 1850.

Genotype : *Cicindela analis,* Fabricius* (1801).

Distribution: India and Malaya.

1. **arrowi,** Horn, D. E. Z., p. 362, 1900[1]; Gen. Ins., p. 204, 1910; Fowler, F. B. I., p. 313, 1912[2]; Dover, & Rib., R. I. M., xxv, p. 351, 1923.[3]

   Type : Brit. Mus.

   N. Chin hills[1, 2, 3]
   Rangoon (B. M.)
   Tavoy[2, 3]

2. **eugenia,** Chaudoir, Cat., Coll., pp. 20, 56, 1865 (*Cicindela*)[1]; Gemm. & Har., Cat., p. 15, 1868 (*Cic.*)[2]; Atkinson, Cat., p. 7, 1889 (*Cic.*)[3]; Gestro, Ann. Gen., p. 87, 1889,[4] p. 367, 1893 (*Cic.*)[5]; Fleut., Cat., p. 128, 1892[6]; A. S. E. F., p. 496, 1893[7], p. 569, 1902[8]; B. S. E. F., p. 49, 1917[9]; Lesne, Mission Pavie, Indo-Chine, iii, p. 60, 1904[10]; Horn, Gen. Ins., p. 204, 1910[11]; A. S. E. B., p. 364, 1913[12]; Fowler, F. B. I., p. 313, 1912[13]; Dover & Rib., R. I. M., xxv, p. 351, 1923.[14]

   Type : Mus. Paris.

   Cochin China[1, 2, 3, 6, 7, 8, 11, 12, 14]
   Bhamo[4, 5, 13, 14]
   Between Yenang-Yaung and Mandalay[14]
   Mandalay [4, 5, 13, 14]
   Teinzo[4, 13, 14]
   Tharawaddy [13, 14]

   Penang[11, 14], Siam[11, 14]
   Vientian[9]
   Battambang[10]
   Tale Sap[10]

3. **kraatzi,** Horn, D. E. Z., p. 54, 1899[1]; Gen. Ins., p. 204, pl. 13, f. 9, 1910[2]; Annand. & Horn, List, p. 7, 1909[3]; Fowler, F. B. I., p. 312, 1912[4]; Dover & Rib., R. I. M., xxii, p. 723, 1921[5]; xxv, p. 352, 1923.[6]

   Type : Coll. W. Horn.

   Sikkim[1, 2]
   Above Tura, 3,000–3,500ft. [5, 6]
   Darjiling[4, 6]
   Ghumti, 1,500-5,000 ft.[6]
   Khasi hills[2, 4, 6]
   Mungphu, 3,000-4,000 ft.[3, 4, 6]
   Pashok, 2,000 ft.[5]
   Singla, 1,500 ft.[5]

4. **nodicollis,** Bates, Ent. Mon. Mag., x, p. 267, 1872 (*Pronyssa*)[1]; Atkinson, Cat., p. 13, 1889 (*Pro.*)[2]; Fleut., Cat., p. 126 1892 (*Pro.*)[3]; A. S. E. F., p. 497, 1893; B. S. E. F., p. 369, 1917[4]; Horn., D. E.

   Darjiling[1, 2, 3, 8, 11]
   Above Sureil, 5,500 ft.[11]
   Khasi hills[5, 8, 9, 10]

---

* Ridley, J. Str. Br. As. Soc. 38, p. 130, 1902, records, *H. analis* from Bombay, but this is obviously an error.

Z., p. 99, 1895; Gen. Ins., p. 204, pl. 13, f. 10, 1910[5]; Maindron, B. S. E. F., p. 264, 1904[6]; Lefroy, Ind. Ins. Life, pl. 16, f. 1, 1909; Annand. & Horn, List, p. 7, 1909[7]; Fowler, F. B. I., p. 311, f. 141, 1912[8]; Codina, Barcel, Cat., p. 400, 1917[9]; Dover & Rib., R. I. M., xxii, p. 722, 1921[10]; xxv, p. 352, 1923.[11]

Type : Coll. R. Oberthur.

Kurseong, 5,000 ft.[6, 7, 8, 9, 11]
Lebong, 6,000 ft. [11]
Mungphu[2, 8, 11]
Pashok, 5,000 ft.[11]
Sureil, 5,000 ft.[11]
Shillong[11], Sikkim[2]

Hagiang[4]
Tonkin[10, 11]

5. **pulchella,** Hope, Gray's Zool. Miscell., p. 21, 1831 (*Cicindela*)[1]; Dohrn., S. E. Z., p. 457, 1879 (*Cic.*)[2]; Atkinson, Cat., p. 11, 1889 (*Cic.*)[3]; Fleut., Cat., p. 118, 1892[4]; Horn., D. E. Z., p. 193, 1898; p. 363, 1900; Gen. Ins., p. 204, 1910[5]; A. S. E. B., p. 364, 1913[6]; Annand. & Horn, List, p. 7, 1920[7]; Fowler, F. B. I., p. 312, 1912[8]; Codina, Barcel, Cat., p. 399, 1917[9]; Dover & Rib., R. I. M., xxii, p. 723, 1921[10]; xxv, p. 352, 1923.[11]

Heptodonta hopci, Parry, A. M. N. H., xiv, p. 454, 1844, Trans. Ent. Soc., iv, p. 84, 1845[12]; Schmid-Gœbel, Faun. Col. Birm., p. 10, 1846[13]; Chaud., Cat., Coll., p. 20, 1865; Gemm. & Har., Cat. p. 18, 1868 (*Cic.*); Atkinson, Cat., p. 9, 1889; Horn, D. E. Z., p. 193, 1889, p. 66, 1902.[14]

Aenictomorpha variipes, Chaud., B. S. N. M., xxiii, (1), p. 11, 1850.

H. yunnana, Fairmaire, A. S. E. B., p. 88, 1887.[15]

Type : Brit. Mus.

Nepal [1, 2, 3, 4, 8, 11]
Darjiling[8, 10]
Gopaldhara[11]
Harkar (A.)
Karen hills[8, 11]
Kurseong[9]
Khasi hills [12, 13]
Kumaon
Manipur (B.M.)
Mt. Mauson[6, 14]
Mungphu[7, 8, 11]
Mussoorie
Naga hills (B. M.)
N. Chin hills (B. M.)
Patkai Hills (B. M.)
Rungeet valley[11]
Rungbong valley[11]
Simla[5, 9, 11]
Tura[10, 11]
Tongon, 1,000 ft.

Yunnan[5, 8, 9, 11, 15]

6 ***pulchella** var. **ferrarii,** Gestro, Ann. Gen., p. 366, 1893[1]; Horn., D. E. Z., p. 363, 1900; Fleut., B. S. E. F., p. 369, 1917[2]; Dover & Rib., R. I. M., xxii, p. 723, 1921,[3] xxv, p. 351, 1923.[4]

Karen hills[1, 3, 4] (Cheba, 400-1,100 metres Ghecu, 1,300-1,400 metres)

* Dr. Horn thinks this is only a blackish *H. pulchella.* In difference to his opinion we have given it as a variety of *H. pulchella,* though in Dover and Riberiro's "List" it is given specific rank. In Col. Cat., Cic., p. 124, 1926, it is only recorded as an aberration of *H. pulchella.*

Heptodonta, pulchella, Hope, Horn, Gen. Ins., p. 204, 1910; Schilder, Ent. Blatt. Berl., p. 176, 1911[5]; Fowler, F. B. I., p. 312, 1912 (all *ex parte*).

Type : Mus. Genova.

N. Shan States3, 4
Ruby mines (B. M.)
Tura 1,000-1,200 ft.[3]

Cambodia[4]
Hongkong[5]
Laos[3, 4]
Macao[5]
S. Formosa[5]
Tonkin[4]
Tranninh[2]

Genus : **CICINDELA,** Linnæus (1758).

Cicindela, Linnæus, Syst. Nat., ii, p. 657, 1735; Syst. Nat., ed. 10, i, p. 407, 1758; de Geer, Mem. Hist. Ins., iv, p. 108, pl. 4, f. 1-8, pl. 17, f. 23, 1774; Olivier, Ent., ii, p. 1, 1790; Encycl. Method, v, p. 723, 1790, Fabricius, Syst. Eleuth, i, p. 231, 1801; Herbst, Käfer, x, p. 146, 1806; Clairville, Ent. Helvet, ii, p. 153, pl. 24, 1806; Schonherr, Synon, Ins., i, p. 238, 1806; Dejean, Cat. Col., p. 1, 1821, p. 2, 1833, p. 2, 1837; Spec. Col., i, p. 17, 1825, ii, p. 413, 1826, v, p. 208, 1831; Icon, i, p. 10, pl. 2-6, 1829; Sturm, Cat. Ins., Samml., i, p. 117, 1826; Cat. Kafer-Sal., p. 1, 1843; Wilson, Encycl. Metro., ix, p. 112, pl. 221, f. 2, 1842; Cristoferi & Jan., Cat., Kerum, Nat. Mus., iii, Ent. I, p. 1, 1832; Klug, Jahr. Ins., i, pp. 14-36, 1834; Brulle, Hist. Nat. Ins., iv, p. 49, 1834; Gistl., Syst. Ins., i, pp. 14, 87, 1837; Castelnau, Hist. Nat. Ins. Col., i, p. 15, 1840; Lacordaire, Mem. Soc. Sc. Liege, p. 105, 1843; Gen. Col., i, p. 17, 1854; Le Conte, Trans. Amer. Philos. Soc., p. 27, 1856; Classif. Col. N. Amer., i, p. 21, f. 1, 2, 5-8, 10, 1861-62; Schaum, Erich. Nat. Ins., Deuts., i, p. 8, 1856; B. E. Z., v, p. 68, 1861; vi, p. 122, 1862; Journ. Ent., ii, p. 57, 1863; Jacquelin Daval, Gen. Col. Eur., i, p. 2, 57, 1857; Thomson, Monogr. Cic., i, p. 5, 1857 (life history); Arcana Nat., p. 85, 1859 (life-history); Motschulsky, Etud. Ent., ii, p. 22, 1862; Chaudoir, Cat. Coll., p. 21, 1865; Gemm. & Har., Cat., p. 8, 1868; Dokhturow, Rev. Meus. Ent., i, p. 66, 1883; Atkinson, Cat., p. 3, 1889; Horn & Roeschke, Monogr. Pal. Cic., p. 1, 1891 (life-history); Ganglbauer, Kafer Mittel., i, pp. 5, 534, 1892; Fleut., Cat., pp. 26, 38, 1892; Horn., D. E. Z., p. 332, 1893, p. 3, 1898, pp. 34, 46, 1899, p. 192, 1903, pp. 24, 56,

1905; Sharp, Camb. Nat. Hist., vi, Insect, ii, p. 201, 1901; Kuhut, Ent. Rund., xxvi, p. 6, f. 1-6, 1909; Fowler, F. B. I., pp. 314-328, 1912 (Life history, key to groups of spp.); Horn., Gen. Ins., pp. 11, 68, 232, 417, 1915 (complete bibliography, synonymy and discussion); Col. Cat., Cic., p. 127, 1926.

Abroscelis, Hope, Col. Man., ii, pp. 19, 28, 1838.

Collytron, Gistle, Natgesch. d. Thierreichs hohere Schulen, Stuttgart, ed. i (1848), ed. 2 (1851), p. 111.

Calochroa, Hope, Col. Man., ii, p. 19, 1838.

Calochroma, Motschulsky, Etud. Ent., p. 22, 1862.

Calomera, Motschulsky, Eutd. Ent., p. 22, 1862.

Calostola, Motschulsky, Eutd. Ent., p. 22, 1862.

Catoptria, Guerin, Rev. Mag. Zool., (2), i, p. 146, 1849.

Chœtostyla, Ganglbauer, *Käfer Mitteleuropa*, i, pp. 11, 18, 1892.

Cylindera, Westwood, Ann. Sc. Nat., 22, p. 300, 1831; Jard. Mag. Zool. Bot., i, p. 251, 1837; Mod. Class. Ins., i, p. 54.

Cylindrostoma, Motschulsky, in litt., Etud. Ent., p. 25, 1859; B. S. N. M., xxxiv (1), p. 98, 1861.

Eumecus, Motschulsky, Kafer, Russ., v. 1850; Etud. Ent., p. 22, 1862.

Habroscelis, Motschulsky, Etude Ent., p. 22, 1862.

Hypœtha, Le Conte, Trans. Amer. Phil. Soc., xi, p. 28, 1860.

Jansenia, Chaudoir, Cat. Coll., pp. 15, 53, 1865.

Myriochile, Motschulsky, Etud. Ent., p. 22, 1862.

Parmecus, Motschulsky, B. S. N. M., xxxvii (3), p. 172, 1864.

Podabra, Hope, Col. Man, ii, p. 19, 1838.

Thopeutica, Schaum-Chaud., B. E. Z., p. 72, 1861.

Tetreurytarsa, Horn, D. E. Z., p. 94, 1892; Fowler, p. 336, 1912.

Genotype : *Cicindela campestris,* Linnæus (1735).

Distribution : Cosmopolitan.

1. **agnata,** Fleutiaux, Comp. Rend., S. E. B., p. 168, 1890[1]; Cat., p. 56, 1892; Horn. D. E. Z., p. 223, 1892; Gen. Ins., p. 292, 1915; Maindron & Fleut., A. S. E. F., p. 13, 1905[2]; Annand. & Horn, List, p. 13, 1909 (habits)[3]; Fowler, F. B. I., p. 367, f. 160, 1912 (habits)[4]; Gravely, R. I. M., vii, p. 208, 1912 (habits); Codina, Barcel. Cat., p. 426, 1917[5]; Dover & Rib., R. I. M., xxv, p. 352, 1923.[6]

   Type : Coll. E. Fleutiaux.

Asansol[1, 4, 6]
Balighai [3, 4, 6]
Behrampur[2, 6]
Caragala (H.)
Chakradharpur[6]
Chounambur [2, 6]
Cuttack (bank of river Mahanadi)[6]
Kurseong[4, 6]
Malabar[2, 6]
Kallinour[2, 6]
Ramnad[4, 6], Sur Lake[6]

Pontianak[5]

2. **albina,** Wiedemann, Zool. Mag., i, (3), p. 169, 1819[1]; Gistl., Syst. Ins., i, p. 63, 1937[2]; Schaum, Journ. Ent., ii, p. 67, 1863; Chaud., Cat. Coll., p. 32, 1865[3]; Gemm. & Har., Cat., p. 9, 1868[4]; Atkinson, Cat., p. 4, 1889[5]; Fleut., Comp. Rend., S. E. B., p. 169, 1890[6]; Cat., p. 82, 1892[7]; Horn, Gen. Ins., p. 307, pl. 20, f. 10, 1915[8]; Annand. & Horn, List, p. 29, 1909 (habits)[9]; Folwer, F. B. I., p. 427, f. 191, 1912[10]; Gravely, R. I. M., vii, p. 208, 1912 (habits); Codina, Barcel, Cat., p. 441, 1917[11]; Dover & Rib., R. I. M., xxii, p. 726, 1921 (habits)[12]; xxv, p. 352, 1923.[13]

   C. albida, Dejean, Spec. Col., i, p. 125, 1825; Brulle, Rev. Ent., Silb., ii, p. 99, 1834.

   Type : Not known to the authors.

Bengal[1–7]
Asansol[6, 10, 13]
Balighai [9, 12, 13]
Berhampur[5]
Darjiling[8, 10, 11, 13]
Lohardaga[9, 12, 13]
Rawalpindi (bank of river Sohan)[12, 13]

3. **albopunctata,** Chaudoir, B. S. N. M., xxv (1), p. 10, 1852[1]; Cat., Coll. p. 24, 1865; Schaum, Journ. Ent., ii, p. 68, 1863; Gemm. & Har., Cat., p. 9, 1868; Atkinson, Cat., p. 4, 1889[2]; Notes Zool., N.-W. P., p. 117, 1882; Fleut., Cat., p. 48, 1892; Annand. & Horn, List, p. 17, 1909[3]; Horn, Gen. Ins., p. 297, 1915[4]; Fowler, F. B. I., p. 378, 1912[5]; Codina, Barcel. Cat., p. 430, 1917[6]; Dover & Rib., R. I. M., xxv, p. 352, 1923.[7]

   C. olivia, Bates, Cist. Ent., ii, p. 33, 1878[8]; Atkinson, Cat., p. 11, 1889[9]; Fleut., Cat., p. 73, 1892.

Simla[1–5, 7]
Bhutan[4, 5, 6, 7]
Chaumusuri[5, 7, 8]
Darjiling[5, 7]
Dehra Dun (F.)
Doiphang valley[7]
Ghoom, 7,500 ft.[3,5,7]
Gopaldhara (A.)
Kangra valley[5, 7]
Kurseong[5, 7]
Moradabad[5, 7,8, 9]

Type : Mus. Paris.

Mungphu[5, 7]
Mussoorie[3, 5, 7]
Soondrigal[3, 7]
Sikkim [3, 6, 7]

4. **andersoni,** Gestro, Ann. Gen., p. 83, 1889[1]; p. 356, 1893[2]; Atkinson, Cat., p. 24, 1889[3]; Fleut., Cat., p. 60, 1892; B. S. E. F., p. 49, 1917[4]; Horn, D. E. Z., p. 223, 1892; A. S. E. B., p. 139, 1914[5]; Gen. Ins., p. 310, 1915; Fowler, F. B. I., p. 437, 1912[6]; Dover & Rib., R. I. M., xxv, p. 352, 1923.[7]

Type : Mus. Genova.

Teinzo[1, 2, 3, 6, 7]
Karen hills[2, 6, 7]
(Cheba to Biapo, 400-900 m.)
N. Chin hills [6, 7]
Rangoon (A.)

Bien-Hoa[5]
Ban-Sop-Mom[4]
Vientian[4]

5. **andrewesi,** Horn, D. E. Z., p. 171, pl. 3, f. 1, 1894 (*Calochroa*)[1]; p. 23, 1907; Gen. Ins., p. 302, 1915[2]; Fowler, F. B. I., p. 392; 1912[3]; Dover & Rib., R. I. M., xxv, p. 352, 1923.[4]

Type : Coll. H. E. Andrewes.

N. Kanara[1–4]

6. **andrewesi,** var. **mauritii,** Horn, D. E. Z., p. 23, 1907[1]; Gen. Ins., pp. 302, 421, 1915[2]; Fowler, F. B. I., p. 392, 1912[3]; Dover & Rib., R. I. M., xxv, p. 352, 1923.[4]

Type : Coll. W. Horn.

Ind. merid occid.[1]
Coorg (H.)
Dera Ismail Khan (H.)
N.-W. Frontier Prov.[2–4]
S. w. India[3]
Sidapur[4]

7. **andrewesi,** var. **unica,** Fleutiaux, B. S. E. F., p. 245, 1895[1]; Fowler, F. B. I., p. 393, 1912[2]; Horn, Gen. Ins., p. 302, 1915; Dover & Rib., R. I. M., xxx, p. 352, 1923.[3]

Calochroa flavoguttata, Horn, D. E. Z., p. 359, 1895.[4]

Type : Coll. E. Fleutiaux.

India[1, 2, 3]
S. India[3, 4]

8. **angulicollis,** Horn, D. E. Z., p. 209, 1900[1]; Gen. Ins., p. 304, pl. 18, f. 2, 1915[2]; Maindron, B. S. E., p. 264, 1904[3]; Fowler, F. B. I., p. 410, 1912[4]; Dover & Rib., R. I. M., xxv, p. 352, 1923.[5]

Type : Coll. W. Horn.

Dacca[1, 4, 5]
Kodaikanal[2, 3, 5]
Madura[2, 4, 5]
Palni Hills (H.)
Tope[5]

9. **angulata,** Fabricius, Ent. Syst. Suppl., p. 62, 1798[1]; Syst. Eleuth., i, 243, 1801[2]; Herbst, Kafer, x, p.

Tranquebar[1–6]
Asansol[7, 10, 14]

176, pl. 171, f. 11, 1806[3]; Dejean, Spec., Col. i, p. 80, 1825; Gistl. Syst. Ins., i, p. 50, 1837[4]; Brulle, Rev. Ent. Silb., ii, p. 99, 1834; Schmidt-Gœb., Faun. Col. Birm., p. 4, pl. 1*f*, 8, 8*a*, 1846; Schaum, S. E. Z., p. 51, 1847; Journ. Ent., ii, p. 62, 1863; Mots., Etud. Ent., p. 31, 1855[5]; Chaud., Cat. Coll., p. 34, 1865; Gemm. & Har., Cat., p. 9, 1868; Atkinson, Cat., p. 4, 1889[6]; Fleut., Comp. Rend., S. E. B., p. 169, 1890[7]; Cat., p. 87, 1892; Maindron & Fleut., A. S. E. F., p. 19, 1905[8]; Annand. & Horn, List, p. 15, 1909, (habits)[9]; Fowler, F. B. I., p. 370, 1912 (habits)[10]; Gravely, R. I. M., vii, p. 209, 1912 (habits); Horn, Gen. Ins., p. 296, 1915[11]; Codina, Barcel. Cat., p. 429, 1917[12]; Dover & Rib., R. I. M., xxii, p. 724, 1921[13], xxv, p. 352, 1923.[14]

C. designata, Dejean, in litt., Cat., ii, p. 1, 1821.

C. sexatilis, Gistl. Syst. Ins., i, p. 49, 1837.

C. latipennis, Parry, A. M. N. H., xiv, p. 454, 1844; T. E. S., iv, p. 84, 1845[15]; Schmidt-Gœb., Faun. Col. Birm., 10, 1846.[16]

Type : Not known to the authors, but presumably in Sweden.

Assam (H.)
Berhampur[6]
Calcutta[6, 9, 10, 14]
Cuttack (bank of R. Mahanadi)[13, 14]
Chota-Nagpur[10, 14]
Coromandel[14]
Dacca[10, 14]
Damukdia[9, 10, 14]
Karachi[10, 11, 12, 14]
Khasi hills[6, 15, 16]
Maldah[6, 9, 10, 14]
Mungphu[10, 14]
Nepal (B. M.)
Orissa (H.)
Pankabari[9, 10, 14]
Pegu[6, 9, 10, 14]
Pondicherry (and environs)[8, 14]
Ponear[8, 14]
Sikhim[6, 9, 11, 12,]
Sitong[13, 14]
Sittaung R.[9, 14]
Sitang valley[6]
Tetara[7, 10, 14]
Toungoo (F.)
Tura, 1,000 ft. [13, 14]

Annam[10, 11, 12, 14]
Cambodia[14]
Hainan[10, 11, 12, 14]
(Five Finger Mts., B. M.).
Penang (B. M.)
Phucsong[12]

10. **angulata,** var. **plumigera,** Horn, D. E. Z., p. 86, 1892[1]; p. 461, 1909[2]; Mem. 1st Congress, Int. Ent., p. 314, 1911[3]; A. S. E. B., lvii, p. 365, 1913[4]; lviii, p. 440, 1914[5]; Gen. Ins., p. 296, 1915[6]; Fowler, F. B. I., p. 370, 1912[7]; Codina, Barcel. Cat., p. 429, 1917[8]; Fleut. B. S. E. F., p. 49, 1917[9]; Dover & Rib., R. I. M., xxv, p. 352, 1923.[10]

Trichinopoly[1, 7, 8, 10]
Burma[2]
Bank of Mahanadi (H.)
Bengal[2], Karachi[2]
Mysore[3, 6, 7, 10]
Nilgiri hills[7, 10]
Pontianak[8]

Type : Coll. W. Horn.

Shembaganur[8]
Sikhim[2]
Shimoga (N.)
Ponear

Annam[2]
Cambodia[4, 5, 8, 10]
Hainan[2], Kang-Kiet[9]
S. Formosa
Sumatra[2, 3, 8, 10]

11. **assamensis,** Parry, A. M. N. H., xiv, p. 454, 1844[1]; T. E. S., iv, p. 84, 1845[2]; v, p. 80, pl. 11, f. 1, 1848[3]; Schmidt-Gœb., Faun. Col. Birm., p. 10, 1846[4]; Mots, Etud., Ent., xi, p. 22, 1862 (*Calostala*); Schaum, Journ. Ent., ii, p. 64, 1863[5]; Chaud., Cat. Coll., p. 38, 1865[6]; Gemm. & Har., Cat., p. 9, 1868[7]; Atkinson *Notes Zool. N.-W. P.*, p. 117, 1882[8]; Cat., p. 4, 1889[9]; Fleut., Cat., p. 117, 1892[10]; A. S. E. F., p. 492, 1893[11]; Horn, D. E. Z., p. 193, 1898; Gen. Ins., p. 302, 1915[12]; Annand. & Horn, List, p. 23, 1909 (habits)[13]; Fowler, F. B. I., p. 395, 1912 (habits)[14]; Codina, Barcel. Cat., p. 436, 1917[15]; Dover & Rib., R. I. M., xxii, p. 725, 1921 (habits)[16]; xxv, p. 353, 1923.[17].

C. horsfield, Hope, in litt., Thomson, Arch. Ent., i, p. 325, 1851.

Type : Coll. R. Oberthur.

Assam[1, 5, 6, 7, 8, 10, 12, 15]
Arrakan[13, 14, 17]
Above Tura, 3,000 ft.[16, 17], Ceylon[2]
Chandkhira[15]
Darjiling[11, 12, 15, 17]
Gopaldhara[17]
Khasi hills[2, 4, 14, 17]
Margherita (H.)
Mungphu[13, 14, 17]
Naga hills[9, 13, 17]
Nemotha[13, 17]
N. Manipur[14]
Pashok[16, 17]
Patkai hills [14, 17]
Sikkim[9, 12, 13, 15, 17]
Sibsagar[13, 14, 17]
Siliguri[13, 17]
Sylhet[11, 14, 17]
Sukna[13, 14, 17]
Rungbong valley[16 17]
Tura[16, 17]
? Tibet [15]

Penang[14, 17]

12. **atkinsoni,** Gestro, Ann. Mus., p. 357, 1893[1]; Fowler, F. B. I., p. 422, 1912[2]; Horn, Gen. Ins., p. 306, 1915; Dover & Rib., R. I. M., xxv, p. 353, 1923.[3]

Type : Mus. Genova.

Karen hills,[1, 2, 3]
(Chela–Biapo, 300-1,000 met.)
Rangoon[2, 3]
Tharrawaddy (H.)

13. **atelesta,** Chaudoir, B. S. N. M., xxvii (1), p. 115, 1854[4]; Cat., Coll., p. 24, 1865[2]; Schaum., Journ, Ent.

Indes Or.[1, 2, 3, 4, 11, 13]
Asansol[6, 7, 9]

II, p. 68, 1863[3]; Gemm. & Har., Cat., p. 9, 1868[4]; Atkinson, Cat., p. 4, 1889[5]; Fleut., Comp. Rend., S. E. B., p. 168, 1890[6]; Fowler, F. B. I., p. 358, 1912 (var.)[7]; Horn, Gen. Ins., p. 294, 1915[8]; Dover & Rib., R. I. M., xxv, 353, 1923[9].

Cicindela charactera, Sturm, in litt., Cat. Ins. Samml., p. 117, 1826; Fleut., B. S. E. F., p. 205, 1895.

C. imperfecta, Chaudoir, nec Le Conte, B. S. N. M., p. 8, 1852[10]; Fleut., Cat., p. 45, 1892[11]; Lefroy, Ind. Ins. Life, pl. 16, f. 6, 1909; nec Fowler, F. B. I., p. 357, 1912; Dover & Rib., R. I. M., xxii, p. 724, 1991.[12]

C. despecta [ex parte], Fleutiaux, Comp. Rend., S. E. B., p. 88, 1886; Cat. p. 48, 1892[13]; Atkinson, Cat., p. 7, 1889.

C. prinsepi, Fleutiaux, B. S. E. F., p. 205, 1895.

Type : Mus. Paris.

Calcutta[7, 8, 9]
Chota-Nagpur[8, 9]
Kurseong[7, 9]
"Ganges" (A.)
Mungphu[7, 9]
Mysore[9]
Nagpur[9]
N. Bombay[8]
Nowatoli[6, 7, 9]
Ranchi[9, 12]
Simla[5, 9, 10]
Surat[9, 12]
Sylhet[7, 9]
Tetara,[6]

14. **aulica,** Dejean, Spec. Col., v, p. 250, 1831[1]; Gist. Syst. Ins., I, p. 57, 1837[2]; Chaud., Cat. Coll., p. 31, 1865[3]; Gemm. & Har., Cat., p. 9, 1868[4]; Horn, Mon. Pal. Cic., p. 157, 1891[5]; D. E. Z., p. 332, 1891; Munch. Kol. Zeit., I, pp. 340, 342, 1903; Gen. Ins., p. 297, 1915[6]; Bedel, Cat., Rais, Col. N. Afric., I, p. 5, 1895[7]; Maindron, A. S. E. F., pp. 379, 380, 1889[8]; Fleut., Cat., p. 75, 1892[9]; Tschitscherin, Hor. Soc. Ent. Ross., pp. 11, 14, (footnote 10), 1903[10]; Fowler, F. B. I., p. 374, f. 164, 1912 (habits)[11]; Dover & Rib., R. I. M., xxv. p. 353, 1923.[12]

var. orienticola, Tschitscherin, Hor., Soc. Ent. Ross., p. 14, (footnote 19), 1903.

Type : Mus. Paris.

Senegal[1–5, 8–12]
Karachi[6, 8, 11, 12]
Manora (N.)

Abyssinia[11, 12]
Angola[12]
Arabia[6, 11, 12]
Baghdad[6, 12]
Bay of Tadjourah[8]
Benguela[11, 12]
Brit. Somaliland[11, 12]
Cape Verde Is.[11, 12]
Djibouti[8], Egypt[11, 12]
Greece[3, 11, 12]
Jemen[6], Kiamari Id.[8]
Nubia[5], Obeck[8]
Qued Atela (pits)[8]
St. Vincent [11, 12]
Suakim[5], Tunis[8, 12]

15. **aulica** var. **diania,** Tschitscherin, Hor. Soc. Ross., p. 10, 1903[1]; Horn, Munch. Kol. Zeit., I, p. 344, 1903; Annand. & Horn, List, p. 16, 1909; Fowler, F. B. I.,

Hudian (Bazmaul environs).[1]
Baluchistan[5]

p. 375, 1912[3]; Horn, Gen. Ins., p. 297, 1915[4]; Dover & Rib., R. I. M., xv, p. 353, 1923[6].

Type : Mus. Petrograd.

———————
sarhad[1, 4, 5]
(Hurmuk, Duz-ab, Podahi)
Bushire[2, 4, 5]
Persian Gulf[3]

16. **aurofasciata,** Dejean, Spec. Col., v, p. 224, 1831[1]; Gistl., Syst. Ins., p. 27, 1837[2]; Guerin, in Delessert, Souv. Voyage Ind., II, p. 33, 1843[3]; Tennant, Ceylon, etc., London, 2nd ed., p. 274, 1859[4]; Chaudoir, Cat., Coll. p. 39, 1865[5]; Atkinson, Cat., p. 5, 1889[6]; Fleut., Cat., p. 117, 1892[7]; Maindron, A. S. E. F., p. 17, 1905; Lefroy, Ind. Ins. Life, pl. 16, f. 7, 1909; Annand. & Horn, List, p. 25, 1909[8]; Fowler, F. B. I., p. 407, f, 174, 1912 (habits)[9]; Horn, Gen. Ins., p. 304, 1915; Codina, Barcel. Cat., p. 438, 1917[10]; Dover & Rib., R. I. M., xxv, p. 353, 1923.[11]

Calochroa crucigera, Hope, Col., Man., II, p. 162, pl. 1, f. 2, 1838.

Type : Mus. Paris.

Indes Or.[1, 3, 7]
Bangalore
Bombay (B. M.)
Ceylon (coast counry or plains)[4]
Cochin[9, 11]
Coondoor, 6,000 ft.[11]
Coorg Jungle (A.)
Coromandel[5, 6]
Darjiling[9, 11]
Deccan[3], Kanara[9, 11]
Mhow (B. M.)
Mysore[9, 10, 11]
Ootacamund, 5,000-7000 ft.[8, 9, 11]
Pondichery (H.)
Shimoga[9, 11]
Siliguri (H.)
Travancore[9, 11]

17. **aurofasciata** var. **goryi,** Chaudoir, B. S. N. M. I., p. 3, 1852[1]; Gem. & Har., Cat., p. 17, 1868[2]; Atkinson, Cat., p. 8, 1889[3]; Horn, D. E. Z., p. 193, 1898[4]; p. 59, 1965[5]; Gen. Ins., p. 304, fig. 98 (on p. 33), 1915.[6]

Cicindela lepida, Gory (nec. Dejean, 1831) Mag. Zool., pl. 96, 1833[7]; as var. Annand. & Horn, List, p. 26, 1909 (habits)[8]; Fowler, F. B. I., p. 408, 1912 (habits)[9]; Codina, Barcel. Cat., p. 438, 1917[10]; Dover & Rib., R. I. M., xxv, p. 353, 1923.[11]

C. lichtensteini, Hope, in litt., Col. Man., II, p. 163, 1838.[12]

Type : Mus. Paris.

Deccan[1, 2, 3, 7]
Bangalore[5, 8, 9, 11]
Belgaum[11]
Gunji (A)
Mysore (Gudalur to Tippukada)[9, 11]
N. Kanara[9, 11]
Ouchterlony Valley, 2,500-3,000 ft. (A.)
Poona
Shimoga[9, 10,11]

19. *****aurofasciata** var. **seminigra,** Fowler, F. B. I., p. 408, 1912[1]; Dopver & Rib., R. I. M., xxv, p. 353, 1923.[2]

India[1, 2]

* This form is relegated to the synonymy of *C. aurofasciata* by Horn (Col. Cat., Cic., p. 185, 1926).

Type : Brit. Mus.

20. **aurulenta,** Fabricius, Syst. Eleuth., I, p. 239, 1801[1]; Herbst., Kafer, x, p. 204, 1806[2]; Schonherr, Syn. Ins., p. 244, 1806; Dejean, Spec. Col., I, p. 46, 1825[3]; Brulle, Hist. Nat. Ins., iv, Col. I, p. 63, 1834; Schmidt-Gœbel, Faun., Col. Birm., p. 2, 1846[4]; Motschulsky, Etud. Ent., p. 29 1855[5]; Schaum, Journ. Ent., II, p. 64, 1863[6]; Chaudoir, Cat., Coll., p. 38, 1865[7]; Gemm. & Har., Cat., p. 10, 1868[8]; Atkinson, Cat., pp. 5, 23, 1889[9]; Gestro, Ann. Gen., p. 81, 1889[10]; p. 360, 1893[11]; Bates, Ent. Suppl., p. 7, 1891[12]; Fleut., Comp. Rend., S. E. B., p. 169, 1890[13]; Cat., p. 116, 1892[14]; A. S. E. F., p. 490, 1893[15]; Heyden, Cat. Col., Sib., (in Deuts. Ent. Gessll.), p. 5, 1893[16]; Horn, Ann. Gen., p. 682, 1894[17]; D. E. Z., p. 97, 1895, p. 59, 1897[18]; p. 66. 1902[19] p. 461, 1909[20]; Mem. Ist Congress, Int. Ent., p. 314, 1911[21]; Gen. Ins., p. 300, 1915[22]; Ridley, J. Str. Br. As. Soc., 38, p. 130, 1902 (habits)[23]; Robinson, Fasc. Malay. Zool., I, p. 179, 1903 (habits); Annand. & Horn, List, p. 19, 1909 (habits)[24]; Schilder, Ent. Blatter, Berl., p. 178, 1911[25]; Fowler, F. B. I., p. 315 (habits), p. 383, 1912[26]; Codina, Barcel. Cat., p. 433, 1917[27]; Bryant, Ent. Mo. Mag., (3), v, p. 73, 1919[28]; Dover & Rib., R. I. M., xxii, p. 725, 1921[29]; xxv, 353, 1923[30].

Sumatra[1, 2, 9, 22, 28]
Asansol[13]
Bhamo hills, 4,000 ft. (B. M.)
Ceylon[20, 21, 22, 26, 27, 30]
Chota Nagpur[20, 22, 26, 27, 30]
Dacca (B. M.)
Dachyny[4]
Karen hills[11, 26, 30]
Kulu[12], Limboi[4]
Malda[9], Meetan [10, 11]
Malewoon[11]
Houngdaran[11]
Mungphu[24, 26, 30, 40, 41] 3-4,000 ft.
Naga hills[9]
N. Chin hills (B. M.)
N. Shan States, [29, 40]
Nepal[9], Palon[11]
Pegu (B. M.)
Salvern[4], Sibsagar[9]
Sikhim[9, 20, 21, 22, 24, 27, 20, 36]
Sylhet (B. M.)
Tavoy[15, 24, 26, 35]
Tharawaddy (A.)
Thagata[9, 10, 12]
Teinzo[9, 10]
Toungoo (B.M.)

Cicindela aurantiaca, Fleutiaux, A. S. E. F., p. 491, 1893[31]; Horn, Arch. Nat., lxxix, A, heft 11, p. 28, 1913.

C. flavomaculata, Chevrolat, Rev. Zool., p. 95, 1845[32]; Chaudoir, Cat. Coll., p. 38, 1865[33]; Horn, Arch, Nat., lxxix, A, heft. 11, p. 28, 1913; as var. Fleutiaux, Cat., p. 116, 1892; A. S. E. F., p. 491, 1893, p. 569, 1902[34]; B. S. E. F., p. 49, 1917[35]; Lesne, Mission, Pavie, Indo-Chine (iii), p. 59, 1904[36]; Horn, D. E. Z., p. 97, 1895, p. 59, 1897[37]; Annand. & Horn, List, p. 20, 1909[38]; Schilder, Ent. Blatter. Berl., p.

Alahanpandjang[17] (A.)
Battambang[36]
Borneo[5, 20 ,21, 26, 27]
Balighe[17]
Banguey Id.[20, 22, 26, 27, 30]
Bedgin[27]
Cambodia[20, 22, 26, 27, 30, 31]
China (Io-Kien)[3, 6, 8, 9, 20, 21, 25, 26, 27, 36]

178, 1911[39]; Fowler, F. B. I., p. 384, 1912[40]; Dover, & Rib., R. I. M., xxv, p. 353, 1923[41].

Type : ?Mus. Stockholm.

Chiem Hoa[19]
Cochin China[9]
Chantabuum[36]
Chusan Is.[26], Deli[17]
Fort de Kock[17]
Formosa[20, 21, 26, 30]
Hainan Id. (B. M.)
Hobart (B. M.)
Hongkong[9, 30, 33, 38, 39, 40, 41]
Huang-By[35]
Java[3, 6, 8, 9, 15, 18, 20, 21, 22, 26, 27, 37]
Keng-Kiet[35]
Kuching (A.)
Kontei (A.), Laokay[34]
Lombok[27, 30]
Macao[30, 32, 40, 41]
Mou-Pin[36]
Mœara Labœ[17]
M. Matang. 1,000 ft.[28]
Nias Id. [20, 22, 26, 27, 30]
Pak-Lay (A.)
Pangl-araug Penang[17]
Perak (hills near Taiping)[29, 30]
Penang[5, 17] (W. hill, 3,000 ft.)[23]
Pontianak (A.)
Prov. Wellesley[23]
Siam[20, 22, 26, 27, 30, 31]
Sandakan[24, 30]
Sarawak[26, 30]
Sinkep Id. [22, 24, 27, 30]
Sanga Sanga (B. M.)
Singapore[9, 15, 20,22, 23, 24, 26, 27, 30]
Sukabumi[27]
Solok[17, 27]
Sunda Is.[7, 36]
Sze-tschuan[16]
Tban-Moi1[9]
Tamansari, 1,600 ft.
Turakhorn[35]

20. **aurulenta,**

21. **aurulenta** var. **batesi,** Fleutiaux, A. S. E. F., p. 491, 1893[1]; B. S. E. F., p. 49, 1917[2]; Horn, Arch. Nat. lxxix, A, heft. 11, p. 28, 1913; D. E. Z., p. 97, 1895[3]; A. S. E. B., p. 365, 1913; Gen. Ins., p. 300, 1915[4]; Schilder, Ent. Blatter. Berl., p. 178, 1911[5]; Codina, Barcel. Cat., p. 434, 1917[6]; Dover & Rib., R. I. M., xxii, p. 725, 1921[7]; xxv, p. 353, 1923[8].

Type : Coll. E. Fleutiaux.

Formosa[6]
Doiphang Valley[7, 8]
Pegu[4, 6, 8]

Chapa[2]
S. China[4, 5, 6]
Hainan (H.)
Java[3]
Kosempo (A.)
Pitani[6]
Taihiorinsho (A.)
Than Moi (H.)

22. **aurulenta** var. **virgula,** Fleutiaux, A. S. E. F., p. 491, 1893[1]; Horn, D. E Z., p. 97, 1895[2]; Arch. Nat., xxix, A, heft. 11, p. 28, 1913; A. S. E. B., p. 365, 1913[3]; Gen. Ins., p. 300, 1915[4]; Annand. & Horn, List, p. 20, 1909[5]; Fowler, F. B. I., p. 384, 1912[6]; Codina, Barcel. Cat., p. 434, 1917[7]; Dover & Rib., R. I. M., xxii, p. 725, 1921[8]; xxv, p. 353, 1923[9].

Type : Coll. E. Fletuiaux.

Momeit[1]
Bhutan [4, 6, 7, 9]
Buxar[5, 6, 8, 9]
Chandkira (N.)
Chittagong[5, 8]
Darjiling[2], Dejoo[8, 9]
Garobadha[8]
Ghumti[8], [9] 4,000 ft.
Gopaldhara[8, 9]
Kurseong[6, 8, 9]
Margherita[8, 9]
Monda[5, 8]
Mungphu[5, 6]
Naga hills[5, 6, 8, 7]
Nepal[4, 6, 7]
N. Shan States (N.)
Pashok, 2,000 ft.[8, 9]
Pedong[1]
Ruby Mines (H.)
Rungeet valley[8, 9]
Rungpo [6, 8, 9]
Shillong[8, 9]
Sibsagar[5, 6, 8, 9]
Silonibari[8, 9]
Sikkim[5, 7, 9]
Singla[8, 9]
Soom, 2,500-4,000 ft.[8, 9]
Sylhet[6, 8, 9]
Tura, 1,000-1,400 ft.[8, 9]

Teinzo[6, 8, 9]

Cambodia[9]
Chusan Is. [4, 7]
Hongkong[5, 6, 8, 9]
Hainan (Five-Finger Mts. N.)
Lao-Kay[3]
Pron-Penh[3]
Singapore[8]
Shanghai[4, 6, 7, 9]

23. **azureocincta,** Bates, Cist. Ent., II, p. 333, 1878[1]; Atkinson, Cat., p. 5, 1889[2]; Fleut., Cat., p. 66, 1892[3]; Horn, A. S. E. B., p. 537, 1892 (*Euryoda*)[4]; D. E. Z., p. 60, 1905[5]; Gen. Ins., p. 287, 1915[6]; Fowler, F. B. I., p. 330, 1912[7]; Dover & Rib., R. I. M., xxv, p. 354, 1923[8].

Type : Coll. R. Oberthur.

Bombay[1–4]
N. Kanara[5–a]
Kharwar (A.)
Kurseong[4]

24. **bellana,** Horn, D. E. Z., p. 63, 1905[1]; Gen. Ins., p. 309, 1915[2]; Fowler, F. B. I., p. 433, 1912[3]; Dover & Rib., R. I. M., xxv, p. 354, 1923.[4]

C. ruda, Horn., D. E. Z., p. 38, 1905[5]; as var. Fowler, F. B. I., p. 433, 1912[6]; Dover & Rib., R. I. M., xxv, p. 354, 1923.[7]

Type : Coll. H. E. Andrewes.

Karachi[1–7]
Fao[2, 3, 4]

25. **belli,** Horn, D. E. Z., p. 173, 1894[1]; Gen. Ins., p. 129, 1915; Maindron, A. S. E. F., p. 12, 1905[2]; Fowler, F. B. I., p. 344, 1912[3]; Dover & Rib., R. I. M., xxv, p. 354, 1923.[4]

Type : Coll. H. E. Andrewes.

N. Kanara[1, 3, 4]
Anamalai hills
Basti (A.)
Belgaum[1, 3, 4]
Nilgiri hills[2]
Wallardi (A.)
Mahe[2, 3, 4]

26. **belli** var. **umbropolit** Horn, D. E. Z., p. 61, 1905[1]; Gen. Ins., p. 291, f. 85, (on p. 32), 1915[2]; Fowler, F. B. I., p. 345, 1912[3]; Dover & Rib., R. I. M., xxii, p. 723, 1921; xxv, p. 354, 1923.[4]

Type : Coll. H. E. Andrewes.

Nilgiri hills[1, 2, 4]
Coonoor[4]
Kanara (? north)[2]
Karkur Ghat[4]
Margalli[4]
Pollibetta[4]
Ouchterlony valley (2,500-5,000 ft.)[3, 4]

Sidapur[4]
S. Mysore (A.)
Teppukadu (2,500-3,000ft.)[4]

27. **bicolor,** Fabricius, Spec. Ins., i, p. 283, 1781[1]; Ent. Syst., i, p. 170, 1792[2]; Syst. Eleuth., i, p. 233, 1801[3]; Olivier, Ent., ii, p. 11, pl. 2, f. 14, 1790[4]; Enc. Meth., p. 728, pl. 174, f. 7, 1790[5]; Herbst, Kafer, x, p. 163, pl. 170, f. 11, 1806[6]; Dejean, Spec. Col., i, p. 43, 1825[7]; Westermann, Rev. Ent. Silb., i, p. 106, 1833 (habits); Gistl., Syst. Ins., i, p. 26, 1837[8]; Mots., Etud. Ent., p. 29, 1855[9]; Chaud., Cat., Coll., p. 38, 1865[10]; Gemm. & Har., Cat., p. 10, 1868[11]; Atkinson, Cat., p. 5, 1889[12]; Fleut., Comp. Rend., S. E. B., p. 170, 1890[13]; Cat., p. 118, 1892[14]; Gestro, Ann. Gen., p. 365, 1893[15]; Annand. & Horn, List, p. 24, 1909 (from young rice fields)[16]; Fowler, F. B. I., p. 400, f. 171, 1912[17]; Horn, Gen. Ins., p. 303, 1915[18]; Codina, Barcel. Cat., p. 437, 1917[19]; Dover & Rib., R. I. M., xxv, p. 354, 1923[20].

C. evanea, Fabr., Fleut., Cat., p. 118, 1892 (in error).

Type : Probably in one of the Swedish museums, but not known to the authors.

Indes Or.[18, 10, 11, 14]
Asansol[13, 17, 20]
Berhampur[16, 17, 20]
Birbhum[16, 17, 20]
Burma[18, 19, 20]
Calcutta and environs[9, 16, 17, 20]
Chapra (F.)
Dehra Dun [18, 19, 20]
Karen hills (Cheba to Biapo)[15]
Khasi hills[17, 20]
Maldah[12, 16, 17, 20]
Murshidabad[17, 20]
Poona[16, 17, 18, 19]
Pubna[20]
Sahibgang,[17, 20]
Simla[17, 18, 19, 20]

28. **bicolor** var. **atavus,** Horn, Ark. Zool. Stockholm, xiii, No. 11, p. 20, 1920[1]; Dover & Rib., R. I. M., xxv, p. 354, 1923.[2]

Type : Coll. W. Horn.

Lashuwala[1, 2]

29. **bigemina,** Klug. Jahrb. Ins., i, p. 30, 1834[a]; Gistl., Syst. Ins., i, p. 52, 1837; Chaud., Cat. Coll., p. 22, 1865[1]; Gemm. & Har., Cat., p. 10, 1868[2]; Atkinson, Cat., p. 5, 1889[3]; Fleut., Comp. Rend., S. E. B., p. 168, 1890[4]; Cat., p. 44, 1892[5]; Horn., A. S. E. B., p. 537, 1892[6]; Gen. Ins., p. 293, 1915[7]; An and. & Horn., List., p. 10, 1909 (habits)[8]; Fowler, F. B. I., p. 347, 1912 (habits)[9]; Codina, Barcel, Cat., p. 427, 1917[10]; Dover & Rib., R. I. M., xxii, p. 723, 1921[11]; xxv, p. 354, 1923[12].

Indes Or. [a, 1, 2, 5, 13]
Asansol[4], Barway[6]
Berhampur[8, 9, 12]
Calcutta[8, 9, 12]
Cawnpore (B. M.)
Chakradharpur[11, 12]
Darjiling[7, 10, 12]
Gwalior[9, 12]
Hazara[12]
Kashmir (H.)
Kurseong [8, 9, 12]
Maldah[3, 8, 9, 12]
N. Kanara[9, 12]

C. tremula, Audoin & Brulle, Arch. Mus. Hist. Nat. Paris, p. 135, pl. 9, f. 3, 1839.[13]

Type : Mus. Berlin.

Pankabari[3, 8, 9, 12]
N. Punjab[7, 10, 12]
Purneah[8, 9, 12]
Pusa (on R. Bank:A.)
Siliguri [10, 11, 12]
Sara Ghat [8, 9, 12]
Sarai Sateh[12]
Tetara[4]

30. **bigemina** var. **brevis,** Horn, D. E. Z., p. 34, 1905[1]; Gen. Ins., p. 293, 1915[2]; Fowler, F. B. I., p. 348, 1912[3]; Dover & Rib., R. I. M., xxii, p. 723, 1921[4]; xxv, p. 354, 1923.[5]

Type : Coll. W. Horn.

Indes Or. [1, 3, 5]
Chapra[2]
Rawalpindi (bank of R. Sohan)[4, 5]

31. **bigemina** var. **iravadicca,** Gestro. Ann. Gen., p. 351, 1893[1]; Fleut., A. S. E. F., p. 483, 1893; Fowler, F. B. I., p. 348, 1912[2]; Horn, Gen. Ins., p. 293, 1915; Dover & Rib., R. I. M., xxv, p. 354, 1923.[3]

Type : Mus. Genova.

Mandalay[1, 2, 3]
Katha[1, 2, 3]
Teinzo [1, 2, 3]
Papun (B. M.)
Tharawaddy[2, 3]

32. **bigemina** var. **procera,** Horn., D. E. Z., p. 34, 1905[1]; Gen. Ins., p. 293, 1915; Fowler, F. B. I., p. 348, 1912[2]; Dover & Rib., R. I. M., xxv, p. 354, 1923.[3]

Type : Coll. W. Horn.

Indes Or. [1, 2, 3]

33. **biramosa** Fabricius, Spec. Ins., i. p. 286, 1781[1]; Syst. Ent., i, p. 175, 1792[2]; Syst. Eleuth, i, p. 240, 1801[3]; Herbst, in Fuessly, Arch. Insect., v, p. 149, pl. 27, f. 16, 1784; Kafer, x, p. 165, pl. 171, f. 1, 1806[4]; Olivier, Ent., ii, p. 26, pl. 2, f. 16*a*, *b*, pl. 3, f. 29, 1790[5]; Enc. Meth., p. 733, pl. 175, f. 4, 1790[6]; Dejean, Spec., Col., i, p. 133, 1825[7]; Schmidt-Gœb., Faun. Col. Birm., p. 6, 1846[8]; Castelnau, Hist. Nat. Ins., p. 18, 1851[9]; Mots., Etud. Ent., p. 30, 1855[10]; Tennent, Ceylon, etc., London, ed. 2, p. 274, 1859; Chaud., Cat. Coll., p. 25, 1865[11]; Redt., Reise Freg. Navarra Zool. ii, Col., p. 2, 1868; Gemm. & Har., Cat., p. 10, 1808[12]; Bates, A. M. N. H., (2), xvii, p. 70, 1886[13]; Atkinson, Cat., p. 5, 1889[14]; Fleut., Cat., p. 61, 1892; A. S. E. F., p. 488, 1893[15]; Gestro, Ann. Gen., p. 356, 1893[16]; Maindron, A. S. E. F., p. 15, 1905[17]; Fletcher, Spol. Zeyl., v, p. 62, 1907 (habits)[18];

Germany[1]
India[2–7, 9, 11, 12]
Arrakan[14, 19, 27]
? Assam [25, 26, 27]
Andaman Is. [14, 17, 19, 21, 22, 25, 26, 27]
Balighai[19, 27]
Barberyn Id. [19, 27]
Belikeri (A.)
Calcutta[10]
Chandipur[27]
Chittagong[14, 19, 22, 27]
Chounambar[17, 27]
Colombo[13, 14, 24, 26]
Coromandel[28]
Delft[24]
Hughli river[8, 22, 27]

Annand., Mem. As. Soc. Beng., i, p. 205, 1906[a]; Annand. & Horn., List, pp. 13, 14, 29, 1909 (habits)[19]; Gravely, R. I. M., vii, p. 207, 1912 (habits); xvi, p. 395, 1919 (distribution)[20]; Houlbert, Insecta Rennes, ii, p. 169, f. 1-13, 1912 (larva)[21]; Fowler, F. B. I., p. 431, 1912[22]; Horn., D. E. Z., p. 385, 1899 (habits)[23]; Spol. Zeyl., ii, pp. 33, 40, pl. 7, f. 7, 1904[24]; Gen. Ins., p. 308, 1915[25]; Codina, Barcel, Cat., p. 441, 1917[26]; Dover & Rib., R. I. M., xxii, pp. 724, 726, 1921 (habits); xxv, p. 354, 1923.[27]

Kanara[14, 19, 27]
Mandapam[27]
Mormugao[27]
Mahe[17, 27]
Madras (town)[10, 14, 19, 27]
Mt. Lavinia[24]
Nicobar Is. [21, 26, 27]
Pamban[19, 27], Puri[19, 27]
Ramnad [a]
Srayikad[19, 27]
Sunderbunds[14, 19, 22, 27]
Tenasserim[15, 27, 22, 27]
Malewoon[16]
Trichinopoly[19, 27]
Trincomali[21, 23, 24, 27]
Trivandrum[27]
Tuticorin[27], Waltair[27]
Weligama[24]

---

var. dilatata, Fleutiaux, A. S. E. F., p. 488, 1893.

C. tridentata, Thunberg, Dissert, Ent. Nov. Ins. Spec., i, p. 26, f. 40, 1781.[28]

Type : Probably in Sweden, but unknown to the authors.

Malacca[15, 22, 27]
Java[25, 21, 26, 27]
China[15, 22, 27]
Celebes[11]

34. **biramosa** var. **contracta,** Fleutiaux, A. S. E. F., p. 488, 1893[1]; Horn., Gen. Ins., p. 308, 1915; Fowler, F. B. I., p. 431, 1912[2]; Dover & Rib., R. I. M., xxv, p. 354, 1923.[3]

Type : Coll. E. Fleutiaux.

Tenasserim[1, 2, 3]
Rangoon[2, 3]
Victoria Point (H.)

35. **calligramma,** Schaum, B. E. Z., p. 69, pl. 1*b*. f. 1, 1861[1]; Journ. Ent., ii, p. 64, 1863; Chaud., Cat. Coll., p. 38, 1865; Gemm. & Har., Cat. p. 10, 1868; Atkinson, Cat., pp. 5, 23, 1889[2]; Fleut., Cat., p. 116, 1892; Annand. & Horn, List. p. 25, 1909[3]; Fowler, F. B. I., p. 413, 1912[4]; Codina, Barcel. Cat., p. 439, 1917[5]; Horn, Spol. Zeyl., ii, pp. 34, 42, pl. 7, f. 17, 1904[6]; Gen. Ins., p. 305, 1915[7]; Dover & Rib., R. I. M., xxv, p. 354, 1923.[8]

var. *conluens,* Fowler, F. B. I., p. 414, 1912.

Calochroa kraatzi, Horn., D. E. Z., p. 172, pl. 3, f. 4, 1894.[9]

Tranquebar[1, 2]
Ceylon[4–8]
Coimbatore[8]
N. Kanara[4, 8, 9]
Pondicherry[1, 2]
Shimoga[4, 5–8]
Teppukadu, 3,000 ft. (A.)
Tope[8]
Trivandrum[2, 3, 4, 8]
Podanur (H.)

Type : Mus. Herlin.

36. **cardoni,** Fleutiaux, Comp. Rend., S. E. B., p. 169, 1890[1]; Cat., p. 89, 1892; Horn, D. E. Z., p. 223, 1892[2]; A. S. E. B., p. 538, 1892[3]; Spol. Zeyl., ii, p. 34, 42, pl. 7, f. 21, 1904; Gen. Ins., p. 296, 1915[4]; Maindron, A. S. E. F., p. 17, 1905[5]; Annand. & Horn, List, p. 15, 1909[6]; Fowler, F. B. I., p. 372, f. 163, 1912[7]; Codina, Barcel. Cat., p. 429, 1917[8]; Dover & Rib., R. I. M., xxii, p. 724, 1921 (habits)[9]; xxv, p. 355, 1923.[10]

Type : Coll. E. Fleutiaux.

Asansol[1, 5, 7, 10]
Bombay (H.)
Barway[5]
Burma[7]
Chilaw[7, 10]
Chounambar[5, 10]
Chota Nagpur[2, 5, 7, 8]
Ghatkuri, 1,200 ft.[10]
Kanara[7, 10]
Manbhum Dist.[9]
Medha, 2,200 ft. [10]
Manharpur[10]
Mysore[7, 10]
Nowatoli[1, 5, 6, 7, 10]
Pondicerry[7, 10]
Pontianak[8]
Purulia[10]
Sikkim[7, 10]
Satara[4, 8, 9]
Taloshi, 2,000 ft.[10]

37. **catena,** Fabricius, Syst. Ent., p. 226, 1775[1]; Spec. Ins., i, p. 286, 1781[2]; Ent. Syst., i, p. 176, 1792[3]; Syst. Eleuth., i, p. 241, 1801[4]; Thunberg, ex parte, Dissert. Ent. Nov. Ins. sp., i, p. 26, 41-43, 1781; Olivier, ex parte, Ent., ii, p. 20, pl. 1, f. 12, 1790[5]; Enc. Meth., p. 731, pl. 174, f. 3, 18, 1790[6]; Herbst. Kafer, x, p. 170, pl. 171, f. 5, 1806[7]; Hübner, Natur., Stück, xxiv, pp. 50, 52, pl. 3, f. 20, 1789[8]; Dejean, Spec. Col., i, p. 117, 1825[9]; Mots., Etud. Ent. p. 30, 1855[10]; Tennent, Ceylon, etc., London, 2nd ed., p. 274, 1859; Schaum, Journ. Ent. ii, p. 67, 1863[11]; Chaudoir, Cat., Coll., p. 32, 1865[12]; Gemm. & Har., Cat., p. 11, 1868[13]; Redt., Reise Freg. Novar. Zool., ii, p. 1, 1868[14]; Bates, A. M. N. H., p. 69, 1886[15]; Atkinson, Cat., p. 5, 1889[16]; Gestro, Ann. Gen., p. 82, 1889[17]; p. 359, 1893[18]; Fleutiaux, Cat., p. 81, 1892; Horn, in litt. D. E. Z., p. 227, 1899; Spol. Zeyl., ii, pp. 34, 42, pl. 7, f. 20, 1904[19]; Gen. Ins., p. 307, 1915[20]; Maindron, A. S. E. F., p. 16, 1905[21]; Annand. & Horn., List, p. 28, 1909[22]; Fowler, F. B. I., p. 426, f. 180, 1912[23]; Annand. & Dover, R. I. M., xxii, pp. 335, 337, 1921 (habits)[24]; Dover & Rib., R. I. M., xxv, p. 355, 1923.[25]

Indes Or. [1–6, 9, 11, 13]
Bangalore (B. M.)
Barkuda Id. [25]
Berhampur[16]
Bombay[25]
Calcutta[22, 23]
Chatrapur[22, 23, 25]
Cheria Id.[25]
Chota-Nagpur[23, 25]
Colombo (Victoria Gardens)[15, 16, 19, 21, 23, 25]
Coromandel[21, 25]
Darjiling[20, 23, 25]
Genji[21], Karachi[25]
Kandy[19, 23, 25]
Madras[10], Mahe[25]
Malabar[10, 12]
Matale[19, 23, 25]
Moolay[18], Mysore[20, 23]
Pondicherry[21, 25]
Peradeniya[25]
Quilon[22, 25]

Ranchi[22, 25]
Teinzo[16, 17, 23, 25]
Tindivanam[21, 25]
Trichur[25], Tope[25]
Trincomali (B. M.)
Tranqebar[7]
Villapuram[21, 25]

Cicindela capensis, Sulser, Abgekürzte Geschichte Ins., i, p. 51, pl. 6, f. 11, 1776[26]; Füessly, Arch. Ins., v, p. 115, p. 27, f. 14, 1784; Wilhelm, Unterhalt. aus. der Natur. Ins., p. 193, pl. 24, f. 178, 1831.

C. cancellata, Horn, (nec. Dejean, 1825), D. E. Z., p. 324, 1891, p. 328, 1893; A. S. E. B., p. 538, 1892; Spol. Zeyl., ii, pp. 34, 42, 1904.

Type : Mus. Stockholm.

Cape of Good Hope[5, 6, 26]
Celebes (N.)
Java (N.)
Hongkong[14]

38. ***catena** var. **cancellata,** Dejean, Spec. Col., i, p. 116, 1825[1]; Gistl., Syst. Ins., i, p. 61, 1837[2]; Schmidt-Gœbel, Faun. Col. Birm., p. 5, pl. 1, f. 4, 1846; Schaum, Journ. Ent., ii, p. 67, 1863[3]; Chaudoir, Cat. Coll., p. 32, 1865[4]; Gemm. & Har., Cat., p. 11, 1865[5]; Atkinson, Cat., p. 5, 1889[6]; Fluet., Comp. Rend., S. E. B., p. 169, 1890[7]; Cat., p. 81, 1892[8]; A. S. E. B., p. 7, 1893; B. S. E. F., p. 369, 1917[9]; Horn, D. E. Z., p. 59, 1897[10]; A. S. E. B., p. 139, 1914[11]; Gen. Ins., p. 306, 1915[12]; Maindron, A. S. E. F., p. 383, 1899[13]; p. 15, 1905[14]; Lesne, Mission-Pavie, Indo-Chine, iii, p. 59, 1904[15]; Annand. & Horn. List, pp. 13, 28, 30, 1909 (habits)[16]; Fowler, F. B. I., p. 424, 1912[17]; Gravely, R. I. M., viii, p. 208, 1912[18]; Codina, Barcel. Cat., p. 440, 1917[19]; Dover & Rib., xxv, p. 355, 1923[29] (treated previously by most authors as a good species).

Cicindela striatifrons, Chaud., Horn, (nec Chaudoir), D. E. Z., p. 217, 1892; A. S. E. B., p. 538, 1892.

C. candei, Chevrolat, Rev. Zool., p. 96, 1845[21]; Schaum, Journ, Ent., ii, p. 67, 1863[22]; Chaud., Cat. Coll., p. 32,

Java[1–6, 8, 10, 12, 15, 17, 19, 20]
Arakan[16, 17, 20]
Asansol[7, 17]
Balighai[16, 20]
Bareilly (H.)
Berhampur[16, 17, 20]
Bhamo hills (B. M.)
Bombay[16, 17, 19]
Cheria Id.[20]
Chounambar[14, 20]
Garobadha[20]
Karachi1[7, 20]
Killinour[14]
Kurseong[17, 20]
Leo, 1,170 ft.[20]
Mahe[14, 17, 20, 28]
Malabar[14, 20]
Murshidabad[17]
N. Kanara[20]
N. Chin Hill[17, 20]
Nowatoli[1, 7, 20]
Pondicherry[20, 26, 28]

* The synonymy given here was suggested by Dr. Horn when our M. S. was submitted to him. In Junk and Schenkling's *Cudeoplerorum Catalogue,* Pars 86, *Corabida : Ciceridelinæ,* p. 188, 1926, however, he gives *cancellata* specific rank with *candci* as a variety.

1865[23]; Gemm. & Har., Cat., p. 11, 1868[24]; Atkinson, Cat., p. 5, 1889[25]; Fleut. Cat., p. 81, 1892; Maindron, A. S. E. F., p. 16, 1905; Schilder, Ent. Blatter, Berl., p. 178, 1911; as var. of *cancellata* : Fowler, F. B. I., p. 424, 1912[26]; Horn, Gen. Ins., p. 307, 1915[27]; Dover & Rib., R. I. M., xxv, p. 355, 1923[28].

Type : Mus. Paris.

Quilon[20], Puri[20]
Sikhim[12, 19], Siliguri[20]
Shimoga (A.)
Sukna, 500 ft.[16, 17, 20]
Sur Lake[20]
Travancore[17, 20]
Trichinopoly[17, 20, 26, 28]
Tuticorin[17, 20]
Tharawaddy[12, 20]

Hongkong[22, 23, 25, 26, 27, 28]
Hainan[11]
Macao[21, 24, 25, 26]
Pontianak[19]
Tathom[9]

39. **ceylonensis,** Horn, D. E. Z., p. 87, 1892[1]; pl. 3, f. 3, 1899; Spol. Zeyl., ii, pp. 33, 42, pl. 7, f. 18, 1904[2]; Gen. Ins., p. 305, 1915[3]; Fowler, F. B. I., p. 414, 1912[4]; Codina, Barcel. Cat., p. 439, 1917[5]; Dover & Rib., R. I. M., xxv, p. 355, 1923.[6]

Cicindela ceylonica, Horn, Fleutiaux, Cat., p. 117, 1892 (in error).[7]

Type : Coll. W. Horn.

Ceylon.[1, 5, 7]
Trincomali[2, 4]
Wellawaya[4, 6]

40. **ceylonensis** var. **diversa,** Horn, Spol., Zeyl., ii, pp. 33, 37, pl. 7, f. 19, 1904[1]; Gen. Ins., p. 305, 1915; Fowler, F. B. I., p. 415, f. 179, 1912[2]; Dover & Rib., R. I. M., xxv, p. 355, 1923.[3]

Type : Coll. W. Horn.

Trincomali[1]
Dambooila[2, 3]

Cicindela chinensis de Geer. This species has been recorded doubtfully from Simla by various authors, but is certainly not Indian, through two examples in the Indian Museum are labelled "Simla." (*Cf.* Dover & Rib., 1921, p. 725.)

41. **chlorida,** Chaudoir, Cat. Coll., pp. 15, 56, 1865 (*Jansenia*)[1]; Gemm. & Har., Cat., p. 33, 1868 (*Jan.*)[2]; Atkinson, Cat., p. 14, 1889 (*Jan.*)[3]; Fleutiaux, Cat., p. 38, 1892 (*Jan.*)[4]; Fowler, F. B. I., p. 334, f. 150, 1912[5]; Horn, Gen. Ins., p. 294, 1915[6]; Dover & Rib., R. I. M., xxv, p. 355, 1923.[7]

Malabar Coast[1, 2, 3, 4, 5, 7]
Padanur[5, 6, 7]
Palni hills (H.)
Tope[7]
Shembaganur (H.)

Type : Mus. Paris.

42. **chloris,** Hope, in Gray's Zool. Miscell., p. 21, 1831[1]; Gemm. & Har., Cat., p. 12, 1868[2]; Moore, in Anderson, Zool. Res. Exped. West., Yun., i, p. 907, 1878[3]; Dohrn, S. E. Z., p. 457, 1879[4]; Atkinson, Cat., p. 6, 1889[5]; Fleutiaux, Cat., p. 73, 1892; Gestro, Ann. Gen. p. 83, 1889[6]; p. 357, 1893[7]; Annand. & Horn, List, p. 16, 1909[8]; Fowler, F. B. I., p. 376, f. 165, 1912[9]; Horn, Gen. Ins., p. 296, pl. 21, f. 2, 1915[10]; Codina, Barcel. Cat., p. 429, 1917[11]; Dover & Rib., R. I. M., xxii, p. 725, 1921[12]; xxv, p. 355, 1923[12].

Cicindela, himaleyica, Redtenbacher, in Hügel, Kashmir, iv (2), p. 497, pl. 23, f. 1, 1848; Chaudoir, Cat. Coll., p. 30, 1865.

Type : Brit. Mus.

Nepal[1, 2, 4, 5, 10, 11]
Amangarh[8, 9, 13]
Bajaura (A.)
Benikhola[8, 9, 13]
Bindukhera[13]
Bijrani1[3]
Bhogpur[13]
Buxar[13]
Darrang Dist.[12]
Dejoo (A.)
E. Duars (A.)
Gilgit, 4,400 ft.[8, 9, 12]
Hathikund[9, 13]
Horai[13], Jalaban[9, 13]
Jumnagwar[9, 13]
Jumoo (A.)
Kaladhungi[13]
Kathgodam1,200 ft.[13]
Khasi hills (N.)
Kiari[13], Kumaon[9, 13]
Kumdhik[8, 9, 13]
Mangaldai Dist.[13]
Moolay[7], Patair[8, 9, 13]
Pegu[5]
Rawalpindi (bank of R. Sohan)[13]
Rudpur[13]
Rungpo, 1,400 ft.[9, 13]
Rungeet valley[12]
Sikkim[5], Sibsagar5
Siliguri1[3]
Soondrijal[8, 9, 13]
Srinagar, 6,000 ft.[13]
Teinzo[6], Tenasserim[5]

Yunnan[3, 5]

43. **chloropleura,** Chaudoir, Cat. Coll., pp. 19, 59, 1865[1]; Gemm. & Har., Cat., p. 12, 1868; Atkinson, Cat., p. 6, 1889[2]; Fleut., Cat., p. 66, 1892; Fowler, F. B. I., p. 327, 1912[3]; Horn, Gen. Ins., p. 287, 1915[4]; Dover & Rib., R. I. M., xxv, p. 355, 1923[5].

N. Hindostan[1, 2]
Darjiling[3]
Dehra dun (F.)
Kalimpong, 600-4,500 ft.[5]
Kashmir[4, 5]
Kumaon (F.)
Pashok, 2,500-3,000 ft.[5]

Thopeutica chlœ, Dokhturow, A. S. E. B., p. 156, 1887 (Indes Or.); Atkinson, Cat., p. 6, 1889 (*Cicindela*); Fleutiaux, Cat., p. 132, 1892 (Indes Or.).

Type : Mus. Paris.

Pashok Spur[5]
Rungeet Valley, 700-800 ft.[5]
Simla[3, 5], Sylhet[3, 5]
Sikkim[4]

44. **cognata,** Wiedemann, Zool. Mag., ii, (1), p. 66, 1823[1]; Schaum, Journ. Ent., ii, p. 67, 1863[2]; Chaudoir, Cat. Coll., p. 27, 1865; Gemm. & Har., Cat., p. 12, 1868[3]; Atkinson, Cat. p. 6, 1889[4]; Fleutiaux, Comp. Rend., S. E. B., p. 169, 1890[5]; Cat., p. 59, 1892; A. S. E. F., p. 487, 1893[6]; Gestro, Ann. Gen., p. 356, 1893[7]; Horn, D. E. Z., p. 96, 1895; A. S. E. B., p. 366, 1913[8]; Gen. Ins., p. 293, 1915[9]; Lesne, Mission Pavie, Indo-Chine, iii, p. 59, 1904[10]; Fowler, F. B. I., p. 364, f. 158, 1912[11]; Dover & Rib., R. I. M., xxii, p. 724, 1921[12]; xxv, p. 355, 1923[13].

Cicindela triramosa, Kollar, Ann. Hof. Mus. Wien., p. 330, 1836[14]; Atkinson, Notes, Zool. N. W. P., p. 117, 1832.

Type : Mus. Vienna.

Bengal[1–4, 6, 9]
Asansol[5, 11, 13]
Bhamo[7]
Berhampur[11, 13]
Cuttack (bank of R. Mahanadi)[12, 13]
Chota Nagpur[11, 13]
Darjiling[11, 13]
Goalbathan[12, 13]
Madras[3, 4, 9, 13, 14]
Palon[7], Pusa (A.)
Teinzo[7], Tikekee[7]

Cambodia[8, 10, 13]

45. ***copulata**, Schmidt-Gœbel, Faun. Col. Birm., p. 9, 1846[1]; Schaum, Journ. Ent., ii, p. 69, 1863; Gemm. & Har., Cat., p. 13, 1868[2]; Atkinson, Cat. p. 6, 1889[3]; Fleutiaux, Cat., p. 81, 1892[4]; B. S. E. F., p. 370, 1917 (in error); p. 252, 1919; Maindron, A. S. E. F., p. 383, f. 1, 1899[8]; Horn., D. E. Z., p. 130, 1899[9]; Gen. Ins., p. 308, 1915[10]; Zool. Jahrb. Syst., 47, p. 325, 1923; Fowler, F. B. I., p. 429, f. 193, 1912[11]; Codina, Barcel. Cat., p. 442, 1917[12]; Dover & Rib., R. I. M., xxii, p. 726, 1921 (habits; note on Calcutta record)[13]; xxv, p. 355, 1923[14].

Type : Mus. Prague.

Cossipore near Calcutta[1, 3, 10, 11], *Calcuta*[2, 3, 4]
Karachi[8, 10, 11, 12, 13, 14]
Manora[8]

46. **corbetti,** Horn, D. E. Z., p. 53, 1899[1]; Gen. Ins., p. 303, 1915; Fowler, F. B. I., p. 402, 1912[2]; Fleutiaux, B. S. E. F., p. 252, 1919; Dover & Rib., R. I. M., xxv, p. 355, 1923[3].

Type. : Brit. Mus.

Tharawaddy[1, 2, 3]

---

*As Dover and Ribeiro (1921) and Horn (1923) have shown, the records of this species from Calcutta are erroneous.

47. **corticata,** Putzeys, Comp. Rend., S. E. B., p. 69, 1875 (*Euryoda*)[1]; (separ.), (2), 14, p. 5, 1875[2]; Atkinson, Cat., p. 6, 1889 (*Cicindela*)[3]; Fleutiaux, Cat., p. 38, 1892 (*Jansenia*)[4]; Horn., Spol. Zeyl., ii, pp. 32, 36, 40, pl. 7, f. 6, 1904[5]; D. E. Z., p. 60, 1905; Gen. Ins., p. 289, 1915[6]; Annand. & Horn, List, p. 8, 1909[7]; Maindron, A. S. E. F., p. 10, 1905[8]; Fowler, F. B. I., p. 335, 1912[9]; Codina, Barcel. Cat., p. 423, 1917[10]; Dover, & Rib., R. I. M., xxv, p. 356, 1923[11]; Horn, Ent. Rec., xxxvii, p. 166, 1925[12].

Type : Mus. Bruxelles.

Ceylon[1, 3, 4, 5, 6, 8, 9, 10, 11]
Coonoor Ghat, 1, 250-3, 500 ft. (A.)
Horawapotana[12]
Kambakkam Hills (H.)
Mahe[8, 11]
Pondicherry[8, 11]
Ramnad[8, 9, 11]
S. Mysore[10]
Shembaganur[10]
Trichinopoly[9, 11]
Trivandrum[7, 8, 9, 11]
Tope[11]

48. ***corticata** var. **læticolor,** Horn, Spol. Zeyl., ii, p. 36, 1904[1]; Gen. Ins., p. 289, 1915[2]; Maindron, A. S. E. F., p. 10, 1905[3]; Fowler, F. B. I., p. 336, 1912[4]; Dover & Rib., R. I. M., xxv, p. 356, 1923[5].

lacunosa race *lacticolor,* Horn, Ent. Rec., xxxvii, p. 166, 1925.[6]

Type : Coll. W. Horn.

Ceylon[1–5]
Trincomali[6]

49. **crassipalpis,** Horn, R. I. M., ii, p. 410, 1908[1]; Gen. Ins., p. 294, pl. 17, f. 8, 1915[2]; Fowler, F. B. I., p. 332, f. 148, 1912[3]; Dover & Rib., R. I. M., xxv, p. 356, 1923[4].

Type: Coll. W. Horn.

Podanur, 1,000 ft.[1–4]

50. **cyanea,** Fabricius, Mant. Ins., i. p. 185, 1787[1]; Ent. Syst., i, p. 170, 1792[2]; Olivier, Enc. Meth., p. 727, 1790[3]; Herbst, Kafer, x, p. 199, 1806[4]; Schaum, B. E. Z., p. 73, 1861[5]; Gemm. & Har., Cat., p. 13, 1868[6]; Atkinson, Cat., p. 6, 1889[7]; Fleutiaux (ex parte) Cat., p. 118, 1892; B. S. E. F., p. 147, 1898[a]; Horn, D. E. Z., p. 87, 1892[8]; Gen. Ins., p. 303, 1915[9]; Annand. & Horn, List, p. 25, 1909[10]; Fowler, F. B. I., p. 406, 1912[11]; Codina, Barcel. Cat., p. 437, 1917[12]; Dover & Rib., R. I. M., xxv. p. 356, 1923[13].

Type : ? Mus. Stockholm.

Indes Or.[1–7]
Balasore (H.)
Palkot[10–13]
Sahibganj[8, 11, 13]
Chota Nagpur[11, a]

China[5]

* In giving the first definite record for this form Horn (1925) seems to regard it as a race of the close allied *C. lacunosa* Putz. In his catalogue, however, (*Coleopterorum Catalogus*, Pars. 80, p. 166, 1926) he adopts the synonymy given us.

51. **cyanea** var. **dejeani,** Hope, in Gray's Zool. Miscell., p. 21, 1831[1]; Dohrn., S. E. Z., p. 457, 1879[2]; Atkinson, Cat., p. 6, 1889[3]; Fleutiaux, Cat., p. 118, 1892[4]; Horn, D. E. Z., p. 223, 1892[5]; p. 193, 1898 (var.); Gen. Ins., p. 303, pl. 18, f. 1, 1915[6]; Annand. & Horn, List, p. 25, 1909[7]; Fowler, F. B. I., p. 406, 1912[8]; Dover & Rib., R. I. M., xxv., p. 356, 1923[9].	Nepal[1–5] Berhampur[7, 8, 9] Chota Nagpur[6, 8, 9] Murshidabad[8, 9, 10]
var. *obliquevittata,* Fleutiaux, B. S. E. F., p. 147, 1898[10].	
Type : Brit. Mus.	
52. **devisoni,** Gestro, Ann. Gen., p. 89, 1889[1]; p. 366, 1893[2]; Atkinson, Cat., p. 24, 1889[3]; Fleut., Cat. p. 131, 1892 (*Euryoda*); Horn, D. E. Z., p. 223, 1892; S. E. Z., p. 121, 1908; Gen. Ins., p. 289, 1915[4]; Fowler, F. B. I., p. 359, 1912[5]; Dover & Rib., R. I. M. , xxv, p. 346, 1923[6].	Thagata (village, 400-500 metres)[1–6]
Type : Mus. Genova.	
53. **decempunctata,** Dejean, Spec. Col., i, p. 145, 1825[1]; Schaum, Journ. Ent., II, p. 66, 1863[2]; Chaud., Cat. Coll., p. 24, 1865[3]; Gemm. & Har., Cat., p. 13, 1868[4]; Atkinson, Cat., p. 6, 1889[5]; Fleut., Cat., p. 47, 1892[6]; A. S. E. F., p. 486, 1893[7]; B. S. E. F., p. 49, 1917[8]; Lesne, Mission-Pavie, Indo-Chine, iii, p. 60, 1904[9]; Annand. & Horn, List, p. 10, 1909[10]; Horn, A. S. E. B., p. 365, 1913[11]; Gen. Ins., p. 292, 1915[12]; Fowler, F. B. I., p. 353, 1912[13]; Codina, Barcel. Cat., p. 425, 1917[14]; Dover & Rib., R. I. M., xxv, p. 356, 1923[15].	Indes Or. [1–7] Assam[12, 14, 15] Birbhum[13, 15] Dacca (B. M.) Dejoo (N.) Lakhimpur (A.) Murshidabad[10, 13, 15] Palon[13, 15, 16] Rajmahal[10, 13, 15] Rangoon[13, 15] Tharawaddy (H.) ————————
C. modica, Gestro, Ann. Gen., p. 354, 1893[16].	Cambodia[9, 11–15] Laos. Tonkin[11, 15] (Kieng-Kiet[8]; Haobin : A.) Penang (B.M.)
Type : Mus. Paris.	
54. **decempunctata,** var. **obscure-dilatata,** Horn, Arch. Nat. 79, A, ii, p. 28, 1913[1]; Gen. Ins., p. 292, 1915; Dover & Rib., R. I. M., xxii, p. 723, 1921[2]; xxv, p. 356, 1923[3].	Delhi[1, 3] Lahore (river side)[1, 2]
Type : Coll. W. Horn.	
55. **delavayi,** Fairmaire, Le Naturaliste, viii, p. 223, 1886[1]; A. S. E. F., p. 303, 1886[2]; Atkinson, Cat., p. 6, 1889[3]; Fleutiaux, Cat., p. 44, 1892[4]; A. S. E. F., p. 484,	Yunnan[1–3, 5, 7–10] Mungphu[10] ————————

1893[5]; Horn., D. E. Z., p. 94, 1895; A. S. E. B., p. 364, 1913[6]; Gen. Ins., p. 289, 1915[7]; Schilder, Ent., Blatter Berl., p. 177, 1911[8]; Codina, Barcel. Cat., p. 423, 1917[9]; Dover & Rib., R. I. M., xxv, p. 356, 1923[10].

Type : ? Mus. Paris.

China[4]
Laokay[6]

56. ***delavayi** var. **funebris,** Schmidt-Gœbel, Faun. Col. Birm., p. 8, 1846[1]; Chaudoir, Cat. Coll., p. 23, 1865; Gemm. & Har., Cat., p. 16, 1868[2]; Atkinson, Cat., p. 8, 1889[3]; Fleutiaux, Cat., p. 44, 1892; A. S. E. F., p. 483, 1893[4]; B. S. E. F., p. 49, 1917[5]; Gestro, Ann. Gen., p. 352, 1893[6]; Horn, A. S. E. B., p. 537, 1892[7]; Gen. Ins., p. 289, 1915[8]; Annand. & Horn, List, p. 9, 1909[9]; Schilder, Ent. Blatter, Berl., p. 177, 1911[10]; Fowler, F. B. I., p. 341, 1912[11]; Codina, Barcel. Cat., p. 423, 1917[12]; Dover & Rib., R. I. M., xxv, p. 356, 1923[13].

Cicindela dolens, Fleutiaux, B. S. E. F., p. 111, 1886[14]; Cat., p. 45, 1892; Atkinson, Cat., p. 7, 1889[15].

Type : Mus. Prague.

Burma[1–3]
Bombay[14, 15]
Cheba, Biapo, 900-1,100m.
Gopaldhara (A.)
Karen Hills[4, 6, 11, 13]
Khasi hills (A.)
Kurseong[7, 9, 11, 13]
Maymyo (A.)
Mungphu[9, 11, 13]
Nagpur (H.)
Naga hills[11]
N. Manipur 4,000 ft.[11, 12]
Pedong[4]
Punjab[8, 11, 12, 13]
Ruby mines[4]
Sikkim[9, 13]

Chapa[5]
Chusan Is.[10]

57. **†despectata,** Horn, D. E. Z., p. 86, 1892[1]; p. 339, 1893; Mem. 1st Congress, Int. Ent., p. 315, 1911[2]; Gen. Ins., p. 296, pl. 17, f. 9, 1915[3]; Fleut., ex parte, Cat., p. 88, 1892; Annand. & Horn, List, p. 14, 1909[4]; Dover & Rib., R. I. M., xxii, p. 724, 1921[5]; xxv, p. 356, 1923[6].

Type: Coll. W. Horn.

Philippine Is.[1, 2, 3, 5, 6]
Tavoy[6]

N. & S. Luzon (Albay. N.)
Perak[2–6]

---

* Horn, Col. Cat., Cic., p. 167, 1926, gives the name of this variety as *doleres* Fleut. but without any reason for the change. It may be because of the form of *C. campestris* described by Sturm in 1827 as *junebris,* but as Sturm's *junebris* is usually treated as a synonym of *C. campestris,* we do not see any reason for changing the name of the Indian variety.

† Dr. Horn doubted the occurrence of this species within our limits, omitting the Burmese locality in his recent catalogue, but the fact that it is not uncommon in the Malay Peninsula (judging from material in the F. M. S. Museum) proves that its occurrence in Tavoy, the fauna of which is almost Malayan, is not remarkable.

58. **discrepans,** Walker, A. M. N. H. (3) II, p. 202, 1858[1]; Tennant, Ceylon, etc., 2nd Ed., London, p. 274, 1859[2]; Gemm. & Har., Cat., p. 14, 1868[3]; Atkinson, Cat., p. 7, 1889[4]; Fleut., Cat., p. 118, 1892[5]; Horn, Spol. Zeyl., II, pp. 33, 41, pl. 7, f. 12, 1904[6]; Gen. Ins., p. 301, f. 130, (on p. 35), 1915[7]; Maindron, A. S. E. F., p. 18, pl. 1, f. 3, 1905[8]; Fowler, F. B. I., p. 389, f. 168, 1912[9]; Dover & Rib., R. I. M., xxv, p. 356, 1923[10].

Type : Brit. Mus.

Ceylon "Coast country or plains)"[1–5, 7]
Colombo[6, 8, 9, 10]
Kandy[8, 9, 10]
Nalanda[6, 8, 9, 10]
Perivipancheram (B. M.)

59. **discrepans** var. **lacrymans,** Schaum, Journ. Ent. Lond., p. 57, 1863[1]; Chaudoir, Cat. Coll., p. 39, 1865[2]; Gemm. & Har., Cat. p. 19, 1868[3]; Horn, Spol. Zeyl., II, pp. 33, 41, pl. 7, f. 13, 1904[4]; Mem. 1st Congress, Int. Ent., p. 315, 1911[5]; Gen. Ins., p. 301, 1915[6]; Ark. Zool. Stockholm, xiii, No. 11, p. 21, 1920[a]; Fowler, F. B. I., p. 390, 1912[7]; Dover & Rib., R. I. M., xxv, p. 356, 1923[8].

Cicindela stillicidium, Dohrn, in litt., S. E. Z., p. 173, 1889[9].

Type : Mus. Berlin.

Ceylon (South)[1–5, 7]
Kandy[4, 7, 8]
Ombolata[5, 6]

Nias[3, 5, 6]

60. **discreta,** Schaum, Journ. Ent. Lond., p. 59, 1863[1]; Chaudoir, Cat. Coll., p. 27, 1865[2]; Gemm. & Har., Cat., p. 14, 1868[3]; Fleutiaux, Cat., p. 60, 1892[4]; Horn, 9, 1925[a].

Cicindela indigna, Chaudoir, in litt., Cat. Coll., p. 27, 1865[13]; Fleutiaux, Cat., p. 60, 1892[14].

C. elaphroides, Dokhturow, Rev. Ent. Caën, p. 276, 1882[15]; Bær, A. S. E. F., p. 98, 1886[16]; Fleut., Cat., p. 39, 1892[17].

C. subfasciata, Horn, D. E. Z., p. 370, 1892[18] (var.).

Type : Mus. Berlin.

Mindanao[1]
Assam[9, 11, 12]
Tura (H.)

Batchian[13, 14]
Borneo[9, 11]
Buru[9, 11, 12, a]
Cabanatuan[10]
Cambodia[8, 9, 11, 12]
Celebeg[1–4, 9, 11, 16]
Ceram (N.)
Java[9, 11, 12,] Kapit[12]
10 mls. S. of Kuching[12], Lohago[11]
Malang[6], Manilla[16, 17]
Mindanao (Aguson river)[14]
Moluccas (N.)
New Guidea[9, 11, 12]
N. Queensland[9, 11, 12]
Nias Id.[9, 11, 12]
Nueve Ecija[10]
Philippines[9, 11, 12]

Sados[11], Sarawak[12]
Serdang[5]
Simalur Id.[9, 11, 12]
Sinabang[7]
Soekaranda[11]
Sukabumi, 2,000 ft.[5]
Sumatra[9, 11, 18]
Tœ-Kang-Besi Is.[9, 11, 12]
Tandjong-Morawa[5]

61. **discreta,** var. **reductula,** Horn. Gen. Ins., p. 298, 1915[1]; Codina, Barcel. Cat., p. 431, 1917[2]; Dover & Rib., R. I. M., xxv, p. 356, 1923[3].

discreta var. *reducta,* Horn, nec. Mots. 1859, D. E. Z., p. 370, 1892[4]; Fowler, F. B. I., p. 361, 1912[5]; Dover & Rib., R. I. M., xxii, p. 724, 1921[6].

Type : Coll. W. Horn.

Sumatra[1–5]
N. Chin Hills[3, 5]
Rangoon (B.M.)
Tura, 1,000-1,500 ft.[3, 5]

Amboyna[1, 2, 3]
Borneo[1, 2]
Sarawak[3, 5]

62. **dissimilis** var. **singalensis,** Horn, Deuts. Ent. Nat. Bibl., p. 53, 1911[1]; Gen. Ins., p. 293, 1915[2]; Dover & Rib., R. I. M., xxv, p. 356, 1923[3].

Type : Coll. W. Horn.

Ceylone[1, 2, 3]
Chapra (H.)

63. **distinguenda,** Dejean, Spec. Col., I, p. 92, 1825[1]; Gistl., Syst. Ins., I, p. 51, 1837[2]; Schaum, Journ. Ent., II, p. 68, 1868[3]; Chaud., Cat. Coll., p. 24, 1865[4]; Gemm. & Har., Cat., p. 14, 1868[5]; Atkinson, Cat., p. 7, 1889[6]; Fleutiaux, Cat., p. 48, 1892[7]; Horn, Spol. Zeyl., II, pp. 34, 43, pl. 7, f. 26, 1904[8]; Gen. Ins., p. 294, 1915[9]; Maindron, A. S. E. F., p. 13, 1905[10]; Fowler, F. B. I., p. 358, 1912[11]; Codina, Barcel, Cat., p. 427, 1917[12]; Annand. & Dover, R. I. M., xxii, p. 336, 1921[13]; Dover & Rib., R. I. M., xxii, p. 724, 1921[14]; xxv, p. 356, 1923[15].

Myriochila dohrni Motschulsky, Etud., Ent., vi, p. 109, 1857[16].

Type : Mus. Paris.

Indes Or.[1–7, 16]
Barkuda Id.[14, 15]
Ceylon (H.)
Kollinour[10]
Madras[9, 12, 15]
Mahe[10, 15]
Nilgiri hills (H.)
Pondicherry[11, 12, 15]
Puttalam[8]
Rambha (H.)
Tranquebar[4]

S. Borneo[12]

64. ***distinguenda** var. **lunatula,** Horn, D. E. Z., p. 35, 1905 (Beilage)[1]; p. 21, 1907; Fowler, F. B. I., p. 359,

Madras[1, 2, 3]
Elephant Pass[4]

* This form is not given definite varietal rank in Col. Cat., Cic., p. 173, 1926, but is regarded as an aberrational form.

1912[2]; Dover & Rib., R. I. M., xxv, p. 356, 1923[3]; Horn, Ent. Rec., xxxvii, p. 166, 1925[4].

Type : Call. W. Horn.

Pondicherry (H.)

65. **dives,** Gory, Mag. Zool., p. 97, 1833[1]; Schaum, Journ. Ent., II, p. 64, 1868; Chaudoir, Cat., Coll., p. 38, 1865; Gemm. & Har., Cat., p. 14, 1868[2]; Atkinson, Notes Zool. N. W. P., p. 117, 1882; Cat., p. 7, 1889[3]; Fleutiaux, Cat., p. 116, 1892; Fowler, F. B. I., p. 413, f. 177, 1912[4]; Horn, Gen. Ins., p. 305, 1915[5]; Codina, Barcel. Cat., p. 438, 1917[6]; Dover & Rib., R. I. M., xxv, p. 356, 1923[7].

Cicindela ambitiosa, Dupont, in litt., Dejean, in litt., Cat., ed. 3, p, 3, 1837.

Type : Mus. Paris.

Deccan[1, 2, 3]
Belgaum[4, 7]
Bengal[4, 7]
Dharwar (N.)
Mhow[4, 7]
Mysore[4, 6, 7]
N. Kanara[4, 7]
Sikkim[4, 7]
S. Bombay[5, 6]

66. **dormeri,** Horn, D. E. Z., p. 198, 1898[1]; Spol. Zeyl., II, pp. 33, 41, pl. 7, f. 10, 1904[2]; Gen. Ins., p. 288, 1915; Fowler, F. B. I., p. 325, f. 144, 1912[3]; Codina, Barcel. Cat., p. 422, 1917; Dover & Rib., R. I. M., xxv, p. 356, 1923[4]; Horn, Ent. Rec., xxxvii, p. 166, 1925[5].

Type : Coll. W. Horn.

Kandy[1–4]
Lady Horton[5]
Peradeniya (A.)

67. **dromicoides,** Chaudoir, B. S. N. M., xxv (1), p. 21, 1852[1]; Cat. Coll., p. 23, 1865; Schaum, Journ. Ent., II, p. 68, 1863; Gemm. & Har., Cat., p. 14, 1868; Atkinson, Notes Zool., N. W. P., p. 117, 1882; Cat. p. 7, 1889[2]; Fleutiaux, Comp. Rend., S. E. B., p. 168, 1890[3]; Cat., p. 38, 1892 (*Jansenia*)[4]; Horn, A. S. E. B., p. 537, 1892[5]; D. E. Z., p. 94, 1895; Gen. Ins., p. 290, 1915[6]; Maindron, B. E. E. F., p. 264, 1904 (*Euryoda*)[7]; Annand. & Horn, List, p. 9, 1909[8]; Fowler, F. B. I., p. 341, f. 151, 1912[9]; Dover & Rib., R. I. M., xxii, p. 723, 1921 (habits)[10]; xxv, p. 356, 1923[11].

Cicindela picta, Motschulsky, B. S. N. M., xxxvii (3), p. 173, 1864; Gemm. & Har., Cat., p. 24, 1868; Atkinson, Cat., p. 11, 1889; Fleutiaux, Cat., p. 38, 1892 (*Jansenia*).

Type : Mus. Paris.

Nepal[1, 2, 9, 11]
Barway[5], Bombay[6, 11]
Chota Nagpur[6, 9, 11]
Coimbatore (H.)
Darjiling[9, 11]
Ghumti 4,000 ft.[10, 11]
Gopaldhara (A.)
Khasi hills[9, 11]
Kousanie, 6,075 ft.[11]
Kumaon[8–11]
Kurseong[3, 7, 8, 9, 11]
Kumbir[3, 9, 11],
Nowatoli Lebong (A.)
Mungphu[8, 9, 11]
Mussoorie
Ranikhet (A.)
Sikkim[6], Simla[6, 8, 11]
Soom to Darjiling, 4,500-7,000 ft.[11]
Suriel, below, 4,800 ft.[11]

68. **duponti,** Dejean, Spec. Col., II, p. 419, 1826[1]; Gistl., Syst. Ins., I, p. 26, 1837[2]; Atkinson, Cat., p. 7, 1889[3]; Fleutiaux, A. S. E. F., p. 138, 1889[4]; Comp. Rend, S. E. B., p. 169, 1890[5]; Cat., p. 116, 1892; A. S. E. F., p. 490, 1893[6]; B. S. E. F., p. 49, 1917[7]; Maindron, A. S. E. F., p. 17, 1905[8]; Horn, D. E. Z., p. 59, 1905; A. S. E. B., p. 365, 1913[9]; Gen. Ins., p. 300, 1915[10]; Annand. & Horn, List, p. 19, 1909[11]; Fowler, F. B. I., p. 382, f. 167, 1912[12]; Fletcher, J. B. N. H. S., xxiii, p. 379, 1914[13]; Codina, Barcel, Cat., p. 433, 1917[14]; Dover & Rib., R. I. M., xxii, p. 725, 1921[15]; xxv, p. 357, 1923.[16]

var. *barmanice,* Gestro, Ann. Gen., p. 360, 1893[17]; Horn, D. E. Z., p. 96, 1895; Gen. Ins., p. 440, 1915; Annand. & Horn, List, p. 19, 1909; Fowler, F. B. I., p. 383, 1912[18]; Codina, Barcel, Cat., p. 433, 1917[19]; Dover & Rib., R. I. M., xxii, p. 725, 1921[20]; xxv, p. 357, 1923.[21]

Cicindela indica, Fleutiaux, A. S. E. F., p. 490, 1893[22]; Horn, Gen. Ins., p. 440, 1915.

Type : Mus. Pari.

Cochin China[1, 2, 3, 6, 9]
Arracan[11, 12, 16]
Bengal[10, 16, 15, 22]
Cheba, 400-1,100 m.
Cheerapunji[11, 12, 16]
Chota Nagpur[8, 12, 16]
Dacca, Kangra[12]
Karen Hills[16, 17, 18, 21]
Khasi hills
Kumbir Nowatoli[5]
Maflong[16], Mahe[12, 26]
Margherita (H.)
Maupun[11, 12, 16]
Merchiston State[11, 16]
Naga hills (H.)
Nongpriang[11, 12, 16]
N. Manipur[18, 21]
N. Chin hills[12, 16]
N. Shan States[16, 20, 21]
Palon[17]
Patkai hills[12, 16]
Pollibetta[13]
Shillong[11, 12, 16]
Shimoga[16]
S. Bombay[10]
Sylhet[16, 18, 21]
Talewadi[16]
Teinzo[17]
Tharawaddy[16, 18, 21]
Travancore[10]
Trichur[16]
Tura, 1,000-1,400 ft.[15, 16]
Upper Maymyo (N.)
Wynaad (B. M.)

Annam[6, 10, 14, 18, 19]
Nape[7]
Penang[12, 16]
Plinc Son[19]
Tonkin[15, 16]
Yunnan[10, 12, 16]

69. **erudita,** Wiedemann, Zool. Mag., II, p. 68, 1823[1]; Schaum, B. E. Z., p. 74, 1861; Chaudoir, Cat. Coll., p. 27, 1865[2]; Gemm. & Har., Cat., p. 14, 1868[3]; Atkinson, Cat. p. 7, 1889[4]; Bates, Entom. Suppl., p. 7, 1891[5]; Fleutiaux, Cat., p. 59, 1892[6]; A. S. E. F., p. 487, 1893[7]; Horn, A. S. E. B., p. 538, 1892[8]; Gen. Ins., p. 293, 1915[9]; Annand. & Horn, List, p. 12, 1909[10]; Fowler, F. B. I., p. 362, f. 156, 1912[11]; Dover & Rib., R. I. M., xxv, p. 357, 1923[12].

Cicindela amabitis, Dejean, Spec. Col., v, p. 228, 1831; Gistl., Syst. Ins., I, p. 40, 1937.

C. chloropus, Audouin & Brullé, Arch. Mus. Hist. Paris, p. 134, pl. 9, f. 2, 1839[13].

C. erythropus, Brullé, in litt., Schaum, in litt., Journ., Ent. Lond., p. 63, 1863.

Type : Unknown to the authors.

Bengal[1–4, 6, 7, 9, 10, 13]
Agra[10, 11, 12]
Allahabad[11, 12]
Barway[8]
Chapra (H.)
Chota Nagpur[11, 12]
Kashmir[1, 11, 12]
Katihar[12]
Kierpur[12]
Kulu[5, 11, 12]
Maldah[10, 11, 12]
Moghalsarai (B. M.)
Naoshera (H.)
Punjab[9], Pusa[11, 12]
S. India[12]
Tenasserim[12]

Java[4]

70. **fastidiosa,** Dejean, Spec. Col., I, p. 95, 1825[2]; Gistl. Syst. Ins., I, p. 52, 1837[2]; Schaum, Journ, Ent., II, p. 68, 1863[3]; Chand., Cat. Col., p. 25, 1865[4]; Gemm. & Har., Cat., p. 15, 1868[5]; Atkinson, Cat., pp. 8, 23, 1889[6]; Fleut., Comp. Rend. S. E. B., p. 168, 1890[7]; Cat., p. 49, 1892[8]; Horn, A. S. E. B., p. 537, 1892[9]; D. E. Z., p. 354, 1896; Spol. Zeyl., ii, p. 34, 44, pl. 7, f. 25, 1904[10]; Gen. Ins., p. 294, 1915[11]; Maindron & Fleut., A. S. E. F., p. 13, 1905[12]; Annand. & Horn, List, p. 11, 1909[13]; Fowler, F. B. I., p. 352, 1912[14]; Codina, Barcel, Cat., p. 427, 1917[15]; Annand. & Dover, R. I. M., xxii, pp. 335, 336, 1921 (habits)[16]; Dover & Rib., R. I. M., xxii, p. 723, 1921[17]; xxv, p. 357, 1923[18].

C. litigiosa, Dejean, Spec. Col., i, p. 97, 1825[19]; Gistl., Syst., Ins., p. 53, 1837[20].

C. despecta, Fleut., ex parte, Comp. Rend. S. E. B., p. 88, 1886; Cat., p. 48, 1892; Atkinson, Cat., p. 7, 1889.

Type : Mus. Paris.

Indes Or.[1–6, 8, 19, 20]
Andaman Is. (H.)
Anuradhapura[18]
Assam[6, 11, 13, 14, 15, 18]
Barway[9]
Barkuda Id.[16, 17, 18]
Chota Nagpur[18]
Dehra Dun[11, 15, 18]
Ganges[15], Genji[12]
Kashmir[14, 15, 18]
Kurseong[11, 14, 18]
Madras? Town (A.)
Manipur (H.)
Mhow[14, 18]
Moghalsarai (B.M.)
Mysore[14, 18]
Nowatoli[7, 14, 18]
Pegu[11, 13, 14, 15, 18]
Pondicherry[12]
Punjab (H.)
Puttalam[10], Rambha[18]
Ranchi (A.)
Rawalpindi[14, 18]
Shembaganur (H.)
Trincomali[14, 18]

71. **fastidiosa** var. **leucoloma,** Chaudoir, B. S. N. M., xxv, (1), p. 12, 1852[1]; Cat. Coll., p. 25, 1865; Gemm. & Har. Cat., p. 20, 1868[2]; Atkinson, Notes Zool., N. W. P., p. 117, 1882; Cat., p. 9, 1889[3]; Fleutiaux, Cat. p. 50, 1892; Maindon & Fleutiaux, A. S. E. F., p. 13, 1905; Fowler, F. B. I., p. 352, f. 154, 1912[4]; Horn, Gen. Ins., p. 294, 1915[5]; Zool. Med. Leiden, vii, p. 109, 1922; Dover & Rib., R. I. M., xxv, p. 359, 1923[6].

Type : Mus. Paris.

Simla[1, 3, 4, 6]
Agra (H.)
Ceylon[6]
East India[6]
India[2, 6]
N. India[5, 6]

72. **fleutiauxi,** Horn, Gen. Ins., p. 299, 1915[1]; Codina, Barcel. Cat., p. 432, 1917[2]; Dover & Rib., R. I. M., xxv, p. 357, 1923[3].

C. berthuri, Fleutiaux, B. S. E. F., p. 316, 1893[4]; Annand. & Horn, List, p. 18, 1909[5]; Fowler, F. B. I., p. 380, f. 166, 1912[6].

Type : Coll. E. Fleutiaux.

Bengal[1 2, 3, 6]
Bihar (H.)
Darjiling (H.)
Ghumti, 1,500-5,000 ft.[3]
Gopaldhara[3]
Kurseong[2, 3, 6]
Mungphu[3, 5, 6]
Rungbong valley[3]
Sikkim[1, 2, 3, 4]
Singla 1,500 ft.[3]

73. **foveolata,** Schaum, Journ, Ent., p. 59, 1863[1]; Chaud., Cat. Coll., p. 22, 1865[2]; Gemm. & Har., Cat., p. 16, 1868[3]; Atkinson, Cat., p. 24, 1889[4]; Fleut., Cat., p. 40, 1892[5]; A. S. E. F., p. 568, 1903[6]; B. S. E. F., p. 49, 1917[7]; Gestro, Ann. Gen., p. 87, 1889[8]; p. 351, 1893[9]; Horn, D. E. Z., p. 60, 1897, p. 66, 1902; Mem. 1st Congress, Int. Ent., p. 315, 1911[10]; A. S. E. B., p. 365, 1913[11]; Gen. Ins., p. 295, 1915[12]; Fowler, F. B. I., p. 345, 1912[13]; Schultze, Phil. Journ. Sci., D. xi, p. 7, 1916[14]; Codina, Barcel. Cat., p. 428, 1917[15]; Dover & Rib., R. I. M., xxii, p. 723, 1921[16]; xxv, p. 357, 1923[17].

Celebes[1–5, 8, 10, 12, 13, 15, 17]
Dacca[13, 17]
Dehra Dun[12, 15]
Karen hills (Cheba to Biapo, 400-1,100 m.)[9, 13, 17]
Kamara[10, 12, 13, 15]
Karkur Gate (H.)
Malabar[10, 12, 15, 17]
Maymyo[17]
Mysore[10, 12, 15]
Rangoon (F.)
Teinzo[4, 8, 9, 10, 15, 17]
Teppukadu, 3,000, ft.[17]
Tharawaddy[13, 17]
Tura, 12,00 ft.[16, 17]

Buru[15]
Chiem Hoa[9, 11]
Cochin China[11, 12, 15]

Manilla[14]
Philippine Is.[10, 12, 15, 17]
Sumatra[10, 12, 13, 15, 17]
Vientian[7]
Yen-Bay (Jobit)[6, 11]

Type : Mus. Berlin.

Cochin China[1, 2, 3, 6, 7, 8, 15, 18, 20, 22, 23]

74. **fuliginosa,** Dejean, Spec. Coll., II, p. 415, 1826[1]; Gistl., Syst. Ins., I, p. 25, 1837[2]; Schmidt-Gœbel, Faun. Col. Birm., p. 5, pl. 1, f. 6, 1846[3]; Schaum, Journ. Ent., II, p. 67, 1863[4]; Chaudoir, Cat. Coll., p. 32, 1865[5]; Gemm. & Har. Cat., p. 16, 1868[6]; Atkinson, Cat., pp. 8, 24, 1889[7]; Gestro, Ann. Gen., p. 82, 1889[8]; p. 359, 1893[9]; Fleutiaux, Cat., p. 79, 1892[10]; B. S. E. F., p. 369, 1917[11]; Horn, Ann. Gen., p. 681, 1894[12]; D. E. Z., p. 59, 1897[13]; p. 66, 1902,[14] p. 461, 1909[15]; Horn, Mem. 1st Congress, Int. Ent., p. 314, 1911[16]; A. S. E. B., p. 366, 1913[17]; Gen. Ins., p. 306, 1915[18]; Ridley, J. Str. Br. As. Soc., 38, p. 130, 1902[19]; Lesne, Mission Pavie, Indo-Chine, III, p. 61, 1904[20]; Annand. & Horn, List, p. 28, 1909[21]; Fowler, F. B. I., p. 422, f. 186, 1912[22]; Codina, Barcel. Cat., p. 440, 1917[23]; Dover & Rib., R. I. M., xxv, p. 3557, 1923[24].

Type : Mus. Paris.

Bhamo[8, 9, 22, 24]
Ceylon[15, 16, 18, 22, 23, 24]
Maymyo (A.)
Okkyi (H.)
Pegu[7, 22, 24]
Rangoon (A.)
Shewego (H.)
Teinzo[8, 9, 22, 24]

? Banka[18, 23, 24]
Borneo[5, 7, 8, 10, 15, 16, 18, 20–24]
Cambodia[15, 17, 18, 20, 22, 23, 24]
Chiemhoa[14, 17], Deli[12]
Java[13, 16, 18, 22, 23, 24]
Jelebu[21]
Laos (Tathom)[11, 15, 18, 23, 24]
Malacca[4, 5, 7, 8, 18, 23, 24]
Negri Sembilan[21]
Penang[19]
Shanghai[15, 16, 18, 20, 22, 24]
Singapore[15, 19]
Sumatra[15, 18, 20, 23, 24]
Perak (Prov. Wellesley)[19, 23]

75. **funerea,** McLeay, Annul. Jav., ed. i, p. 12, 1825[1]; ed. ii, p. 107, 1833[2]; Schaum, S. E. Z., p. 74, 1861[3]; Journ., Ent., II, pp. 62, 67, 1863[4]; Chaudoir, Cat. Coll., p. 30, 1865[5]; Gemm. & Har., Cat., p. 16, 1868[6]; Atkinson, Cat., p. 8, 1889[7]; Fleutiaux, Comp. Rend. A. S. B., p. 169, 1890[8]; Cat., p. 73, 1892[9]; B.

Java[1–6, 7, 9, 15–19, 21, 28, 29, 31, 32, 35]
Allanmyo[18, 21]
Amherst[24]
Barway[11, 18, 21]
Darjiling[21, 27]

S. E. F., p. 49, 1917[10]; Horn, A. S. E. B., p. 538, 1892[11]; D. E. Z., p. 59, 1897[12]; A. S. E. B., p. 365, 1913[13]; p. 140, 1914[14]; Gen. Ins., p. 297, 1915[15]; Lesne, Mission Pavie, Indo-Chine, III, p. 61, 1904[16]; Annand. & Horn, List., p. 16, 1909 (habits)[17]; Fowler, F. B. I., p. 377, 1912[18]; Codina, Barcel. Cat., p. 429, 1917[19]; Dover & Rib., R. I. M., xxii, p. 725, 1921[20]; xxv, p. 357, 1923[21].

Cicindela marginepunctata, Dejean, Spec. Col., II, p. 428, 1826[22]; Gistl., Syst. Ins., I, p. 55, 1837[23]; Schm.-Goeb., Faun. Col. Birm., p. 5, 1846[24].

C. assimilis, Hope. In Gray's Zool. Miscell., p. 21, 1831; Horn, A. S. E. B., p. 538, 1892 (var.)[25]; D. E. Z., p. 96, 1892, p. 193, 1898; A. S. E. F., p. 140, 1914[26]; Fleutiaux, Cat., p. 73, 1892; Annand. & Horn, List, p. 17, 1909 (var.)[27].

C. opigrapha, Dejean, Spec. Col., V, p. 347, 1831[28]; Gistl., Syst. Ins., I, p. 55, 1837[29]; Chaudoir, Cat. Coll., p. 30, 1865 (var.)[30]; Gemm. & Har., Cat., p. 16, 1868 (var.)[31]; Atkinson, Cat., p. 8, 1889 (var.); Fleutiaux, Cat., p. 73, 1892[32]; Horn, Ann. Gen. p. 681, 1894 (var.)[33]; D. E. Z., p. 59, 1897.

Cicindela marginesignata, Schonherr, in litt., Dejean, in litt., Cat., ed. 3, p. 4, 1837[34].

Type : Brit. Mus.

Doiphang valley[21]
Ghatkuri 1,200 ft.[21]
Gopaldhara[21]
Kalimpong 600-4,500 ft.[21]
Khasi hills (N.)
Kurseong[18, 21, 25]
Manharpur[21]
Mungphu[18, 21, 27]
Mysore[15, 18, 19, 21]
Nowatoli[8]
N. Shan States[21]
Pankabari[18, 21, 27]
Pegu[17, 18, 21, 27]
Poona[18, 21, 27]
Rungbong valley[21]
Rungeet valley 700-800ft.[21]
Rungpo, 900 ft.[21]
Sadon, 5,000 ft.[21]
Sibsagar[12, 21, 27]
Sikkim[15, 16, 17, 19, 21, 27]
Simla[15, 18, 19, 21]
Singla, 1,500 ft.[21]
Sukna[18, 21, 27]
Sundrijal[21, 27]
Sylhet[18, 21]
Taung-gno (A.)
Teinzo[18, 21]
Tenasserim[17, 18, 21, 27]
Tharawaddy[18, 21]
tura 1,000 ft.[20, 21]

Annam (Keng Trap)[13, 15, 19, 21]
Ban-Sop-Vi[10]
Ban-Sop-Tia[10]
Bao-Læ[16]
Belighe[33]
Cambodia[16]
Celebes[4, 5, 7, 9, 15, 16, 18, 19, 21, 30]
Cochin China[6, 7, 13, 15, 18, 19, 22, 23, 24]

Hainan[18, 21]
Keng-Kiet[10]
Lago Toba[33]
Lampan[21]
Laos (Vientian)[14]
Lombok[5]
Moeara Labœ[33]
Perak[15, 17, 18, 19, 21]
Patalung[21]
Sukubumi, 2,000 ft.[12]
Sumatra[15, 16, 19, 21]
Talé Sap.[20]

75*a*. **ganglbaueri,** Horn, D. E. Z., p. 95, 1892[1]; p. 339, 1893; Spol. Zeyl., II, pp. 33, 41, pl. 7, f. 11, 1904[2]; Gen. Ins., p. 288, pl. 17, f. 5, 1915[3]; Fleut., Cat., p. 85, 1892[4]; Fowler, F. B. I., p. 324, 1912[5]; Dover & Rib., R. I. M., xxv, p. 357, 1923[6].

Type : Coll. W. Horn.

Ceylon[1–6]

76. *****germanica** var. **kirilovi,** Fischer, B. S. N. M., p. 7, pl. 1, f. 3, 1844; Gebler, Bull. Acad. Sc. St. Petersbourg, III, p. 97, 1845 (ed. 3, VI, 1844), B. S. N. M., XXXIII (2), p. 1, 1860; Chaudoir, Cat. Coll., p. 22, 1865; Gemm. & Har., Cat., p. 19, 1868; Solsky, Reise Turk, Fedt., Cols. I, St. Petersbourg, p. 9, 1874; Marseul, L'Abeille, XIX, p. 20, 1880; Heyden, Cat. Coll., Siberia, (Deuts. Ent. Gesell.), p. 3, 1880; p. 5, 1893; Dokhturow, Hor. Soc. Ent. Ross., p. 255, 1885; Horn, Mon. Pal. Cic., p. 88, pl. 3, f. 3a, 1891 (var. of subsp. *obliquefasciata,* Adams); D. E. Z., p. 28, 1905 (sub-sp.); Fleutiaux, Cat., p. 43, 1892 (var. of subsp. *obliquefasciata,* Adams); Fowler, F. B. I., p. 354, 1912; Horn, Gen. Ins., pp. 292, 346 (for Pal. synonymy see Horn) 1915; Dover & Rib., R. I. M., xxv, p. 358, 1923.

Type : Mus. Dresden.

Songoria Rossica
(Drus Krælin)
(Gilgit 5,000 ft)

Altai Mts. Altalan,
Barkul
Dauria, Dsungaria
Kolyvan dist.
Kuldscha (B. M.)
Sarbad, Seistan
S. Russia.
Transcaspia
Tschemkent
W. Siberia

---

* The footnote to this form regarding the locality "Barkul" in Dover and Ribeiro's "List" (1923) the proofs, of which I did not see was inserted without my knowledge and should not be considered. The locality "Barkul" here given does not refer to the Indian locality at all, but to the Russian one! It has not been possible to indicate the source of the localities in dealing with this Palaratic form. The locality Gilgit in Kashmir was recorded by Fowler. C.D.

77. **grammophora,** Chaudoir, B. S. N. M., xxv (1), p. 7, 1852[1]; Cat. Coll., p. 27, 1865; Gemm. & Har., Cat., p. 17, 1868[2]; Atkinson, Notes Zool., N. W. P., p. 117, 1882; Cat. pp. 8, 24, 1889[3]; Fleutiaux, Comp. Rend. S. E. B., p. 169, 1890[4]; A. S. E. F., p. 487, Cat., p. 59, 1829[5]; Lefroy, Ind. Ins. Life, p. 261, pl. 16, f. 12, 1909; Annand. & Horn, List, p. 12, 1909 (habits)[6]; Fowler, F. B. I., p. 363, f. 157, 1912[7]; Horn, Gen. Ins., p. 293, 1915; Codina, Barcel. Cat., p. 426, 1917; Dover & Rib., R. I. M., xxii, p. 724,[8] 1921; xxv, p. 358, 1923.[9]

Type : Mus. Paris.

Hindostan[1, 2, 3, 5]
Asanso[13, 7, 9]
Calcutta & environs[7, 9]
Chota Nagpur[7, 9]
Damukdia[6, 7, 9]
Dehra Dun (A.)
Goalbathan[7, 9]
Kaladhungi[8, 9]
Katihar[9]
Mahanadi[3, 6, 9]
Maldah[6, 7, 9]
Patna dist.[7]
Punjab (H.), Pusa[7, 9]
Sara Ghat[6, 7, 9]
Saran[9], Siripur[9]

78. *granulata var. **stoliczkana,** Bates, P. Z. S., p. 713, 1878[1]; Sci. Rep. 2nd, Yarkand, Mission, p. 1, 1891[2]; Stoliczka, II, Yarkand Miss. Col., p. 1, pl. Geodeph., f. 8, 1890, Dokht., Hor. Soc. Ent. Ross,. p. 278, 1885; Annand. & Horn, List, p. 31, 1909[3]; (as var. of *burmeisteri*); Fowler, F. B. I., p. 439, 1912 (as var. of *burmeisteri*); Horn, Gen. Ins., p. 338, 1915[4]; Dover & Rib., R. I. M., xxv, p. 354, 1923[5] (as var. of *burmeisteri).*

Type : Indian Museum, Calcutta.

N. W. Kashmir or E. Turkestan (Type specimen)[3]
Jhelum valley[3, 5]
Kuen Lun (n. slopes)[1, 2]
Tarbagatai Mts.[2]

79. **guttata,** Wiedemann, Zool. Mag., II, p. 63, 1823[1]; Schaum, Journ. Ent., II, p. 65, 1863[2]; Chaud., Cat. Coll., p. 38, 1865[3]; Gemm. & Har., Cat., p. 17, 1868[4]; Atkinson, Cat., p. 8, 1889[5]; Fleut., Comp. Rend. S. E. B., p. 170, 1890[6]; Cat., p. 117, 1892[7]; Annand. & Horn, List, p. 25, 1909[8]; Fowler, F. B. I., p. 412, f. 176, 1912[9]; Horn, Gen. Ins., p. 305, 1915[10]; Codina, Barcel. Cat., p. 438, 1917[11]; Dover & Rib., R. I. M., xxv, p. 358, 1923.[12]

Type : Unknown to the authors.

Bengal[1–5, 7, 10, 11]
Chapra[12]
Chota Nagpur (H.)
Dehra Dun[10, 11, 12]
Ganjam dist.[10, 11, 12]
Lohardaga[8, 9, 12]
Kunbir[6], Nowatoli[11]
Palkot (N.)
Ranchi[8, 9, 12]
Shimoga[9, 12], Tetara[6]

---

*This species is not described by Fowler as he understood that the Kashmir locality was erroneous, but there seems to be no reason for doubting the validity of the specimen from Jhelum in the Indian Musem though Dr. Horn omits it from his recent catalogue. The true type locality of the species is probably the northern slopes of the Kuen-Lun, but the exact locality is not given.

80. **gyllenhali,** Dejean, Spec., Col., I, p. 143, 1825[1]; Schaum, Journ. Ent. Lond., p. 62, 1863; Chaud., Cat. Coll., p. 27, 1865[2]; Gemm. & Har., Cat., p. 17, 1868[3]; Atkinson, Cat., p. 9, 1889[4]; Fleutiaux, Cat., p. 60, 1892[5]; Annand. & Horn, List, p. 31, 1909[6]; Fowler, F. B. I, p. 438, f. 197, 1912[7]; Horn, Gen. Ins., p. 310, 1915[8]; Codina, Barcel. Cat., p. 443, 1917[9]; Dover & Rib., R. I. M., xxv, p. 358, 1923.[10]

Cicindela marginella, Gyllenhal, in litt.; Dejean, in litt., Cat., ed. 3, p. 6, 1837.[11]

var. immarginata, Horn, D. E. Z., p. 81, 1892; Fleut., Cat. p. 60, 1892[12]; Fowler, F. B. I., p. 439, 1912.

Type : Mus. Paris.

Indes Or.[1–5, 11, 12]
Bandra[7, 10]
Bassein[8, 9, 10]
Bombay (H.)
E. Indies[6, 10]
Karachi[7–10]
Manora (N.)
? N. Kanara (H.)

81. ****hæmorrhoidalis,** Wiedemann, Zool. Mag., II, p. 63, 1823[1]; Gistl., Syst. Ins., I, p. 27, 1837; Schaum, B. E. Z., p. 74, 1861; Journ. Ent., II, p. 65, 1863[2]; Chaudoir, Cat. Coll., p. 38, 1865[3]; Gemm. & Har., Cat., p. 17, 1868[4]; Atkinson, Cat., p. 9, 1889[5]; Fleutiaux, Comp. Rend. S. E. B., p. 170, 1890[6]; Cat., p. 117, 1892[7]; Maindron & Fleutiaux, A. S. E. F. p. 18, 1905; Horn, A. S. E. B., p. 538, 1892[8]; D. E. Z., pp. 234, 395, (footnote) 1899 (termitophily); Spol. Zeyl., II, pp. 33, 42, pl. 7, f. 16, 1904[9]; Gen. Ins., p. 303, 1915; Zool. Med. Leiden, VII, p. 107, 1922; Wasmann, Zool. Jahrb. Syst., xvii, p. 158, 1902 (termitophily); Annand. & Horn., List, p. 24, 1909[10]; Fowler, F. B. I., p. 402, 1912[11]; Fletcher, Journ. B. N. H. S., xxiii, p. 379, 1914 (habits)[12]; Codina, Barcel. Cat., p. 437, 1917[13]; Annandale & Dover, R. I. M., xxii, pp. 335, 337, 1921 (termitophily)[14]; Annandale, R. I. M., xxv, p. 241, 1993 (termitophily); Dover & Rib., R. I. M., xxv, p. 358, 1923.[15]

Cicindela quadrimaculata, Sturm, Cat. Ins. Samml., p. 55, pl. 1, f. 1, 1826; Audouin, Mag. Zool., pl. 18, 1832.

C. flavopunctata, Audouin, Mag. Zool., pl. 18, 1832; Tennant, Ceylon, etc., London, 1859, 2nd ed., p. 274.[16]

Type : Unknown to the authors.

Bengal[1–5, 7], Angul[15]
Anuradhapura[9]
Asansol[6]
Barkuda Id.[15]
Barkul[15] 0-1,000 ft.
Barway[8], Bombay[13]
Calcutta[5, 10, 11, 15]
Ceylon (plain of coast country)[11, 13, 16]
Ganges River[11]
Ganjam[12], Gantasila[15]
Jubbulpur[15]
Karkur Ghat[11, 15]
Lohardaga[10, 15]
Malabar[15], Mysore[11, 15]
Nilgiri hills[11, 15]
N. Kanara[11, 15]
Nowatoli (N.)
Pollibetta[12]
Puttalam (H.)
Ranchi[10, 16]
Rawalpindi[15]
Sibsagar[5, 10, 15]
Trivandrum[11, 15]
Wellawaya[15]

Canton[15]
China[10, 11]

* Dr. Horn (Col. Cat., Cic., p. 184, 1926) considers this a subspecies of *C. bicolor* Fabr., but it seems to us more convenient to keep it distinct.

82. **hæmorrhoidalis** var. **xanthospilota,** Fowler, F. B. I., p. 403, 1912[1]; Horn, Gen. Ins., p. 303, 1915; Dover & Rib., R. I. M., xxv, p. 358, 1923.[2] — Madras[1, 2]; Karkur Ghat[3]; Merchiston State[3]

hœmorrhoidalis var. Annand. & Horn., List, p. 25, 1909.[3]

Type : Brit. Mus.

83. **hamiltoniana,** Thomson, Arch. Ent., I, p. 323, 1857[1]; Horn, D. E. Z., p. 223, 1892[2]; p. 193, 1898, p. 60, 1905[3]; Gen. Ins., p. 301, f. 138, (on p. 36), pl. 17, f. 10, 1915[4]; Maindron, B. S. E. F., p. 264, 1904[5]; Annand. & Horn., List, p. 23, 1909[6]; Fowler, F. B. I., p. 390, f. 169, 1912 (habits)[7]; Fletcher, Journ. B. N. H. S., xxiii, p. 379, 1914 (habits)[8]; Dover & Rib., R. I. M., xxii, p. 725, 1921[9]; xxv, p. 358, 1923.[10] — Madras[1, 2]; Bandipur[6]; Coromondel Coast,[11, 12, 13]; Malabar (F.); Merchiston State[6, 10]; Mysore[4, 7, 10]; Nadgani[9, 10]; Nilgiri hills[3, 7, 10]; Pollibetta[8, 9, 10]; Wallardi[5]

Cicindela flavovittata, Chaudoir, Cat. Coll. pp. 39, 61, 1865[11]; Gemm. & Har., Cat., p. 15, 1868[12]; Atkinson, Cat., p. 8, 1889[13]; Fleutiaux, Cat., p. 118, 1892.

Type : Brit. Mus.

84. ***helferi,** Schaum, Journ. Ent., II, p. 69, 1863[1]; Gemm. & Har., Cat., p. 17, 1868[2]; Atkinson, Cat., p. 9, 1889[3]; Fleutiaux, ex parte, Cat., p. 38, 1892; (*Jansenia*)[4]; Horn, Gen. Ins., p. 309, 1915[5]; Zool. Jahrb. Syst., xlviii, p. 325, 1923; Dover & Rib., R. I. M., xxv, p. 356, 1923.[6] — ? Burma[1, 2, 3, 5, 6, 8]; ? Bengal[4]

Cicindela limbata, Schmidt-Gœbel, nec. Say, 1823; Faun. Col. Birm., p. 7, 1846[7]; Horn, D. E. Z., p. 131, 1899; Fowler, F. B. I., p. 431, 1912.[8]

Type : Mus. Prague.

85, **henryi,** Horn, Ent. Rec. xxxvii, p. 165, 1925.[1] — Horawapotana (Ceylon)[1]

Type : Coll. Brit. Mus.

86. #**histrio,** Tschitscherin, Hor. Soc. Ent. Ross., XXXVI p. 16, 1903[1]; Fowler, F. B. I., p. 425, f. 188, 1912[2]; Horn, Munch. Kol. Zeits., p. 344, 1903[3]; Gen. Ins., p. 307, 1915[4]; Dover & Rib., R. I. M., xxv, p. 358, 1923.[5] — Karachi[2–5], Manora[2, 5]; ———; Bampur[1, 2, 5]; Bamrud[1]; Duz-ab (to Podalix)[1]

Type : Mus. Petrograd.

---

* The chact locality of this species is not known (see Horn. 1923).

The type is from a locality in Persia, we cannot say exactly which.

Hurmuk[1]
Kuh-i-Murhak[1]
Nasirabad[1]
Neh-i-Bendan (environs of Bendan)[1]
Rarra[1]
Shur-ab[1]

87. **holosericea,** Fabricius, Syst. Eleuth., I, p. 243, 1801[1]; Herbst., Käfer, X, p. 210, 1806[2]; Motschulsky, Etud. Ent., p. 31, 1855[3]; Schaum, B. E. Z., p. 74, 1861; Chaudoir, Cat., Coll., p. 23, 1865[4]; Gemm. & Har., Cat., p. 28, 1868 (as var. of *triguttata*, Herbst)[5]; Atkinson, Cat., p. 13, 1889 (as var. of *viduata*, Fabr.)[6]; Fleutiaux, Comp. Rend. S. E. B., p. 168, 1890[7]; Horn., Ann. Gen., p. 680, 1894[8]; D. E. Z., p. 95, 1895, p. 56, 1897[9]; Gen. Ins., p. 290, 1915[10]; Lesne, Mission Pavie, Indo-Chine, iii, p. 60, 1904[11]; Fowler, F. B. I., p. 345, 1912[12]; Codina, Barcel. Cat., p. 424, 1917[13]; Dover & Rib., R. I. M., xxv, p. 358, 1923.[14]

Cicindela viduata, Fabricius, Syst. Eleuth., I, p. 242, 1801[15]; Herbst. Käfer, X, p. 209, 1806; Gistl., Syst. Ins., I, p. 64, 1837[16]; Schaum, S. E. Z., p. 51, 1847; Journ. Ent., II, p. 66, 1863[17]; Atkinson, Cat., p. 13, 1880; Fleutiaux, A. S. E. F., p. 485, 1893.

C. triguttata, Dejean ex parte, Spec. Col. I, p. 146, 1825.

C. parvula, Schönherr, in litt., Dejean, in litt., Cat., ed. 3, p. 6, 1837.

? *C. myrrha,* Thomson, Arch. Ent., I, p. 129, 1857.[18]

C. stygica, Chaudoir, in litt., Cat. Coll., p. 23, 1865; Fleutiaux, Cat. Coll., p. 44, 1892.

Type: Mus. Stockholm.

Java[1–6, 10, 13, 14, 16, 17]
Asansol[7, 12, 14]
Karen hills[12, 14]
Khasi hills[12, 14]
Nowatoli[7, 12, 14]

Battambang[11]
Borneo[6, 18]
Celebes[4, 6, 17]
Deli[8]
Dœsœn Tengah[8]
Malacca[4], Palabuan[9]
Palawan[13]
Philippine Is,[10, 11, 13, 14]
Preanger Reg.[9]
Sarawak[18]
Siam[4]
Singapore[4]
Sœrœlangœn[8]
Sukabumi 2,000 ft.[9]
Sumatra[6, 10, 13–17]
Yunnan (H.)

88. **humillima,** Gestro, Ann. Gen., p. 353, 1893[1]; Fleutiaux, A. S. E. F., p. 485, 1893; Annand. & Horn, List, p. 10, 1909[2]; Fowler, F. B. I., p. 355, f. 155, 1912 (habits)[3]; Horn, Gen. Ins., p. 292, 1915; Codina, Barcel. Cal., p. 425, 1917; Dover & Rib., R. I. M., xxv, p. 358, 1923[4].

Type : Mus. Genova.

R. Irawaddy[1]
Bhamo[1, 3, 4], Katha[1, 3, 4]
Paurigde (A.)
Rangoon[1–4]
Schwegoo[1, 3, 4]
Teinzo[1, 3, 4]
Tharawaddy[3, 4]
Tyyaing (N.)

89. **indica,** Fleutiaux, A. S. E. F., p. 484, 1893[1]; Fowler, F. B. I., p. 342, f. 152, 1912[2]; Horn, Gen., Ins., p. 289, 1915[3]; Dover & Rib., R. I. M., xxv, p. 358, 1923.[4]

India[1]
Bombay[2, 3, 4]

Cicindela dolens, Chaudoir, in litt., Fleutiaux, A. S. E. F., p. 484, 1893.

Type : Coll. E. Fleutiaux.

90. **intermedia,** Chaudoir, B. S. N. M., xxv (1), p. 6, 1852[1]; Schaum, Journ., Ent., II, p. 64, 1863[2]; Chaudoir, Cat. Coll., p. 38, 1865[3]; Gemm. & Har., Cat., p. 19, 1868[4]; Atkinson, Notes Zool. N. W. P., p. 117, 1882; Cat., p. 9, 1889[5]; Bates, Sci. Reports 2nd Yarkand Mission, p. 2, 1891[6]; Fleutiaux, Cat., p. 116, 1892[7]; Annand. & Horn, List, p. 18, 1909[8]; Fowler, F. B. I., p. 379, 1912[9]; Horn, Gen. Ins., p. 299, 1915[10]; Codina, Barcel. Cat., p. 432, 1917[11]; Dover & Rib., R. I. M., xxii, p. 725, 1921[12]; xxv, p. 358, 1923.[13]

Type : Mus. Paris.

Indes Or. [1–7]
Assam[9, 13]
Cambellpore (B. M.)
Jhelum valley[5, 6, 8, 9, 13]
Kathgodam 1,200 ft.[13]
Kousanie, 6,075 ft.[13]
Kumaon, 1,200-6,075 ft.[11, 12]
Matheran, 800 metres (H.)
N. Punjab to Simla[10, 13]

91. **interrupto-fasciata,** Schmidt-Gœbel, Faun. Col. Birm., p. 2, pl. 1, f. 1, 1846[1]; Schaum, Journ. Ent., II, p. 65, 1863[2]; Gemm. & Har., Cat., p. 19, 1868[3]; Atkinson, Cat., p. 9, 1889[4]; Gestro, Ann. Gen., p. 80, 1889[5]; p. 361, 1893[6]; Fleutiaux, Cat., p. 117, 1892[7]; A. S. E. F., p. 494, fig. 1893[8]; Horn, D. E. Z., p. 217; 1892, p. 98, 1895, p. 131, 1899, p. 66, 1902[9]; A. S. E. B., p. 366, 1913[10]; Gen. Ins., p. 302, f. 137 (on p. 36), 1915[11]; Schilder, Ent. Blatter Berl., p. 178, 1911[12]; Fowler, F. B. I., p. 399, 1912[13]; Codina, Barcel. Cat., p. 436, 1917[14]; Dover & Rib., R. I. M., xxv, p. 358, 1923.[15]

Cicindela flavolineata, Chaudoir, Cat. Coll., pp. 38, 60, 1865[16]; Gemm. & Har., Cat., p. 15, 1868[17]; Atkinson, Cat., p. 8, 1889[18]; Fleutiaux, Cat., p. 117, 1892[19]; A. S. E. F., p. 495, 1893[20]; Gestro, Ann. Gen., p. 361, 1893[21]; Fowler, F. B. I., p. 400, 1912 (var.).

C. ditissima, Bates, Ent. Mo. Mag., IX, p. 49, 1872[22]; Atkinson, Cat., p. 7, 1889[23]; Fleutiaux, Cat., p. 117 1892[24].

Type: Mus. Prague.

Burma[1–4, 7, 8, 11, 14, 20]
Maymyo[15]
Pegu[4, 15]
Ruby Mines[13]
Teinzo[5, 6, 13, 15]

Cambodia[10–16, 19]
Hong-Kong[11, 13 14, 15, 20–24]
Laos[11, 14, 16, 17, 18]
New Guinea[16]
S. China[12, 14]
Than-moi[9, 10]

92. **kaleea,** Bates, P. Z. S., p. 340, 1866[6]; T. E. S., p. 323, 1873[2]; Gemm. & Har., Cat., p. 19, 1868[3]; Atkinson, Cat., p. 9, 1889[4]; Fleutiaux, Cat., p. 46, 1892[5]; A. S. E. F., p. 485, 1893[6]; Horn, D. E. Z., p. 327, 1891, p. 95, 1982, p. 333, 1893, p. 66, 1900[7]; A. S. E. B., p. 365, 1913[8]; p. 140, 1914[9]; Gen. Ins., p. 298, 1915[10]; Codina, Barcel. Cat., p. 431, 1917[11]; Dover & Rib., R.I.M., xxv, p. 358, 1923.[12]

Type: Coll. R. Oberthür.

Formosa[1, 3, 4, 5, 10, 11]
Assam[10, 11]
Margherita (H.)
Sikkim[10, 11]

Australia[6], Bao-Lac[9]
Chiem-Hoa[7, 8]
China[2, 4, 5, 11]
Fuchow[2, 4], Japan[6]
Kaichou[10, 11]
Kiantschu[10, 11]
Shanghai[2, 4]
Tuyen-Quan[9]
Yunnan[10, 11]

93. **lacunosa,** Putzeys, Comp. Rend. S.E. B., p. 68, 1875[1]; Atkinson, Cat., p. 9, 1889; Fleutiaux, Cat., p. 38, 1892 (*Jansenia*); Horn, Spol. Zeyl., II, pp. 34, 44, pl. 7, f. 28, 1904[2]; Gen. Ins., p. 295, 1915; Fowler, F. B. I., p. 325, 1912[3]; Dover & Rib., R. I. M., xxv, p. 358, 1923.[4]

Type: Mus. Bruxelles.

Jesso[1]
Habarane[3, 4]
Kandy (H.)
? Madura (H.)
Puttalam[2, 3, 4]

94. **lefroyi,** Horn. R. I. M., II, p. 409 1908[1]; Gen. Ins., p. 305, pl. 18, f. 3, 1915[2]; Fowler, F. B. I., p. 418, f. 182, 1912[3]; Dover & Rib., R. I. M., xxv, p. 358, 1923.[4]

Type: Coll. H. E. Andrewes.

Chapra,[1–4]
Pusa[1, 3, 4]

95. **limosa,** Saunders, T. E. S., pp. 61 (habits), 64, pl. 7, f. 6, 1834[1]; Schmidt-Gœbel, Faun. Col. Birm., p. 7, 1846[2]; Schaum, B. E. Z., p. 74, 1861; Journ. Ent., II, pp. 62, 68, 1863; Chaudoir, Cat. Coll., p. 27, 1865; Gemm. & Har., Cat., p. 20, 1868; Atkinson, Cat., p. 9, 1889[3]; Fleutiaux, Cat., p. 60, 1982, B. S. E. F., p. 204, 1985; Horn, D. E. Z., p. 354, 1896, p. 461, 1909[4]; Spol. Zeyl., II, pp. 34, 42, pl. 7, f. 23, 1904[5]; Mem. 1st Congress, Int. Ent., p. 314, 1911[6]; Gen. Ins., p. 310, 1915[7]; Annand. & Horn. List, p. 31, 1909[8]; Fowler, F. B. I., p. 436, 1912;[9] Dover & Rib., R.I.M., xxii, p. 726, 1921 (habits)[10]; xxv, p. 359, 1923[11].

Cicindela cinctella, Chevrolat, Le Naturaliste, ii, p. 93, 1882[12]; Atkinson, Cat., p. 6, 1889[13]; Fleut., Cat., p. 60, 1892.[14]

Diamond Harbour[1, 2, 3]
R. Hughli[1]
Andaman Is.[4, 6, 7, 9, 11–14]
Burma[4, 6, 7]
Ceylon[4, 6, 7, 9, 11]
Chandipur[10, 11]
Durgapur (salt lakes)[10, 11]
Iranativu[5]
Mungphu[8, 9, 11]
Nicobar Is.[4, 6, 7, 8, 9, 11]
Pegu[8, 9, 11], Puttalam[5]
Sibpur[3], Sikkim[4, 6, 7]

Chusan is, [4, 6, 7, 9, 11]

C. *nivicinctoides*, Horn, D. E. Z., p. 81, 1892, Fleutiaux, Cat., p. 60, 1892.[15]

Type: Unknown to the authors.

Indo-China[15]
Malaya[15]
Shanghai[6]

96. **lunulata,** Fabricius, Spec. Col., I, p. 284, 1781[1]; Ent. Syst., I, p. 172, 1792; Syst. Eleuth., I; p. 236, 1801; Olivier, Ent., II, No. 33, p. 17, pl. 2, f. 24, 1790[2]; Enc. Meth., p. 730, pl. 174, f. 14, 1790; Herbst, Käfer, X, p. 187, pl. 172, f. 10, 1806; Horn, Monog. Pal. Cic., pp. 157, 159, 161, pl. 6, f. *7b*, 24, 1891[3]; Munch, Kol. Zeit., I, pp. 340, 346, 1903; Gen. Ins., pp. 297, 356 (palæarctic syn.), 1915[4]; Fleutiaux, Cat., p. 74, 1892; Bedel, Cat. Rais, Col., N. Afric., I, pp. 4, 8, 1895; Tschitscherin, Hor. Soc. Ent. Ross., xxvi, pp. 12-16, 1903; Grandi. Rivesta Col. Ital., pp. 85-105, pl. 1, f. 1, 12, 20, pp. 220-227, 1906; Semenow, Rev. Russ. Ent., pp. 307, 309, 1908; Barthe, Miscell. Ent., XVI, (1-2), Beilage, Cic., pp. 4-7, 1908; Annand. & Horn., List, p. 15, 1909; Dupins, A. S. E. B., p. 193, 1900, Ferrer, Buttl. Ins. Cat. Hist. Nat., (2) viii, p. 83, pl. 2, f. 6, 14, pl. 4, f. 11, 1911; Schilder, Ent. Blatter Berl., p. 178, 1911; Fowler, F. B. I., p. 375, 1912[5]; Dover & Rib., R. I. M., xxv, p. 359, 1923[6].

Type : Brit. Mus. (Bank's Coll.)

"Cape of Good Hope"[1, 2]
Nushki dist.[4, 5, 6]

Bagdad[4]
Kashgar
Khirgiz Steppes[3]
Mongolia[3]
Pekin[4]
Siberia to Atlai[3]

Widely distributed in Southern Europe, N. Africa.

97. ***lunulata** var. **nemoralis,** Olivier, Ent., II, No. 33, p. 13, pl. 3, f. 36, 1790[1]; Enc. Méth., p. 729, pl. 174, f. 10, 1790[2]; Gené, Mem. Accad. Torino, p. 166, 1836; Girard, B. S. E. F., p. 186 1873 (Bionomics); Horn, Monog. Pal. Cic., pp. 159-163, pl. 6, f. *7a*, *e-h*, 1891; Gen. Ins., pp. 297, 356, (palæarctic syn.) 1915[3]; Grandi, Riv. Col. Ital., p. 93, pl. 1, f. 2, 10, pp. 220-227, 1906; Annand. & Horn, List, p. 15, 1909[4]; Schilder, Ent. Blatter, Berl., p. 203, 1911; Fowler, F. B. I., p. 375, 1912[5]; Codina, Barcel. Cat., p. 483, 1917; Dover & Rib., R. I. M., xxv, p. 359, 1923[6].

Type : Unknown to the authors.

Provence[1, 2]
Perso-Baluch Frontier[4, 6]

Also from S. Europe and Western Asia.
Pamir[3, 6]
Khirgiz steps to Kolywan & Altai[3]
Kulu Nor.[3]
Mongolia[6,]
N. China[30]
N. Tibet, 14,000 ft.[3]
Seistan[3, 5, 6]
Turkestan to Pamir[3]

---

* *Cicindela lunulata* and its var. *nemoralis* are strictly Palæarctic forms, included here as the region covered by these catalogues is a political rather than a zoogeographical one.

98. **maindroni,** Horn, Ent. Nachr., p. 98, 1897[1]; Gen. Ins., p. 309, fig. (on p. 34), 1915[2]; Maindron, A. S. E. F., p. 380, f., p. 381, 1899[3]; Fowler, p. 432, 1912[4]; Dover & Rib., R. I. M., xxv, p. 359, 1923.[5]

Karachi[1, 2, 4, 5]
Baluchistan[4, 5]
Kiameri Id.[3, 4, 5]

Type : Coll. W. Horn.

99. **malabarica,** Fleutiaux & Maindron, B. S. E. F., p. 72, 1903[1]; Maindron, A. S. E. F., p. 14, pl. 1, f. 5, 1905[2]; Fowler, F. B. I., p. 438, f. 196, 1912[3]; Horn, Gen. Ins., p. 311, 1915[4]; Codina, Barcel. Cat., p. 443, 1917[5]; Dover & Rib., R. I. M., xxv, p. 359, 1923.[6]

Mahe[1–6]
Kalyan (H.)

Type : Coll. E. Fleutiaux.

100. **mariæ,** Gestro, Ann. Gen., p. 361, 1893[1]; Fleutiaux, A. S. E. F., p. 493, 1893[2]; Horn, D. E. Z., p. 97, 1895; Gen. Ins., p. 303, f. 129, (on p. 35), 1915[3]; Annand. & Horn, List., p. 24, 1909[4]; Fowler, F. B. I., p. 401, 1912[5]; Dover & Rib., R. I. M., xxv, p. 359, 1923.[6]

Karen hills (Cheba to Biapo)[1, 2, 5, 6]
N.Chin. Hills[5, 6]
Pegu[3–6], Rangoon[5, 6]
Taung-ngu[1, 5, 6]
Tharawaddy[5, 6]
Tenasserim[3, 5, 6]
Victoria Pt. (B. M.)

Type : Mus. Genova.

101. **melancholica,** Fabricius, Ent. Syst. Suppl., p. 63, 1798[1]; Syst. Eleuth., I, p. 236, 1801[2]; Herbst., Käfer, X, p. 200, 1806[3]; Gistl., Syst. Ins., I, p. 51, 1837[4]; Schaum, S. E. Z., p. 50, 1847; Kuster, Käfer Europa, XIX, No. 2, 1849; Chaudoir, Cat. Coll., p. 25, 1865[5]; Gemm. & Har., Cat., p. 22, 1868[a]; Solsky, Reise, Turk. Fedt., Col., I, Petersburg, p. 6, 1874; Dokhturow, Hor. Soc. Ent. Ross., p. 261, 1885; Atkinson, Cat., p. 10, 1889; Horn., Monog. Pal. Cic., p. 130, pl. 4, f. 8*a*, pl. 5, f. 40, 1891[6]; D. E. Z., p. 63, 1896, p. 62, 1905; Gen. Ins., p. 294, 264, (African syn.), 354 (palæarctic distrib.), 1915[7]; Fleutiaux, Cat., p. 49, 1892; Heyden, Cat. Col. Sib. (Deuts. Ent. Gissel), p. 6, 1893; Péringuey, Trans. S. African Philos. Soc., p. 47, 1893; Bedel, Cat. Rais. Col. N. Afric., I, p. 13, 1895; Tschitscherin, Hor. Soc. Ent. Ross., p. 3, 1903[8]; Maindron, A. S. E. F., p. 14, 1905; Annand. & Horn, List, p. 11, 1909[9]; Schilder, Ent. Blatter Berl., p. 177, 1911[10]; Fowler, F. B. I., p. 356, 1912[11]; Codina, Barcel. Cat., pp. 403, 481, 1917[12]; Dover & Rib., R. I. M., xxii, p. 723, 1921[13]; xxv, p. 359, 1923.[14]

Guinea Dom[1–4]—
Sierra Leone[a]
Ambala [13, 14]
Bandra[11, 14]
Benga[17, 11, 14]
Chapra (A.)
Kaladhungi[13, 14]
Kalka, 2,400 ft.[13, 14]
Karachi1[1, 12, 14]
Nagpur[11, 14]
Punjab to Bengal[7, 12]
Pusa[14]

Arabia[6, 7]
Ausellabad (environs)[8]
Bushire[9, 13, 14]
Hongkong[6], Java[5, 6]
Kenduk[8]
Mt. Kuhi-Hodghe[8]
Nasirabad[8]

Type : ? Mus. Stockholm.

Also found in S. Europe, the whole Continent of Africa, and Western Asia.

102. **minuta,** Olivier, Ent., II, No. 33, p. 31, pl. 2, f. 13*a*, *b*, 1790[1]; Enc. Meth., V, p. 735, pl. 175, f. 13, n. 2, 1790[2]; Fabricius, Ent. Syst., I, p. 178, 1792;[3] Herbst, Käfer, X, p. 196; p. 173, f. 8, 8*a*, 1806[4]; Motschulsky, Etud. Ent., p. 31, 1855[5]; Chaudoir, Cat. Coll., p. 27, 1865[6]; Gemm. & Har., Cat., p. 22, 1868; Atkinson, Cat., p. 10, 1889[7]; Gestro, Ann. Gen., p. 85, 1889,[8] p. 356, 1893[9]; Fleutiaux, A. S. E. F., p. 138, 1880 (mimicry); Comp. Rend. S. E. B., p. 169, 1890[10]; Cat., p. 59, 1892; B. S. E. F., p. 49, 1917[12]; Horn, Ann. Gen., p. 680, 1894[13]; D. E. Z., p. 58, 1897[14]; p. 66, 1902[15]; A. S. E. B., p. 365, 1913[16]; Gen. Ins., p. 292, 1915[a]; Lesne, Mission Pavie, Indo-Chine, iii, p. 61, 1904[17]; Maindron, A. S. E. F., p. 12, 1905[18]; Annand. & Horn, List, p. 12, 1909[19]; Gravely, R. I. M., vii, p. 209, 1912[20]; Fowler, F. B. I., p. 366, 1912[21]; Codina, Barcel. Cat., p. 426, 1917[22]; Dover & Rib., R. I. M., xxii, p. 724, 1921[23]; XXV, p. 359, 1923.[24]

Cicindela baltimorensis, Herbst, Käfer, X, p. 181, pl. 172, f. 3, 1806[25] (the locality given is surely an error).

C. *tremebunda,* MacLeay, Ann. Jav., ed., p. 12, 1825, ed. 2, p. 107, 1833[26]; Schaum, S. E. Z., p. 74, 1861.

C. *pumila,* Dejean, Spec. Col., II, p. 425, 1826[27]; Schmidt-Gœbel, Faun., Col. Birm. p. 5, 1846.[28]

C. *prinsepii,* Saunders, T. E. S., Lond., p. 62, (habits), p. 65, pl. 7, f. 7, 1834[29]; Schaum, Journ. Ent., II, p. 62, 1863.

C. *acuminata,* Kollar, Ann. Hofmus, Wien., p. 331, 1836, Gistl., Syst. Ins., I, p. 53, 1837; Maindron, A. S. E. F., p. 13, 1905.

C. *discreta,* Koningsberger, Med., Lands, Plant., No. 44, p. 113, pl. 6, f. 22, 1901.[30]

Type : Unknown to the authors.

Indes Or.[1–5]
Ackra[7, 20]
Asansol[10]
Behrampur[21, 23]
Calcutta[19, 21, 23]
Catepaleom (A.)
Chakradharpur[23]
Chapra[21, 23]
Cheriadongra (A.)
Chota Nagpur[21, 23]
Cuttack (Bank of R. Mahanadi)[23]
Dacca[21, 23]
Darjiling[23], Delhi
Dhanli, Garobadha
Mahe[17, 18], Malda[7]
Margherita (H.)
Myawadi
Myeen Kyan[8, 9]
Myingyan
Nerbudda-Bank
N. Kanara (A.)
Pashok, 1,000 ft.
Patan, 2,050 ft.
Pondicherry[18, 21]
Pegu, Pusa
Sikhim[17]
Siliguri (A.)
Teinzo[8, 9]
Tharawaddy[21]
Trichur
Tura, 1,000 ft.

? *Baltimore*[25]
Batavia
Batu Pahat (B.M.)
Bedagi, Borneo[17]
Cape Engano[13]
Chein Hoa[15]

China[11]
Cochin China[17, 18]
Deli[13], Hue[16]
Java[6, 7, 18, 26, 27, 28, 30]
Kaju Tanam[13]
Keng-Kiet[12]
Kuching (B.M.)
Kukub (B. M.)
Martaban[28]
Misauw[18]
Mœara Labœ[13]
Mt. Tenger, 4,000 ft.[14]
Mt. Tjikorai, 4,000 ft.[14]
Nias[17], Perak[17, 19]
Philippine Is.[18]
Pontianak
Prom-Penh[16]
Sarawak (B.M.)
Sukabumi, 2,000 ft.[14, 18]
Sumbawa[17]
Taynuh[18]
Yunnan (A.)

103 **motschulskyana,** Horn, Gen. Ins., p. 289, 1915[1]; Dover & c. Rib., R. I M., xxv, p. 359, 1923[2].

Cicindela mostchulskyi, Horn, D. E. Z., p. 198, 1893[3]; Fowler, F. B. I., p. 342, 1912.[4]

Type : Col. W. Horn.

N. Kanara[1, 2, 4]
Indes Or.[3]

104 ***mouhoti** var. **anometallescens,** Horn, Ent. Nachr., p. 140, 1893[1]; A. S. E. B., p. 140, 1914[2]; Gen. Ins., p. 302, 1915; Fleutiaux, A. S. E. F., p. 492, 1893; Fowler, F. B. I., p. 398, 1912[3]; Codina, Barcel. Cat., p. 436, 1917[4]; Dover & Rib., R. I. M., xxv, p. 359, 1923.[5]

Type : Coll. W. Horn.

Momeit[1, 3, 4, 5]
Gokteik (A.)
Maymyo[3, 5]
Ruby Mines[3, 5]
Teinzo (B. M.)

Frontier of Tonkin and Laos[2]
Yunnan[5]

105. **mouhoti** var. **bramani,** Dokhturow, Rev. Ent. Caën, p. 262, 1882[1]; Atkinson, Cat., p. 5, 1889[2]; Fleut.,

Pegu[6]

China[6, 7]

* The typical *mouhoti* is a Siamese species.

Cat., p. 116, 1892; Fowler, F. B. I., p. 396, 1912[3]; Horn, A. S. E. B., p. 366, 1913[4]; Gen. Ins., p. 302, 1915[5]; Dover & Rib., R. I. M., xxv, p. 359, 1923.[6]

Cicindela interruptofasciata, Schmidt-Gœbel, Fleutiaux (nec. Gœb.), A. S. E. F., p. 494, f. 1893[7]; Horn, D. E. Z., p. 89, 1904.

Type : Unknown to the authors.

Cambodia[3]
Cochin China [2, 3, 5, 7]
Lakhon[6, 7]
Laos[5, 6]
Taynish[4]
Tonkin[4, 5 6]

106. **mouhoti** var. **cariana,** Gestro, Ann. Gen., p. 363, 1893[1]; Fleutiaux, A. S. E. F., p. 493, f., 1893[2]; Horn, D. E. Z., p. 89, 1904; Gen. Ins., p. 302, 1915; Fowler, F. B. I., p. 398, f. 170, 1912 (habits)[3]; Dover & Rib., R. I. M., xxii, p. 725, 1921[4]; p. 355 (as species in error), p. 359, 1923.[5]

Type : Mus. Genova.

Karen hills[1, 2, 3, 5] (Cheba, 900-1,100 metres)[1]
Ghecu, (1,300-1,400 metres)[1]
N. Shan States[4, 5]
Tenasserim[3, 5]

107. **mouhoti** var. **gœbeli,** Horn, D. E. Z., p. 92, 1895[1]; p. 89, 1904, Gen. Ins., p. 302, 1915; Fowler, F. B. I., p. 397, 1912 (habits)[2]; Codina, Barcel. Cat., p. 436, 1917[3]; Dover & Rib., R. I. M., xxv, p. 359, 1923.[4]

Cicindela tritoma, Gestro, nec. Schm.-Gœb., Ann. Gen., p. 81, 1889[5]; p. 361, 1893; Fleutiaux, B. S. E. F., p. 246, 1895; Horn, D. E. Z., p. 97, 1895.

C. anometallescens, ex parte, Fleutiaux, A. S. E. F., p. 492, f., 1893[6]; Horn, D. E. Z., p. 97, 1895.

Type : Coll. W. Horn.

Teinzo[1–6]
Bhamo[2–5]
Momeit[6]
Rangoon[2, 4, 5]

108. **multiguttata,** Dejean, Spec. Col., I, p. 109, 1825[1]; Gistl., Syst. Ins., I, p. 58, 1937[2]; Schaum, Journ. Ent., II, p. 67, 1863[3]; Chaud., Cat. Coll., p. 31, 1865; Gemm. & Har., Cat., p. 22, 1868[4]; Atkinson, Cat., p. 10, 1889[5]; Fleutiaux, Comp. Rend. S. E. B., p. 169, 1890[6]; Cat., p. 76, 1892; A. S. E. F., p. 488, 1893; Annand. & Horn, List, p. 27, 1909[7]; Fowler, F. B. I., p. 417, 1912[8]; Horn, Gen. Ins., p. 305, 1915; Dover & Rib., R. I. M., xxv, p. 359, 1923.[9]

Type : Mus. Paris

Indes Or.[1–5]
Asansol[6, 8, 9]
Calcutta[7, 8, 9]
Dehra Dun (H.)
Dejoo (A)
Maldah (H.)
Nepal[8, 9]
Patkai hills[8, 9]
Sara Ghat[7, 8, 9]

109. ***mutata,** Fleutiaux, A. S. E. F., p. 486, 1893[1]; Fowler, F. B. I., p. 365, 1912[2]; Horn, A. S. E. B., p. 365,

Palon[1, 6]
Bhamo[2, 5, 6]

* Horn (Col. Cat., Cic., p. 171, 1926) regards this form as a variety of *C. cognata* Wied.

1913[3]; Gen. Ins., p. 293, 1915[4]; Dover & Rib., R. I. M., xxv, p. 259, 1923.[5]

Cicindela cognata, Gestro, nec. Wiedemann, Ann. Gen., p. 356, 1893.[6]

Type : Coll. E. Fleutiaux.

Papun (B.M.)
Teinzo[1, 2, 5, 6]
Tharawaddy[2, 5]
Tikekee[6]

Laos (H.)
Tonkin[3, 4, 5]
Yunnan (A.)

110. **nietneri,** Horn, D. E. Z., p. 220, 1894[1]; Spol. Zeyl., II, pp. 34, 44, 1904[2]; Gen. Ins., p. 291, 1915[3]; Maindron, A. S. E. F., p. 12, 1905[4]; Fowler, F. B. I., p. 351, 1912[5]; Dover & Rib., R. I. M., xxv, p. 360, 1923.[6]

Type : Coll. W. Horn.

Ceylon[1–6]

111. **octogramma,** Chaudoir, B. S. N. M., xxv (i) p. 4, 1852[1]; Cat., p. 38, 1865; Gemm. & Har., Cat., p. 23, 1868; Atkinson, Notes Zool. N. W. P., p. 117, 1882, Cat., p. 10, 1889[2]; Fleutiaux, Cat., p. 117, 1892; Fowler, F. B. I., p. 404, f. 173, 1912[3]; Horn, Gen. Ins., p. 304, 1915; Codina, Barcel. Cat., p. 438, 1917[4]; Dover & Rib., R. I. M., xxv, p. 360, 1923.[5]

Type : Mus. Paris.

Dinapore[1, 2, 3, 5]
Bengal (H.)
Darpoor (B.M.)
Dharhar[3, 5]
Kangra valley[3, 5]
Pegu[2], Sibsagar[2]
Tenasserim[2]

Sinkep Id.[2]
Yunnan[4, 5]

112. **octogramma** subsp. **fabriciana,** Horn, Gen. Ins., p. 303, 1915[1]; Zool. Med., Leiden, VII, p. 107, 1922; Codina, Barcel. Cat., p. 437, 1917[2]; Dover & Rib., R. I. M., xxv, p. 356, 1923.[3]

Calochroa fabricii, Horn, D. E. Z., p. 171, 1894[4]; Fowler, F. B. I., p. 403, f. 172, 1912.[5]

Type : Coll. H. E. Andrewes.

N. Kanara[1–5]
Bombay (H.)

113. **octonotata,** Wiedemann., Zool. Mag., I, p. 168, 1819[1]; Gray, in Griffith, Anim. King., XIV, p. 264, pl. 29, f. 3, 1832; Dejean, Spec. Col., I, p. 45, 1825; Gistl., Syst. Ins., I, p. 27, 1837[2]; Schaum, Journ. Ent., II, p. 64, 1863[3]; Chaud., Cat. Coll., p. 38, 1865[4]; Gemm. & Har., Cat., p. 23, 1868[5]; Atkinson, Notes Zool. N. W. P., p. 117, 1882; Cat., pp. 10, 24, 1889[6]; Gestro, Ann. Gen., p. 81, 1889,[7] p. 360, 1893[8]; Fleutiaux, Comp. Rend. S. E. B., p. 169, 1890[9]; Cat., p. 116, 1892[10]; Stebbing, Man. Elem. Forest, Zool., Calcutta,

Bengal[1–7, 10, 13, 14]
Agra[11–14, 16]
Asansol[9, 12, 16]
Chandkhira (B. M.)
Chittagong Hill Tracts[16], Dacca[12, 16]
Darjiling[15, 16]
Dhargaly (B.M.)
Ganges river[12]
Garobadha[15, 16]
Jamayi[12, 16]

p. 84, pl. 28, f. 148, 1908[a]; Annand. & Horn, List, p. 18, 1909 (habits)[11]; Fowler, F. B. I., p. 381, 1912 (habits)[12]; Horn, Gen. Ins., p. 299, 1915[13]; Codina, Barcel. Cat., p. 432, 1917[14]; Dover & Rib., R. I. M., xxii, p. 725, 1921[15]; xxv, p. 360, 1923.[16]

Cicindela solonotata, Westermann, in litt., Gistl., in litt., Syst. Ins., I, p. 27, 1837.

Type : Unknown to the authors.

Khasi Hills[12]
Lushai hills[12, 15]
Maldah[11, 12, 16]
Margherita (H.)
Miani Mukh (at junction of Miani & Kassalong rivers)[16]
Minhla[6, 7, 8]
Murshidabad[12, 16]
N. Chin hills[12]
N. Manipur[12]
Pankabari[11, 12, 16]
Pegu[12, 16], Pusa[12, 16]
Riv. Reidak[a]
Riv. Sunkos[a]
Sibsagar[12, 16]
Sikhim[14], Simla[12, 16]
Sukna, 500 ft.[11, 15, 16]
Sylhet[12, 16], Teinzo[6, 7, 8]
Tharawaddy[12, 16]
Tura, 1,200 ft.[15, 16]
Zamayi (A.)

Penang (B.M.)

114. **ornata,** Fleutiaux, B. S. E. F., p. 146, 1898[1]; Maindron, A. S. E. F., p. 383, fig. 1899[2]; Fowler, F. B. I., p. 429, f. 192, 1912[3]; Horn, Gen. Ins., p. 308, f. 134, (on p. 36), 1915[4]; Dover & Rib., R. I. M., xxv, p. 360, 1923[5].

Type : Coll. E. Fleutiaux.

Karachi[1–5]
Manora[2]
? N. Kanara (H.)

115. **phalangioides,** Schmidt-Gœbel, Faun. Col. Birm., p. 8, pl. 1, f. 5, 1846[1]; Schaum, Journ. Ent., II, p. 68, 1863[2]; Chaud., Cat. Coll., p. 27, 1865[3]; Gemm. & Har., Cat., p. 23, 1868[4]; Atkinson, Cat., p. 11, 1889[5]; Fleutiaux, Cat., p. 60, 1892[6]; Annand. & Horn, List, p. 31, 1909[7]; Fowler, F. B. I., p. 435, 1912[8]; Horn, Gen. Ins., p. 311, pl. 18, f. 5, 1915[9]; Dover & Rib., R. I. M., xxv, p. 360, 1923[10].

Type : Mus. Prague.

Burma[1–5]
Karen hills[10]
N. Chin hills[16]
Pegu[5, 7–10]
Teinzo[10]

116. ***princeps,** Vigors, Zool. Journ., I, p. 413, pl. 15, f. 1, 1825[1]; Gistl., Syst. Ins., I, p. 27, 1837; Parry, T. E. S., p. 80, 1848 (*Calochroa*)[2]; Schaum, Journ.

Madras (neighbourhood)[1, 6, 8]

---

* Dr. Horn suggests that the localities "Nilgiri Hills", "Pondicherry" and "Nowatoli" may belong to the var. *ducalis*.

Ent., II, p. 64, 1863; Chaud., Cat. Coll., p. 39, 1865[3]; Gemm. & Har., Cat., p. 24, 1868; Atkinson, Notes Zool. N. W. P., p. 117, 1882; Cat., p. 11, 1889[4]; Fleutiaux, Comp. Rend. S. E. B., p. 170, 1890[5]; Cat., p. 117, 1892; Fowler, F. B. I., p. 409, f. 175, 1912[6]; Horn, Gen. Ins., p. 304, 1915[7]; Dover & Rib., R. I. M., xxv, p. 360, 1923[8].

Type : Brit. Mus.

Chingleput (H.)
Ceylon[2]
Coromondel[4]
C. India[6, 8], Deccan[4]
Kamhakhan hills (H.)
? Nilgiri hills[4]
? Nowatoli[5]
? Pondicherry[3]
Surada[7, 8]

117. **princeps** var. **ducalis,** Horn, D. E. Z., p. 254, 1897[1]; Gen. Ins., p. 304, 1915[2]; Annand. & Horn, List, p. 26, pl. I, f. 3, 1909[3]; Fowler, F. B. I., p. 410, 1912[4]; Codina, Barcel, Cat., p. 438, 1917[5]; Dover & Rib., R. I. M., xxv, p. 360, 1923[6].

Type : Coll. W. Horn.

Chota Nagpur[1–6]
Nowatoli (B.M.)
Palkot[4, 6]

118 **prothymoides,** Horn, S. E. Z., p. 120, 1908[1]; Gen. Ins., p. 289, 1915[2]; Fowler, F. B. I., p. 359, 1912[3]; Dover & Rib., R. I. M., xxv, p. 360, 1923[4].

Type : Coll. H. E. Andrewes.

Karkur Ghat[1–4]
Malabar[4]

119 **quadrilineata,** Fabricius, Spec. Ins., I, p. 285, 1781[1]; Ent. Syst., I, p. 175, 1792[2]; Syst. Eleuth., I, p. 239, 1801[3]; Herbst, in Fuessly, Arch. Ins., V, p. 115, pl. 27, f. 15, 1784; Käfer, X, p. 162, pl. 170, f. 10, 1806; Olivier, Ent., II, no. 33, p. 25, pl. I, f. 4, 8*a, b,* 1790; Enc. Meth., p. 733, pl. 175, f. 3, 1790; Leach, Zool. Miscell. Lond., II, p. 44, pl. 115, 1815; Dejean, Spec. Col., I, p. 132, 1825; Hope, Mag. Nat. Hist., p. 171, 1840 (ethnographical note) ; Schm.-Gœbel, Faun., Col. Birm., p. 6, 1846[4]; Castelnau, Hist. Nat. Ins., I, p. 18, 1851; Mots., Etud. Ent., p. 30, 1855[5]; Tennant, Ceylon, etc., London, 2nd ed., p. 274, 1859; Chaud., Cat., p. 28, 1865; Gemm. & Har., Cat., p. 24, 1868; Bates, A. M. N. H., xvii (5), p. 70, 1886[6]; Atkinson Cat. p. 11, 1889; Fleut., Cat., p. 60, 1892 Indes; Maindron, A.S. E. F., p. 381, 1899, p. 15, 1905; Horn, D. E. Z., p. 62, 1905, p. 174, pl. 1, f. 22-25, 1906; Gen. Ins., p. 307, f, 142-144 (on p. 36); 1915; Annand. & Horn, List, p. 29, 1909 (habits); Lefroy, Ind. Ins. Life, p. 261, 1909; Fowler, F. B. I., p. 434, f. 194, 1912; Codina, Barcel. Cat., p. 441, 1917; Gravely, R. I. M., xvi, p. 395, 1919 (habits); Dover & Rib., R. I. M., xxii, p. 726, 1921 (habits); xxv, p. 360, 1923; Smedley, Journ. F. M. S. Museums (in

India[1, 2, 3]
Arrakan
Baluchistan
Cacara Bay
Chandipore
Channel Creek nr. Mud Point
Chounambar
Colombo[4, 6]
Dabhol
Gangetic Deita
Harni, Karachi
Koltharen, Madras[5]
Mahe, Manora
Martaban[4]
Pondicherry
Pontianak
Rangoon (B.M.)
Ratnagiri dist.
Sunderbunda
Tenasserim

press).

Type : Mus. Stockholm.

Singapore[5]
Morib. W. Coast, Malay Pen. (C.D.)

120. **quadrilineata,** var. **renei,** Horn, Ann. Gen., p. 273, 1896[1]; Spol. Zeyl., II, pp. 33, 40, pl. 7, f. 8, 1904[2], D. E. Z., p. 174, pl. 1, f. 26, 1906; Gen. Ins., p. 308, 1915[3]; Maindron, A. S. E. F., p. 381, 1899; Fowler, F. B. I., p. 435, 1912[4]; Codina, Barcel. Cat., p. 441, 1917[5]; Dover & Rib., R. I. M., xxii, p. 726, 1921[6]; xxv, p. 360, 1923[7].

Type : Coll. W. Horn.

Baluchistan[1, 3, 5, 7]
Colombo[2]
Karachi[4, 7]
Manora (N.)
Pamben[6, 7]
Shembaganur[5]
Trichinopoly[4, 7]

121. **rhytidopteroides,** Horn, Mem. Dept. Agric. Ind., viii, No. 9, p. 89, 1924[1].

Type : Coll. W. Horn.

Bengal[1]

122. **rugosiceps,** Chaudoir, Cat. Coll., pp. 18, 57, 1865[1]; Gemm. & Har., Cat., p. 25, 1868[2]; Atkinson, Cat., p. 11, 1889[3]; Fleut., Cat., p. 66, 1892; Maindron, A. S. E. F., p. 10, pl. 1, f. 4, 1905[4]; Horn, D. E. Z., p. 60, 1905[5]; Gen. Ins., p. 287, 1915; Fowler, F. B. I., p. 333, f. 149, 1912 (habits)[6]; Codina, Barcel. Cat., p. 422, 1917[7]; Dover & Rib., R. I. M., xxv, p. 360, 1923[8].

Type : Mus. Paris.

Malabar coast[1, 2, 3]
Coonoor, 2,500-3,500 ft.[4, 8]
Madras (B.M.)
Mysore[6, 8]
Nilgiri hills[5, 6, 8]
Pillur[8], Ramnad[6, 8]
Shembaganur[7]

123. **seriepunctata,** Horn, D. E. Z., p. 80, 1892[1]; Gen. Ins., p. 292, 1915[2]; Fleut., Cat., p. 47, 1892[3]; Fowler, F. B. I., p. 351, 1912[4]; Dover & Rib., R. I. M., xxii, p. 723., 1921; xxv, p. 360, 1923[5].

Type : Coll. W. Horn.

Darjiling[1, 3, 4]
Gopaldhara (A.)
Kalimpong, 600-4,500 ft.[5]
Kurseong[4, 5]
N. Bengal[2, 5]
Pashok, 2,000 ft.[5]
Rungeet valley, 700-800 ft.[5]
Sikhim[2]
Singla, 1,500 ft.[5]

124. **sexpunctata,** Fabricius, Syst. Ent., p. 226, 1775[1]; Spec. Ins., I, p. 285, 1781[2]; Ent. Syst., I, p. 175, 1792[3]; Syst. Eleuth., I, p. 239, 1801[4]; Herbst, in Fuessly, Arch. Insect, VIII, p. 172, pl. 46, f. 1, 1786; in Fuessly, Arch. Ent. p. 149, pl. 46, f. 1, 1794; Käfer, x, p. 164, pl. 170, f. 12, 1806; Olivier, Ent., II, No. 33, p. 24, pl. 1, f. 6, 1790[5]; Enc. Meth., p. 732, pl. 175, f. 2, 1790[6]; Dejean, Spec. Col., I, p. 47,

Littora Malabariaca[1, 2, 3, 4, 5, 6, 7]
Alipur Duars
Andaman Is.[12]
Arrakan[12], Asansol[16]

1825; Westermann, Rev. Ent. Silberm., I, p. 106, 1833 (habits), Saunders, T. E. S. p. 61, 1834 (habits); Gistl., Syst. Ins., I, p. 28, 1837[7]; Schm.-Gœbel, Faun. Col. Birm., p. 4, 1846[8]; Guérin, Rev. Mag. Zool., p. 84, 1849; Schaum, Journ. Ent., II, p. 67, 1863; Chaud., Cat. Coll., p. 38, 1865[9]; Bates, Proc. Zool. Soc., p. 342, 1866[10]; Gemm. & Har., Cat., p. 26, 1868; Bær, A. S. E. F., p. 98, 1886[11]; Atkinson, Cat. p. 11, 1889[12]; Gestro, Ann. Gen., p. 82, 1889[13]; p. 359, 1893[14]; Cotes, Ind. Mus. Notes, I, p. 59, 1889[15]; II, p. 148, 1893; III, pp. 17, 118, fig., 1893; Fleut., Comp. Rend. S. E. B., p. 569, 1890[16]; Cat., p. 114, 1892; A. S. E. F., p. 489, 1893[17]; p. 569, 1902[18]; B. S. E. F., p. 49, 1917[19]; Horn, A. S. E. B., p. 538, 1892[20]; D. E. Z., p. 96, 1895, p. 66, 1902; Spol. Zeyl., II, pp. 33, 41, pl. 7, f. 14, 1904; Arch., Nat. Jahrg., LXXVIII A, heft. II, p. 83, 1912; A. S. E. B., p. 365, 1913; Gen. Ins., pp. 266, 301, 1915; Barlow, Ind. Mus. Notes, IV, p. 46, 1896 (preying on *Leptocorisa varicornis*); Peal, Ind. Mus. Notes, V, pp. 40, 48, 1900 (preying on *Hispa armigera*); Lesne, Mission, Pavie, Indo-Chine, III, p. 61, 1904; Maindron, A. S. E. F., p. 17, 1905; Lefroy, Ind. Ins. Pests, Calcutta, p. 118, f. 134, 1906 (preying on *Leptocorisa varicornis*); Ind. Ins. Life, p. 261, fig., p. 259, 1909; Annand. & Horn, List, p. 21, 1909 (habits); Schilder, Ent. Blatter Berl., p. 178, 1911; Fowler, F. B. I., p. 385, 1912 (habits); Vuillet, Insecta II, p. 85, f. 1 (on p. 84), 1912 (habits); Fletcher, Some South Indian Insects, p. 199, 1914; Poyser, Planter's Chron., XI, p. 622, 1916; Schultze, Philipp. J. Science, D. XI, p. 8, 1916; Codina, Barcel, Cat., p. 434; 1917; Dover & Rib., R. I. M., xxv, p. 360, 1923.

Cicindela flavomaculata, Hope, in Gray's Zool. Miscell., p. 21, 1831; Mag. Nat. Hist., p. 171, 1840; Dohrn, S. E. Z., p. 456, 1879; Atkinson, Cat., p. 8, 1889; Fleut., Cat., p. 115, 1892; Horn, D. E. Z., p. 223, 1892.

C. *tripunctata,* Buquet, in litt., Dejean, in litt., Cat., ed. 3, p. 2, 1837 (var.); Bær, A. S. E. F., p. 98, 1886; Fleut., Cat., p. 115, 1892 (var.); A. S. E. F., p. 490, 1893.

C. *risi,* Helfer, in litt., Schm.-Gœb., Faun. Col. Birm., p. 4, 1846.

C. *sexignata,* Buquet, in litt., Chaudoir, in litt., Cat. Coll.,

Bangalore[12]
Barway[20], Basanti
Berhampur
Bhamo[13, 14]
Birbhum, Bombay
Bonaventure Estate
Shevaroy Hills
Bosondhar
Calcutta and environs[7, 8, 12]
Champaran
Chota Nagpur
Coorg, Damukdia
Diamond, Harbour[15]
Farm caves nr. Moulmein
Gilgit
Irrawaddy (Delta)
Kandy (B.M.)
Karen hills (B.M.)
Kawkareik
Kharagpur
Maldah[12], Mahe
Moghul Sarai (B.M.)
Mungphu
Nepal (Raxaul)
N. Kanara (A.)
Pegu, Peradeniya
Pondicherry[12]
Poona, Purneah
Pusa, Ranchi
Rangoon[13, 14]
Raniganj, Sadiya
Sara Ghat, Shillong
Shimoga
Shivegoo[13, 14]
Sikhim[12]
Sittaung river[12]
Sunderbunds
Sylhet (N.)
Trichur, Teinzo[13, 14]
Tetara[16]
Tharawaddy (A.)

p. 38, 1865. — Wellawaya

C. javanica, Vries van Doesburgh, Mem. de Mexico, 1893; Horn, D. E. Z., p. 96, 1895, p. 59, 1897 (as var.).

C. punctata, Stebbing, Man. Elem. Forest Zool., Calcutta, p. 84, pl. 28, f. 147, 1908 (preying on *Leptocorsis varicornis*).

Type : Mus. Stockholm.

Annam, Bangkok
Cambodia
Cameroon
Chiem-Hoa
S. China
Formosa[10]
(Kagi : B. M.)
Gerona, Hanoi[18]
Hongkong
Java (Modjokerto)
Bet. Chantaboun and Battambang
Kompong-Than[19]
Manilla[11], Marliban[8]
Phuc Son, Senegal[9]
Singapore[17]
Tarlac, Tourakhon[19]
Vientian[19], Siam[17]
Senegambia[9, 13]

125. ***sexpunctata** var. **aurovittata,** Audouin & Brullé, Arch. Mus. Hist. Nat. Paris, p. 127, pl. 8, 1839[1]; Schaum, Journ. Ent., II, pp. 62, 67, 1863; Gemm. & Har., Cat., p. 26, 1868[2]; Atkinson, Notes, Zool. N. W. P., p. 117, 1882; Cat., p. 11, 1889; Fleut., Cat., p. 115, 1892[3]; A. S. E. F., p. 490, 1893[4]; Horn, D. E. Z., p. 96, 1895[5]; p. 461, 1909[6]; Spol. Zeyl., II, pp. 33, 41, pl. 7, f. 15, 1904[7]; Mem. 1st Congress Int. Ent., p. 314, 1911[8]; Gen. Ins., p. 303, 1915[9]; Fowler, F. B. I., p. 386, 1912[10]; Annand. & Dover, R. I. M., xxii, pp. 335, 337, 1921 (habits)[11]; Dover & Rib., R. I. M., xxv, p. 353, 1923[12].

Type : Mus. Paris

Pondicherry[1, 2, 4, 6, 8, 9, 10, 12]
Andaman Is.[5, 6, 9, 10, 12]
Barkuda Id.[12]
Batticaloa (B. M.)
Calcutta and environs[9, 12]
Ceylon[6, 7, 8, 9, 10, 12]
Nicobar Is.[4, 6, 8, 9, 10, 12]
Rangoon[4, 6, 8, 9, 10, 12]

Centr. Japan[6, 9, 10, 12]
Indo-China[3]
Philippine Is.[6, 8, 9, 10, 12]

126. #**shivah,** Parry, T. E. S., V, p. 80, pl. 21, f. 2, 1848; (*Calochroa*)[1]; Schaum, Journ. Ent., II, p. 65, 1863[2]; Chaud., Cat., p. 39, 1865; Gemm. & Har., Cat., p. 26, 1868[3]; Atkinson, Notes Zool. N. W. P., p. 117, 1882;

Assam[1, 2, 6, 7]
Nepal[5, 7]
N. Kanara[5, 7]

---

* Dr. Horn (Col. Cat., Cic., p. 183, 1926) treats this as a good species.

\# Dr. Horn (Col. Cat., Cic., p. 185, 1926) queries the locality "Assam," giving S. Bombay as the distribution.

Cat., p. 12, 1889[4]; Fleut., Cat., p. 118, 1892; Fowler, F. B. I., p. 411, 1912[5]; Horn, Gen. Ins., p. 304, 1915[6]; Dover & Rib., R. I. M., xxv, p. 361, 1923[7].

Nilgiri hills[8]
Sylhet[3, 4]

Cicindela boisduvali, Dupont, in litt., Dejean, in litt., Cat., ed. 3, p. 2, 1837.

C. erichsoni, Hope, in litt., Col. Man., II, p. 163, 1838[8]; Horn, D. E. Z., p. 193, 1898; Fowler, F. B. I., p. 411, 1912.

Type : Brit. Mus. (Coll. Nevinson).

127. **sinica,** Fleutiaux, A. S. E. F., p. 137, 1889[1]; Cat., p. 46, 1892[2]; B. S. E. F., p. 49, 1917[3]; Gestro, Ann. Gen., p. 352, 1893[4]; Horn, D. E. Z., p. 66, 1902[5]; A. S. E. B., p. 365, 1913[6]; Gen. Ins., p. 293, 1915[7]; Fowler, F. B. I., p. 355, 1912[8]; Dover & Rib., R. I. M., xxv, p. 365, 1923[9].

Type : Coll. E. Fleutiaux.

Saigon[1, 6, 8, 9]
Ceylon[5, 6]
Maymyo (A.)
Taik-gyi[8, 9]
Tikekee in Pegu[4]

Bangkok[7, 8], China[8]
Chiem Hoa[5, 6, 8, 9]
Cochin China[2]
Hoabin (A.)
Hongkong[7, 9]
Java[7, 9]
Kompong-Thom.[3]
Siam (H.)
Vientian[3]

128. **spinolai,** Gestro, Ann. Gen., p. 85, 1889[1], p. 355, 1893[2]; Fleutiaux, Comp. Rend. S. E. B., p. 168, 1890[3]; Cat., p. 47, 1892; A. S. E. F., p. 486, 1893[4]; Horn, D. E. Z., pp. 97, 224, 1892[5]; p. 95, 1895[6]; A. S. E. B., p. 364, 1913[7]; Gen. Ins., p. 292, 1915[8]; Annand. & Horn, List, p. 10, 1909 (habits)[9]; Fowler, F. B. I., p. 346, f. 153, 1912[10]; Codina, Barcel. Cat., P., 425, 1917[11]; Dover & Rib., R. I. M., xxii, p. 723, 1921[12]; xxv, p. 361, 1923[13].

Cicindela demangei, Fleutiaux, A.S.E.F., p. 569, 1902[14]; B.S.E.F., p. 369, 1917 (var.)[15]; Horn, A.S.E.B., p. 364, 1913 (var.)[16].

Type : Mus. Genova.

Meetan[1]
Asansol[3, 10, 13]
Bhamo[1, 2, 10, 13]
Chota Nagpur[5, 10, 13]
Dacca[10, 13]
Darjiling[10, 13]
Dehra Dun[8, 11, 13]
Gokteik (A.)
Gopaldhara (A.)
Karen hills (Cheba, 900-1,100 m.)[2, 10, 13]
Margherita (H.)[11]
Momeit[4, 10, 13]
N. Chin. hills[10, 13]
Pashok, 1,000-5,000 ft.[13]

128. **spinolai,—**

Patkai hills[10, 13]
Pegu (B.M.)
Rangamati[12, 13]
Rangoon[10, 13]
Ruby Mines[10, 13]
Rungeet valley, 700-800 ft.[13]
Singla[13]
Sukna, 500 ft.[9, 13]
Sylhet[10, 13]
Teinzo[1, 2, 10, 13]
Tenasserim[10, 13]
Tharawaddy[10, 13]
Tura, 1,000 to 1,500 ft.[12, 13]

Cochin China[7, 8, 10, 11]
Haobin (A.)
Laokay[14, 16]
Luang-Prabang[15]
Trinninh[15]
Yunnan[13]

129. **striatifrons,** Chaudoir, B. S. N. M., XXV, (1), p. 12, 1852[1]; Cat. Coll., p. 32, 1865[2]; Schaum, Journ. Ent., II, p. 67, 1863[3]; Gemm. & Har., Cat., p. 27, 1868[4]; Atkinson, Notes Zool. N. W. P., p. 117, 1882; Cat., pp. 12, 24, 1889[5]; Fleut., Cat., p. 80, 1892[6]; Horn, A. S. E. B., p. 538, 1892[7]; Gen. Ins., p. 307, 1915; Annand. & Horn, List, p. 28, 1909[8]; Fowler, F. B. I., p. 426, f. 190, 1912 (habits)[9]; Dover & Rib., XXV, p. 361, 1923[10].

Type : Mus. Paris.

Indes Or.[1–4]
Barway[7]
Berhampur[5, 8, 9, 10]
Murshidabad[8, 9, 10]
N. Kanara[9, 10]

130. **striolata,** Illiger, Arch. Zool. and Zootomie (Wied.), I, (2), p. 114, 1800[1]; Schaum, B. E. Z., p. 173, 1862; Journ. Ent., II, p. 64, 1863 (as var. of *semivittata,* Fabr.); Gemm. & Har., Cat., p. 27, 1868[2]; Bær A. S. E. F., p. 98, 1886[3]; Atkinson, Cat., p. 12, 1889[4]; Gestro, Ann. Gen., p. 80, 1889[5], p. 357, 1893[6]; Fleutiaux, Comp. Rend. S. E. B., p. 169, 1890[7]; Cat., p. 76, 1892[8]; A. S. E. F., p. 488, 1893[9]; B. S. E. F., p. 49, 1917[10]; Horn, A. S. E. B., p. 538, 1892[11]; Ann. Gen., p. 681, 1895[12]; D. E. Z., p. 96, 1895, p. 59,

Sumatra[1, 16, 18, 21, 23, 24, 25, 26, 27, 28]
Asansol[7]
Barway[11]
Bhamo[5, 6]
Calcutta[18, 23]
Chota Nagpur[18, 23]
Coromandel[27]
Dehra Dun[16, 21, 23]

1897[13]; p. 66, 1902[14]; p. 60, 1905, S. E. Z., p. 174, 1896; A. S. E. B., p. 366, 1913[15]; Gen. Ins., p. 306, 1915[16]; Maindron, A. S. E. F., p. 15, 1905[16]; Annand. & Horn, List, p. 27, 1909[17]; Schilder, Ent. Blatter, Berl., p. 178, 1911; Fowler, F. B. I., p. 419, f. 183, 1912[18]; Fletcher, J. B. N. H. S., XXIII, p. 379, 1914 (habits)[19]; Schultze, Philipp. Journ. Sci., D, XI, p. 8, 1916[20]; Codina, Barcel. Cat., p. 439, 1917[21]; Dover & Rib., R. I. M., XXII, p. 725, 1921 (habits)[22]; XXV, p. 361, 1923[23].

Cicindela semivittata, Fabricius, Syst. Eleuth., I, p. 237, 1801[24]; Herbst, Käfer, X, p. 201, 1806[25]; Gistl., Syst. Ins., I, p. 25, 1837[26]; Schm.-Gœb., Faun. Col. Birm., p. 3, pl. 1, f. 2, 1846[27]; Schaum, B. E. Z., p. 74, 1861; Journ. Ent., II, p. 64, 1863; Chaud., Cat. Coll., p. 39, 1865[28]; Bates, Proc. Zool. Soc., p. 342, 1866[29]; Horn, Ann. Gen., p. 682, 1895 (var.)[30].

C. vigorsi, Dejean, Spec. Col., V, p. 223, 1831.

C. multiguttata, Fleutiaux, A. S. E. B., p. 169, 1890, p. 7, 1893.

C. dorsolineata, Chevrolat, Rev. Zool. p. 95, 1845; as var.; Schaum, B. E. Z. p. 173, 1862; Gemm. & Har., Cat., p. 27, 1868; Bær, A. S. E. F., p. 98, 1886[33]; Atkinson, Cat., p. 12, 1889; Fleutiaux, A. S. E. F., p. 489, 1893[34].

Type : Mus. Berlin.

Kanara[18, 23]
Karen Hills (Cheba to Biapo, 900-1,100 m.)[6]
Karkar Ghat (A.)
Khasi hills[18, 23]
Kokareit[5, 6]
Mahe[18, 23]
Malda[17, 18, 23]
Malewoon[6]
Mungphu[17, 18, 23]
N. Chin Hills (B. M.)
Ouchterlony valley, 3, 500-4,000 ft.
Pashok, 3,000 ft.[23]
Pegu[4, 17, 18, 23]
Pollibetta[19, 23]
Rangamati[22, 23]
Sikhim[4, 16, 21]
Soom, 3,000-3,500 ft.[23]
S. Mysore (A.)
Taung-ngu[6]
Teizo[5, 6, 18, 23]
Tenasserim[18, 23]
Tharawaddy[18, 23]
Trivandrum[18, 23]
Wallardy[16]

Borneo[16, 18, 23]
Benquet[20]
Bagino[20]
Celebes (Pangie: A.)[2, 4, 8, 9, 10, 16, 18, 23, 28, 29]
Chien Hoa[14, 15]
China[21, 34]
Diampejam[21]
N. China[9, 16, 18]
Formosa (Yentemps: A.)[16, 18, 21, 23, 29]
Hongkong[4, 28]
Java[2, 4, 9, 13, 16, 18, 21, 23, 28]
Laos (Xieng Khonang : A.)

130. **striolata,**–

Luzon[26]
Macao[4, 33]
Mt. Mauson[15]
Olongapo[20]
Perak[21]
Philippines[4, 8, 9, 10, 16, 18, 21, 23, 29, 33]
Romblou[20]
Singtan[16, 23]
Sumatra (Deli)[2, 3, 4, 9]
Than-Moi[14, 15]
Vientian[10]
Zambales[20]

131. **striolata** var. **lineifrons,** Chaudoir, Cat. Coll., pp. 39, 62, 1865[1]; Gemm. & Har., Cat., p. 20, 1868[2]; Atkinson, Cat., pp. 9, 24, 1889[3]; Gestro, Ann. Gen., p. 79, 1889[4]; p. 357, 1893[5]; Fleut., Cat. p. 77, 1892; A. S. E. F., p. 489, 1893[6]; Horn, D. E. Z., p. 337, 1893; p. 96, 1895, p. 66, 1902[7]; A. S. E. B., p. 366, 1913[8]; Gen. Ins., p. 306, 1915[9]; Annand. & Horn, List, p. 28, 1909[10]; Fowler, F. B. I., p. 421, f. 184, 1912[11]; Dover & Rib., R. I. M., XXII, p. 726, 1921[12]; XXV, p. 361, 1923[13].

Cicindela interrupta, Fleutiaux, A S. E. F., p. 569, 1902 (as var.)[14]; Horn., D. E. Z., p. 66, 1902[15].

Type : Mus. Paris.

Cambodia[1, 2, 3, 6, 8, 9, 11, 13]
Above Tura 2,500 ft.[12, 13]
Bhamo (B.M.)
Bhamo hills, 4,000 ft. (B.M.)
Dehra Dun[9, 13]
Kanara[11]
Karen hills (Cheba 400-1,100; Ghecu, 13-1,400 m.)[5, 11, 13]
Maymyo (A.)
Meetan[4, 5]
Moolay[5]
Nilgiri hills[11, 13]
N. Chin Hills[11, 13]
Pègu[3, 10, 11, 13]
Teinzo[4, 5, 11, 13]
Tenasserim[3, 11, 13]
Thagata[4, 5]
Tharawaddy[11, 13]
Tura[9]

Cochin China[8, 9]
Lao-Kay[14]
Mt. Mauson[7, 8]
Perak[9, 11, 13]
Luang-Pranang (A.)

132. ***striolata** var. **parvimaculata,** Fowler, F. B. I., p. 421, f. 185, 1912[1]; Horn, Gen. Ins., p. 305, 1915; Dover & Rib., R. I. M., XXV, p. 361, 1923.[2]

Type : Brit. Mus.

Dhargeely[1]
Darjiling[2]

133. **sublacerata** var. **balucha,** Bates, Cist. Ent., II, p. 332, 1878[1]; Atkinson, Cat., p. 5, 1889[2]; Fleut., Cat., p. 50, 1892[3]; Horn, D. E. Z., p. 334, 1893, p. 11, 1894[4]; p. 354, 1896; Munch., Kol. Zeit., I, p. 340, 1903; Gen. Ins., pp. 295, 355, 1915[5]; Semenow, Hor. Soc. Ent. Ross., XXX, p. 231, 1895-96[6]; Fowler, F. B. I., p. 368, 1912[7]; Dover & Rib., R. I. M., XXV, p. 361, 1923[8].

Type : Coll. R. Oberthür.

Baluchistan[1–8]
Skardo[7, 8]
————
Armenia[8]
Caucasus[8]
Perso-Baluch Frontier[8]
Trans-caspia[8]
Turkestan (to Ferghana and the Pamirs)[8]

134. **sumatrensis,** Herbst, Käfer, X, p. 179, pl. 172, f. 1, 1806[1]; Dejean, Spec. Col., I, p. 88, 1825; Gistl., Syst. Ins., I, p. 49, 1837[2]; Schm.-Gœb., Faun. Col. Birm., p. 4, 1846[3]; Schaum., Journ. Ent., II, p. 67, 1863[4]; Chaud., Cat., Coll., p. 34, 1865[5]; Gemm. & Har., Cat., p. 27, 1868[6]; Bær, A. S. E. F., p. 98, 1886; Bates, A. M. N. H., XVII (5), p. 69, 1886[7]; Atkinson, Cat., pp. 12, 24, 1889[8]; Gestro, Ann. Gen., p. 82, 1889[9]; p. 359, 1893[10]; Fleutiaux, Comp. Rend. S. E. B., p. 169, 1890[11]; Cat., p. 88, 1892; B. S. E. F., p. 49, 1917[12]; Horn, A. S. E. B., p. 538, 1892[13]; Ann. Gen., p. 681, 1894[14]; D. E. Z., p. 58, 1897, p. 66, 1900[15]; Spol. Zeyl., II, pp. 34, 42, pl. 7, f. 22, 1904[16]; A. S. E. B., p. 365, 1913[17], p. 140, 1914[18]; Gen. Ins., p. 295, 1915[19]; Robinson, Fasc., Malay, Zool., I, p. 180, 1903 (habits); Lesne, Mission Pavie, Indo-Chine, III, p. 61, 1904[20]; Maindron & Fleut., A. S. E. F., p. 16, 1905[21]; Annand. & Horn, List, pp. 14, 30, 1909 (habits)[22]; Childer, Ent. Blatter Berl., p. 178, 1911[23]; Fowler, F. B. I., p. 371, f. 162, 1912[24]; Gravely, R. I. M., VII, p. 209, 1912; Schultze, Philipp. Journ. Sci., D, XI, p. 8, 1916[25]; Codina, Barcel. Cat., p. 428, 1917[26]; Annand. & Dover, R. I. M., XXII, pp. 335-337, 1921[27]; Dover & Rib., R. I. M., XXII, p. 724, 1921; xxv, p. 361, 1923[28].

Sumatra[1–5, 8, 14, 24, 28]
Asansol[11, 38, 39]
Amherst[3]
Barkuda Id.[28]
Barway[13, 26]
Bhamo[8, 9, 10, 24, 28]
Bhogpur[28]
Cacara Bay[28]
Calcutta[22, 24, 28]
Chakradharpur[28]
Cheria Id.[28]
Colombo[7, 16]
Cuttack[28]
Damukdia [22, 24, 28]
Karachi[19, 21, 24, 28]
Kashmir[19, 26, 28]
Khandesh[24, 28]
Killinour (A.)
Kumdhik[22, 24, 28]
Mahe[21, 24, 28]
Mahe[24, 28]
Malda[24, 28]
Martaban[3]
Mergui[8]
Nerbudha bank (A.)

---

* Dr. Horn thinks this a good species.

Cicindela catena, var. No. 3, Thunberg, Dissert, Nov.-Ins., sp., I, p. 27, pl. 1, f. 43, 1781[29].

C. catena, Thunberg, Nov. Ins. Spec., p. 26, figs. 41-43, 1784[30].

C. arcuata, Kollar, Ann. Hofmus. Wien, p. 330, 1836[31].

C. catena var.♂, Gistl. Syst. Ins., I, p. 61, 1837.

C. westerhauseri, Gistl., Syst. Ins., I, p. 61, 1837.

C. leguilloni, Guérin, Rev. Zool., p. 120, 1841[32]; as var.; Bær, A. S. E. F., p. 98, 1886[33]; Fleut., Cat., p. 88, 1892[34].

C. boyeri, Blanchard, Voy. Pole Sud. Zool., IV, p. 4, pl. 1, f. 2, 1853[35].

C. niponensis, Bates T. E. S., p. 216, 1883[36]; as var.; Atkinson, Cat., p. 12, 1889; Fleut., Cat., p. 88, 1892.

C. renardi, Fleutiaux, Comp. Rend. S. E. B., p. 169, 1890 (var.)[38]; Cat., p. 88, 1892 (var.); Fowler, F. B. I., p. 972, 1912 (var.).

Type : Mus. Berlin.

N. Chin hills[24, 28]
N. Kanara (B. M.)
N. Shan States (N.)
Nowatoli[11]
Papun (H.)
Patan, 2,050 ft.[28]
Pegu[22, 24, 28]
Pondicherry[24, 26, 28]
Pontianak[26]
Purulia[28]
Ranchi[24, 28]
R. Ganges[28]
Siliguri[28]
Tavoy[24, 28]
Teinzo,[9, 10, 24, 28]
Telicherry[21, 28]
Tetara[11]
Tharawaddy[24, 28]
Trivandrum[22, 24, 28]
Tura, 1,000 ft.[28]

Auckland[20]
Ban-Sop-Tia[12]
Bao-Læ[18]
Borneo[6, 8, 19, 24, 26, 28, 32]
Cambodia[20]
Cebu[25]
Centr. China[8, 28, 20]
Chiem Hoa[15]
Cochin China[17]
Formosa[19, 24, 26, 28]
Hainan (f-finger Mts. N.)[24, 28]
Japan (Nippon)[8, 19, 20, 21, 23, 24, 26, 28, 36, 37]
Java[2, 4, 5, 8, 19, 26, 28]
Keng Kiet[12]
Kiang Si[26]
Lombok
Luzon (Rizal)[25]
Malacca[4]
Mindanao[5, 6, 33, 34]
Pachim dist.

134. **sumatrensis,**–

Pak-Lay[12]
Perak[24, 28]
Philippine Is.[4, 8, 9, 19, 20, 21, 24, 26, 28]
Quang-Tang[28]
Sinkep Id.[22, 28]
Sumbawa[19, 26, 28]
Sunda Is.[9, 21]
Toledo[25]
Yunnan[19, 26, 28]

135. ***sumatrensis** var. **fowleri,** nom. nov. *sumatrensis* var. *imperfecta*, Horn (nec. Le Conte, 1851), D. E. Z., p. 173, 1894[1]; A. S. E. B., p. 538, 1892; Gen. Ins., p. 298, 1915 ; Annandale & Horn, List, p. 14, 1909; Fowler, F. B. I., p. 372, 1912[2]; Dover & Rib., R. I. M., XXII, p. 724, 1921; XXV, p. 361, 1923[3].

Type : Col. H. E. Andrewes.

N. Kanara[1, 2]
Castle Rock[3]
Helvek, 2,000 ft.[3]
Jubbalpore[3]
Medha, 2,200 ft.[2]
Mussoorie[3]
Nerbudda Bank (B.M.)
Patan, 2,050 ft.[3]
Pimpli[3,] Ranchi[3]
Shimoga[2, 3,] Surat[3]
Taloshi, 2,000 ft.[3]

136. **tetragrammica,** Chaudoir, Cat. Coll., pp. 18, 58, 1865[1]; Gemm. & Har., Cat., p. 27, 1868[2]; Atkinson, Cat., pp. 12, 24, 1889[3]; Fleut., Cat., p. 66, 1892; Horn, D. E. Z., p. 60, 1905[4]; Gen. Ins., p. 295, 1915[5]; Annand. & Horn., List, p. 8, 1909[6]; Fowler, F. B. I., p. 331, f. 147, 1912[7]; Codina, Barcel. Cat., p. 428, 1917[8]; Dover & Rib., R. I. M., XXV, p. 362, 1923[9].

Type : Mus. Paris.

Malabar Coast[1, 2, 3, 7]
Anamalai hills, 3,000-5,500 ft.[4, 7, 9]
Coorg (H.)
Ouchterlony valley (A.)
Pykara, 6,000 ft.[9]
Ramnad[7, 9]
Shembaganur[7, 8, 9]
S. Bombay[5, 8, 9]
Tavoy (H.)
Trichinopoly[6, 7, 9]

137. **tetrastacta** Wiedemann, Zool. Mag., II, p. 65, 1823[1]; Schaum, B. E. Z., p. 74, 1861; Journ. Ent., II, p. 64, 1863[2]; Gemm. & Har., Cat., p. 27, 1868[3]; Atkinson, Cat., p. 12, 1889[4]; Fleut., Comp. Rend. S. E. B., p. 168, 1890 (*Euryoda*)[5]; Cat., p. 38, 1892 (*Jansenia*)[6];

Bengal[1, 2, 3, 6]
Asansol[5, 10, 14]
Barkuda Id.[12, 13, 14]
Birbum[4, 9, 10, 14]

---

* In his Col. Cat., Cic., p. 174, 1926, Horn, sinks this form as a "dlc—F" form of *sumatrensis.* Even in this case, however, the insect appears to require a new name.

Horn, D. E. Z., p. 94, 1892 (*Tetreurytarsa*); A. S. E. B., p. 537, 1892 (*Tetreur.*)[7]; D. E. Z., p. 33, 1905 (Beiheft: *Euryoda*); Gen. Ins., p. 289, 1915[8]; Annand. & Horn., List, p. 8, 1909[9]; Fowler, F. B. I., p. 337, 1912[10]; Codina, Barcel. Cat., p. 423, 1917[11]; Annand. & Dover, R. I. M., XXII, pp. 336, 337, 1921[12]; Dover & Rib., R. I. M., XXII, p. 723, 1921[13]; XXV, p. 362, 1923[14].

Cicindela colon, Klug, Jahrb., I, p. 11, 1834; Audouin & Brullé, Arch. Mus. Hist. Nat. Paris, p. 123, pl. 7, f. 9, 1839; Anderson, Notes Zool., N. W. p. 117, 1882[15].

Type : Unknown to the authors.

Calcutta[9, 10, 14]
Chota Nagpur[10, 14]
Dharwar[10, 14]
Ganjam[8, 11, 14]
Gantasila[12, 13, 14]
Godag (A.)
Kurseong[7]
Nowatoli[9, 10, 11, 14]
R. Ganges Valley[9, 14]
Surada (H.)
N. W. Prov.[15]

138. **triguttata,** Herbst, Käfer, X, p. 182, pl. 172, f. 5, 1806[1]; Dejean, ex parte, Spec. Col., I., p. 146, 1825[2]; Schm.-Gœb., Faun. Coll. Birm., p. 8, 1846[3]; Schaum, Journ., Ent., II, p. 66, 1863[4]; Redt., Reise Freg. Novr. Zool., II, p. 2, 1868[5]; Gemm. & Har., Cat., p. 28, 1868[6]; Atkinson, Cat., ex parte, p. 13, 1889[7]; Fleut., Cat., p. 46, 1892[8]; A. S. E. F., p. 484, 1893[9], p. 137, 1889[10]; p. 569, 1903[11]; B. S. E. F., p. 49, 1917[12]; Gestro, Ann., Gen., p. 352, 1893[13]; Horn, Ann. Gen., p. 680, 1894[14]; D. E. Z., p. 95, 1895, p. 56, 1897[15]; A. S. E. B., p. 364, 1913[16]; Gen. Ins., p. 290, 1915[17]; Annand. & Horn, List, p. 9, 1909[18]; Schilder, Ent. Blatter Berl., p. 177, 1911[19]; Fowler, F. B. I., p. 343, 1912[20]; Schultze, Philipp. Journ. Sci., D, XI, p. 8, 1916[21]; Codina, Barcel. Cat., p. 424, 1917[22]; Dover & Rib., R. I. M., XXII, p. 723, 1921[23]; XXV, p. 362, 1923[24].

Cicindela sexmaculata, Dejean, in litt., Spec. Col., I, p. 146, 1825[25].

C. chlorochila, Chaudoir, B. S. N. M., XXV (1), p. 25, 1852[26]; Cat., p. 23, 1865[27]; Gemm. & Har., Cat., p. 12, 1868; Akinson, Notes Zool. N. W. P., p. 117, 1882; Cat., pp. 6, 23, 1889[28]; Gestro, Ann. Gen., p. 85, 1889[29], p. 353, 1893[30]; Fleut., Cat. p. 46, 1892[31]; A. S. E. F., p. 485, 1893[32]; Horn., D. E. Z., p. 213, 1892, p. 95, 1895, p. 57, 1897[33]; p. 66, 1902[34]; Robinson, Fasc. Malay, Zool., I, p. 180, 1903[35]; Schilder, Ent. Blatter Berl., p. 177, 1911 (var.)[36]; Fowler, F. B. I., p. 343, 1912 (var.)[37].

Type : Mus. Berlin.

N. America[1]
Indes Or.[2, 8, 25]
Banka (N.)
Bhamo[24, 29, 30, 37]
Calcutta[17, 18, 20, 24]
Dehra Dun[17, 22, 24]
Darjiling dist., 600-4,500 ft.[24]
Kalimpong, 600-4,500 ft.[24]
Karen hills (Cheba, 900-1,100 m.)[13, 20, 24]
Maymyo (A.)
Margherita (H.)
Meetan[29, 30]
Pegu[18, 20, 24, 30, 37]
Rangamati[23, 24]
Rangoon[30]
Siliguri[23, 24]
Taung-ngu[24, 37]
Teinzo[20, 24, 28, 29, 30, 37]
Tharawaddy[24, 37]
Tikekee[30]
Tura, 1,300 ft.[23, 24]

Ban-Sop-Mom[12]
Borneo[7, 17, 20, 24]
Celebes[4, 7, 17, 22, 24, 37]
Chiem-Hoa[34]
Hoabin (A.)

138. **triguttata,**—

Hong Kong[19, 24, 26, 27, 28, 31, 32, 36, 37]
Ban Sai Kan[35]
Java[4, 5, 7, 15, 33]
Komblon (H.)
Kompong-Thom.[12]
Lao-Kay[11]
Malacca (N.)
Malay Archipelago[8, 9, 20, 24] Malaya[8, 9, 16]
Nawngchik[35]
Philippine Is.[17, 20, 24]
Penang (N.)
Saigon[10]
Singapore[17, 24]
Sumatra (Deli)[4, 6, 7, 14, 17, 24]
Sumbawa[17, 24]
Shanghai[17, 24]
Tœkang Besi[17]
Vientian[12]
Zambales & Olonapo in Luzon[21]
Occidental negros and Bago in Negros[21]

139. ***tritoma,** Schmidt-Gœbel, Faun. Col. Birm., p. 3, pl. 1, f. 3, 1846[1]; Schaum, Journ. Ent., II, p. 65, 1863[2]; Gemm. & Har., Cat., p. 28, 1868[3]; Atkinson, Cat., p. 13, 1889[4]; Fluet., Cat., p. 117, 1892[5]; A. S. E. F., p. 493, fig., 1893[6]; Horn, D. E. Z., p. 131, 1899; Gen. Ins., p. 302, 1915[7]; Fowler, F. B. I., p. 394, 1912[8]; Dover & Rib., R. I. M., XXV, p. 362, 1923[9].

Cicindela lauræ, Gestro, Ann. Gen., p. 364, 1893[10]; Fleutiaux, A. S. E. F., p. 493, 1893[11]; Fowler, F. B. I., p. 394, 1912[12].

Type : Mus. Prague.

Burma[1, 2, 3, 5, 6, 7]
Karen Hills[9, 10, 12]
Pegu[4] -------------
Cochin China[6]
Laos (Luang-Prabang:A.)

140. **undulata,** Dejean, Spec. Col. I, p. 94, 1825[1]; Gistl., Syst. Ins., I, p. 52, 1837[2]; Chaud., Cat. Coll., p. 24, 1865[3]; Gemm. & Har., Cat. p. 28, 1868[4]; Atkinson, Cat. p. 13, 1889[5]; Fleutiaux, Comp. Rend. S. E. B., p. 168, 1890[6];

Indes Or.[1-4, 7]
Allahabad[14, 18]
Asansol[6, 14, 18]
Balugaon[18]

* Dr. Horn doubts all the exact localities given for this species as it has been mixed with *C. mariæ,* gestro, etc.

Cat., p. 48, 1892[7]; B. S. E. F., p. 369, 1917[8]; Horn, Ann. Gen., p. 680, 1894[9]; Spol. Zeyl., II, pp. 34, 43, pl. 7, f. 24, 1904[10]; Gen. Ins., p. 294, 1915[11]; Maindron, & Fleutiaux, A. S. E. F., p. 13, 1905[12]; Annand. & Horn, List, p. 11, 1909[13]; Fowler, F. B. I, p. 356, 1912[14]; Codina, Barcel. Cat., p. 427, 1917[15]; Annand. & Dover, R. I. M., XXII, p. 336, 1921[16]; Dover & Rib., R. I. M., XXII, p. 724, 1921[17]; XXV, p. 362, 1923[18]. Type : Mus. Paris.	Barküda Id.[17, 18] Calcutta[13, 14, 16] Ceylon (H.) Coimbatore (A.) Dacca[14, 18] Darjiling[15] Dehra Dun (H.) Genji[12] Gopkuda Id.[13, 18] Gulistan[18] Gwalior[14, 18] Lohardage[13, 14, 18] Margherita[15] Maldah[4, 13, 16] Nowatoli[6, 14, 18] Mysore[11, 14, 18] Mahe[4, 16, 18] Puttalam[10] Ramnad[14, 18] Samal Id.[18] Sylhet[14, 18] ---------- Annam (Ten-Sinh)[8] China[5] Hongkong[5, 14, 18] Japan[5.] Java[2] Lœbœ Gedang[9] Perak[18.] Silago[9]
141. **undulata** var. **dubia**, Horn, D. E. Z., p. 80, 1892[1], p. 87, 1904, Gen. Ins., p. 294, 1915[2]; Fleut., Cat., p. 48, 1892[3]; Fowler, F. B. I., p. 357, 1912[4]; Dover & Rib., R. I. M., XXII, p. 724, 1921[5]; XXV, p. 362, 1923[6]. Type : Coll. W. Horn.	? India[1, 3, 4] Darjiling[2, 6] Mandalay[5, 6] Margherita (H.) Myingyan[6] N. Assam[2, 6] Naga hills (H.) Pakokku, 180 ft.[6] Singla, 1,500 ft.[5, 6]
142. **venosa,** Kollar, Ann. Hofmus. Wien., p. 331, 1836[1]; Horn, A. S. E. B., p. 365, 1913[2], p. 140, 1914[3]; Gen. Ins., p. 292, 1915[4]; Gravely, R. I. M., VII, 1912[5]; Fleut., B. S. E. F., p. 49, 1917[6]; Codina, Barcel. Cat., p. 427, 1917[7]; Dover & Rib., R. I. M., xxv, p. 362, 1923 (in error).	*Georgia Americœ*[1] Bengal[4, 7, 8, 9, 10, 11, 14, 15] Asansol[13, 16, 18] Bhutan[16, 18] Berhampur (bank of R. Bhagirati)[17]

Cicindela nitida, Wiedemann, nec Licht, 1796, Germar's, Mag. Ent., IV, p. 117, 1821[8]; Dejean, Spec. Col., II, p. 91, 1825; Gistl., Syst. Ins. I., p. 50, 1837[9]; Schaum, Journ. Ent., II, p. 67, 1863[10]; Chaud., Cat., p. 26, 1865[11]; Gemm. & Har., Cat., p. 22, 1868; Atkinson, Cat., p. 24, 1889[12]; Fleutiaux, Comp. Rend. S. E. B., p. 168, 1890[13]; Cat., p. 56, 1892[14]; Lesne, Mission-Pavie, Indo-Chine, III, p. 61, 1904[15]; Horn, D. E. Z., p. 292, 1908; Fowler, F. B. I., p. 366, f. 159, 1912[16]; Dover & Rib., R. I. M., xxii, p. 724, 1921 (habits and mimicry)[17]; xxv, p. 360, 1923[18].

Type : Mus. Vienna.

Bombay[17]
Calcutta[16]
Chota Nagpur[16, 18]
Cuttack (bank of R. Mahanadi)[18]
Darjiling[16, 18]
Juggenathgunj (R. Brahmaputra)[18]
Kaladhungi[18]
Karachi[4, 7, 18]
Kashmir[4, 7]
Kurseong[16, 18]
Maldah[12, 16, 18]
Patna dist.[16]
Pegu[4, 7, 12, 16, 18]
Pondicherry[16, 18]
Rajmahal[16, 18]
Ramnad[16, 18]
Sara Ghat[16, 18]

Cambodia[2, 4, 7, 15, 16, 18]
Lao-Kay[3] Pak-lay[6]

143. **venus,** Horn, D. E. Z., p. 22, 1907[1]; Gen. Ins., p. 287, pl. 17, f. 6, 1915[2]; Fowler, F. B. I., p. 328, f. 146, 1912[3]; Dover & Rib., R. I. M., XXV, p. 362, 1923[4].

Type : Coll. H. E. Andrewes.

Nilgiri Hills[1-4]
Nadgani[2, 4]

144. **venus** var. **stuprata,** Horn, A. S. E. B., p. 446, 1909[1]; Gen. Ins., p. 287, 1915[2]; Dover & Rib., R. I. M., XXV, p. 362, 1923[3].

Type : Coll. W. Horn.

Wallardi[1]
Travancore[2, 3]

145. **vigintiguttata,** Herbst, Käfer, X, p. 174, pl. 171, f. 9, 1806[1]; Dejean, Spec., Col., I, p. 108, 1825[2]; Westermann, Rev. Ent. Silber, I, p. 106, 1833 (habits)[3]; Gistl., Syst. Ins., I, p. 58, 1837; Schaum, Journ. Ent., II, p. 67, 1863[4]; Chaud., Cat. Coll., p. 31, 1865[5]; Gemm. & Har., Cat., p. 29, 1868[6]; Atkinson, Cat., p. 13, 1889[7]; Fleut., Cat., p. 76, 1892[8]; Horn, D. E. Z., p. 60, 1897[9]; Gen. Ins., p. 305, f. 96, (on p. 33), 1915[10]; Annand. & Horn, List, p. 26, 1909 (habits)[11]; Fowler, F. B. I., p. 416, f. 180, 1912 (habits)[12]; Codina,

Ostindien[1-6, 8]
Barkul[14, 15]
Berhampur[11, 12, 15]
Bhutan[10, 11, 12, 13, 15]
Birbhum[11, 12, 15]
Dacca[12, 15]
Damukdia Ghat[11, 12, 15]
Kurseong, 5,000 ft.[11, 12, 15]
Maha River (N.)

Barcel. Cat., p. 439, 1917[13]; Dover & Rib., R. I. M., XXII, p. 725, 1921[14]; XXV, p. 362, 1923[15].

Type : Mus. Berlin.

Mungphu, 3,000-4,000 ft.[11, 12, 15]
Murshidabad[11, 12, 15]
Pankabari[7, 11, 12, 15]
Pusa (A.)
Rajmahal[12, 15]
Sara Ghat[11, 12, 15]
Sikkim[7, 10, 11, 13, 15]

Java[5, 8, 9]

146. **viridicincta,** Horn, D. E. Z., p. 173, 1894[1]; Gen. Ins., p. 287, 1915[2]; Fowler, F. B. I., p. 328, f. 145, 1912[3]; Fletcher, J. B. N. H. S., XXIII, p. 379, 1914[4]; Dover & Rib., R. I. M., XXII, p. 723, 1921[5]; XXV, p. 362, 1923[6].

azureocincta var. *Kurseong,* Horn, A. S. E. B., p. 537, 1892.

Type : Coll. H. E. Andrewes.

N. Kanara[1, 2, 3, 6]
Barway (H.)
Chota Nagpur[2, 3, 6]
Dehra Dun[2, 6]
Mysore[2, 6]
Pollibetta[4, 5, 6]
S. Mysore (H.)
Teppukadu 3,000 ft. (A.)
Nilgiri Hills[3, 6]

147. ***viridilabris,** Chaudoir, B. S. N. M., XXV (1), p. 24, 1852[1]; Cat., p. 23, 1865[2]; Schaum, Journ. Ent., II, p. 66, 1863[3]; Gemm. & Har., p. 29, 1868[4]; Atkinson, Notes Zool., N. W. P., p. 117, 1882[5]; Cat., p. 13, 1889[6]; Bates, A. M. N. H., XVII (5), p. 70, 1886[7]; Fleut., Cat. p. 47, 1892[8]; A. S. E. F., p. 486, 1893[9]; Horn., D. E. Z., p. 61, 1905; Gen. Ins., p. 291, 1915[10]; Zool. Mus. Leiden, VII, p. 109, 1922; Annandale & Horn, List, p. 9, 1909; Fowler, F. B. I., p. 349, 1912[11]; Dover & Rib., R. I. M., XXII, p. 723, 1921[12]; XXV, p. 363, 1923[13].

Type : Mus. Paris.

Indes Or.[1-4, 8]
N. India (H.)
Bengal[10, 13]
Dehra Dun (H.)
Kalka, 2,400 ft.[12, 13]
Pusa[13], Simla[10, 13]

E. Indies[11, 13]

148. **viridilabris** var. **fallaciosa,** Horn, D. E. Z., p. 57, 1897[1]; Gen. Ins., p. 291, 1915; Zool. Med. Leiden, VII, p. 109, 1922; Fowler, F. B. I., p. 343, 1912[2]; Dover & Rib., R. I. M., XXV, p. 363, 1923[3].

Cicindela viridilabris, Gestro, nec. Chaudoir, Ann. Gen., p. 355, 1893[4].

Teinzo[1-4]
Karen hills[2-4]

---

* This species has been recorded from the Karen hills, Madras, various localities in Ceylon, etc., but Dr. Horn says that all these records are wrong, as these specimens belong to the variety *labiocenea,* Horn.

C. chlorochila, Fleutiaux, ex parte, A. S. E. F., p. 485, 1893.

Type : Coll. W. Horn.

149. **viridilabris** var. **fuseo-euprascens,** Horn, D. E. Z., p. 60, 1905[1]; Gen. Ins., p. 291, 1915; Fowler, F. B. I., p. 350, 1912[2]; Dover & Rib., R. I. M., XXV, p. 363, 1923[3].

Type : Coll. H. E. Andrewes.

Nilgiri hills, 1,250 ft.[1, 2, 3]
Coimbatore (H.)
Kallar (B.M.)
Trichinopoly[2, 3]

150. **viridilabris** var. **labioænea,** Horn, D. E. Z., p. 79, 1892[1]; p. 61, 1905[2]; Spol. Zeyl., II, pp. 34, 44, pl. 7, f. 27, 1904[3]; Gen. Ins., p. 291, 1915[4]; Fleut., Cat., p. 47, 1892[5]; Maindron & Fleutiaux, A. S. E. F., p. 11, 1905[6]; Annand. & Horn, List, p. 9, 1909 (habits)[7]; Fowler, F. B. I., p. 349, 1912[8]; Dover & Rib., XXV, p. 363, 1923[9]; Horn, Ent. Rec., XXXVII, p. 166, 1925[a].

Cicindela viridilabris, Bates, A. M. N. H., XVII (5), p. 70, 1886[10]; Fleutiaux, A. S. E. F., p. 486, 1823; Codina, Barcel. Cat., p. 424, 1917.

Type : Coll. W. Horn.

Ceylon[1, 2, 4, 5, 10]
Colombo[3, 6, 9]
Dehiwela (A.)
Hatton[3]
Horawapotana[a]
Kandy[3, 6, 7, 8, 9]
Madulsima (B.M.)
Mahe[6], Nalanda[3]
Negombo[3, 6, 9]
Puttalam[3, 6, 9]
Peradeniya[3, 6, 9]
Weligama[3, 6]
Wellawaya[3, 6]
Yatiyantota[3]

151. **viridilabris** var. **severini,** Horn, A. S. E. B., p. 537, 1892[1], p. 364, 1913[2]; D. E. Z., p. 61, 1905[3]; Gen. Ins., p. 291, 1915[4]; Maindron & Fleutiaux, A. S. E. F., p. 12, 1905[5]; Fowler, F. B. I., p. 350, 1912[6]; Dover & Rib., R. I. M., XXV, p. 363, 1923[7].

Type : Coll. W. Horn.

Barway[1]
Bengal[6, 7]
Betchery Id.[5]
Chota Nagpur[4, 7]
Karkur Ghat (A.)
Kanara[6], Mahe[5, 7]
Ouchterlony valley, 2,500-5,000 ft. (A.)
Pollibetta[7]
Pandakel Bridge[5]
Teinzo[3, 6, 7]
Travancore[5, 6, 7]
Trichur (H.)
Wallardi (A.)

Tonkin[2, 3, 4, 6, 7]
S. China[4]

152 **vittigera,** Dejean, Spec. Col., I, p. 107, 1825[1]; Gistl., Syst. Ins., p. 58, 1837[2]; Chaud., Cat., Coll., p. 31, 1865; Redt., Reis. Freg. Novar. Zool., II, (1), p. 1,

Indes Or.[1, 2, 4]
Berhampur[9, 10]
Calcutta[9, 10]

1868[3]; Gemm. & Har., Cat., p. 29, 1868[4]; Atkinson, Cat., p. 13, 1889; Fleut., Cat., p. 76, 1892; A. S. E. F., p. 488, 1893[5]; Horn, D. E. Z., p. 60, 1897[6]; Gen. Ins., p. 305, 1915[7]; Annand. & Horn, List, p. 27, 1909[8]; Fowler, M. B. I., p. 417, f. 181, 1912[9]; Dover & Rib., R. I. M., XXV, p. 363, 1923[10].

Type : Mus. Paris.

Dacca[9, 10]
Damukdia Ghat[7, 9, 14]
Dinajpur[5]
Goalbathan[9, 10]
Maldah[7, 9, 10]
Pusa (H.)

Hongkong[3]
Java[2, 4]

153. **waterhousei,** Horn, D. E. Z., p. 206, 1900[1]; Spol. Zeyl., II, pp. 33, 41, pl. 7, f. 9, 1904[2]; Gen. Ins., p. 288, 1915[3]; Fowler, F. B. I., p. 325, 1912[4]; Dover & Rib., R. I. M., XXV, p. 363, 1923[5].

Type : Coll. W. Horn.

Ceylon[1-5]
Minneri (B.M.)

154. ***westermanni**, Schaum, B. E. Z., p. 75, 1861 (*Dromica*)[1]; Chaud., Cat., Coll., p. 15, 1865 (*Jansenia*); Gemm. & Har., Cat., p. 33, 1868[2]; Atkinson, Cat., p. 14, 1889[3]; Fleut., Cat., p. 38, 1892 (*Jan.*); Fowler, F. B. I., p. 332, 1912[4]; Horn, Gen. Ins., p. 290, pl. 17, f. 7, 1915[5]; Dover & Rib., R. I. M., XXV, p. 363, 1923[6].

Type : Mus. Berlin.

Tranquebar[1, 4, 6]
Coromandel[4, 5, 6]
Madras[1, 2, 3, 4, 6]
Minneri (B.M.)

155. **whithilli,** Hope, Col. Man., II, p. 23, 1838[1]; Chaud., Cat. Coll., p. 38, 1865[2]; Dohrn., S. E. Z., p. 310, 1881 (*Witchilli*)[3]; Dokhturoff, B. S. E. F., p. 151, 1887[4]; Atkinson, Cat., pp. 13, 24, 1889[5]; Fleut., Cat., p. 115, 1892; Maindron & Fleutiaux, A. S. E. F., p. 17, 1905[6]; Horn, D. E. Z., p. 59, 1905[7]; p. 154, 1908 (habits); Gen. Ins., p. 301, 1915[8]; Lefroy, Ind. Ins. Life, pl. 16, f. 2, 1909; Annand. & Horn, List, p. 22, 1909 (habits)[9]; Fowler, F. B. I., p. 385, 1912 (habits)[10]; Dover & Rib., R. I. M., XXV, p. 363, 1923.[11]

Type : Hope Museum, Oxford.

Bombay[1, 4, 5]
Anaimalai Hills, 3,000-4,000 ft.[6, 7, 10, 11]
Belgaum[10, 11]
Coonoor (environs)[6, 11]
Coorg[6]
Koondah range (B.M.)
Khanapur[10, 11]
Madras[3], Mahe[10, 11]
Nepal[2]
Nilgiri hills[6, 10, 11]
Ootacamund[10, 11]
Permaad (B.M.)
Poona[5, 8, 9, 10, 11]
Talewadi[11]
Tharrawaddy[10, 11]

---

* Dr. Horn remarks that all the localities of this species require confirmation.

155. **whithilli,**— Trichur[11]
Travancore Tea Co. (B.M.)
Wallardi[6, 11]
Wynaad[9, 10, 11]
Ghats merid. R.
Bretandeau[6]
Malabar[6]
N. Malabar[11]

156. **willeyi,** Horn, Spol. Zeyl., II, pp. 33, 36, pl. 7, f. 4, 1904[1]; Gen. Ins., p. 288[2]; Fowler, F. B. I., p. 326, 1912[3]; Dover & Rib., R. I. M., XXV, p. 363, 1923[4]. — Ceylonn[1, 2, 3, 4]

Type : Coll. W. Horn.

Genus : **APTEROESSA**, Hope (1838).

Apteroessa, Hope, Col. Man., II, pp. 13, 16, 159, 1838; Guérin, Rev. Zool., p. 13, 1834; Lacordaire, Mem. Soc. Sc. Liege, p. 114, 1843; Rev. Zool., p. 60, 1843; Gemm. & Har., Cat., p. 32, 1868; Atkinson, p. 14, 1889; Fleutiaux, Cat., p. 37, 1892; Horn, D. E. Z., p. 47, 1899, p. 46, 1905; Gen. Ins., pp. 210, 232, 413, 1915; Fowler, F. B. I., pp. 300, 440, 1912 Horn, Col. Cat., Cic., p. 313, 1926.

Cicindela, Fabricius (ex parte), Spec. Ins., I, p. 282, 1781.

Genotype : *Cicindela grossa* Fabricius (1781).

Distribution : Coromandel Coast.

1 ***grossa**, Fabricius, Spec. Ins., I., p. 282, 1781; Mant. Ins., I, p. 185, 1787; Ent. Syst., I, p. 170, 1792; Syst. Eleuth., I, p. 231, 1801; Olivier Ent., II, no. 33, p. 8, pl. 2, f. 23, 1790; Enc. Meth., p. 727, pl. 174, f. 4, 1790; Herbst, Käfer, X, p. 184, pl. 172, f. 7, 1806; Hübner, Natur forscher, Stück XXIV, p. 48, pl. 2, f. 18, 1789; Wilhelm, Unterhaltungen aus der Natur., Ins., I, p. 192, pl. 24, f. 175, 1834; Hope, Col. Man., II, p. 159, pl. 1, f. 1,[a-d] 1838; Motschulsky, — Tranquebar

* As this is known only from Tranquebar on the Coromandel Coast it has not been necessary to use numbers against the references. No perfect specimen of this species exists, and it has not been taken since it was first described. In his catalogue (1926) Horn adds the locality Ammayanayakanur, Madura, to the distribution of this species, but on what authority we do not know.

Etud. Ent., p. 33, 1855, Chaudoir, Cat., p. 54, 1865; Gemm. & Har., Cat., p. 32, 1868; Atkinson, Cat., p. 14, 1889; Fleutiaux, Cat., p. 37, 1892; Horn, D. E. Z., p. 47, 1899, p. 46, 1905; Gen. Ins., p. 414, pl. 20, f. 6, 1915; Fowler, F. B. I., p. 440, f. 198, 1912; Dover & Rib., R. I. M., XXV, p. 363, 1923.

Type : Brit. Mus. (Bank's Cabinet).

In poor condition.

Sub-family : MEGACEPHALINÆ, Horn (1907).

Megacephalina, Horn, Gen. Ins., pp. 8, 126, 1910 (key to genera).

Megacephalides, Lacordaire, Gen. Col., I, pp. 5, 9, 1854.

Megacephalites, Thomson, Monog. Cic., p. 17, 1857; Horn, D. E. Z., p. 466, 1907.

Megacephalidœ, Horn, D. E. Z., p. 16, 1898,

Megacephalini, Horn, D. E. Z., p. 466, 1907; Gen. Ins., pp. 116, 117, 1910.

Genus : MEGACEPHALA, Latreille (1802).

Megacephala, Latreille, Hist. Nat. Ins., III, p. 79, 1802; VIII, p. 191, pl. 71, f. 4, 1804; Latreille & Dejean, Hist. Nat. Icon. Col. Eur., ed., I, p. 36, 1822; Dejean, Spec. Col., I, p. 6, 1825; Icon. Hist. Nat. Col. Eur., ed. 2, p. 6, 1829; Brullé, Hist. Nat. Ins., IV, Col. I, p. 38, 1834; Castelnau, Rev. Ent. Silberm., II, p. 27, 1834; Etud. Ent., I, p. 33, 1835; Hist. Nat. Col., I, p. 10, 1840; Trans. Roy. Soc. Victoria, pp. 30, 36, 38, 1867; Lacordaire, Mem. Soc. Sciences, Liege, p. 100, 1843; Gen. Col., I, p. 11, 1854; Chaudoir, B. S. N. M., p. 67, 1850; Westwood, T. E. S., p. 57, 1852; Thomson, Monog., Cic., p. 18, 1857; Dokhturow, Spec. Cic., I, pp. 26-32, 1882; Horn, D. E. Z., p. 326, 1893, pp. 94-97; 1904, p. 105, 1905, pp. 263-271, 1907, p. 163, 1909; Fleut., Cat., p. 18, 1892; Rev. Ent. Cæn, p. 123, 1894, p. 285, 1896, p. 45, 1899; Sloan, Proc. Linn. Soc., N. S. Wales, p. 311, 1906; Horn,

Gen. Ins., p. 130, 1910; Fowler, F. B. I., p. 441, 1912; Horn, Col. Cat., Cic., p. 61, 1926.

Tetracha, Hope, ex parte, Col. Man., II, p. 6, 1838; Lacordaire, Mem. Soc. Sc., Liege, p. 100, 1843; Gen. Coll., I, p. 12, 1854; Chaud., B. S. N. M., p. 6, 1850; XX (4), p. 334, 1860; Jacquelin-Duval, Gen. Col., Eur., I, p. 121, f. 4, 1861-62; Brown, T. E. S., p. 351, 1860; Gemm. & Har., Cat., p. 3, 1868; G. Horn, Trans. Amer. Ent. Soc., p. 234, 1876; Dokhturow, Spec. Cic., I, pp. 33-73, 1882; Schaupp, Bull. Brookl. Ent. Soc., p. 8, 14, 1878; p. 77, 1884; Atkinson, Cat., p. 24, 1889; Fleutiaux, Cat., p. 19, 1892.

Grammognatha, Motschulsky, Käfer, Russlands, pl. 5, p. 5, 1850.

Genotype : *Cicindela senegalensis,* Linneaus.

Distribution : Cosmopolitan.

1. **euphratica,** Latreille & Dejean, Hist. Nat. Col. Eur., V, p. 37, pl. 1, f. 4, 1822; Castelnau, Rev. Ent. Silberm., II, p. 28, 1834; Hist. Nat. Ins., I, p. 11, 1851; Dejean, Spec. Col., I, p. 7, 1823; Ménétriés, Catal. Rais. Obj. Zool. Caucase, p. 93, 1832 (bionomics); Fischer, B. S. N. M., (4), p. 431, 1832; Brullé, Hist. Nat. Ins., IV, Col., I, p. 43, 1834; Gistl., Syst. Ins., I, p. 5, 1837; Blanchard, Rev. Zool., p. 160, 1846; Græęlls, B. S. E. F., p. 109, 1847 (bionomics); Westood, T. E. S., p. 57, 1852; Jacquelin-Duval, Gen. Col. Eur., I, pp. 1, 57, pl. 1, f. 1, 1*a*, 1857; Thomson, Monog. Cic., p. 29, pl. 4, figs. 7, 8, 1857; Coquerel, A. S. E. F., p. 616, pl. 14, figs. 1, 1*a-f*, 1859 (metamorphosis); Cotty, A. S. E. F., p. 327, 1860 (bionomics); Mem. Soc. Linn. N. Fr., I, pp. 159-162, 1866 (bionomics); Chaud., Cat. Coll., p. 42, 1865; Gemm. & Har., Cat., p. 4, 1868 (*Tetracha*); Mulsant & Mayet, Mem. Acad.-Lyon, XIX, pp. 314-318, 1872 (metamorphosis); Opuse, XV, p. 66, 1873 (metamorphosis); Fleut., Cat., p. 19, 1892 (*Tetracha*); Heyden, Cat. Col. Sib., (Deuts. Ent. Gesell), p. 5, 1893 (*Tetr.*); Atkinson, Notes Zool. N. W. P., p. 117, 1882; Cat., p. 24, (*Tetracha*);

Agra
Karachi
Kathiawar
Cutch
Salsette Id.

Amu Daria
Banks of Euphrates & Nile.
Neh-i-Bendan
Sarhad (Fontaime Tshaashei)
Also found in Spain, Levant, N. Africa & Western Asia.

Bedel, Cat. Rais. Col. N. Africa, I, p. 2, 1895; Maindron, A. S. E. F., p. 379, 1899; Tschitscherin, Hor. Soc. Ent. Ross., p. 2, 1903; König, Wien. Ent. Zeit., p. 140, 1904; Annand. & Horn, List, p. 5, 1909; Horn, Gen. Ins., p. 136, 1910; Fowler, F. B. I., p. 442, f. 199, 1912; Dover & Rib., R. I. M., XXV, p. 363, 1923.

Tetracha algeriana, Guérin, Rev. Zool., p. 428, 1846; B. S. E. F., p. 103, 1846.

T. syriaca, Gistl., in litt., Syst. Ins., I, p. 11, 1837.

Grammognatha euphratica var. *œegypten,* Motschulsky, Käfer, Russlands, p. 5, 1850.

Type : Mus. Paris.

2. ***euphratica** var. **armeniaca,** Castelnau, Rev. Ent. Silberm, II, p. 28, 1834; Brullé, Hist. Nat. Ins., IV, Col., I, p. 44, 1834; Dokhturow, Hors. Soc. Ent. Ross., p. 247 1885; Fleut., Cat., p. 20, 1892; Annand. & Horn, List, p. 4, pl. 1, f. 5, 1909; Horn, Gen. Ins., p. 136, 1910; Fowler, F. B. I., p. 443, pl. 199, 1912; Dover & Rib., R. I. M., XXV, p. 363, 1923.

Agra
Afghan-Baluchistan Frontier

Armenia
Siestan
Transcaspia to Amu Daria

Type : Unknown to the authors, but presumably in the Paris Museum.

* This form is sunk as a synonym of the typical form by Horn (1926).

PART XIV-PALPICORNIA

SYSTEMATICS. Palpicornia [Heer (1841), Fauna Col. Helv., I, p. 471, *excl. Georissida*] - Bach (1851), Käfer fn. Nord u. Mitteldeutschl. I, p. 4 — Fuss. (1859), Archiv. Ver. Siebenb. Landes K. n. F. IV (3), p. 103 — Seidlitz (1872), Fauna Baltica, ed. I, p. 31, 10 Bedel (1881), Fne. Col. Bassin Seine, I, p. 289 (*Hydrophilidoe* + *Sphæridiidoe*) — Lameere (1900), Ann. Soc. Ent. Belg., XLIV, p. 365—Ganglbauer (1904), Kaf. Mitteleur., IV, I, p. 151—de Peyerimhoff (1907), Bull. Soc. Entom. Fr., p. 125, 127 renv. (1)—Handlirsch (1908), Die foss. Insekt., p. 1276—Reitter (1909), Fna. Germ., II, p. 2—Kolbe (1911), Mem. 1 er Congrès intern. d'Entom., II, p. 62—d'Orchymont (1916), Ann. Soc. Ent. Fr., LXXXV, p. 91; (1919), Revue Zool. Afric. VI, p. 163.

Hydrophiles De Geer (1774), Mem. Hist. Ins., IV, p. 365.

Sphoeridiota Latreille (1894), Hist. Nat. Crust. Ins., X, p. 48 (Hydrophiliens) + Spheridiotes s. str.).

Hydrophilii + *Sphoeridiots* Latreille (1806-1807), Gen. Crust. Ins. I, p. 240, II, p. 61, 70.

Hydrophilides + *Sphoeridides* Leach (1815), in Brewster's Edinb. Encycl., IX, p. 95, 96.

Hydrophilii + *Sphoerididea* Leach (1817), Zool. Misc., III, p. 90, 95.

Palpicornes Latreille (1817), in Cuvier, Le Règne Animal. III, p. 269.

Hydrophiliens Lamarek, ex. p. (1817), Hist. Nat. Anim. Sans. Vert., IV, p. 533—Chenu (1851), Encycl. Hist. Nat. Col., I, p. 226 (*Hydrophiliens. Hydrophilii*) (*—Elmides, Hétérocérides*).

Helophoridoe + *Hydrophilidoe* + *Sphoeridiidoe* Stephens (1829), Ill. Brit. Ent. Mandib., Ill, p. 109, 118, 125.

Hydrophilides + *Hélophorides* Solier (1834), Ann. Soc. Ent. Fr., III. p. 302, 303.

Hydrophilen Erichson (1837), Käf. Mark Brandenburg, I. p. 193.

Hydrophili Redtenbacher (1845), Gatt. deutsch. Käfer-fna, p. 51, 70.

Hidrofiloideos Solier (1849), in Gay, Hist. fis. y polit. de Chile, Fauna, Zoologia, IV, p. 295.

Hydrophilidœ Fairmaire & Laboulbène (1855), Fne Entom. Francaise, I p. 223—Bose (1859), in Gutfleisch, Kaf. Dautschl., p. 5, 64—J.L. Leconte (1861 & 1862), Class. Col. North. Am., I, p. xxiii, 43—Packard (1870), Guide Study Ins., p. 437—Solsky (1876), Fedtsch. Reise Turkest., II, p. 3—Schlosser Klekovski (1877), Fna. Kornjasah, I, p. 106—Provancher (1877), Fna Ent. Canada, I, p. 208—Sharp (1882), Biol. Centr. Amer., I, 2, p. 53—J. Lec. et Horn (1883) Class. Col. North-Amer., p. xxx, 69—Seidlitz (1888), Fna Balt., ed. II, Famil., p. xli, 21; (1889), Fna. Transsylv., Famil., p. xli, 21—Comstock (1895), Man. Study Ins., p. 507, 527—Wickham (1895), Canad. Ent. XXVII, p. 181, 213—Kolbe (1898), Deutsch ost-Afrika, IV, Kaf., p. 125—Everts (1899), Col. Neerl. I, p. 16, 635; (1903), Suppl. p. 1; (1922), III, p. 283—Kliment (1899), Cesti Brouci, p. 109—Stierlin (1900), Col. Helv., p. 5, 211—Kolbe (1901) Archiv. Naturg., LXVII, Beih., p. 139—Ganglbauer (1904) Käfer Mitteleur., IV, 1, p. 141—Reitter (1909) in Brauer Süsswasser fna. Deutschl., 3-4, p. 7, 51; (1909), Fna Germ. II, p. 346—Blatchley (1910) Col. Indiana, Bull. Indiana Dep. Geol., I, p. 247—Jacobson (1911) Käfer Russl. p. 838—Fowler (1912) Fna. Brit. Ind. (Gen. Intr.) p. 96, 128—Kuhnt (1913), Ill. Best. tab. Käf. Deutschl., p. 320—Des Gozis (1916), Misc. Ent., XXIII, p. 1—Tullgren et Wahlgren (1920), Svenska Ins. (2) p. 200, 232.

Hydrophilides Imhoff (1856) Einf. Stud. Kol., p. 98.

Hydrophiliden Kuwert (1890), Verh. Naturf. Ver. Brünn, XXVIII, 1889, p. 3.

Palpicornios Germain (1901), Annales de la Universidad de Santiago de Chile, CIX p.? (1902), Apuntes Entomolojicos, Heloforidos Chilenos, p. 3.

CATALOGUES—

General. Gemminger et Harold (1868), Cat. Col. II, p. 475—Zaitzev (1908), Horœ Soc. Ent. Ross., XXXVIII, p. 324—Knisch (1924), Col Catal. W. Junk & S. Schenkling, Pars 79, Hydrophilidae, p. 1—306.

Indian. Atkinson (1890), Jourl. Asiat. Soc. Bengal, LIX (2), Suppl. p. 164.

ONTOGENESIS. *General.* Beutenmüller (1891), Journ. New York Microscopic 1 Soc., VII, p. 5-6 Brocher (1913), L'Aquarium de Chambre p. 210—Ganglbauer (1904), Die Käf. von Mittelleur., IV, 1, p. 146—MacGillivray (1903), New York State Mus. Bull. LXVIII (Ent. 18) p. 289—d'Orchymont (1913), Ann. Biol. lacustre, VI, p. 173, 204—Reitter (1909) in Brauer, Susswasserfna. Deutschl., 3-4 p. 51—Richmond (1920), Bull., American Mus. Nat. History, XLII, Art. I, p. 1, Pl. 1-XVI—Rupertsberger (1880), Biol. Käfer Eur. p. 269; (1894), Nachtr., p. 279—Schiödte (1861-1862), De Metamorphosi Eleutheratorum observations. Naturh. Tidsskr., I, pp. 202-203, Pl. III-VII; (1872) l.c. VIII, pp. 211-221, Pl. IX.

GENERAL PHYLOGENY AND CLASSIFICATION.

Houlbert (1894) Rapp. Nat. Col., p. 63—Lameere (1900) Ann. Soc. Ent. Belg., XLIV, p. 365—Ganglbauer (1904), Käfer, Mitteleur., IV, I, p. 151—Handlirsch (1906-1908) Die foss. Ins., p. 1276—d'Orchymont (1913), Ann. Biol. lacustre, VI, p. 181; (1916), Ann. Soc. Ent. France, LXXXV, p. 91-106; (1916), Bull. Soc. Ent. France, p. 139-141; (1916), l. c., p. 235-240; (1919), Revue Zool. Africaine, VI, p. 163-168; (1919), Ann. Soc. Ent. Belg., LIX, p. 84-85; (1919), Ann. Soc. Ent. France., LXXXVIII, p. 105-168—Richmond (1920), Bull. American Mus. Nat. History, XLII, Art. I, p. 77.

KEY TO FAMILIES.

1. The anterior transverse suture of head (clypeo-frontal) is not in relation in the middle posteriorly with a sagittal and the region of head before this

transverse suture, behind the labrum is constituted by the clypeus. Club of antennae composed of five[1] pubescent[2] joints. When only apparently pentarticulate by turgescence of the hairy second joint, the club appears strangled and the whole antenna has but 6 joints. First four joints of tarsi very short or more or less subequal in length, the fifth elongate nearly as long or longer than the four preceding joints taken together.

Larvae with feet, with inclined head, distinctly separated gular sutures, with well developed gula, broadly in contact with the occipital foramen, mouth parts inserted in a furrow of the undersurface of head, with primitive maxillæ. Antennae inserted nearer the externo frontal angle of head than are the mandibles. Ocular areas in groups of five single eyes, not aggregated. HYDRAENIDÆ

1. The anterior transverse suture of head (antenno frontal), when present, is directed angularly backwards in the middle and connected here with a sagittal, the whole forming a Y-suture. The region of head before the anterior branches of this Y, behind labrum is constituted by the prefrons. Clypeus sometimes a partly present but reduced then to the preclypeus as a sort of connective membrane, frequently of clearer colour, between labrum and prefrons, lodged in an emargination of the latter (*Hydrous, Neohydrophilus*, etc.). Club of antennae always composed of less than five pubescent joints. Only five ventral segments on abdomen by reduction of the 3rd morphological sternite. If it happens that there occurs a 6th segment (9th sternite), then this sclerite is membraneous (*Hydrochus*) or more or less retractile under the fifth (*Berosus, Laccobius*, etc.). Hind wings with cantharidiform venation. HYDROPHILIDÆ[3]

Larvae with horizontal frequently elevated head, gular sutures contiguous, exceptionally on a single point or not contiguous usually on a long space and

---

1. The African genus *Prosthetops* has only a 3-articulate antennal club, but the postfrons bears two conspicuous ocelli.
2. The antennæ of the New-Zealand *Orchymontia* have nearly entirely glabrous antennæ and are not conspicuously clubbed, but ony gradually increassate from the fifth joint onwards.
3. In a restricted sense.

pushing away the gula which is then very reduced and lying at a great distance from occipital foramen; mouthparts inserted on the extreme anterior under side of head or exceptionally in a furrow, very salient, with palpiform maxillae and labrum coalesced with clypeus to a labro-clypeus before prefrons. Ocular areas in groups of six single eyes, aggregated to a more or less dense group appearing sometimes (*Cercyon* f. i.) to form only one eye.

FAM. **HYDRÆNIDÆ**

Hydrænidae **A.** d'Orohymont (1919), Rev. Zoologique Africaine, VI, p. 166.

ONTOGENESIS AND MORPHOLOGY. **A.** d'Orchymont (1916), Ann. Soc. Ent. Fr., LXXXV, pp. 100, 104, 105 (*Palpicornia*, Categorie **I**).

KEY TO SUB FAMILIES.

1. At least six ventral segments on abdomen by non-reduction of the 3rd (morphological) sternite; the 6th ventral segment is thus the eighth sternite. Venation of hind wings staphyliniform. 2.

Larvae with 9 complete urites, the 9th with biarticulated procerci[1]; holopneustic type; eight pairs of annular uncovered adbominal spiracles. Clypeus well developed behind labrum and before prefrons and separated therefrom by a suture. Body covered with long and stiff hydrofugal setae, the teguments glabrous on the interspaces. Maxillae with an internal and an external lobe. Tenth write salient.

1' Only five ventral segments by reduction of the 3rd sternite. Prothorax not placed intimately against

---

1. According to Berlese, Gli Insetti, I, 1909, p. 333, the appendices of 9th tergite must be called procerci and are the true cerci. Richmond calls these true cerci, mesocerci, and reserves the term procerci for the appendices of 8th segment. It seems generally accepted however that the 8th segment bears no appendices and that when such are nevertheless present in appearance, when reduction of the last segments has taken place, they belong really to the following reduced segments. The mesocerci are for Berlese appendices of the 10th and, for him also, the acrocerci are dependant on the 11th articulation of abdomen (9th for Richmond). The prostyli belong according to Berlese to the 9th sternite; for Richmond to the 10th segment. New investigations seem necessary to settle there different points.

hind body, but free and usually narrowed behind. Labrum concealed under the projecting clypeus. Short, very convex form with projecting elytra. First abdominal segment more or less occupied by the posterior coxal cavities. Elytra often provided with tubercles. Eyes not divided by a canthus. Gular sutures not contiguous. Anterior coxal cavities open or not behind. Scutellum tolerably large as a long triangle. Venation of hind wings cantharidiform. Larvae with 8 complete urites with very reduced procerci; metapneustic type: abdominal spiracles reduced to the eighth pair, which arises on the ground of a stigmatic atrium. No labrum. Maxillae wih only an internal lobe. Body covered with a very short and dense, not hydrofugal pubescence mingled with longer setae. Segments less chitinous. 3. *Spercheinoe.*

2. Prothorax not placed intimately against hind body, but free and usually narrowed behind with ordinarily fovei-form impressions on the disc.

Larvae (*Hydroena, Ochthebium*): labrum little or not widened in its first half; first anterior seta of labrum (one on each side) curved to the axis of body and ramous at extremity or at least bifid; urotergites without distinct trace of an anterior transversal row of setigerous pores. *Hydrœninœ*

2′ Prothorax placed intimately against hind body, distinctly widened behind nearly as wide or of the same width as the base of elytra without impressions on the disc. Elytra truncate leaving the apex of abdomen uncovered. Seven ventral segments.

Larvae: labrum distinctly widened in the middle; first anterior seta covered also to the body but single, not even bifid, at extremity; urotergites with an anterior row of setigerous pores. Procerci widely separated proximally and nearly parallel. Setae on the clypeus (six in number) not placed at the anterior margin and equidistant even in the middle. Lobes of maxillae strongly separated apically. Mentum rounded on the sides, not very much widened in the middle, no wider than base of submentum. *Limnebiinœ*

SUB-FAM. **HYDRAENINAE.**

Hydroeninoe [Ganglbauer (1904), Käf. Mittel eur., IV, 1, pp. 152, 175, excl. *Hydrochus*]—d'Orchymont (1916), Ann. Soc. Ent. Fr., Vol. 85, p. 105; (1919), Revue Zool. Afric. VI, p. 165–Everts (1922), Col. Neerl. III. p. 284.

Hydroenaires Mulsant (1844) Hist. Nat. Col. Fr., Palpic., pp. 27, 50.

Ochrebiidoe Thomson (1859), Skand. Col., I, p. 15; (1860), l. c., II, p. 70.

Hydrenaires Rey (1885), Ann. Soc. Linn. Lyon XXXI, p. 353; (1886), l. c., XXXII, p. 14.

Hydroenitoe Kuwert (1890), Verh. Naturf. Ver. Brünn, XXVIII, p. 23.

Hydroenini Everts (1899), Col. Neerl., I, pp. 638, 640—Ganglbauer (1904), Kaf. Mitteleur., IX, 1, p. 176, 180.

Hydroenina [Zaitzev (1908), Hor.–Soc. Ent. Ross., XXXVIII, p. 336 (Catal.)]–Jacobson (1911), Käf. Russl., p. 840.

ONTOGENESIS. A. d'Orchymont (1919), Ann. Soc. Ent. Belg., LIX, p. 84-85—Richmond (1920), Bull. Amer. Museum Nat. Hist., XLII, Art. I, p. 21.

KEY TO GENERA.

1. Maxillary palpi longer, with an ultimate joint longer than the penultimate and not conspicuously more slender. Pronotum not bordered at some places with a membrane. 2.

1'. Maxillary palpi short, at most as long as antennae, with ultimate joint shorter and narrower than the more or less incrassate penultimate articulation. Antennae nine jointed with a pentarticulate club and the second glabrous joint elongate, oblong or obconic. Head with more or less conspicuous ocelli, one near each eye. Pronotum bordered at least in part with a translucid membrane. Third joint of tarsi longer than the two preceding joints. Labrum horizontal.

Larvae: Procerci nearly contiguous proximally and very divergent. First anterior seta of labrum bifid at extremity. Antennae shorter, with no prominent inner swellings and with the satae all distally inserted on the second joint. Pectinated process of posterior piece of mandibles (lacinia mobilis Richmond) narrower. Lobes of maxillae not separated apically. Mentum not conspicuously widened in the middle, not wider or not much wider than base of submentum. Gula more developed in the longitudinal direction. Setae of clypeus placed near anterior border forming two ternal groups very widely separated in the middle. Setae of the integument finer and less stiff. 3. *Ochthebius.*

2. Maxillary palpi shorter reaching anterior angles of pronotum; the latter transverse, widened on the sides and narrowed behind. Elytra irregularly punctured.

Larvae unknown. 1. *Loeliœna.*

2'. Maxillary palpi extraordinarily elongated, reaching far over anterior angles of pronotum; the latter more or less hexagonal, rarely transverse. Elytra ordinarily —not always–punctured in rows.

Larvae: Procerci widely separated proximally although they are borne in life in somewhat divergent manner. First enterior seta of labrum ramous or multifurcate at extremity. Antennae longer; second joint with two inner swellings of which the first placed in the middle or nearer the base, bears a long seta; third joint without inner swelling. Lacinia mobilis broader. Lobes of maxilla very slightly separated apically. Mentum conspicuously widened in the middle, very much wider than base of submentum. Gula more transverse less developed in the longitudinal direction. Setae of clypeus placed in the middle, far from anterior border, and equidistant even in the middle. Setae of integument more robust and stiff. 2. *Hydroena.*

1. Læliæna J. Sahlberg.

Læliæna J. Sahlberg (1900), Õfvers, Finsk Vet. Soc. Forh., XLII, p. 203. Kumaon (W. Almora)[1];

sahlbergi G.C. Champion (1920), Ent. Mo. Mag. LVI, p. 169. — Turkestan (Kendyk Taue).

sparsa J. Sahlberg (1900), l., c., p. 203.

2. **Hydraena** Kugelann.

SYSTEMATCS. *Hydroena* Kugelann (1794), in Schneid. Mag., I, 5, p. 578–Latreille (1817) in Cuvier, Règne Anim. III, p. 273; (1829), nouv. ed. p. 520; (1825) Families naturelles du Règne Anim. p. 366–Leach (1815), in Brewster's Edinb. Encycl., IX, p. 96; (1817), Zool. Misc. III, p. 90–Stephens (1829), Ill Brit. Ent. Mandib., II, p. 117–Curtis (1830), Brit. Ent., VII, p. 307 Text–Solier (1834), III, p. 303–Brullé (1835), Hist. Nat. Ins. Vol. Vbis, p. 308–Sturm (1836), Deutschl. Fna. Ins., X, p. 69, t. 224–Erichson (1837), Kãf. Mark Brandenb., I, p. 200–Laporte de Castelnau (1840) Hist. Nat. Ins., II, p. 407–Heer (1841), Fna. Col. Helv., I., p. 478–Mulsant (1844), Hist. Nat. Col. Fr., Palp., p. 50, 73–Redtenbacher (1845), Gatt. Deutsch. Käferfna. p. 70; (1849), Fna. Austr., p. 129; (1858), ed. II, p. 112; (1874), ed. III, II, p. 122–Kiesenwetter (1849), Linnaea Ent., IV, p. 156–Bach (1851), Käferfna. Nord u. Mitteldeutschl., I, p. 122, 127, 391–Chenu (1851), Encycl. Hist. Nat., I, p. 239–Lacordaire (1854), Gen. Col. I, p. 465, 469–Fairmaire et Laboul-bène (1854), Fne. Ent. France. I, p. 245–Leconte (1855), Proc. Ac. N. Sc. Philad., p. 357; (1861–1862), Class. Col. North. Amer; I, p. 45–Jacquelin Duval (1857), Gen. Col. I, p. 93, 98–Kuhn (1858) Käf. Südbayer. Flachl., p. 38–Bose (1859), in Gutfl. Käf. Deutschl., p. 69–Fuss (1859), Archiv. Ver. Siebenb-Landesk., n. F., IV (3), p. 120–Thomson C.G. (1859), Skand. Col. I, p. 15, (1860), l., c., II, p. 70; (1867), l. c., IX, p. 114–Seidlitz (1872), Fna. Balt., p. 11; (1888), ed. II, p. (24) (Genera), 121 (species); (1888), Fna. Transsylv. p. (24) (Genera), 121 (species)–Dalla Torre (1877) Jahresb. Ver. Naturk. Linz. (8), p. 67–Provancher (1877), Fne. Ent. Canada, I, p. 212–Schlosser Klek (1877), Fna. Kornjašah, I, p. 107–Bedel (1881), Fne. Col. Bass. Seine, I, p. 291–Sharp (1882) Biol. Centr. Amer., I, 2, p. 99–Leconte et Horn (1883) Class. Col. North-Amer., p. 71–Rey (1886), Ann. Soc. Linn. Lyon XXXII, (1885), P. 14, 68–Fowler (1887),

Col. Brit. Isl. I, p. 241, 247–Kuwert (1888), Deuts. Ent. Zeitschr., XXXII, p. 113; (1890), Verh. Naturf. Ver. Brünn, XXVIII, p. 23, 278–Wickham (1895), Canad. Ent., XXVII, p. 181–Acloque (1896), Fne. France, p. 95–Everts (1899), Col. Neerl., I, p. 640, 664–Kliment (1899), Cešti brouci, p. 124–Lameere (1900), Man. Fne. Belge, II, p. 325–Stierlin (1900), Col. Helv., p. 215–Sainte Claire Deville (1902), Echange, XVII, p. 76-78, 83-85; (1905), l. c., XXI, Annexe, p. 1–Ganglbauer (1904), Käf. Mitteleur; IV, 1, p. 180, 197–Reitter (1909), in Brauer, Süsswasserfna Deutschl., 3-4, p. 57, 61, fig. 41; (1909), Fna. Germ., II, p. 350, 353–Blatchley (1911), Col. Indiana, Bull. Indiana Dep. Geol. I (1910), p. 249–Jacobson (1911), Käf. Russl., p. 844–Kuhnt (1913), Ill. Best. Tab. Kãf. Deutschl. p. 417, 426, f. 7–Schaufuss (1916) in Calwer's Käferb., ed. 6 (1913), p. 325, 329–Des Gozis (1918), Misc. Ent., XXIV, p. 97–d'Orchymont (1925), Ann. & Bull. Soc. Ent. Belg. LXV, p. (45).

Amphibolus + *Hydroena* G.R. Waterhouse (1833), Ent. Mag., I, p. 292.

NTOGENESIS. A. d'Orchymont [1919 (4th July)], Ann. Soc. Ent. Belg., LIX, p. 84-85 (egg-case, larva)–Richmond [1920 (20th Sept.)], Bull. Amer. Museum Nat. Hist., XLII, Art. I, p. 82 (egg-case), 84 (larva).

Sub-gen. **Hydraena** s. str. Stierlin.

Hydroena s. str. Stierlin (1990), Col. Helv. p. 230–Ganglbauer (1904), Käf. Mitteleur., IV, I, p. 197, 199–Sainte Claire Deville (1906) Abeille, XXX, p. 283–Reitter (1909), Fna. Germ., II, p. 354–Kuhnt (1913), Ill. Best.-Tab. Kaf. Deutschl., p. 427–Schaufuss (1916), in Calwer. Käferb., ed. 6 (1913), p. 329–Des Gozis (1919), Misc. Ent., XXIV, p. 154–Barthe (1920), l. c., p. 180–Everts (1922), Col. Neerl., III, p. 302–d'Orchymont (1923), Ann. Soc. Ent. Belg., LXIII, p. 33; (1925), Ann. et Bull. Soc. Ent. Belg., LXV, p. 46.

Hoplydroena+*Hydroena* in sp. Kuwert (1888), Deutsch. Ent. Zeitschrift., p. 115; (1890), Verh. Naturf. Ver. Brünn, XXVIII, p. 279–Seidlitz (1888), Fna. Balt., ed. II, p. 121; (1888), Fna Transsylv., p. 128–Porta (1900), Misc. Entom., VIII, p. 7.

bihammata G.C. Champion (1920), Ent. Mo. Mag., Vol. LVI, p. 70, fig. 2 (p. 69). ♂[1]. — Kumaon (W. Almora)[1]; Chakrata, Mussoorie.

cirrata G.C. Champion (1920), Ent. Mo. Mag., Vol. LVI, p. 69, fig. 1 ♂[1]. — Kumaon (W. Almora); Chakrata, Mussoorie.

dimorpha A. d'Orchymont (1922), Records Ind. Mus., VIII, 1922, p. 623, fig. 1♂♀. — Assam (Dibrugarh).

indica A. d'Orchymont (1920), Ann. Soc. Ent. Belg., LX, p. 18. — Khandesh.

* **maculicollis** G.C. Champion (1920), Ent. Mo. Mag., Vol. LVI, p. 70. — Kumaon (W. Almora).

orientalis Breit (1916), Col. Rundschau, V, p. 54-55. — Buchara, Persia.

scabra A. d'Orchymont (1925) subsp. ♂ *knischi* nov.** — Tonkin (Lac Tho).[1]

? scabra A. d'Orchymont (1925), Ann. Bull. Soc. Ent. Belg., LXV, p. 261.[1]

3. Ochthebius Leach.

Ochthebius Leach (1815), in Brewster's Edinb. Encyclopedia, IX, p. 95; (1817), Zool. Misc., III, p. 90–Latreille (1829) in Cuvier, Règn. Anim., nouv. ed. IV, p. 520–Curtis (1830), Brit. Ent., VII, p. 291–Solier (1834), Ann. Soc. Ent. Fr., III, p. 303–Brullé (1835), Hist. Nat. Ins., Vol. 5bis, p. 307–Sturm (1836),

* A. Knisch in his Catalogue (1924) has this species under the subgenus *Hadrenya=Hœnydra)* perhaps because of Champion saying that *maculicollis* is not unlike *atricapilla* Waterh. But there are at least 8 series of punctures on elytra between scutellum and shoulder. It is thus a true *Hydrœna* s. str.

** When working on new material from the Philippines and from Sumatra communicated by the late Mr. Alfred Knisch and on re-preparing therefore the specimen from Tonkin (Lac Tho) attributed by me with doubt to *scabra*, I have found that this specimen is a ♂ and that it possesses near the last third of inner side of the anterior, not curved, tibiæ a tolerably well developed mucron, accompanied by some much finer spines. Nothing of the sort could be observed in the type of *scabra*, which is a ♂ (not ? female as printed in of Philippines Journal of Science 30, 1926, p. 363), nor in two other males, one from Mount Maquiling (Philippines), the second from Fort de Kock (Sumatra). The Indo-Chinese specimen is otherwise wholly identical to these specimen, with perhaps the exception of the medio-lateral angulation of sides of pronotum which is more apparent. It represents apparently a ♂ sub-species of *scabra*. I have named it in hounour of my regretted colleague, in mind of the aid he has given me to interpret the species.

Deutschl. Fna., Ins., X, p. 53, t. 221–Erichson (1837), Käf. Mark Brand. I, p. 199–Laporte de Castelnau (1840), Hist. Nat. Ins., II, p. 48–Heer (1841) Fna. Col. Helv., I, p. 477–Mulsant (1844), Hist. Nat. Col. Fr. Palp., p. 50-51–Redtenbacher (1845), Gatt. Deutschl. Käferna, p. 71; (1849), Fna. Austra., p. 128; (1858), l., c., ed. II, p. 110–Bach (1851), Käferfna Nord-ü Mittel Deutschl., I, p. 122, 125–Chenu (1851), Encycl. Hist. Nat. Col. I, p. 237–Wollaston (1854), Ins. Mader., p. 90–Lacordaire (1854), Gen. Col., I, p. 465, 468–Fairmaire et Laboulbène (1854) Fne Fr. Col. I. p. 235, 239–Leconte (1855), Proc. Acad. N. Sc. Philad., p. 357–*Jacquelin Duval* (1857), Gen. Col. Eur., I, p. 93, 98–Kuhn (1858) Käf. Südbayer, Flachl., p. 38–Bose (1859), in Gutfl. Käf. Deutschl., p. 68–Fuss (1859), Archiv. Ver. Siebenb, Landesk., n.f., IV, (3), p. 119–Leconte (1861-1862) Classif. Col. North. Amer., p. 45–Seidlitz (1872), Fna. Balt., p. 11; (1888), id., ed. II, Gattung., p. 24–Solsky (1876), Fedtsch. Reise Turkestan, II, p. 21–Leconte and Horn (1883), Class. Col. North-Amer., p. 71–Kuwert (1887), Deutsch. Ent. Zeitschr., p. 369; (1890), Verh. Naturf. Ver. Brünn, XXVIII, p. 23, 234–Horn G. H. (1890), Trans. Am. Ent. Soc., XVII, p. 17–Wickam (1895), Canad. Entom., XXVII, p. 181, 184–Acloque (1896), Fne. France, p. 95–Everts (1899), Col. Neerl., I, p. 640, 661; (1922), l. c., III, p. 300–Kliment (1899), Cešti Brouci, p. 123–Lameere (1900), Fne Belg., II, p. 325–Stierlin (1900), Col. Helv., p. 215–Ganglbauer (1904), Käf. Mitteleur-IV, 1, p. 180–Reitter (1909), in Brauer, Süsswasserfna Deutschl., 3-4, p. 57, 58, fig. 40; (1909), Fna. Germ., II, p. 350, 351–Blatchley (1911), Col. Indiana, Bull. Indiana Dept. Geol. I. (1910), p. 249, 253–Jacobson (1911), Käf. Russl., p. 844–Kuhnt (1913), Ill. Best. Tab. Käf. Deutschl., p. 417, 424. fig. 8–Schaufuss (1916), in Calwer Käferb., ed. 6 (1913), p. 325, 327–Des Gozis (1918), Misc. Ent., XXIV, p. 96–Tullgren et Wahlgren (1920), Svenska Ins., (2), p. 233.

Hydroena Illiger (1798) (nec. Kugelann, 1794), Verz. Käf, Preuss., p. 275.

Ochtebius+Enicocerus Stephens (1829), Ill. Brit. Ent. Mandib., II, p. 194-196.

Ochtebius+Asiobates C.G. Thomson (1859) Skand. Col. I, p. 15; (1860), id., II, p. 73; (1867), id., IX, p. 114.

Ochthebius+Calobius Wollaston (1854), Ins. Mader., p. 90, 92.

Henicocerus+Ochthebius Bedel (1881), Fne. Col. Bass. Seine, I, p. 291, 294.

Henicocerus+Ochthobius+Calobius Rey (1886), Ann. Soc. Linn. Lyon, XXXII, p. 14, 15, 22, 63.

Henicocerus+Ochthebius Fowler (1887), Col. Brit. Isl., I, p. 241, 243.

ONTOGENESIS. Ganglbauer (1904), Käf. Mitteleur., IV, I, p. 182-183–A. d'Orchymont (1913), Ann. Biol. lac., VI, p. 182, 207–Richmond (1920), Bull. Amer. Nat. Hist., XLII, Art. 1, p. 21, 22, 82, 84, Pl. 1, f. 4.

Sub-gen. **Asiobates C.G. Thomson.**

Asiobates C.G. Thomson (1859), Skand. Col., I, p. 15; (1860), id., II, p. 73–Everts (1899), Col. Neerl., I, p. 662; (1922), id., III, p. 300–Ganglbauer (1904), Käf. Mitteleur., IV, 1, p. 184–Reitter (1909), in Brauer, Süsswasserfna Deutschl., 3-4, p. 60; (1909), Fna. Germ., II, p. 352–Schaufuse (1916), in Calwer, Käferb., ed. 6 (1913), p. 328.

Trymochthebius Kuwert (1887), Deuts. Ent. Zeitschr., XXXI, p. 380, t. I, fig. 12.

Asiobathes Kuhnt (1913), Ill. Best. Tab. Käf. Deutschl., p. 424, fig. 49[a].

amplicollis G.C. Champion (1925), Ent. Mo. Mag., Vol. LXI, p. 172.	Nainital.
atricapillus Reitter (1901), Deutsch. Ent. Zeitschr., p. 66.	Buchara.
depressus J. Sahlberg (1900), Ofvers. Vet. Soc. Forh., XLII, p. 195.	Turkestan.
jagthanae G.C. Champion (1921), Ent. Mo. Mag. Vol. LVII, (1925), l. c., Vol. LXI, p. 260.	India; Kumaon (Jagthana)
minimus G.C. Champion (1925), l. c., Vol. LXI, p. 173.	Bodiar (Haldwani).
kosiensis G.C. Champion (1920), Ent. Mo. Mag., LVI, p. 169.	Kumaon (W. Almora).
laevisculptus Reitter (1901), Deuts. Ent. Zeitschr., p. 65.	Buchara (Kulab).

Sub-gen. **Homalochthebius Kuwert.**

Homalochthebius Kuwert (1887), Deuts. Ent. Zeitschr., Vol., XXXI, p. 383, t. I, fig. 15; (1890), Verh. Naturf. Ver. Brünn, XXVIII, p. 237–G.H. Horn (1890) Trans. Amer. Ent. Soc. XVII, p. 18–Everts (1899), Col. Neerl. I, p. 663; (1922), id., III, p. 300–Ganglbauer (1904), Käfer Mitteleur., IV, 1, p. 184–Reitter (1909), in Brauer, Süsswasserfna Deutschl., 3-4, p. 59; (1909), Fna. Germ., II, p. 352–Kuhnt (1913), Ill. Best. Tab. Käf. Deutschl., p. 424, fig. 49*b*.–Schaufuss C. (1916), in Calwer, Käferb., ed. 6 (1913), p. 328.

Asiobates Rey (1886) (nec. C.G. Thomson, 1859), Ann. Soc. Linn. Lyon, XXXII, p. 25.

coomani A. d'Orchymont (1925), Ann. & Bull, Soc. Ent. Belg., LXV, p. 265. — Tonkin.

yunnanensis A. d'Orchymont (1925), l. c., p. 263. — Yunnaz.

Sub-gen. **Hymenodes** Mulsant.

Hymenodes Mulsant (1844), Hist. Nat. Col. Fr., Palp., p. 68–Chenu (1851), Encycl. Hist. Nat. Col., I, p. 239–Rey (1886), Ann. Soc. Linn. Lyon, XXXII, p. 53–Seidlitz (1888), Fna. Balt., ed. II, p. 119; (1888), Fna. Transsylv., p. 126–Stierlin (1900), Col. Helv., p. 229–Ganglbauer (1904), Käf. Mitteleur; IV, 1, p. 184–Reitter (1909) in Brauer, Süsswasserfna Deutschl., 3-4, p. 60; (1909), Fna. Germ; II, p. 353–Everts (1912), Tijdschr. Ent., LV, p. 292–Kuhnt (1913), Ill. Best. Tab. Käf. Deutschl., p. 424, f. 57–Schaufuss, C. (1916) in Calwer, Käferb., ed. 6 (1913), p. 328–Everts (1922), Col. Neerl., III, p. 301.

Cheilochthebius+*Camptochthebius* (ex p.) + *Eccoptochthebius* Kuwert (1887), Deuts. Ent. Zeitschr., XXXI, p. 374, 377, 379, t. I, fig. 6, 9, 10; (1890), Verh. Naturf. Ver. Brünn, XXVIII, p. 235, 236.

cupricollis G.C. Champion (1925), Ent. Mo. Mag., Vol. LXI, p. 172 [praeocc.: *cupricollis* J. Sahlberg (1902-1903), Öfv. Finska Vet. Soc. Forh. XLV, n°. 10, p. 6, Kopet Dagh.] — Kumaon (Haldwani).

foveolatus Germar (1823), Ins. Spec. nov. (1824), p. 90[1]–Sturm (1836), Deutschl. Fna, Käf. X, p. 58, t ccxxi, f.m. — Germany.[1] I. Sylt.[7] Europe med.[4]

M.–Laporte de Castelnau (1840), Hist. Nat. Ins., II, p. 48–Heer (1841), Fna. Col. Helv., I, p. 477–Mulsant (ex p.) (1844), Palp., p. 70, var. B–Redtenbacher (1849), Fna. Austr., p. 129; (1858), id., ed. 3, II, p. 121–Bach (1851), Käferfna Deutschl., I, p. 127–Fairmaire et Laboulbène (1854), Fne. Ent. Fr., I, p. 244, var B.–Bose (1859), in Gutfl., Käf. Deutschl., p. 69–Fuss (1859), Archiv. Ver. Siebenb. Landesk n. F., IV, (3); p. 120-Dalla Torre (1877), Jahresber. Ver. Naturk. Linz (8), p. 71–Schlosser Klek (1877), Fna. Kornjasah, I, p. 114–Bedel (1881), Fne, Col. Bases. Seine, I, p. 294, 318–Rey (1886), Ann. Soc. Linn. Lyon, XXXII, p. 54, 60 (*Ochthobius*)[2]–Kuwert (1887) Deuts. Ent. Zeitschr., XXXI, p. 375, t. II, fig. 18; (1890), Verh. Naturf. Ver. Brünn, XXVIII, p. 247[3]–Seidlitz (1888), Fna. Balt., ed. 2, p. 121; (1888), Fna. Transsylv, p. 128–Acloque (1896), Fne Fr., p. 96–Kliment (1899), Cešti brouci, p. 124,–Stierlin (1900), Col. Helv., p. 229–Ganglbauer (1904), Käf. Mitteleur., IV, I, p. 192[3, 4]–Reitter (1909), in Brauer, Süsswasserfna Deutschl., 3-4, p. 60; (1909), Fna. Germ., II, p. 353, t. 78, fig. 11–J. Müller (1909), Zool. Bot-Ges. Wien, LLX, p. 475–Kuhnt (1913), Ill. Best-Tab. Käf. Deutschl., p. 426, f. 57–J. Sahlberg (1913), Ofv. Finsk. Vet. Soc. Forh., LV, A, n°. 19, p. 55[5]–Des Gozis (1919), Misc. Ent., XXIV, p. 152–G.C. Champion (1920), Ent. Mo. Mag., LVI, p. 168[6]–Everts (1922), Col. Neerl., III, p. 299, 302.[7] — France.[2, 6] Italy.[3] Dalmatia.[4] Bosnia.[4] Herzegovina.[4] Caucasus.[6] Anatolia.[5] Kumaon[6] (W. Almora Ranikhet)

brunneus Dalla Torre (1877), Jahresber. Ver. Nat. Linz, p. 71.

niger Dalla Torre (1877), l. c.

nitidipennis G.C. Champion (1920), Ent. Mo. Mag., LVI, p. 166. — Kumaon (W. Almora)

opacipennis G.C. Champion (1920), l. c., p. 167[1]–A. d'Orchymont, (1925), Ann. & Bu 1. Soc. Ent. Belg., LXV, p. 266.[2] — Kumaon (W. Almora).[1] Tonkin.[2]

rivalis G.C. Champion (1920), l. c., p. 166. — Kumaon (Riv. Sarda Gorge).

scintillans G.C. Champion (1920), l. c., p. 168. — Kumaon (W. Almora).

sexfoveatus G.C. Champion (1920), 1. c., p. 165. — Kumaon (W. Almora, Ranikhet).

strigosus G.C. Champion (1921), Ent. Mo. Mag., LVII, p. 177. — Kumaon (Ranikhet).

Sub-gen. **Bothochius** Rey.

Bothochius Rey (1886), Ann. Soc. Linn. Lyon, XXXII, p. 53–Ganglbauer (1904), Käf. Mitteleur., IV, 1, p. 184–Reitter (1909), in Brauer, Süsswasserfna. Deutschl., 3-4, p. 59; (1909), Fna. Germ., II, p. 353–Kuhnt (1913), Ill. Best. Tab. Käf. Deutschl., p. 424, fig. 55B–Schaufuss C. (1916), in Calwer, Käferb., ed. 6 (1913), p. 328–Everts (1912), Tidschr. Ent., LV, p. 291; (1922), Col. Neerl., III, p. 301.

Odontochthebius+Camptochthebius (ex. p.)+*Colpochthebius* Kuwert (1887), Deutsch. Ent. Zeitschr., XXXI, p. 377, 379, t. I, f. 8, 9, 11; (1890), Verh. Naturf. Ver. Brünn, XXVIII, p. 236.

Colpochthebius+Camptochthebius (ex. p.) Seidlitz (1888), Fna. Balt., ed. 2, p. 119; (1888), Fna. Transsylv., p. 126.

Odontochthebius G.H. Horn (1890), Trans. Amer. Ent. Soc., XVII, p. 18.

nobilis Villa (1835), Suppl. Col. Europ. Dupl., p. 48 (Helophorus)–Heer (1841), Fn. Col. Helv., I, p. 478–de Marseul (1882), Abeille, XX, p. 175[1]–Rey (1886), Ann. Soc. Linn. Lyon, XXXII, p. 53. 54, (*Ochthobius*)–Kuwert (1887), Deutsch. Ent Zeitschr., XXXI, p. 377, t. IV, f. 78; (1890), Verh. Naturf. Ver. Brünn, XXVIII, p. 251[3]–Ganglbauer (1904), Käf. Mitteleur., IV, 1, p. 190[4]–Reitter (1909), in Brauer, Süsswasserfna Deutschl., 3-4, p. 59[5]; (1909), Fna. Germ., II, p. 353[5]–Kuhnt (1913), Ill. Best. Tab. Käf. Deutschl., p. 425, f. 55B–Des Gozis (1919), Misc. Ent., XXIV, p. 138–G.C. Champion (1920), Entom. Mo. Mag., LVI, p. 167.[6]

Switzerland.[1, 3]
Italy.[1, 2, 3, 4]
France.[2], Spain.[2]
Bavaria.[5]
Dalmatia.[4]
Greece.[3, 4]
Naxos.[4]

Kumaon (W. Almora, Ranikhet), Riv. Sarda Gorge, Tanakpur)[6]

hybernicus Sturn (1836) (nec. Curtis, 1829), Deutschl. Fna. Käf., X, p. 62, t. CCXXII, f. c. C.

punctatus Redtenbacher (1849) (nec Stephens, 1829), Fna. Austr., p. 129.

hibernicus Redtenbacher (1858) (nec Stephens, 1829),

Fna. Austr., ed. 2, p. 111; (1874) id., ed. 3, p. 121.

fluviatilis Guillebeau (1896), Bull. Soc. Ent. Fr., p. 240 (*Ochthobius*). Turkestan (Mt. Alai).

turkestanus Kuwert (1892), Deutsch. Ent. Zeitschr., p. 102.

SUB-FAM. **LIMNEBIINÆ.**

Limnebiinoe A. d'Orchymont (1913), Ann. Biol. lac., VI, p. 182; (1916), Ann. Soc. Ent. Fr., Vol. 85, p. 105; (1919), Rev. Zool. Afric., VI, p. 165–Everts (1922), Col. Neerl. III, p. 285.

Limnebiaires Mulsant (1844), Hist. Nat. Col. Fr., Palp., p. 88.

Limnebiidoe C.G. Thomson (1859), Skand. Col., I, p. 14; (1860), id., II, p. 67; (1867), id., III, p. 113.

Limnobiini Bedel (1881), Fne. Col. Bass. Seine, I, p. 302.

Limnobiates Rey (1885), Ann. Soc. Linn. Lyon, XXXI, p. 314.

Limnebiitoe Kuwert (1890), Verh. Naturf. Ver. Brünn, XXVIII, p. 19, 87.

Limnebiini Everts (1899), Col. Neerl., I, p. 638,–Ganglbauer (1904), Käf. Mitteleur. IV, 1, p. 220, 257.

Limnebiina Zaitzev (1908), Hor. Soc. Ent. Ross., XXXVIII, p. 395 (Catal).

Limnobiina Jakobson (1911), Käf. Russl., p. 840.

ONTOGENESIS. A. d'Orchymont (1913), Ann. Biol. lac., VI, p. 181–Richmond (1920), Bull. Amer. Mus. Nat. Hist., XLII, Art. 1, p. 20.

KEY TO GENUS.

Comprises only the genus *Limnebius* Leach.

4. **Limnebius** Leach.

Limnebius Leach (1815), in Brewster Edinb. Encycl., IX, p. 96; (1817), Zool. Misc., III, p. 92–Latreille (1829) in

Cuvier, Règne Anim., nouv. ed., IV, p. 524–Stephens (1829), Ill. Brit. Ent. Mandib., II, p. 119; (1839), Man. Brit. Col., p. 88–Erichson (1837), Käf. Mark Brandenb. I, p. 201–Laporte de Castelnau (1840), Hist. Nat. Ins., II, p. 56–Westwood (1840), Introd. Mod. Class. Ins. II, Synopsis, p. 9–Heer (1841), Fna. Col. Helv., I, p. 481–Mulsant (1844), Hist. Nat. Col. Fr., Palp., p. 88–Redtenbacher (1845), Gatt. Deutschl. Käferfna, p. 71; (1849), Fna. Austr., p. 130; (1858), id, ed. 2, p. 106; (1874), id., ed. 3, p. 115–Chenu (1851), Encycl. Hist. Nat. Col., I, p. 241–Bach (1851), Käferfna Nord Mitteldeutschl., I, p. 122, 130–Lacordaire (1854), Gen. Col., I, p. 455, 460–Fairmaire et Laboulbène (1854), Fne. France, I, p. 227, 232–Wollaston (1854), Ins., Mader., p. 93–Jacquelin Duval (1857), Gen. Col. Eur., I, p. 89, 98–Kuhn (1858), Käf. Südbayer. Flachl., p. 38–Bose (1859), in Gutfl., Käf. Deutschl., p. 67–Fuss (1859), Archiv. Ver. Siebenb. Landesk. n. F., IV, (3), p. 113,–C.G. Thomson (1860), Skand. Col., II, p. 68–Seidlitz (1872), Fna. Balt., p. 11; (1888), ed. 2., p. 21; (1888), Fna. Transsylv., p. 21–Solsky (1876), Fedtsch. Reise Turkest., II, p. 21–Dalla Torre (1877), Jahresber. Ver. Naturk. Linz (8), p. 66–Schlosser Klek (1877), Fna. Kornjašah, I, p. 107–Leconte et G.H. Horn (1883), Class. Col. North-Amer., p. 72–Rey (1883), Rev. Ent. Caen, II, p. 84–Fowler (1887), Col. Brit. Isl. I, p. 222, 230–Kuwert (1890), Verh. Naturf. Ver. Brünn, XXVIII, p. 19–Acloque (1896), Fne. Fr., p. 92–Everts (1899), Col. Neerl., I, p. 639, 648–Kliment (1899), Cešti brouci, p. 114–Lameere (1900), Man. Fne. Belg., II, p. 329–Casey (1900), Journ. N. York Ent. Soc., VIII, p. 51–Stierlin (1900), Col. Helv., p. 211–Ganglbauer (1904), Käf. Mitteleur., IV, I, p. 258–Reitter (1909), in Brauer, Süsswasserfn. Deutschl., 3-4, 64, 76, fig. 52; (1909), Fna Germ, II, p. 357, 365–de Peyerimhoff (1909) L'abeille, XXXI, p. 121–Kuhnt (1913) Ill. Best. Tab, Käf. Deutschl. p. 417, 435, f. 9, 11^{A}–Schaufuss C. (1916), in Calwer Käferb., ed. 6 (1913), p. 345–Des Gozis (1916), Misc. Ent., XXIII, p. 3–Tullgren et Wahlgren (1920), Svenska Ins., (2), p. 234, 235–A. d'Orchymont (1920), Ann. Soc. Ent. Fr., LXXXIX, p. 33, T III, fig. 17 (Wing venation); (1921), Ann. Soc. Ent. Belg., LXI, p. 272, fig. 14 (Wing-venation)–Everts

(1922), Col. Neerl., III, p. 291.

Limnobius Bedel (1881), Fne. Col. Bass. Seine, I, p. 314–Rey (1885), Ann. Soc. Linn. Lyon, XXXI, p. 314.

Limnocharis G.H. Horn (1872), Trans. Amer. Ent. Soc., IV, p. 144; (1873), Proceed. Amer. Philos. Soc., XIII, p. 118–Sharp (1882), Biol. Centr. Amer., I, 2, p. 85–Casey (1886), Bull. Cal. Acad. Sci., II, p. 167.

ONTOGENESIS. A. d'Orchymont (1913), Ann. Biol. lac., VI, p. 181, 207–Richmond (1920), Bull. Amer. Mus. Nat. Hist., XLII, Art. 1, p. 82, 84.

Sub-gen. **Limnebius** s. str. Rey.

Limnebius in sp. Rey (1883), Rev. Ent. Caen, II, p. 87–Seidlitz (1888), Fna. Balt., ed. 2, p. 106; (1888), Fna. Transsylv., p. 113–Ganglbauer (1901) Käf. Mitteleur., IV, 1, p. 259–Reitter (1909), in Brauer, Süsswasserfna Deutschl., 3-4, p. 76; (1909), Fna. Germ., II, p. 365–de Peyerimhoff (1909), L'abeille, XXXI, p. 125–Kuhnt (1913), Ill. Best. Tab. Käfer Deutschl., p. 435, f. 92,–Des Gozis (1916), Misc. Ent., XXIII, p. 4–Everts (1922), Col. Neerl., III, p. 291.

Tricholimnebius + *Embololimnebius* + *Odontolimnebius* + *Limnebius* in sp. Kuwert (1890), Verh. Naturf. Ver. Brünn, XXVIII, p. 88-99–Everts (1899), Col. Neerl. I, p. 648-650.

almoranus Knisch. (1924) Wien Ent. Zeit., XLI, p. 29.[1]

rufipennis Règimbart (1903), Ann. Soc. Ent. Fr., LXXII, p. 336. — Kumaon[1] (Ranikhet, C. Almora). Pusa.

singularis Knisch (1924), Wien. Ent. Zeits., XLI, p. 31. — Pondicheri, Genji.

Sub-gen. **Bilimneus** Rey.

Bilimneus Rey (1883), Rev. Ent. Caen, II, p. 88. — Kumaon (W. Almora).

Bolimnius Rey (1885), Ann. Soc. Linn. Lyon, XXXI, p. 318–Seidlitz (1888), Fna. Balt., ed. 2, p. 107; (1888), Fna. Transsylv., p. 114–Everts (1899) Col. Neerl., I, p. 648–Ganglbauer (1904), Käf. Mitteleur., IV, 1, p. 259–de Peyerimhoff (1909), L'abeille, XXXI, p. 125–Reitter (1909) Fna. Germ., II, p. 366–Kuhnt

(1913), Ill. Best. Tab. Käf. Deutschl., p. 435, fig. 93[a]–Des Gozis (1916), Misc. Ent., XXIII, p. 4–Everts (1922), Col. Neerl., III, p. 291.

Crepilimnebius Kuwert (1890), Verh. Naturf. Ver. Brünn, XXVIII, p. 105.

distinctus Knisch (1924) Wien. Ent. Zeit., VII, p. 31. (*Limnebius-Bolimnius*). Kumaon (W. Almora).

SUB-FAM. **SPERCHEINÆ.**

Spercheinæ Ganglbauer (1904), Käf. Mitteleur., IV, 1, p. 153, 215–Reitter (1909), in Brauer, Süsswaserfna Deutschl., 3-4, p. 52, 63; (1909), Fna. Germ., II, p. 346, 356–Fowler (1912), Fna. Brit. Ind. (Gen. Intr.), p. 130–Kuhnt (1913) Ill. Best. Tab. Käf. Deuts., p. 417, 428–Schaufuss C. (1916), in Calwer, Käfb., ed. 6 (1913), p. 331, 332–A. d'Orchymont (1916), Ann. Soc. Ent. Fr., Vol. 85, p. 105; (1919), Rev. Zool. Afric., VI, p. 165–Tullgren et Wahlgren (1920), Svenska Ins. (2), p. 233–Everts (1922), Col. Neerl., III, p. 284.

Spercheini Erichson (1837), Käf. Mark Brand; I, p. 193.

Spercheida Heer (1841), Fna. Col. Helv., I, p. 473.

Sperchéens Mulsant (1844) Hist. Nat. Col. Fr. Palp., p. 23, 24.

Sperchini Bach (1851), Käferfna Nord und Mitteldeutschl., I, p. 121, 123.

Spercheii Fairmaire & Laboulbène (1854), Fne. Fr., I, p. 224, 234.

Spercheides Lacordaire (1854) Gen. Col., I, p. 447, 462.

Spercheites Jacquelin Duval (1857), Gen. Col. Eur., I, p. 90, 97.

Spercheidoe Thomson (1859), Skand. Col., I, p. 16; (1860), id, II, p. 83.

Spercheina Fowler (1887), Col. Brit. Isl. I, p. 219, 232.

Spercheitoe Kuwert (1890), Verh. Naturf. Ver. Brünn, XXVIII, p. 23.

Sperchèiens Lameere (1900), Man. Ene. Belg., II, p. 324.

ONTOGENESIS, Schiödte (1872), Nat. Tidskr. (3) VIII, p. 217 (*sperchei*)–Ganglbauer (1904) Käf. Mittleur. IV, I, p. 216–Richmond (1920), Bull. Amer. Mus. Nat. Hist., XLII, Art. I, p. 30.

KEY TO GENUS.

Comprises only the genus *Spercheus* Kugelann.

5. **Spercheus** Kugelann.

Spercheus Kugelann (1798), in Illiger's Verz. Käf. Preuss., p. 241–Fabricius (1801), Syst. Eleuth., I, p. 248–Latreille (1804), Hist. Nat. Crust. Ins., X, P. 69; (1807), Gen. Crust. Ins., II, p. 63; (1817), in Cuvier, Règne Anim., III, p. 272; (1829), id, nouv. ed., IV, p. 521–Duftschmid (1805) Fna. Austr., I, p. 294–Gyllenhal (1808), Ins. Suec., I, p. 124–Leach (1815), in Brewster's Edinb. Encycl., IX, p. 96; (1817) Zool. Misc., III, p. 92,–Lamarck (1817), Hist. Nat. Anim. Sans Vert., IV, p. 536–Stephens (1829), Ill. Brit. Ent. Mandib., II, p. 124; (1839), Man. Brit. Col., p. 88–Curtis (1832), Brit. Ent. IX, p. 394–Solier (1834), Ann. Soc. Ent. Fr., III. p. 302, 317—Sturm (1835), Deutschl Fna. Ins. IX, p. 92, f. 214—Erichson (1837), Käf. Mark Brandenb., I, p. 193—Laporte de Castelnau (1840), Hist. Nat. Ins. II, p. 57—Westwood (1840), Intr. Mod. Class. Ins., II, Synopsis, p. 9—Heer (1841), Fna. Col. Helv., I, p. 473— Mulsant (1844), Hist. Nat. Col. Fr., Palp., p. 24—Redtenbacher (1845), Gatt. Deutsch. Käferfn., p. 71; (1849), Fna. Austr., p. 127; (1858) ed. 2, p. 107; (1874) ed. 3, II, p. 116— Chenu (1851), Encycl. Hist. Nat. Col., I. p. 243—Bach (1851), Käferfn. Nord u. Mitteldeutschl., I, p. 121, 123—Fairmaire et Laboulbène (1854), Fne. Ent. Fr., I, p. 234—Lacordaire (1854), Gen. Col., I, p. 464—Imhoff (1856), Einf. Stud. Kol., p. 99, t. 11, f. 16—Leconte (1855) Proc. Ac. Philad., p. 356—Jacquelin Duval (1857) Gen. Col. Eur., I, p. 91, 98—Kuhn (1858) Käf. Südbayer. Flachl., p. 38—Bose (1859), in Gutfleisch, Käf. Deutschl. p. 67—Fuss (1859), Archiv. Ver Siebenb. Landesk. n. F., IV, (3), p. 114—Thomson C.G. (1860), Skand Col., II, p. 83—Seidlitz (1872), Fna. Balt., p. 11; (1888), ed. 2, p. 23 (Gatt.); (1888), Fna. Transsylv., p. 23 (Gatt.)—Dalla Torre (1877), Jahresber. Ver. Naturk. Linz (8) p. 67—Bedel (1881), Fne. Col. Bass. Seine, I, p. 301—Rey (1885), Ann. Soc. Linn.

Lyon, XXXI, p. 349—Fowler (1887), Col. Brit. Isl. I, p. 232.—Kuwert (1890), Verh. Naturf. Ver. Brünn, XXVIII, p. 23, 179—Everts (1899) Col. Neerl., I, p. 640, 656—Kliment (1899) Cesti bronci, p. 121—Lameere (1900), Man. Fne. Belg. II, p. 328—Sterlin (1900), Col. Helv., p. 214—Ganglbauer (1904), Käf., Mitteleur., IV, I, p. 216—Reitter (1909), in Brauer, Süsswasserfna Deutschl., 3-4, p. 64, fig. 42, 43; (1909), Fna. Germ., II, p. 356—Kuhnt (1913), Ill., Best. Tab. Käf. Deutschl., p. 417, 428, f. 2 B, 4—Schaufuss C. (1916) in Calwer, Käferb., ed. 6 (1913), p. 332—A. d'Orchymont (1916), Ann. Soc. Ent. Fr., LXXXV, p. 97, f. 4 (morphology) of head)—Des Gozis (1918), Misc. Ent., XXIV, p. 94—Tullgren and Wahlgren (1920), Svenska Ins. (2), p. 234.

Sperchus Scholsšer Kelk (1877), Fna. Kornjašah, I, p. 106.

ONTOGENESIS, Sturm (1835), Deutschl. Fna. Ins., IX, p. 95—C.G. Thomson (1860) Skand. Col. II, p. 84—Schiödte (1862), Nat. Tidskr. (3) I, p. 233—A. d'Orchymont (1913), Ann. Biol. lac, VI, p. 208—Richmond (1920), Bull. Amer. Mus., Nat. Hist., XLII, Art., 1, p. 83, 84.

belli G.C. Champion (1919), Ent. Mo. Mag., LV, p. 239.	Karachi.
gibbus G.C. Champion (1919), Ent. Mo.♀ Mag., LV, p. 238. (♂)[1]—A. d' Orchymont (1923), Mem. Dep. Agric. India, a, VIII, p. 1 (♂♀).[2]	Bengal.[2], Sarda.[1] Bihar.[2] Sunderbans.[3]
binodulus G.C. Champion (1919), Ent. Mo. Mag., LV, p. 239. ( ).[3]	
stangi Schwarz & Barber (1918), Proc. Ent. Soc. Wash., XIX (1917), p. 133[1]—A. d'Orchymont (1919), Ann. Soc. Ent. Belg., LIX, p. 70.[2]	Philippines.[1] Cambodia.[2] Cochin-China.

FAM. **HYDROPHILIDÆ.***

Hydrophilidæ [Fairmaire et Labou'bène, excl. *Hydrænidæ*], A. d'Orchymont (1919), Rev. Zool. Afric., VI, p. 166.

ONTOGENESIS & MORPHOLOGY. A. d'Orchymont (1916), Ann. Soc. Ent. Fr., LXXXV, p. 100, 104, 105 (*Palpicornia,* Categorie II).

---

* Restricted sense. For the use of the name in a wider sense, according to Famaire et Laboulebéne, *see* under "Palpicornia".

KEY TO SUB-FAMILIES.

1. Pronotum with five longitudinal sulci: a median, two inner and two outer ones, enclosing two median, two inner and two outer more or less granulate intervals; disc very transverse wider than head with eyes. Body contour not uniformly curved, the prothorax conspicuously separate from the hind part. Antennae short and nine-jointed with pubescent three-jointed and not compact articulated club, sixth joint without inner process. Eyes not divided by a canthus, moderately convex. Scutellum small, not elongate. Anterior coxal cavities open posteriorly. Abdomen with five segments. Fifth joint of posterior tarsi shorter than the four preceding taken together. General form more or less elongate, not very convex, with elytra not hanging over the hind body.

Larvae with nine complete urites and tri-articulate promonent procerci. Holopneustic type: each urite bears a pair of very small spiracles. Gula very small[1], very far from occipital foramen by confluence over a long space of the gular sutures. With feet. Antennae inserted outside the base of mandibles. 4. *Helophorina.**

1. Pronotum on anterior border with a triangular overhanging process which hides a great part of head; the disc unequal with impressions and obtuse carinae. Body contour not uniformly curved. Antennae nine-jointed with pubscent three-jointed and compact club, sixth joint without inner process. Eyes divided in part by a canthus. Scutellum small, somewhat longer than vide. Anterior coxal cavities open behind. Abdomen with five segments. Fifth joint of posterior tarsi a little shorter than joints one to hour. General form short, very convex, with elytra hanging over the hind body. First ventral segment more or less invaded by the hind coxal cavities. Elytra often with tubercles. Galea of maxillae not falciform.

---

* As stated in 1924 (Bull. Soc. Ent. Belg., VI, p. 93) it seems probable that the New Zealand genus *Horelophus* (without thoracic sulci) belongs to the beginning of the tribe *Rygmodini*, in the sub-family *Spharidiinae.* So the arrangment of the key can be simplified.

Larvae unknown. 5. *Rpimetopinæ*

1″. Pronotum otherwise formed . . .

Larvae with only eight normal urites and reduced cerci, at most bi-articulate . . . .

Body contour not uniformly curved and body not regularly convex, the prothorax narrower than the hind body and conspicuously separated therefrom, nearly as wide as head with eyes. The latter very prominent. Antennae short, seven-jointed with three-articulate pubescent club, the fourth joint very asymetrical, cuplike and with an inner spine-like process. Maxillary palpi never very long. Scutellum small and not elongate. Anterior coxal cavities closed behind. Last joint of hind tarsi as long as the four preceding joints, the first joint very short. Abdomen with five chitinous segments and a sixth more or less membraneous behind the fifth; 2nd to 5th segments on the base with tuberculiform swellings. Pronotum with more or less conspicuous impressions.

Larvae with feet. Gular sutures not contiguous or continguous at a single point, the gula in contact, at least at this point, with the occipital foramen. Metapneustic type: abdominal functional spiracles reduced to the eighth pair, which arises on the ground of a very large, very well developed stigmatic atrium. Antennae with their points of insertion nearer the externo-frontal angles than those of the mandibles (as in *Helophorinœ*). 6. *Hydrochinæ*.

2′. Body-contour uniformly curved and body regularly convex, the prothorax not narrower than the hind body and not conspicuously separated therefrom; with the exception of some *Berosini*, but then the scutellum is elongate, and some *Rygmodini* (New Zealand forms), but then the 8 or 9-jointed antennae are very much longer than the maxillary palpi. Last joint of hind tarsi much shorter than the preceding joints taken together except in *Helobata* (American) and *Acidocerus* (African) but then the maxillary palpi are very long. Pronotum without impressions or grooves.

Larve with gular sutures contiguous on a long space and gula lying at a great distance from occipital foramen. Antennae with their points of insertion farther from the externo-frontal angles of head than those of the madibles, inserted thus within the space separating the latter. 3.

3. Antennae normally longer than maxillary palpi, which are never very long the last glabrous joint obconic, fitted more or less tightly against first joint of pubescent club, the latter laxly articulated or very compact in extreme genera, sometimes in appearance (truly ?) four articulate, Posterior and intermediate tarsi with first joint sometimes shorter than the following but then antennae are much longer than maxillary palpi (*Rygmodini*), sometimes as long but ordinarily longer. Labrum very often concealed or retractile under anterior border of prefrons. Always five ventral segments.

Larvae (of *Rygmodini* not known): ocular areas round, usually small and more closely aggregated; legs reduced or absent pleural lobes not prominent; abdomen truncate; abdominal segments following the eighth reduced forming a stigmatic atrium; metapneustic type. (Always ? or characters only good for specialized genera?)

More or less terrestrial forms. 7. *Sphæridiinæ*.

3. Antennae shortened, as long as and often shorter than maxillary palpi, the last glabrous joint more or less asymetrical, cup-like, embracing the first joint of pubescent always tri-articulate club, which is never compact. First joint of posterior and intermediate tarsi much shorter than following, except in *Helobata* (American) and *Acidocerus* (African) but here the maxillary palpi are very elongate, much longer than antennae. Labrum very often exposed. Sometimes a sixth ventral segment is visible but it is more or less retractile under the fifth.

Larvae: ocular areas oval, larger, with eyes more distant; legs well developed, visible from above except in *Paracymus* and *Anacoena*; pleural lobes usually prominent; abdomen not noticeably truncate, more or less elongate; metapneustic type: abdominal segments following the eighth reduced forming a stigmatic atrium; or type apneustic (*Berosini*): abdominal segments following the eighth very much reduced not forming stigmatic atrium, first to seventh urites each with a pair of very prominent tracheal gills.

Aquatic forms . . . . 8. *Hydrophilinæ*.

Sub-fam. **HELOPHORINÆ.**

Helophorinæ Ganglbauer (1904), Käf. Mitteleur., IV, 1, p. 152, 153—Reitter (1909), in Brauer, Süsswasserfna. Deutschl., 3-4 p. 52; (1909), Fna. Germ; II, p. 346—Fowler (1912), Fna. Brit. Ind. (Gen. Intr.). p. 130—Kuhnt (1913), Ill. Best-Tab. Käf. Deutschl., p. 417, 420—Schaufuss C. (1916), in Calwer, Käferb., ed. 6 (1913), p. 321-322—A. d'Orchymont (1916), Ann. Soc. Ent. Fr. LXXXV, p. 105—Tullgren & Wahlgren (1920), Svenska Ins. (2) p. 233—Everts (1922) Col. Neerl., III, p. 285.

Helophoridoe C. G. Thomson (1859), Skand. Col., I, p. 16; (1860), II, p. 77; (1867), IX, p. 113—Bedel (1881), Fne. Col. Bass. Seine, I, p. 290, 297.

Helophorates Rey (1885), Ann. Soc. Linn. Lyon, XXXI, p. 354.

Helophorina Fowler (1887), Col. Brit. Isl., I, p. 219, 233.

Helophorini [Bach (1851), Käferfna Nord & Mitteldeutschl., I, p. 121, 123. Excl. *Hydrochus*] Jakobson (1911), Käf. Russl., p. 839—Sharp (1915), Ent. Mo. Mag, LI, p. 2, 27, 113.

ONTOGENESIS. Ganglbauer (1904), Die Käf.

Mitteleur., IV, I, p. 152, 153—Richmond (1920), Bull, Amer. Mus. Nat. Hist., XLII, Art I, p. 17.

KEY TO GENUS.

Comprises only the genus *Helophorus* Fabricius.

6. Helophorus [Fabricius] emend Leach.

Helophorus [Fabricius] emend. Leach (1815), in Brewster's Edinb. Enc., IX, p. 95; (1817), Zool. Misc. Ill., p. 90—Zetterstedt (1828), Fauna Ins. Lapp. I, p. 199; (1840) Ins. Lapp., p. 123—Stephens (1829) Ill. Brit. Ent. Mandib., II, p. 111; (1839), man. Brit. Col., p. 83—Brullé, (1835), Hist. Nat. Ins., Vbis, p. 303—Erichson (1837), Käf. Mark. Brand, I, p. 194—Westwood (1840) Intr. Mod. Class. Ins. II, Synopsis, p. 9—Heer (1841), Fna. Col. Helv., I, p. 473—Mulsant (1844), Hist. Nat. Col. Fr., Palp., p. 28—Redtenbacher (1845), Gatt. Deutsch. Käferfna, p. 71; (1849), Fna. Austr., p. 127; (1858), ed. 2, p. 107; (1874), ed. 3, II, p. 116— Bach (1851), Käferfna Deutschl. I, p, 122, 123; (1866), Syst. Verz. Käf. Deutschl., p. 467—Lacordaire (1854), Gen. Col., I, p. 465—Leconte (1855) Proc. Acad. Nat. Sc. Philad., p. 357; (1861-1862) Class. Col. North. Amer., p. 45—Jacquelin Duval (1857), Gen. Col. Eur. I, p. 91, 98—Bose (1859) in Gutfl., Käf. Deutschl., p. 67-Fuss (1859) Archiv. Ver. Siebenb. Landesk. n. F. IV, (3), p. 114—Thomson C.G. (1860), Skand. Col. H, p. 77; (1868), X, p. 297—Zimmerman (1868-1869), Trans. Amer. Ent. Soc. II, p. 249—Seidlitz (1872), Fna. Balt., p. 11—Solsky (1876), Fedtsch., Reise Turkest. II, p. 21—Dalla Torre (1877), Jahresber. Ver. Naturk. Linz (8) p. 67—Provancher (1877), Fne. Entom. Canada, I, p. 210—Scholssen Klek, (1877), Fna. Kornjasah I, p. 107—Bedel (1881), Fne. Col. Bass. Seine, I, p. 297—Sharp (1882), Biol. Centr. Amer., I, (2), p. 88—Leconte et Horn (1883), Class. Col. North, Amer., p. 71—Kuwert (1885), Wien. Ent. Zeit., IV, p. 229; (1886), V, p. 221; (1890), Verh. Naturf. Ver. Brünn, XXVIII, p. 23, 180—Fowler (1887) Col. Brit. Isl., I, p. 233—Wickham (1895), Canad. Ent., XXVII, p. 181, 182—Acloque (1896), Fne. Fr., p. 95— Everts (1899), Col. Neerl. I, p. 640, 656—Kliment (1899), Cesti brouci, p. 121—Lameere (1900), Man. Fne. Belg., II,

p. 325—Ganglbauer (1904), Käf. Mitteleur., IV, I, p. 153—Edwards (1908), Ent. Mo. Mag. 2, XIX, p. 218—Reitter (1909), in Brauer, Süsswasserfna Deutschl., 3-4, p. 52, f. 38; (1909), Fna. Germ., II, p. 346—Blatchley (1910), Col. Indiana in Bull. Ind. Dep. Geol., I, p. 249, 250—Kuhnt (19130 Ill. Best. Tab. Käf Deutschl., p. 417, 420, f. 2 A.C. D. f. 43 A. B. C.—A. d'Orchymont (1913), Ann. Soc. Ent. Belg., LVII, p. 100—Schaufuss C. (1916) in Calwer, Käferb., ed. 6 (1913), p. 322—Tullgren & Wahlgren (1920) Svenska Ins. (2), p. 233—A. d'Orchymont (1924) Bull. Soc. Ent. Belg. VI, p. 91; (1926), Ann. Soc. Linn. Lyon; LXXII, p. 111-143.

Elophorus Fabricius (1775), Syst. Ent., p. 66 (ex. p.).

Empleurus + *Helophorus* Rey (1885), Ann. Soc. Linn. Lyon, XXXI, p. 354 & sq.

Empleuri (G. *Empleurus, Megempleurus, Bradempleurus)* + *Trichelophori* (G. *Eutrichelophorus, Trichelophorus)* + *Gephelophorus* + *Meghelophorus* + *Helophorus* + *Atracthelophoras* + *Cyphelophorus* Sharp (1915), Ent. Mo. Mag. LI, p. 130-138, 156-163, 198-204, 233-238, 270-277; (1916), l. c., LII, p. 83-86, 108-112, 125-130, 164-177, 193-200.

ONTOGENESIS. Schiödte (1861-1862), Nat. Tidskr., (3), I, p. 212, 222, 223; (1872), id. VIII, p. 212—Ganglbauer (1904), Käfer Mitteleur.; IV, I, p. 155—Brocher F. (1913), L'aquarium de Chambre, p. 224—A. d'Orchymont (1913), Ann. Biol. lac., VI, p. 208—Richmod (1920), Bull. Amer. Mus, Nat. Hist., XLII, Art. I., p. 17, 82, 84, 87—A. d'Orchymont (1926), Ann. Soc. Linn. Lyon, LXXII, p. 112, 113.

Sub-gen. **Lihelophorus** Zaitzev.

Lihelophorus Zaitzev (1908), Ann. Mus. Zool. Acad. Sc. St. Petersb., XIII, p. 421—Kniz (1910); Zool. Jahrb. Syst. Geogr. Biol., XXIX, p. 451— Sharp (1915), Ent. Mo. Mag., LI, p. 201—A. d'Orchymont (1924), Bull. Soc. Ent. Belg., VI, p. 92; (1926), Ann. Soc. Linn. Lyon, LXXII, p. 116.

lamicola Zaitzev (1908), l. c., p. 421—Kniz (1910), l. c., p. 451. — Tibet (Kuku Noor).

ser Zaitzev (1908), l. c., p. 422—Kniz (1910), l. c., p. 452. — Tibet (Kuku Noor).

Sub-gen. **Meghelophorus** Kuwert.

Meghelophorus Kuwert (1890), Verh. Naturf. Ver. Brünn, XXVIII, 1890, p. 181, 197—Sharp (1915), Ent. Mo. Mag., LI, p. 4, 200—A. d'Orchymont (1924), Bull. Soc. Ent. Belg., VI, p. 91, 92; (1926), Ann. Soc. Linn. Lyon, LXXII, p. 116, 118.

Megahelophorus Kuwert (1886), Wien. Ent. Zeit., V, p. 226 (rectius *Meghelophorus).*

Megelophorus Seidlitz (1888), Fne. Fna. Balt., ed. 2, p. 115; (1888), Fna. Transsylv., p. 122.

Megalelophorus Ganglbauer (1904), Käf. Mitteleur., IV, I, p. 157.

kozlovi Zaitzev (1908), l. c., p. 423. — Tibet or.

Sub-gen. **Trichelophorus** [Kuwert] emend Seidlitz.

Trichelophorus [Kuwert] emend. Seidlitz (1888), ed. 2, p. 115; (1888), Fna Transsylv., p. 122—Ganglbauer (1904), Käf. Mitteleur., IV, I, p. 157—Sharp (1915), Ent. Mo. Mag., LI, p. 4—Schaufuss C. (1916), in Calwer, Käferb., ed. 6 (1913), p. 323—A. d'Orchymont (1924), Bull. Soc. Ent. Belg., VI, p. 91, 92; (1926); Ann. Soc. Linn. Lyon, LXXII, p. 117, 122.

Trichohelophorus Kuwert (1886), Wien. Ent. Zeit., V, p. 223 (ex. p., rectius *Trichelophorus*).

Eutrichelophorus + *Trichelophorus* Sharp (1915), Ent. Mo. Mag., LI, p. 156, 157, 160.

Group of Species Eutrichelophorus Sharp.

micans Faldermann (1835), Nouv. Mem. Soc. Nat. Mosc., IV, p. 234; (1836), Fna. Entom. Transcaucasica, Col. I, p. 234 (*Elophorus)*[3]—Bedel (1881), Bull. Soc. Ent. Fr., P. (86), ex. p. et (102)—de Marseul (1882), L'abeille, XX, p. 157—Rey (1885), Ann. Soc. Linn. Lyon, XXXI, p. 368, nota, ex. p.[2]—Kuwert (1885), Wien. Ent. Zeit., IV, p. 230, ex. p.; — Siberia (Kirgish).[2] Russia.[2, 7] Causcasus.[3, 5] Transcaucasus.[3, 6] Asia Minor.[3] Armenia.[4] Mesopotamia.[4]

(1886), l. c., V, p. 225, ex p. ; (1890), Verh. Naturf. Ver. Brünn, XXVIII, p. 190, ex p.[3]—Ganglbauer (1904), Käf. Mitteleur. ; IV, I, p. 160, ex p.—Sharp (1915), Ent. Mo. Mag., LI, p. 159, Pl. IV, f. 34 (*Eutrichelophorus)*[4]— A. d'Orchymont (1926), Ann. Mus. Civ. Stor. Nat. Genova, (3) X (L), p. 386.[5] — Persia.[4, 5] Arabia.[4] Turkestan.[1,8]

subcostatus Kolenati (1846), Meletemata Ent., V, p. 65.[6]

opalisans Motschulsky (1860), Schrenck's Reise, II, p. 107.[7]

elegans Ballion (1870), Bull. Soc. Nat. Mosc. XLIII, n°. 4, p. 329.[8]

Sub-gen. **Empleurus** Hope.

Empleurus Hope (1838), Col. Man., II, p. 149—Bedel (1881), Fne. Col. Bass. Seine, I, p. 298—Rey (1884), Rev. Ent. Caen, III, p. 268; (1885), Ann. Soc. Linn. Lyon, XXXI, p. 355—Seidlitz (1888), Fna. Balt, ed 2, p. 24; (1888), Fna. Transsylv., p. 24—Everts (1899), Col. Neerl., I, p. 657—Stierlin (1900), Col. Helv., p. 215—Ganglbauer (1904), Käf. Mitteleur., IV, I, p. 156—Reitter (1909), in Brauer, Süsswasserfna Deutschl., 3-4, p. 52; (1909), Fna. Germ., II, p. 347—Kuhnt (1913), Ill. Best. Tab. Käf. Deutschl., p. 420, fig. 33*a*—Sharp (1915), Ent. Mo. Mag., Li, P. 4—Schaufuss C. (1916), in Calwer Käferb., ed. 6 (1913), p. 323—Des Gozis (1918), Misc. Ent., XXIV, p. 97—A. d'Orchymont (1926), Ann. Soc. Linn. Lyon, LXXII, p. 117, 124.

Empleuri (G. *Empleurus* + G. *Megempleurus* + G. *Brадempleurus*) Sharp (1915), Ent. Mo. Mag., LI, p. 130, 131, 134, 136.

Megempleurus in sp. Sharp. (1915), l. c., p. 131, 132.

Simempleurus, Sharp (1915), l. c., p. 132.

crinitis Ganglbauer (1901), Verh. Zool-bot. Ges. Wien, L, p. 312. — Tibet: (Amdo, Ordos, Kuku (Noor).

fausti Kuwert (1887), Wien. Ent. Zeitg., VI, p. 165. — Turkestan.

Sub-gen. **Helophorus** s. str. Ganglbauer.

Helophorus s. str. Ganglbauer (1904), Käf. Mitteleur., IV, I, p. 157—Reitter (1909), in Brauer, Süsswasserfna Deutsschl., 3-4, p. 54; (1909), Fna. Germ., II,

p. 348—Kuwert (1913), Ill. Best. Tab. Käf. Deutschl., p. 421, f. 37B, 42—Sharp (1915), Ent., Mo. Mag., LI, p. 4, 223—Schaufuss C. (1916), in Calwer, Käferb. ed. 6 (1913), p. 323—Everts (1922), Col. Neerl., III, p. 295, 296—A. d'Orchymont (1924), Bull. Soc. Ent. Belg., VI, p. 93, 98; (1926), Ann. Soc. Linn. Lyon, LXXII, p. 117, 128.

Rhopalohelophorus Kuwert (1886), Wien Ent. Zeit., V, p. 247.

Rhopalhelophorus Kuwert (1890), Verh. Naturf. Ver. Brünn, XXVIII, p. 181.

hingstoni A. d'Orchymont (1926), Ann. et Bull. Soc. Ent. Belg., LXVI, p. 95. — Tibet (Pangle).

montanus A. d'Orchymont (1926). l. c., p. 97[1], p. 201[2]. — Tibet (Tingri, Lingka, Shekkar Gyangtse).[1] (Mangzaka Lake).[3] Kumona (laptel).[1] Syr Daja: Aulie Ata.[2, 4, 5]

splendidus A. Kniz (1910), Zool. Jahrb. Geogr. Biol. XXIX, p. 453[3]—(1919), Ent. Blatt., 15, p. 12[4]—A. d'Orchymont (1925), Ann. et Bull. Soc. Ent. Belg., LXV, p. 145[5] (nec. J. Sahlberg, 1880)

auliensis (Reitter, in coll.) (A. d'Orchymont, l. c., 1926, p. 99 nota).

punctatosulcatus Kuwert (1892), Deuts. Ent. Zeitschr., p. 102. — Turkestan (Mt. Alai).

splendidus j. Sahlberg (1880) forma *imaensis* A. d'Orchymont (1926), Ann. & Bull. Soc. Ent. Belg., LXVI p. 97. — Tibet (Phari. Kampa Dzong, Lingka, Tingri).

Sub-gen. **Atracthelophorus** Kuwert.

Atracthelophorus Kuwert (1890), Verh. Naturf. Ver. Brünn, XXVIII, p. 181—Sharp (1915), Ent. Mo. Mag. (3) I (LI), p. 4; (1916), II (LII), p. 193—A. d'Orchymont (1924), Bull. Soc. Ent. Belg., VI, p. 108; (1926), Ann. Soc. Linn. Lyon. LXXII, p. 118, 139.

Atractohelophorus Kuwert (1886), Wien. Ent. Zeit, V, p. 227 (rectius *Atracthelophorus).*

Atractelophorus Ganglbauer (1904), Käf. Mitteleur, IV, I, p. 157.

Atrachthelophorus Zaitzev (1908), Hor. Soc. Ent. Ross., XXXVIII, p. 327 (Catal.).

bang-haasi Kniz (1919), Ent. Blatt., XV, p. 11 (*Atractelophorus*).	Syr Darja: Aulie Ata.
bernhaueri Breit (1916), Col. Rundsch. V, p. 83 (*Atractelophorus*).	Semirjetschensk : Djarkent.
discretus A. d'Orchymont (1925), Ann. & Bull. Soc. Ent. Belg. LXV, p. 147.	Syr. Darja : Aulie Ata.
frater A. d'Orchymont (1926), Ann. et Bull. Soc. Ent. Belg., LXVI, p. 93.	Kumaon (Laptel, Tibetan frontier) Kashmir: Rupschu, Tso Morar.
longipennis Ganglbauer (1901), Verh. Zool. Bot. Ges. Wien, L. p. 315 (*Atractelophorus*).	Syr. Darja : Aulie Ata.
pictus Ganglbauer (1901), l. c., p. 317 (*Atractelophorus*).	Buchara.

Sub-fam. **EPIMETOPINÆ.**

Epimetopinæ A. d'Orchymont (1916), Ann. Soc. Ent. Fr., LXXXV, p. 105—Everts (1922), Col. Neerl., III, p. 285, nota 2.

Epimetopinæ Zaitzev (1908), Hor., Soc. Ent. Ross., XXXVIII, p. 353 (Catalog.).

KEY TO GENUS.

Comprises only the genus *Epimetopus* Lacordaire.

7. **Epimetopus** Lacordaire.

Epimetopus Lacordaire (1854), Trans. Amer. Ent. Soc. V, p. 251—Bedel (1880), Ann. Soc. Ent. Fr. (5) X, p. LXXIII—Sharp (1882), Biol. Centr. Amer. I (2) p. 88—Leconte & Horn (1883), Class. Col. North Amer., p. 71—E. A. Schwarz (1908), Proc. Ent. Soc. Wash. IX, p. 115—E.A. Schwarz & Barber (1917), Proc. Ent. Soc. Wash., XIX, p. 132.

Ceratoderus Mulsant (1851), Mem. Acad. Sc. Lyon I, p. 1 (nec. *Ceratoderus Westwood, 1842, Paussidæ).*

Sepidulum Leconte (1874), Trans. Amer. Ent. Soc. V, p. 47—Sharp (1875), Ent. Mo. Mag. XI, p. 247.

asperatus G.C. Champion (1919), Ent. Mo. Mag., LV, p. 237.	Kumaon (W. Almora) Sarju Valley.

bullatus Sharp (1875), Ent. Mo. Mag, XI, p. 249 sepidulum—Schwarz & Barber (1917), Proc. Ent. Soc. Wash. XIX, p. 132. India.

flavidulus Sharp (1890), Trans. Ent. Soc. London, p. 355—Schwarz & Barber, l. c., p. 132. Ceylon.

maindroni, Regimbart (1903), Ann. Soc. Ent. Fr., LXXII, p. 338—Schwarz & Barber, l. c., p. 132. Ganji.

Sub-fam. **HYDROCHINÆ.**

Hydrochinæ A. d'Orchymont (1916), Ann. Soc. Ent. Fr., LXXV, p. 105—Everts (1992), Col. Neerl., III, p. 285.

Hydrochidæ C. G. Thomson (1859), Skand. Col., I, p. 15; (1860), id, II, p. 75; (1867), id., IX, p. 113.

Hydrochoates Rey (1885), Ann. Soc. Linn., Lyon, XXXI, p. 354; (1886), id., XXXII, p. 1.

Hydrochoini Ganglbauer (1904), Käf. Mitteleur., IV, I, p. 176.

Hydrochoina Zaitzev (1908), Hor. Soc. Ent. Ross., XXXVIII, p. 333 (Catal.) — Jakobson (1911), Käf. Russl., p. 840.

Hydrochoinae Richmond (1920), Bull. Amer Mus. Nat. Hist. XLII, Art. I, p. 27.

Oxiogenesis. A. d'Orchymont (1916), Ann. Soc. Ent. Fr., LXXXV, p. 98, 101—Richmond (1920), Bull. Amer. Mus. Nat. Hist., XLII, Art. I, p. 27 (Hydrochoinae).

KEY TO GENUS.

Comprises only the genus *Hydrochus* Leach.

8. **Hydrochus** Leach.

Hydrochus Leach (1817), Zool. Misc., III, p. 90—Latreille (1829), in Cuvier, Regne Anim., nouv. ed., IV, p. 520—Stephens (1829), Ill. Brit. Ent. Mandib. II, p. 110; (1839), Man. Brit. Col., p. 83—Solier (1834), Ann. Soc. Ent. Fr., III, p. 303—Brulle (1835), Hist. Nat. Ins. V, p. 306—Sturm (1836), Deutschl. Fna. Ins., X, p. 44, t. 220—Erichson (1837), Käf. Mark. Brand., I. p. 197—Laporte de Castenan (1840), Hist.

Nat. Ins., II, p. 46—Westwood (1840), Intr., Mod. Class Ins., II, Synopsis p. 9—Heer (1841), Fna. Helv., I, p. 476—Mulsant (1844), Hist. Nat. Col. Fr., Palp. p. 43—Redtenbacher (1845), Gatt. deutsch. Käferfna p. 70; (1849), Fna. Austr., p. 128; (1858), ed. 2, p. 109; (1874), ed. 3 II, p. 119—Chenu (1851), Encycl. Hist. Nat. Col., I, p. 236—Bach (1851), Käferfna Nord ü Mittel Deutschl., I, p. 121, 390—Lacordaire (1854), Gen. Col. I, p. 465, 466—Fairmaire & Laboulbène (1854), Fne Fr., I, p. 238—Leconte (1855), Proc. Acad. Philad., p. 357; (1861-1862), Class. Col. North Amer., I p. 45—Jacquelin Duval (1857), Gen. Col. Eur., I, p. 92, 98 Kuhn (1858), Käf. Südbayer, Flachl., p. 38—Bose (1859), in Gutfl., Käf. Deutschl., p. 68—Fuss (1859), Archiv., Ver. Siebenb. Landesk. n. F. (3) p. 117—Thomson C. G. (1860), Skand Col., II, p. 75—Seidlitz (1872). Fna. Balt. p. 11; (1888), ed. 2, Gatt. p. 24; (1888), Fna. Transsylv., p. 24 (Gatt.)—Dalla Torre (1877), Jahresber. Ver. Naturk. Linz. (8), p. 67—Schlosser Klek (1877), Fna. Kornjasah, I, p. 107—Bedel (1881), Fne. Col. Bass. Seine, I, p. 291—Sharp (1882), Biol. Centr. Amer. I (2); p. 89—Leconte & Horn (1883), Class. Col. North Amer., p. 71—Fowler (1887), Col. Brit. Isl., I, p. 241—Kuwert (1887), Deutsch. Ent. Zeitz., XXXI, p. 289; (1890), Verh. Naturf. Ver. Brünn. XXVIII, p. 23, 228—Wickham (1895), Canad. Ent., XXVII, p. 181, 183—Acloque (1896), Fne. Fr., p. 95—Blackburn (1898), Trans. Roy. Soc. S. Austr., XXII, p. 227—Everts (1899), Col. Neerl, I, p. 640, 660—Kliment (1899), Cešti brouci, p. 123—Lameere (1900), Man. Fne. Belg., II, p. 325—Stierlin (1900), Col. Helv., p. 215—Germain (1901), Anales Univ. Chile, CIX, p. ; (1902), apuntes Entom. Heloforidos Chilen., p. 8—Reitter (1909), in Brauer, Süsswasserfna Deutschl., 3-4, p. 57; (1909), Fna. Germ., II, p. 350, 351—Blatchley (1910), Col. Indiana, Bull. Ind. Dep. Geol. I, p. 249, 251—Kuhnt (1913), Ill. Best. Tab. Käf. Deutschl., p. 417, 423, f. 5 A B, 6—Schaufuss C. (1916), in Calwer, Käferb., ed. 6 (1913), p. 325, 326—Des Gozis (1918), Misc. Ent., XXIV, p. 96— Tullgren & Wahlgren (1920), Svenska Ins. (2), p. 233.

Hydrocus Provancher (1877), Fne. Ent. Canada. I, p. 211.

Hydrochoiis Bedel (1881), Fne. Col. Bass. Seine, I, p. 316—Rey (1886), Ann. Soc. Linn. Lyon, XXXII, p. 1—Ganglbauer (1904), Käf. Mitteur., IV, I, p. 176 *(Hydrochous)*.

Ontogenesis, Richmod, (1920), Bull. Amer. Mus., Nat Hist., XLII, Art. I, p. 27, 82, 85 (*Hydrochous*).

annamita Regimbart (1903), Ann. Soc. Ent. Fr., LXXII, p. 64[1]—A. d'Orchymont (1922), Rec. Ind. Mus. VIII, p. 624[2]; (1926), Philipp. Journ. Sc., 30, p. 363.[3]
N. E. Assam. Annam.[1] Tenkin.[3] ———— Philippines.[3]

binodosus Motschulsky (1860), Schrenck's Reis., II, p. 104[1]–? A. d'Orchymont (1922), Rec. Ind. Mus., VIII, p. 624.[2]
India Or.[1] N. E. Assam.[2]

lacustris Nietner (1856), Journ., As. Soc. Beng., XXV, p. 537[1]; (1857), Ann. Mag. Nat. Hist. (2), XIX, p. 387[1]—Règimbart (1903), Ann. Soc. Ent. Fr., LXXII, p. 64.[2]
Ceylon.[1, 2], India.[2] ———— Cochin-China.[2]

subsp. *inornatus* A. d'Orchymont (1927), Journ. F.M.S. Mus., XIII, p. 246.
F. M. S. Java

latitans Fairmaire (1888), Ann. Soc. Ent. Fr., (6) VIII, p. 337 (= ? *lacustris* Nietner; cf. A d'Orchymont (1922), Rec. Ind. Mus., VIII, p. 624).
Tonkin.

opacus Motschulsky (1860), Schrenck's Reis., II, p. 103 (= ? *lacustris* Nietner; cf. A. d'Orchymont (1922), Rec. Ind. Mus., VIII, p. 624).
India Or.

violaceomicans Motschulsky (1860), Schrenck's Reis., II, p. 103 (= ? *lacustris* Nietner; cf., A. d'Orchymont (1922), Rec. Ind. Mus. VIII, p. 624).
India Or.

Sub-fam. **SPHAERIDIINÆ.**

Sphaeridiinæ Fowler (1887), Col. Brit. Isl., I, p. 218, 251; (1912), Fna. Brit. Ind. (Gen. Intr.), p. 130—Everts (1899), Col. Neerl., I, p. 637—Ganglbauer (1904), Käf. Mitteleur, IV, I, p. 153, 265—Reitter (1909), in Brauer Süsswasserfna Deutschl., 3-4, p. 52, 77; (1909), Fna. Germ., II, p. 346, 366—Kuhnt (1913), Ill, Best. Tab. Käf. Deutschl., p. 417, 436—Schaufuss C., (1916), in Calwer, Käferb., ed. 6 (1913), p. 322, 346—A. d'Orchymont (1916), Ann. Soc. Ent. Fr., LXXXV, p. 105; (1919), LXXXVIII, p. 105; (1916),

Bull. Soc. Ent. Fr., p. 235, f. 1 (add. *Rygmodini*)—Tullgren & Wahlgren (1920), Svenske Ins. (2), p. 233, 235—Everts (1922), Col. Neerl., III, p. 285.

Sphæridiota Latreille (1804), Hist. Nat. Crust. Ins., X, p. 77.

Sphaeridides Leach (1815), in Brewster's Edinb. Encycl., IX, p. 96.

Sphaerididea Leach (1817), Zool. Misc. III, p. 95.

Sphaerididae W. S. M'Leay (1825), Annul Java., p. 36.

Sphaeridiotes Latreille (1829), in Cuvier, Règne Anim., nouv. ed., IV, p. 525.

Sphaeridiidae Stephens (1829), Ill. Brit. Ent. Mandib., II, p. 135.

Sphèridiens Brullé (1835), Hist. Nat. Ins. (V), II, p. 247, 287.

Sphaeridiini Erichson (1837), Kä. Mark. Brand. I, p. 212.

Sphèridiotes Laporte de Castelnau (1840), Hist. Nat. Ins. II, p. 58.

Spheridida Heer (1841), Fna. Col. Helv. I, p. 486.

Geophilies Mulsant (1844), Hist. Nat. Col. Fr., Palp., p. 23, 145.

Spherididae Chenu (1851), Encycl. Hist. Nat. Col., I, p. 252.

Sphaeridii Fairmaire & Laboulbène (1854), Ene. Ent. France, I, p. 249.

Sophaeridiides Lacordaire (1854), Hist. Nat. Ins. Gen. Col., I, p. 447, 470.

Sphaeridiides Leconte (1854-1855), Proc. Acad. Nat. Sc. Philad., p. 357.

Sphèriiites Jacquelin Duval (1857), Gen. Col. Eur., I, p. 94, 97.

Sphaeridini Dalla Torre (1877), Jahresber. Ver. Naturk. Linz (8), p. 67.

Sphèridiiens Lameere (1900), Man. Fne. Belg., II, p. 328.

Sphaeridiini + *Cyllomina* + *Rygmodus* bis *Tormus* Zaitzev (1908), Hor. Soc. Ent. Ross., XXXVIII, p. 401, 400,

375-376 (Catal.).

ONTOGENESIS. Ganglbauer (1904), Käf Mitteleur., IV, 1, p. 265—Wessenbrg Lund (1915), Insektlivet i ferske vande, p. 299—A. d'Orchymont (1916), Bull. Soc. Ent. Fr., P. 238—Richmond (1920), Bull. Amer. Mus., Nat. Hist. XLII, p. 77.

KEY TO TRIBES

1. Head not narrowed just before the eyes, the outer margin forming no conspicuous angle with the latter, the antennae inserted under a laminated border (antennal sternite) very conspicuous from beneath and concealing the base of antennae from above. This border penetrates sometimes into the eyes, which appear then emarginate. Antennae usually much longer than maxillary palp. Metasternum usually prolonged between the middle coxae, closely united with the mesostital elevation. Metathoracic episterna wide.*

   Posterior tarsi with first joint longer than 2nd, and 3rd usually very long, exceptionally subequal or scarcely longer than 2nd, but then the first ventral segment is carinate along the middle. Intermediate coxae normally separate. Metasternum with sometimes a defined elevated area in the middle or very conspicuously defined femoral abutments.

   Larvae with tarsi (*Coelostoma, Phænonotum*) or tarsi absent (*Sphæridium*). *Spharidiini.*

1′. Head abruptly narrowed before the eyes, the outer margin forming a conspicuous angle with the latter, which are not emarginate; the shorter antennae inserted freely, not under the laminated border of antennal sternite, the insertion quite visible from above. Metasternum not prolonged between the middle coxae except in *Omicrus* (American and Hawaiian). Metathoracics episterna never very wide. 2.

---

* These characters are also those of the exotic *Rygmodini.*

2. Mesostital process longer than wide, as a tablet or lamina. Middle coxae never widely separated. Metathoracic episterna moderately wide. Elytra not clasping the body. Space between femoral abutments of metasternum narrower behind, more or less elevated and often otherwise sculptured. Epipleura of elytra conspicuous, often horizontal. Prostitum only carinate or exceptionally as an elevated flat or tectiform tablet.

   Larvae (in *Cercyon*) footless — *Cercyonini.*

2′. Mosostital process as a flat tablet at least as wide as long and always widely in contact with metasternum.* Middle coxae very widely separated. Metathoracic episterna narrow and partly covered by the margin of elytra. The latter clasp the body; if not the metasternum has no femoral lines and no median elevated part. Gula and mentum not lying in the same horizontal plane, the submentum placed nearly perpendicularly between the two. Prostitum always elevated at middle to the plane of mesostitum as a flat tablet, carinate or not carinate. Epipleura of elytra sometimes indistinct or reduced. Anterior part of prefrons before the eyes having a tendency to be deflected towards the underside.

   Larvæ (in *Megasternum*) with vestigial legs — *Megasternini.*

TRIBE SPHÆRIDIINI.

Sphæridiini A. d'Orchymont (1919), Ann. Soc. Ent. Fr., LXXXVIII, p. 106, fig. 1.

Cyclonotaires + *Sphæridium* Rey (1886), Ann. Soc. Linn. Lyon, XXXII, p. 112, 113, 120, 121.

Cyclonotitæ + *Sphæridium* Kuwert (1890), Verh. Naturf. Ver. Brünn, XXVIII, p. 22.

Cyclonoti + *Sphæridium* G. H. Horn (189), Trans. Ann. Ent. Soc. XVII, p. 281.

Caelostomini + *Sphæridium* Everts (1899), Col. Neerl., I, p. 638.

---

* Except in the north American *Pemelus* not known to me.

Ontogenesis, A. d'Orchymont (1919), Ann. Soc. Ent. Fr., LXXXVIII, p. 107.

KEY TO GENERA.

1. Elytra covering the apex of abdomen. Scutellum short, triangular. 2.

1'. Apex of abdomen not covered. Scutellum as an elongate triangle. Elytra with a sutural stria. First ventral segment not carinate. Antennae 8-jointed. First joint of posterior tarsi longer than second. 13. *Sphæridium.*

2. Club of 9-jointed antennae very long and laxly articulated. Elytra only with a sutural stria, irregularly punctured, without punctures arranged to form series of striae, even on the sides, of a more convex shape, with less planate underside, and tibiae less enlared and flattened. First ventral segment not carinate (only obscurely carinate in *horni*). Prostitum normally developed before anterior coxae, not carinate or with an anteriorly dentate carina (in *horni*). Posterior tarsi with first joint twice as long as second joint. 9. *Cælostoma.*

2'. Club of antennae shorter and its articulation not so lax. Elytra normally with sutural stria and several other striae or seriately arranged larger punctures, at least at the sides; exceptionally with caelostomoid appearance from the upperside, without the least trace of such series, even at the sides, and with sutural stria vanishing but of a more depressed and less convex habit with more planate underside, tibiae more enlarged and flattened and first abdominal segment well developed and ordinarily carinate along its whole length.* 3.

3. Elevated part of mesostitum before middle coxae not arrow-headed with transverse carina accompanied before and behind by a slender longitudinal more or less short carination. Eyes small, without canthus. Taris very short with basal joints subequal. Elytra punctate-striate a least in part. 4.

---

* Not carinate in *Dactylosternum (Coæloflētium) exstriatum* m. from the Philippines. This species the upperside of pronotum and elytra is besides practically without punctuation.

3'. Elevated part of mesostitum arrow-headed in the middle. Eyes some times with a rudimentary canthus. Tarsi less short or also very short, basal joint of variable length, sometimes twice as long, sometimes subequal to second joint. Elytra punctuate-striate, punctate-seriate, or with vanishing series, or even without series, sutural stria and punctuation. 12. *Dactylosternum.*

4. Second and third striæ or series of elytra shortened behind effected here by the fourth which approaches the first. Prefrons not bordered anteriorly: metasternal femoral abutments reaching intermediate coxal cavities.† Metasternum without curved abruptly impressed area on antero-external angles; anteriorly with a slender longitudinal and short carinula. Form tolerably convex transversely and small. First abdominal segment carinate, second carinate or not. 10. *Protosternum.*‡

4'. Second and third striae not shortened behind, reaching apex of elytra just like the first and fourth. Prefrons finely bordered anteriorly. Metasternum with abruptly impressed area on antero-external angles limited by a curved line which reaches the metathoracic episternum. Femoral abutments present from the middle of posterior margin of metasternum to the middle of coxal cavities and meeting here the limit of antero-external depressed area. First and second ventral segments carinate in the middle. 11. *Mucetum.*

9. **Coelostoma** Brullè.

Coelostoma Brullè (1835), Hist. Nat. Ins., V, (2), p. 289, 293—Kuwert (1890), Verh. Naturf. Ver. Brünn, XXVIII, p. 22, 175—Everts (1899), Col. Neerl. I, p. 640, 655—Kliment (1899), Cešti brouci, p. 120—Reitter (1909), in Brauer, Süsswasserfna Deutschl., 3-4, p. 77, 78, f. 53; (1919), Fna. Germ., II, p. 336, 367, f. 70, p. 368—Jakobson (1911), Käf. Russl., p.

---

† These characters are also those of *Blanodium* from Formosa and Sumatra.

‡ Sicne my note on this genus (Ann. and Bull. Soc. Ent. Belg., LXVI, 1926, p, 104. I have received a second species belonging to it and new to science. It has a well developed carina on second vential segment just as in *Mucetum.*

858—Kuhnt (1913), Ill. Best. Tab. Käf. Deutschl., p. 419, 436, f. 28, 29— Schaufuss C. (1916), in Calwer, Käferb., ed. 6 (1913), p. 346, 347—Tullgren & Wahlgren (1920), Svenska Ins. (2), p. 235—Knisch (1921), Archiv. Naturg. (1919), p. 71.

Cyclonotum Erichson (1837), Käf. Mark. Brandebg., I, p. 212.

ONTOGENESIS Richmond (1920), Bull. Amer. Mus. Nat. Hist., XLII, Art., I, p. 87.

æneolum Règimbart (1903), Ann. Soc. Ent. Fr., LXXII, p. 337.
Calicut, Mahè.

fabriciusi Montrouzier emend. (1860), Ann. Soc. Ent. Fr. (3) VIII, p. 245 (*Ochthebius fabricii*)[1]— Fauvel (1883), Rev. Ent. Caen, II, p. 357; (1903), l. c., XXII, p. 355 (*Cyclonotum fabricii*)[2]— A. d'Orchymont (1924), Nova guinea, XV, Zool., Livr. 1, p. 30[3]; (1925), Ann. Bull. Soc. Ent. Belg., LXV, p. 270[4] (cf. *Coelostoma* sp. B, in Ann. Soc. Ent. Belg., LIX, 1919, p. 73[4] bis.
New Caledonia,[1, 2, 4]
Australia.[2 to 6]
Tasmania.[4]
Hawaii.[3, 4, 7]
New Guinea.[3]
Borneo.[4], Sumatra.[4]
Tonkin.[4]
Laos, Annam.[4] bis.

? *mastersi* W. M'Leay (1871), Trans. Ent. Soc. N. S. Wales, II, 2, p. 133 (*Cyclonotum mastersii*).[5]

? *australe* Blackburn (1888), Proc. Linn. Soc. N. S. Wales (2), III, p. 839 (*Cyclonotum*).[6]

extraneum Sharp (1908), Fna Hawaiensis, III, 579 (*Cyclonotum*).[7]

horni Règimbart (1902), Ann. Soc. Ent. Fr., LXXI, p. 474 (*Cyclonotum*)[1]—A. d'Orchymont (1922), Rec. Ind. Museum, VIII, p. 265; (1925),[2] Ann. & Bull. Soc. Ent. Belg., LXV, p. 267.[3]
Ceylon.[1], India.[3]
Nicobars,
Andamans.[3]
Abor Country.[2]

Tonkin.[3]
Sumatra.[3], China.[3]

var. *lazarense* A. d'Orchymont (1925), Ann. & Bull. Soc. Ent. Belg., LXV, p. 200[1]; (1926), Philip. Journ. Sc., 30, 3, p. 364.[2]
Philippines.[1, 2]
Sumatra.[3]
F. M. S. (Selangor, Perak, Pahang).

var. *transcaspicum* Reitter (1906), Wien. Ent. Zeitg., XXV, p. 32[1]—A. d'Orchymont (1925), Ann. & Bull. Soc. Ent. Belg., LXV, p. 267[2]; (1926), Philipp. Journ. Sc., 30, 3. p. 364.[3]
Transcaspia.[1, 3]
Buchara.[1, 3] Arabia.[2]
Asia minor.[2] China[2]
Cambodia.[2]

Tonkin, Malacca.[2]
F. M. S. Selangor.

orbiculare Fabricius (1775), Syst. Ent. p. 229[1]; (1781), Spec. Ins. I, p. 290; (1787), Mant. Ins. I, p. 188; (1792), Ent. Syst. I, p. 184; (1801), Syst. Eleuth. I, p. 252 (*Hydrophilus*)—Brullè (1835), Hist. Nat. Ins., V, (2), p. 294—Erichson (1837), Käf. Mark. Brandeb. I, p. 214 (*Cyclonotum*)— Mulsant (1844), Hist. Nat. Col. Fr. Palp., p. 148 (*Cyclonotum*)— Thomson G.C. (1860), Skand. Col. II, p. 101 (*Cyclonotum*)— Kuwert (1890), Verh. Naturf. Ver. Brünn. XXVIII, p. 178 —Everts (1899), Col. Neerl. I, p. 656—Ganglbauer (1904), Käf. MItteleur, IV, I, p. 268 (*Cyclonotum*)—Reitter (1909), Fna. Germ., II, p. 368, t. 80, f. 13—Tullgren & Wahlgren (1920), Svenska Ins. (2), p. 235.*

Europe.[1]
Siberia.
Japan.[2]
Syr Darja. (Aulie ata).[3]

pilula O. Muller (1776), Zool. Dan. Prodr., p. 69 (*Hydrophilus*).

punctatum Gœze (1777), Entom. Beytr., I, p. 666 (*Hydrophilus*).

immaculatum Rossi (1790), Fna. Etrusc. I, p. 48, pl. 3, f. 5 (*Hydrophilus*).

orbiculatum Rossi (1792), Mant. Ins., I, p. 65 (*Hydrophilus*).

dalmatinum Kuster (1848), Käf. Eur., XIII, n° 40 (*Cyclonotum*).

breve Sharp (1874), Trans. Ent. Soc. London, p. 419 (*Cyclonotum*).[2]

puncticolle Reitter (1906), Wien. Ent. Zeitg., XXV, p. 32.[3]

salvazai A. d'Orchymont (1919), Ann. Soc. Ent. Belg., LIX, p. 73.[1]

Cambodia, Annam.[1]
Cochin-China.

stultum Walker (1858), Ann. & Mag. Nat. Hist., (3), II, p. 209 (*Hydrobius*)[1]—A. d'Orchymont (1919), Ann. Soc. Ent. Belg., LIX, p. 71, 72[2]; (1922), Rec. Ind. Mus., VIII, p. 625[3]; (1926), Philipp. Journ. Sc., 30, 3, p. 363.[4]

simplex Sharp (1874), Trans. Ent. Soc. Lond., p. 419[5]; (1890), l. c., p. 356[6] (*Cyclonotum*).

Ceylon.[1, 6], Burma.[3]
Assam.[3]

Indo-China.[3]
Annam, Tonkin,[2]
Cochin, China.
F. M. S. Selangor.

---

* Principal authors. For complete bibliography see : Knisch, Col. pars. 79, 1924, p. 111-112.

Sumatra., Java.[3] Borneo.[3], formosa Philippines.[4] Japan.[5], China.[5]

vitalisi A. d'Orchymont (1923), Mem. Dep. Agric. India, VIII, p. 2 (in litt); (1923), Treubia, III, 3-4, p. 418[2]; (1925), Ann. & Bull. Soc. Ent. Belg., LXV, p. 269[3] (c.f. *Coelostoma* sp. A, in Ann. Soc. Ent. Belg., LIX, 1919, p. 72).[4] — India.[3, 5, 7], Bihar.[1] ? Genji.[6], ? Cylon.[7]

———

Indo-china.[3, 5] Tonkin.[2, 4]

orbiculare Regimbart (1903), Ann. Soc. Ent. Fr., LXXII, p. 63[5], 337[6] (ex p. ?)—Knisch (1921), Archiv. Naturg. (1919) p. 77 (ex p. ?).[7] (nec Fabricius, 1775). — F. M. S. Malacca.[3] Sumatra.[2, 3, 5, 7] Bengkalis Is.[3] Java.[3, 7], Borneo.[3, 5]

10. **Protosternum** Shap.

Protosternum Sharp (1890), Trans. Ent. Soc. London, p. 356—A. d'Orchymont (1926), Ann. & Bull. Soc. Ent. Belg., LXVI, p. 104.

atomarium Sharp (1890), l. c., p. 356. — Ceylon.

11. **Mucetum** A. d'Orchymont.

Mucetum A. d'Orchymont (1926), Ann. & Bull. Soc. Ent. Belg., LXVI, p. 101, 104.

carinatum A. d'Orchymont (1926), l. c., p. 102. — F. M. S.: Singapore.

12. **Dactylosternum** Wollaston.

Dactylosternum Wollaston (1854), Ins., Mad., p. 99, t. 3, f. 1—Redtenbacher (1874), Fna. Austr., ed. 3, II, p. 124–Bedel (1881), Fne. Col. Bass. Seine, I, p. 336, nota–Rey (1886), Ann. Soc. Linn. Lyon, XXXII, p. 113, 118– Seidlitz (1888), Fna. Balt. ed. 2. Gatt. p. 23; (1888), Fna. Transsylv., Gatt., p. 23–Kuwert (1890), Verh. Naturf. Ver. Brünn, XXVIII. p. 22, 178—G. H. Horn (1890), Trans. Am. Ent. Soc., XVII, p. 281, 282—Ganglbauer (1904), Käf. Mitteleur., IV, I, p. 266, 267—Jakobson (1911), Käf. Russl., p. 858—Schaufuss C., (1916), in Calwer, Käferb., ed. 6 (1913), p. 346—Des Gozis (1917), Misc. Ent., XXIII, p. 64—Knisch (1921), Archiv. Naturgesch. (1919), A. 8, p. 71—A. d'Orchymont (1923), Mem. Depart. Agric. India, VIII, p. 2; (1926), Ann. & Bull. Soc. Ent. Belg., LXVI, p. 104.

Trichopoda Brullè (1835), Hist. Nat. Ins., V, 2, p. 289, 294— Laporte de Castelnau (1840), Hist. Nat. Anim., Art. II, p. 59 (nec Latreille).

Coelostoma Laporte de Castelnau (ex p.) (1840), Hist. Nat. Anim. Art. II, p. 58.

Cyclonotum Mulsant (ex. p.) (1840) Ann. Soc. Agric. Lyon, VII, 167, 169—Sharp (1882) Biol.-Centr. Amer., I, (2), p. 94.

Macrocercyon Alluaud (1899), Bull. Soc. Ent. Fr., p. 379.

abdominale Fabricius (1792), Ent. Syst., I, p. 79; (1801)[1], Syst. Eleuth., I, p. 94 (*Sphæridium*)[2]—Mulsant (1844), Ann. Soc. Agric. Lyon, VII, p. 179 (*Cyclonotum*)[4]—Wollaston (1860), Ann. Mag. Nat. Hist., p. 55; (1865), Col. Atlantidum, p. 76[5]; (1877), Col. Sanctae Helenae, p. 20—Gerstacker (1873), Decken's Reise Ostafr. III (2), p. 79 (*Cyclonotum*)—Redtenbacher (1874), Fna., Austr., ed. 3, II, p. 124—Sharp (1887), Biol. Centr. Amer. I (2), p. 769 (*Cyclonotum*)[6]—G. H. Horn (1890), Trans. Amer. Ent. Soc., XVII, p. 283[7]—Leng & Mutchler (1918), Bull. Amer. Mus. Nat. Hist., XXXVIII, p. 111—A. d'Orchymont (1925), Ann. & Bull. Soc. Ent. Belg., LXV, p. 274[8]; (1926), Philipp. Journ. Sc., 30, 3, p. 365.[9]

insulare Laporte de Castelnau (1840), Hist. Nat. Anim. Art., II, p. 59 (*Coelostoma*)[10]—A. d'Orchymont (1913) Suppl. Ent., II, p. 10.

? *nitidum* Boheman (1851), Ins. Caffr. I, p. 602 (*Cyclonotum*).[11]

rousseti Wollaston (1854), Ins. Mad., p. 100, Pl. 3, f. 1.[12]

? *mulsanti* Murray (1859), Ann. Mag. Nat. Hist. (3), IV, p. 352 (*Cyclonotum*).

? *natalense* Gemminger & Harold (1868), Cat. Col. II, p. 495.

semistriatus Schaufuss L. W. (1887), Hor. Soc. Ent. Ross., XXI, p. 108 (*Hydrobius*).[14]

? *foveonitidum* Kuwert (1890), Verh. Naturf. Ver. Brünn, XXVIII, p. 179.

Brasil.[1], Antilles.[6, 7]
St. Vincent.[2]
Mexico.[6, 7, 8]
Florida, N. Carolina.[7]
Madeira.[5, 6, 7, 12]
Canaries.[5, 8]
Cape Verde I.
TropicalAfrica.[8]
Caffraria.[11], Guinea.[13]
Mauritius.[4, 8, 10]
Seychelles.
Madagascar.[4, 6, 7, 8]
Annobon I.
S.Thome I,
Gibraltar.[8], Tunis.[8]
Algeria.[8], Spain.
france., syria.[15]

Ceylon.[8], India.[8]
Andamans.[8]

Malacca.[8]
Singapore.[14]
Sumatra.[8], Java.[8]
Philippines.[9]
Samoa.[8]
Christmas I.[8], Key I.
Hawaii.[8]
N. Caledonia.
Australia.[8]
Queensland.

	Warmer regions of globe, excl. New Zealand.[8]
bakeri A. d'Orchymont (1925), Ann. & Bull. Soc. Ent. Belg., LXV, p. 149.	Penang I.
coælostomoides A. d'Orchymont (1923), Mem. Depart. Agric. India, VIII, p. 3, 6.	Matheran.
corporaali, A. d'Orchymont (1923), Treubia, III, p. 416.[1]	Sumatra.[1] F. M. S. Pahang.
cycloides Knisch (1921), Archiv. f. Naturg. (1919), A. 8, p. 73.[1]	Sumatra.[1] Mentawei I.[1] F. M. S. Pahang.
dachinabadense A. d'Orchymont (1923), Mem. Depart. Agric. India, VIII, p. 3, 5.	Trichinopoly.
dytiscoides Fabricius (1775), Syst. Ent., p. 67; (1781), Spec. Ins., I, p. 78; (1792), Ent. Syst., I, p. 79; (1801), Syst. Eleuth., I, p. 94 (*Sphaeridium*)—Linnè (1789), Syst. Nat., ed. XIII, Gmel., I, IV, p. 1626 (*Silpha*)—Olivier (1790), Ent., II, No. 15, p. 5, Pl. 2, f. 10 (*Sphaeridium*)—Wollaston (1878), Col. Sanctae Helenae, p. 19 (*Cyclonotum*) Bedel (1881), Bull. Soc. Ent. Fr., p. LXXXI & LXXXVI— Règimbart (1903); Ann. Soc. Ent. Fr., LXXII, p. 63[1]—A. d'Orchymont (1913), Suppl. Ent., II. p. 9, t. 1, f. 5, 6[2]; (1926), Philipp. Journ. Sci., 30, 3, p. 366.[3] *melanopternum* Montrouzier (1857), Ann. Soc. Agr., Lyon, VII, p. 19 (*Sphaeridium*).[4] *cowieyi* Blackburn (1898), Trans. R. Soc. S. Austr., XXII, 2, p. 229 (*Cyclonotum*).[5]	Ceylon.[2] —————— F. M. S. ; Singapore Poulo Pinang.[1, 2] Sumatra. Engano I.[2] Mentawei I.[2] Nias I.[2], Java.[2] Batjan.[2], Borneo.[1, 2] Philippines.[2, 3] Celebes.[2], Buru I. Morotai.[1, 2] New Britain.[1, 2] New Pomerania.[2] New Guinea.[2] Woodlark I.[4, 2] Australia.[5]
fletcheri A. d'Orchymont (1923), Mem. Depart. Agric. India, VIII, p. 3, 6.[1]	Matheran.[1] Kanara.
hydrophiloids W.S. M'Leay (1825), Annul. Javan., p. 36 (*Sphaeridium*)[1]—Erichson (1845), Archiv. f. Naturg. (Wiegman's Archiv.), XI, 2, p. 105 (*Cyclonotum*) —A. d'Orchymont (1914), Ent. Mitteil., III, p. 322; (1923), Truebia, III, p. 417[2]; (1925), Ann. & Bull. Soc. Ent. Belg., LXV, p. 274.	Java.[1, 2, 4, 6] Sumatra.[2, 4] Mentawei I. Buru I., Boreno.[4] Philippines.[2, 4] Formosa.[5] Annam.[4, 5, 7]

capense Mulsant (1844), Ann. Soc. Agric. Lyon, VII, p. 170 (*Cyclonotum*).	Tonkin.[4] Laos-Annam. Penins. Siam. Malacca.[4, 7] Singapore.
rubripes Boheman (1858), Eugenies Resa, Ent. Bidr., I, Col., p. 24 (*Cyclonotum*)[3]. Règimbart (1903), Ann. Soc. Ent. Fr., LXXII, p. 63[4]—A. d'Orchymont (1913), Suppl. Ent., Vol. II, p. 7-9, t. 1, f. 7.[5]	———— Indian.[4], Assam.
? *nitidum* Laporte de Castelnau (1840), Hist. Nat. Anim. Art., II, p. 58 (*Coelostoma*)[6]—Règimbart (1903), Ann. Soc. Ent. Fr., LXXII, p. 63.[7]	
indicum A. d'Orchymont (1923), Mem. Depart. Agric. India, VIII, p. 3.[1]	Shembaganur.[1] N. Kanara
salvazai A. d'Orchymont (1925), Ann. & Bull. Soc. Ent. Belg., LXV, p. 272, 274.	Laos.
seriatum Knisch (1921), Archv. f. Naturg. (1919), A. 8, p. 76.[1]	Sumatra.[1] Mentawei I.[1] Java., Engano. I. Cochinchina, Tonkin.
vitalisi A. d'Orchymont (1925), Ann. & Bull. Soc. Ent. Belg., LVX, p. 271, 274.	Laos. Fokien.

13. **Sphaeridium** Fabricius.

Sphaeridium Fabricius (ex. p.) (1775), Syst. Ent., p. 66; (1787), Mant. Ins., I, p. 43; Ent. Syst. (1792), I, p. 77; (1801), Syst. Eleuth, I, p. 92—Latreille (ex p.) (1796), Prècis car. gèn. Ins., p. 8; (1817), in Cuvier, Règne Anim. III, p. 273; (1829), nouv. ed., IV, p. 525—Leach (1815), in Brewster's Edinb. Enc., IX, p. 96; (1817), Zool. Misc. III, p. 95—Lamarck (1817), Hist. Nat. Anim. Sans vert; IV, p. 541—Stephens (1829), Ill. Brit. Ent. Mandib. II, p. 154; (1839), Man. Brit. Col., p. 93—Curtis (1834), Brit. Ent. XI, nr. 518—Brullè (1835), Hist. Nat. Ins., V (2), p. 289, 291—Erichson (1837), Käf. Mark. Brand. I, p. 214—Westwood (1840). Intr. Mod. Class. Ins, II, Synopsis, p. 10—Heer (1841) Fna Coll. Helv., I, p. 487—Mulsant (1844), Hist. Nat. Col. Fr. Palp., p. 147, 150, t. f. 14, 15, 17—Redtenbacher (1845), Gatt. Deutsch. Käferfna. p. 70; (1849), Fna Austr., p. 133; (1858), ed. 2, p. 114; (1874), ed. 3, II, p. 124—Chenu (1851), Encycl. Hist. Nat.

Col., I, p. 253—Bach (1851), Käferfna Nord u. Mitteldeutschl. I, p. 122, 134—Murray (1853), Ann. Mag. Nat. Hist. (2), XII, p. 74, t. 4, f.—Fairmaire & Laboulbène (1854), Fne. Ent. France, I, p. 250—Lacordaire (1854), Gen. Col. I, p. 471, 472—Wollaston (1854), Ins. Mader., p. 101—Jacquelin Duval (1857), Gen. Col. Eur., I, p. 95, 99—Kuhn (1858), Käf. Südbayer. Flachl., p. 37—Bose (1859), in Gutfl., Käf. Deutschl., p. 71—Fuss (1859), Archiv. ver. Siebenb. Landesk n. F IV, (3), p. 121—Thomson G. C. (1859), Skand Col., I, p. 19; (1860), II, p. 101; (1867), IX, p. 123—Seidlitz (1872), Fna. Balt., p. 11; (1888), ed. 2, Gatt. p. 23; (1888), Fna. Transsylv., Gatt., P. 23—Solsky (1876), Fedtsch. Reise Turkest. II, p. 21—Dalla Torre (1877), Jahresber. Ver. Naturk. Linz (8), p. 67—Schlosser Klek (1877), Fna. Kornjašah, I, p. 108—Bedel (1881), Fne. Col. Bass. Seine, I, p. 334, 335—Leconte & Horn (1883), Class. Col. North. Amer. p. 73—Rey (1886), Ann. Soc. Linn. Lyon, XXXII, p. 120, 121-Fowler (1887), Col. Brit. Isl., I, p. 252—Kuwert (1890), Verh. Naturf. Ver. Brünn, XXVIII, p. 22, 174—Wickham (1895), Canad. Ent., XXVII, p. 182—Acloque (1896), Fne. Fr., p. 98—Everts (1899), Col. Neerl. I, p. 639, 655—Kliment (1899), Cešti brouci, p. 120—Lameere (1900), Man. Fne. Belg., II, p. 335—Stierlin (1900), Col. Helv., p. 213—Ganglbauer (1904), Käf. Mitteleur., IV, I, p. 266, 269— Reitter (1909), Fna Germ., I, p. 366, 367—Blatchley (1910), Col. Indiana in Bull. Ind. Dep. Geol., I, p. 265—Jakobson (1911), Käf. Russl., 858—Kuhnt (1913), Ill. Best. Tab. Käf. Deutschl., p. 419, 436, fig. 22 A. 23, 24—Joy (1914), Ent. Mo. Mag., L, p. 83—Schaufuss c. (1916), in Calwer, Käferb, ed. 6 (1913), p. 346, 347—Sainte Claire Deville (1919), Bull. Soc. Ent. Fr., p. 230—A. d'Orchymont (1919), Ann. Soc. Ent. Fr., LXXXVIII, p. 112, f. 3—Sharp (1919), Ent. Mo. Mag., LV, p. 124—Tullgren & Wahlgren (1920), Svenska Ins. (2), p. 235—Knisch (1921), Archiv. Naturg., LXXXV, (1919), A. (8), p. 70—Menozzi (1921), Bull. Soc. Ent. Italiana LIII, p. 57.

ONTOGENESIS, Mulsant (1844), Hist. Nat. Col. Fr. Palp., p. 147 & 151—Lacordaire (1854), Gen. Col. I, p. 470—Schiödte (1862), Nat. Tidskr. (3) I, p. 220, 223; (1872), VIII (1), p. 216—Ganglbauer (1904),

Käf. Mitteleur; IV, I, p. 270, f. 10—A. d'Orchymont (1913), Ann. Biol. lac., VI, p. 211; (1919), Ann. Soc. Ent., Fr., LXXXVIII, p. 107—Schaufuss C. (1916), in Calwer, Käferb., ed. 6 (1913) p. 348—Richmond (1920), Bull. Amer. Mus. Nat. Hist., XLII, Art. 1, p. 83.

cameroni A. d'Orchymont (1926), Ann. & Bull. Soc. Ent. Belg., LXVI, p. 202.[1] — Mohand Rau, Siwaliks; Dehra Dun.[1]

F. M. S.: Seremban (var.).

dimidiatum Gory (1834-1842), in Guèrin—Mèneville, Icon. Règne Anim. de G. Cuvier, Insectes, VII, p. 73, fig. 15[1]—Laporte de Castelnau (1840), Hist. Nat. Anim. Art, II, p. 60[2]—A. d'Orchymont (1913), Suppl. Ent., II, p. 13, 15[3]; (1920), Ann. Soc. Ent. Belg., LX, p. 19.[4] — Java.[1, 2, 3] Sumatra.[3], Celebes. Philippines.[3] China.[3] Indo-China.[3] India.[3, 4]

kolleri A. d'Orchymont (1925), Ann. & Bull. Soc. Ent. Belg., LXV, p. 155. — Laos.

quinquemaculatum Fabricius (1798), Ent. Syst. Suppl. p. 39; (1801), Syst. El., I, p. 94—Laporte de Castelnau (1840), Hist. Nat. Anim. Art., II, p. 60[1]—Sharp (1890), Trans. Ent. Soc. London, p. 357[2]—A. d'Orchymont. (1913), Suppl. Ent., II, p. 11, 14, t. 1, f. 10, 14.[3] — Java.[1, 3, 4, 5] Sumatra., Formosa.[3] Philippines. China.[2, 3, 7] F. M. S. Indo-China.[3]

India.[2, 3], Indus.[3] Ceylon.[2, 3, 6] Baluchistan.[3]

? *marginatum* W. S. MacLeay (1825), Annul. Jav. p. 36 (nec Fabricius, Scriba) (sec. Knisch., Col. Cat., Pars 79, 1924, p. 124).[4]

? *vicinum* Laporte de Castelnau (1840), Hist. Nat. Anim. Art., II, p. 60.[5]

cricolor Walker (1858), Ann. & Mag. Nat. Hist. (3), II, p. 209 (nec Fourcroy, 1785).[6]

chinense Frivaldsky (1889), Termeszetr. Füzetek, XXII, p. 197.[7]

seriatum A. d'Orchymont (1913), Suppl. Ent., II. p. 12, 14[1]; Treubia (1926), VII, p. 128.[2] — Sumatra.[1], Buru I.[2] Formosa.[1] Philippines Tonkin.[1] Fokien.

severini A. d'Orchymont (1919), Ann. Soc. Ent. Fr., LXXXVIII, p. 116, f. 7[1]; (1923), Mem. Depart. Agric. India, VIII, p. 7[2]; (1923), Treubia, III, p. 418.[3] — Malabar.[1] Khasi Hills.[2] Madura.

Cambodia, Tonkin. F.M.S.: singapore. Sumatra.[1], Java.[3] Hongkong.

vitalisi A. d'Orchymont (1925), Ann. & Bull. Soc. Ent. Belg., LXV, p. 153. — Upper Mekong.

TRIBE CERCYONINI.

Cercyonini A. d'Orchymont (1919), Ann. Soc. Ent. Fr., LXXXVIII, p. 118, fig. 8; (1925), Ann. & Bull. Soc. Ent. Belg., LXV, p. 292 & nota 2.

Cercyones (—Sphaeridium) G.H. Horn (1890), Trans. Amer. Ent., Soc., XVII, p. 281, 287.

ONTOGENESIS. A. d'Orchymont (1919), Ann. Soc. Ent., Fr., LXXXVIII, p. 119.

KEY TO GENERA.

1. Anterior part of disc of head, before antenno-frontal suture, not particularly or very slightly deflected to the underside. Elytra seriate punctate or striate puncatate. Posterior tarsi with first joint longer than second. — 2.

1'. Prefrons very strongly deflected to the underside so that in extreme cases this part of disc of head is nearly in the same plane as ventral side, when the head is closely applied to the prothorax. The eyes are very small and seen from above as sub-angular projections at each side of the head. Pronotum without series of larger punctures at posterior margin. Labrum outstanding. Mentum and gula not in the same horizontal plane, the submentum placed nearly perpendicularly or very obliquely between the two. — 10.

2. First ventral segment not carinate. No antennal cavities on anterior margin of prostitum. Mesostitum declivous, only carinate, not elevated as a tablet or

a lamina. Head, pronotum and scutellum without true punctures, but sculptured with very small perfect circles, more or less closely distributed. 14. *Cycreon.*

2'. First ventral segment carinate along the middle. 3.

3. Prostitum only carinate along the middle not conspicuously differentiated from the lateral antennal cavities, the pubescent slopes of carina descrnding gently in the glabrous bottom of the latter. No series of larger punctures on hind margin of pronotum. 4.

3'. Prostitum carinate along the middle and conspicuously differentiated here from the large lateral antennal cavities, at least by an unevenness or a delicate ridge-like carina; if not (in some *Armostus*) a crest situated on metasternum just behind the middle coxae does not reach the external margin of the cavities of the latter but bends before towards the inner side of metathoracic episterna, and no femoral lines or crests are present between femoral abutments and outer margin of said middle coxal cavities 5.

4. Mesostital elevation not being an extension of the metasternum, in touch with the latter by a single point, or separated therefrom by a large hiatus, or engaged in a small triangular excision of its anterior middle side. Femoral crests on metasternum present or not. 15. *Cercyon.*

4'. Mesostital elevation appearing as a sort of prolongation of disc of metasternum in the shape of a more or less elongate pentagon reunited to this piece by its more or less wide base, a delicate suture being still visible. No crests on metasternum to continue the femoral abutments. 16. *pelosoma.*

5. Differentiated middle part of prostitum not elevated, on a slightly more depressed plane than lateral antennal cavities. Mesostital process elongate posteriorly with an emargination which applies obliquely on the metasternum. Femoral crests are present on the latter. No. series of larger punctures on hind margin of pronotum. Elytral epipleura nearly horizontal near posterior coxae. 17. *peltocercyon.*

5'. Differentiated middle part of prostitum elevated above antennal cavities, the beetle being examined from the underside. 6.

6. Mesostital process very thick, its underside not as a plane tablet, its antero-lateral sides straight and posteriorly divergent, very much and transversely curved afterwards, sometimes nevertheless somewhat plane on the underside and recalling then *Pelosoma,* nearly 2½ times as long as wide, its anterior vertical side forming on the inferior extremity a very conspicuous outstanding swelling; the posterior side of process, rounded, or more or less angular, is tolerably broadly in contact with metasternum and displaced obliquely on the anterior side of the latter, between middle coxae. Sides of metasternum before posterior coxae with some larger punctures partly concealed by pubescence of surface. Elytral epipleura nearly horizontal near posterior coxae. No series of larger punctures on hind margin of pronotum but punctuation of the latter is of two sizes. No femoral crest are present on metasternum. 18. *pseudocercyon.*

6'. Mesostital process as a plane tablet on its underside. 7.

7. Elytral epipleura nearly vertical the surface of metasternum lying very deep in the elytral cavity; sides of the latter with some larger punctures partly concealed by pubescence of surface. Mesostital elevation pentagonal or lance-like, some what longer than wide, more or less widely in touch with middle part of metasternum between coxae. Tenth series of elytral punctures anteriorly far from elytral margin. Differentiated middle part of prostitum as a reversed lozenge with more or less angulate anterior side. 19. *Gillisius.*

7'. Elytral epipleura more horizontal or only oblique along metasternum. Differentiated middle part of prostitum more pentagonal with straight anterior side 8.

8. Metasternum with a fine crest bordering middle coxal cavities and reaching its anterior outer angles (in Asiatic forms). Metasternal femoral abutments prolonged with a crest reaching these angles.

Elytra more or less pubescent. Pronotum with a more or less conspicuous series of larger punctures along its hind margin. Mesostital process usually short oval, sometimes more or less pentagonal. 20. *Oosternum.*

8'. Metasternum with an anterior crest-accompanied often by a groove which originates near teh inner side of middle coxae, bordering then the latter, but not reaching the anterior outer angles; a little before these angles the crest bends backwards to follow then to a certain extents the innerside of metathoracic episterna from which it is sometimes separated by a more or less deep depression. Body rounded oval somewhat narrowed behind not conspicuously pubescent. Pronotum without series of larger punctures along its hind margin. No femoral crests to continue the metasternal femoral abutments. 9.

9. A deep hiatus between mesostital process and metasternum middle elevation, this process oval, longer than wide more or less rounded behind and more acuminate in front. Middle part of postitum somewhat differentiated or not differentiated from large lateral antennal cavities. First joint of posterior tarsi somewhat longer than two following taken together. Elytra punctate-striate. 21. *Armostus.*

9'. No hiatus between mesostital process and metasternum, the former oval, pointed anteriorly, widely truncate behind and closely united here by the whole width of truncature with elevated part of the latter, this elevated part gibbous and more prominent than mesostital tablet. Middle part of prostitum more differentiated from lateral parts. First joint of posterior tarsi subequal to joints 2 and 3 taken together. Elytra striate-sulcate, crenelate in the striae. 22. *Morastus.*

10. First ventral segment not carinate. 11.

10'. First ventral segment carinate. Tarsi very short. Intermediate coxae not very much separated. Mesostital elevtion as a rhomboidal arrow head, but with median part not dectylate longitudinally. First joint of posterior tarsi somewhat longer than second. Side margins of elytra more or less widely explanate. 25. *Paromicrus.*

11. Mesostital lamina cultriform, double, its anterior part concave in the middle, its posterior part alone in touch with metasternal elevation between middle coxae. Posterior tarsi more elongate, their first joint much longer than second. Elytra explanate as in *Paromicrus*. Middle coxae narrowly separate. Antennae 9-jointed (6+30) — 23. *Omicrogiton.*

11′. Mesostital elevation as a wide pentagonal plane, in touch with metasternal process between middle coxae. Tarsi shorter. Side margins of elytra not explanate. Middle coxae more widely separate. Antennae 8-jointed (5+3) — 24. *Psalitrus.*

14. **Cycreon** A. d'Orchymont.*

Cycreon A. d'Orchymont (1919), Ann. Soc. Ent. Fr., LXXXVIII, p. 119.

15. **Cercyon** [Leach] Mulsant.

Cercyon [Leach (1817), Zool. Misc., III, p. 95—Brullé (1835), Hist. Nat. Ins., V (2), p. 290, 293—Erichson (ex p.) (1837), Käf. Mark Brandbg., I, p. 216—Westwood (1840), Intr. Mod. Class. Ins., II, Synopsis, p. 10]—Mulsant (1844), Hist. Nat. Col. Fr. Palp., p. 147, 156, t. f. 19—Redtenbacher (1845), Gatt. Deutsch Käferfn., p. 70; (1849), Fna. Austr., p. 134; (1858), ed. 2, p. 114; (1874), ed. 3, II, p. 124— Chenu (1851), Enc. Hist., Nat. Col., I, p. 254—Bach (1851), Käferfn. Nord Mitteldeutsch., I, p. 122, 134—Murray (1853), Ann. Mag. Nat. Hist. 2, XII, p. 78, t. 4 f. — Fairmaire & Laboulbène (—*Lafertei* Mulsant) (1854), Fne. Ent. Fr., I, p. 251—Lacordaire (1854), Gen. Col., I, p. 471, 473—Wollaston (1854), Ins. Mader., P. 102—Jacquelin Duval (1857), Gen. Col. Eur., I, p. 95, 99—Kuhn (1858), Käf. Südbayer. Flachl., p. 37–Bose in Gutfl. (1859), Käf. Deutschl., p. 71—Fuss (1859), Archiv. Ver. Siebenb. Landesk. n. F. IV (3), p. 122—Thomson (1859), Skand-Col., I, p. 19; (1860), II, p. 103; (1867), IX, p. 123—Seidlitz (1872), Fna. Balt., p. 12—Solsky (1876), Fedtsch. Reise Turkestan, II, p. 21—Dalla Torre (1877), Jahresber. Ver. Naturk. Linz (8), p.

* Introduced in this catalogue on account of an undescribed new species, in my cabinet, from Singapore. This specimen, a ♂ comes very near the Sumatran *Cycreon sculpturatus* m.

67—Bedel (1881), Fne. Col. Bass. [Seine, I, p. 334, 336—Sharp (1882), Biol. Centr. Amer., I, 2, p. 104; (1884), Trans. Ent. Soc. London, p. 466—Rey (1886), Ann. Soc. Linn. Lyon, XXXII, p. 120, 127—Fowler (1887), Col. Brit. Isl. I, P. 252, 253—Kuwert (1890), Verh. Naturf. Ver. Brünn, XXVIII, p. 21, 159—Horn G. H. (1890), Trans. Am. Ent. Soc., XVII, p. 287, 288—Wickham (1895), Canad. Ent., XXVII, p. 182, 214—f. 14, 15, p. 215—Acloque (—*Lafertei* Mulsant) (1896), Fne. Fr., p. 98—Everts (1899), Col. Neerl., I, p. 639, 651–Kliment (1899), Češtibrouci, p. 116—Lameere (1900), Man. Fne. Belge, II, p. 335—Ganglbauer (1904), Käf. Mitteleur., IV, I, p. 266, 272—Reitter (1909), in Brauer, Süsswasserfna Deutschl., 3-4, p. 77, 78; (1909), Fna. Germ. II, p. 367, 368—Jakobson (1911), Käf. Russl., p. 858—Kuhnt (1913), Ill. Best. Tab. Käf. Deutschl., p. 419, f. 30, 31—A. d'Orchymont (1914), Ent. Mitteil., III, p. 325—Schaufuss C. (1916), in Calwer, Käferb., ed. 6, (1913), p. 346, 349—Des Gozis (1917), Misc. Ent. XXIV, p. 65—Leng & Mutchler (1918), Bull. Amer. Mus. Nat. Hist., XXXVIII, p. 113–Knisch (1921), Archiv. Naturg. (1919), A. 8, p. 71—Tullgren & Wahlgren (1920), Svenska Ins. (2), p. 235— Everts (1922), Col. Neerl., III, p. 292.

Cercydion Latreille (1829), Règne Anim., nouv. ed., IV, p. 526.

Cercyon + *Paracercyon* Seidlitz (1888), Fna. Balt., ed. 2, p. 23 (Gatt.); (1888) Fna. Transsylv., p. 23 (Gatt.)—Stierlin (1900), Col. Helv., p. 214.

ONTOGENESIS Thomson (1860), Skand-Col., p. 103—Schiödte. (1862), Nat. Tidskr. (3), I, p. 219, 223; (1872), VIII, p. 216—Ganglbauer (1904), Käf. Mitteleur., IV, I, p. 273-274—A. d'Orchymont (1913), Ann. Biol. lac., VI, p. 212—Schaufuss C. (1916), in Calwer, Käferb., ed. 6 (1913), p. 349—Richmond (1920), Bull. Amer. Mus Nat. Hist., XLII, Art. 1, p. 83 (*Cercyon ?*), 87, 88.

Subg. **Cercyon** s. str. Ganglbauer.

Cercyon s. str. Ganglbauer (1904), Käf. Mitteluer., IV, 1, p. 274—Reitter (-*flavipes* Thunb.) (1909), Fna Germ., II, p. 369—Kuhnt (-*flavipes* Thunbg.) (1913), Ill. Best.

Tab. Käf. Deutschl., p. 437, f. 98 e — Schaufuss C. (1916), in Calwer, Käferb., ed. 6 (1913), p. 350—Des Gozis (—*analis* Payk.) (1917), Misc. Ent., XXIV, p. 70, 72—Everts (—*analis* Payk.) (1922), Col. Neerl, III, p. 293.

Cercyon (—*haemorrhous* Gyll.,—*analis* Payk) + *Cerycon* Rey (1886), Ann. Soc. Linn. Lyon, XXXII, p. 130, 156. *Cercyon* s. str. (—*ustulatus* Preyssl.) + *Cheilocercyon* + *Cercyon* Seidlitz (1888), Fna. Balt., ed. 2, p. 112, Gatt.; (1888), Fna. Transsylv., p. 119, Gatt.

Cercyon (—*litoralis* Gyll.—*depressus* Steph.,—*arenarius* Rey,—*hoemorrhous* Gyll.) + *Epicercyon* Kuwert, Verh. Naturf. Ver. Brünn, XXVIII, 1890, p. 159, 168.

Cercyon s. str. (—*ustulatus* Preyssl) + *Chilocercyon* + *Epicercyon* Everts (1899), Col. Neerl., I, p. 652, 653.

dilutus Règimbart (1903), Ann. Soc. Ent. Fr., LXXII, p. 338—A. d'Orchymont (1926), Ann. & Bull. Soc. Ent. Belg., LXVI, p. 210.
Mahlè.
Nilgiri Hills.

? **hydrophiloides** Motschulsky (1863), Bull. Soc. Nat. Mosc., XXXVI, I, p. 445 (gen. dub.).
Colombo.

indicus A. d'Orchymont (1926), Ann. & Bull. Soc. Ent. Belg., LXVI, p. 205, 211.
Nilgiri Hills.

lineolatus Motschulsky (1863), Bull. Soc. Nat. Mosc., XXXVI, K, p. 444 (*Trichopoda*)[1]—Sharp (1890), Trans. Ent. Soc. Lond., p. 357[2]—A d'Orchymont (1926), Ann. & Bull. Soc. Ent. Belg., LXVI, p. 209.
Ceylon.[1, 2], Bihar.
———
Tonkin.
Philippines.

lividulus A d'Orchymont (1926), Ann. & Bull. Soc. Ent. Belg., LXVI, p. 204, 210.
Nilgiri Hills.

lunulatus Gemminger & Harold (1868), Cat. Col. II, p. 497 (in Catal.) Knisch (1921), Archiv. Naturg., (1919), A. 8, p. 79[1]—A. d'Orchymont (1925), Ann. & Bull. Soc. Ent. Belg., LXV, p. 277[2]; (1926), l. c., LXVI, p. 212.[3]
Ceylon.[1, 3, 4,]
———
Tonkin.[2]
Singapore.
Sumatra.[5] : Painan.

lunigerum Motschulsky (1863), Bull. Soc. Nat. Mosc., XXXVI, I, p. 444 (nec. Mannerheim (1853).[4]

luniger Règimbart (1902), Ann. Soc. Ent., Fr., LXXII, p. 474 (nec. Horn, 1890).[5]

maindroni Règimbart (1903), Ann. Soc. Ent. Fr., LXXII, p. 357[1]—A. d'Orchymont (1926), Ann. & Bull. Soc. Ent. Belg., LXVI, p. 211.

Nilgiris.[1], Coonoor.[3] Madura : Shembaganur

nigriceps Marsham (1802), Ent. Brit., I, p. 72 (*Dermestes*)— Stephens (1829), Ill. Brit. Ent. Mandib., II, p. 151; (1839), Man. Brit. Col., p. 97—Bach (1851), Käferfna Nord und Mittel- deutchl., I, p. 469—Bedel (1881), Fne Col. Bass. Seine, I, p. 339, 343—Fowler (1887), Col. Brit. Isl., I, p. 258—Kuwert (1890), Verh. Naturf. Ver. Brünn, XXVIII, p. 167—G. H. Horn (1890), Trans. Amer. Ent. Soc., XVII, p. 291, 302—Wickhan (1895), Canad. Ent. XXVII, p. 216—Everts (1899). Col. Neerl., I, p. 654—Lameere (1900), Man. Fne. Belg., II, p. 337, 338—Ganglbauer (1904), Käf. Mitteleur., IV, 1, p. 281—Reitter (1909), Fna. Germ., II, p. 370—Blatchley (1910), Col. Indiana in Bull. Ind. Depart. Goel., I, p. 266, 268—Kuhnt (1913), Ill. Best. Tab. Käf. Deutschl., p. 357, f. 99—Schaufuss C. (1916), in Calwer, Käferb., ed 6, (1913), p. 351—Leng & Mutchler (1918), Bull. Amer. Mus. Nat. Hist., XXXVIII, p. 113—Des Gozis (1918), Misc., Ent., XXIV, p. 88—A. d'Orchymont (1926), Ann. & Bull. Soc. Ent. Belg., LXVI, p. 212.

Palaerctic Region
Corsica.
Africa, Madeira, Canaries, Salvage.
Amer. N. & S.
Antilles.
Asia S. & E.
Formosa.
Philippines.

India.
Ceylone.

Indo-China; Tonkin.
Australia
nearly Cosmopolitan.

atricapillus Marsham (1802), Ent. Brit. I, p. 72 (*Dermestes*).

laevis Marsham (1802), l. c., p. 73 (*Dermestes*).

concinnus Marsham (1802), l. c., p. 74 (*Dermestes*).

centrimaculatum Sturm (1807), Deutschl. Fna, H, p. 23, t. 22 f. e. E. (*Sphaeridium*).

atriceps Stephens (1829), Ill. Brit. Ent. Mandib., II, p. 151; (1839), Man. Brit. Col., p. 97.

inustus Stephens (1829), l. c., 152; (1839), l. c., p. 97.

ustulatus, Stephens (1896), l. c., p. 152; (1839), l. c., p. 97.

bimaculatus, Stephens (1829), l. c., p. 152; (1839), l. c., p. 97.

nubilipennis Stephens (1832), l. c., V, p. 401; (1839), Man. Brit. Col., p. 98.

pulchellus Heer (1841), Fna. Helv., I, p. 492.

munulus Melsheimer (1844), Proc. Acad. Philad., II, p. 102.

centromaculatus Rey (1886), Ann. Soc. Linn. Lyon, XXXII, p. 131, 150.

simplex Delahon (1913), Deuts. Ent. Zeitschr., p. 530.

punctigerus Knisch (1921), Archiv. Naturg. (1919), A. 8, p. 79[1]—A. d'Orchymont Ann. & Bull. Soc. Ent. Belg., LXVI, p. 209[2] (prope ? *rufotestaceus* Motschulsky ?).

Ceylon.[1], India.[2] : Trichinopoli.
———————
Tonkin., Singapore.
Engano I.
Sumatra : Medan.

? **rufotestaceus** Motschulsky (1863), Bull. Soc. Nat. Mosc., XXXVI, I, p. 445 (gen. dub.).

Ceylon.

tropicus A. d'Orchymont (1925), Ann. & Bull. Soc. Ent. Belg., LXV, p. 300[1]; (1926), Treubia, VII, p. 129[2]; (1926) Ann. & Bull. Soc. Ent. Belg., LXVI, p. 204.[3]

Buru I.[1, 2]
? Tonkin.[2]

uniformis Sharp (1890), Trans. Ent. Soc. London, p. 357[1]—A. d'Orchymont (1914), Ent. Mitteil., III, p. 323[2], (1926), Ann. & Bull. Soc. Ent. Belg., LXVI, p. 207, 212.[3]

Ceylon.[1]
Nilgiri Hills.[3]
———————
? Tonkin., ?sumatra.
Sychelles I.

vicinalis Walker (1859), Ann. Mag., Nat Hist., (3), III, p. 358—Sharp (1890), Trans. Ent. Soc. London, p. 357[2]—Règimbart (1903), Ann. Soc. Ent. Fr., LXXII, p. 63[3]—A. d'Orchymont (1926), Ann. & Bull. Soc. Ent. Belg., LXVI, p. 209.[4]

nigriceps Motschulsky (1863), Bull. Soc. Nat. Moscou., XXXVI, I, p. 445.[5].

atriceps Gemminger & Harold (1868), Cat. Col., II, p. 496 (Catal.).

India.[1, 3, 4]
Ceylon.[3, 5]
———————
Indo-China.[3]
Tonkin.
Sunda Is.
Sumatra.[3]
Formosa.
Philippines.

wividus A. d'Orchymont (1925), Ann. & Bull. Soc. Ent. Belg., LXV, p. 279.

Tonkin

16. **Pelosoma** Mulsant.*

Pelosoma Mulsant (1844), Hist. Nat. Col. Fr. Palp, p. 147, 184, t. f. 18—Chenu (1851), Encycl. Hist.

* An American genus but known also from New Britain & Buru Island introduced here on account of several specimens, still unstudied, belonging to this group and received from Tonkin (R. P. de Cooman leg.).

Nat. I, p. 254— Murray (1853), Ann. & Mag. Nat. Hist., (2) XII, p. 75, t. 4, f.—Redtenbacher (1858), Fna. Austr., ed. 2, p. 116; (1874) ed. 3, II, p. 127—Sharp (1882), Biol. Centr. Amer., I, 2, p. 109—Rey (1886), Ann. Soc. Linn. Lyon, XXXII, p. 165—Kuwert (1890), Verh. Naturf. Ver. Brünn, XXVIII p. 21—G. H. Horn (1890), Trans. Amer. Ent. Soc., XVII, p. 287, 306—Ganglbauer (1904), Käf. Mitteleur., IV, I, p. 266, 283—Jakobson (1911), Käf. Russl., P. 858— Schaufuss C. (1916), in Calwer, Käferb., ed. 6 (1913), p. 346—Des Gozis (1917), Misc. Ent., XXIV, p. 74—Knisch (1920), Archiv. Naturg. (1919), A. 8, p. 71.

17. **Peltocercyon** A. d'Orchymont.

Peltocercyon A. d'Orchymont (1925), Ann. & Bull. Soc. Ent. Belg., LV, p. 284.

coomani A. d'Orchymont (1925), l. c., 285. Tonkin.

18. **Pseudocercyon** A. d'Orchymont.

Pseudocercyon A. d'Orchymont (1926), Ann. & Bull. Soc. Ent. Belg., LVI, p. 212.

andrewesi A. d'Orchymont (1926), l. c., p. 212. Tonkin. Nilgiri Hills. Carin-Cheba.

19. **Gillisius** A. d'Orchymont.

Gillisius A. d'Orchymont (1925), Ann. & Bull. Soc. Ent. Belg., LV, p. 287.

andrewesi A. d'Orchymont (1926), l. c., LVI, p. 214, 216. Nilgiri Hills.

madurensis A. d'Orchymont (1926), l. c., LV, p. 288; (1926), l. c., LVI, p. 216. India, Madura.

20. **Oosternum** Sharp.

Oosternum Sharp (1882), Biol. Centr. Amer. I (2), p. 112—Jakobson (1911), Käf Russl., p. 585—A. d'Orchymont (1914), Ent. Mitteil., III, p. 324, 325; (1924), Ann. Mus. Civ. St. N. Genova, LI. p. 262, nota—Knisch (1921), Archiv. Naturg. (1919), A. 8, p. 71.

? *Crypteuna* Motschulsky (1863), Bull. Soc. Nat. Moscou, XXXVI, I, p. 448.

? *Notocercyon* Blackburn (1898), Trans. R. Soc. S. Austr., XXII, p. 230.

costatum Sharp (1882), Biol. Centr-Amer., I (2), p. 113, f. 16[1]—Jakobson (1911), Käfer Russl., p. 858—H. Scott (1913), Trans. Linn. Soc. Lond., XVI (2), p. 221-222—A. d'Orchymont (1914), Ent. Mitteil., III, p. 328—Leng & Mutchler (1918), Bull. Amer. Mus. Nat. Hist., XXXVIII, p. 113.[2] — Mexico.[1], C. Amer.[1] Florida.[2] Porto Rico.[2] Porto Rico.[2] Cuba.[2] Var ? Kumaon, Sarju Valley.

? *aequinoxialis* Motschulsky (1855), Etud., Ent. p. 20 (*Cryptopleurus aeqinoxialis*); (1863), Bull. Soc. Nat. Moscou, XXXVI, I, p. 448 (*Crypteuna*).

horni A. d'Orchymont (1914), Ent. Mitteil., III, p. 326, 328, f. 2, p. 327[1]—Knisch (1921), Archiv. Naturg. (1919), A. 8, p. 81.[2] — Formosa.[1] China: Hongkong. Ceylon.[2] Barway Nilgiris.

saundersi A. d'Orchymont (1925), Ann. & Bull. Soc. Ent. Belg., LV, p. 289.[1] — Singapore.[1] Tonkin

21. **Armostus** Sharp.

Armostus Sharp (1890), Trans. Ent. Soc., Lond., p. 358—A. d'Orchymont (1925), Ann. & Bull. Soc. Ent. Belg., LXV, p. 290.*

---

* In this note I have stated that *Armostus* was an aberrant *Cercyonini* related to *Psalitrus, Paromicrus* and *Omicrus.* It is an error. I have attached too great a weight to the statement of Sharp that it makes a considerable approach to *Peratogonus* and had at the time only before me a somewhat abromal specimen from Singapore. Since I have seen a cotype of the genotype *(optatus)* and more material; I can definitely point out that *Armostus* is a true *Ceryonini* with mesostital elevation definitely separated from the metasternal elevated area, just as in *Cercyon tristis* III. *(minutus* Gyllh.) from Europe. Also the differentiated and carinate middle part of prostitum, which is there by subject ot some variation, seems only a slight modification of what is found in several *Cercyon*-species and this may be said in a still more evident manner from *schenklingi* and *fleutiauxi* m. In these two forms this middle part is indeed practically not or only obscurely elevated above and separated from the large lateral antenal cavities. Were it not for the peculiar morphology of crests of sides of metasternum I should now be inclined to think that *Armostus* ought not to be separated from *Cercyon,* in which genus I had formerly placed *schenklingi* and *fleutiauxi.* However, and again, this special morphology can be derived from such form as *Cercyon lumulatus* Gemm. & Har., where the different metasternal crests are still complete the coxal one in touch with, the femoral issued from femoral abutment, the former directed to and reaching the outer margin of coxal cavities, the latter bending before this outer margin and hindwards along the enisternum. These different forms are very troublesome for the systematist.

fleutiauxi A. d'Orchymont (1925), Ann. & Bull Soc. Ent. Belg., LV, p. 276 *(Cercyon)*[1]; (1926), l. c., LVI, p. 208 (Nota). — Cochin-China.[1] Java.[1]

optatus Sharp (1890), Trans. Ent. Soc. Lond., p. 359[1] — Cylon.[1], India.[2, 3]

---

crenulatus Règimbart (1903), Ann. Soc. Ent. Fr., LXXII, p. 337 (*Cercyon*)[2]—A. d'Orchymont (1925), Ann. & Bull. Soc. Ent. Belg., LXV, p. 276 (*Cercyon*)[3]; (1926), l. c., LVI, p. 208 (Nota). — Indo-China.[3] Sumatra.[2, 3] Java.[2]

schenklingi A. d'Orchymont (1914), Entom. Mitteil., III, p. 323 (*Cercyon*)[1]; (1926), Ann. & Bull. Soc. Ent. Belg., LVI, p. 208 (Nota). — Formosa.[1]

22. **Morastus** A. d'Orchymont.

Morastus A. d'Orchymont (1926), Ann. & Bull. Soc. Ent. Belg., LVI, p. 216.

gracilipalpis A. d'Orchymont (1926), l. c., p. 217. — Nilgiri Hills..[3]

23. **Omicrogiton** A. d'Orchymont.

Omicrogiton A. d'Orchymont (1919), Ann. Soc. Ent. Fr., LXXXVIII, p. 121, 126—Knisch (1921), Archiv. Naturg. (1919), A. 8, p. 71.

insularis A. d'Orchymont (1919), l. c., p. 122[1]—G. C. Champion (1925), Ent. Mo. Mag.; LXI, p. 171, f. 3.[2] — Ceylon.[2]

Cochin-China. Tonkin., Engano I.[1] Buru I. Philippines.

24. **Psalitrus** A. d'Orchymont.

Psalitrus A. d'Orchymont (1919), Ann. Soc. Ent. Fr., LXXXVIII, p. 123, 126; (1925), Ann. & Bull. Soc. Ent. Belg., LXV, p. 159.

championi A. d'Orchymont (1925), Ann. & Bull. Soc. Ent. Belg., LXV, p. 160[1]—G. C. Champion (1925), Ent. Mo. Mag., LXI, p. 170[2], 260.[3] — Nilgiri Hills.[1, 2,] Haldwani.[3]

vandenbosscheae A. d'Orchymont (1919), Ann. Soc. Ent., Fr., LXXXVIII, p. 124[1]; (1925), Ann. & Bull. Soc.? Ent. Belg., LXV, p. 159.[2] — Sumatra.[1] Penang I.[2] Cape.[2]

25. Paromicrus H. Scott.

Paromicrus H. Scott (1913), Trans. Linn. Soc. Lond., XVI, (2), p. 206—A. d'Orchymont (1919), Ann. Soc. Ent. Fr., LXXXVIII, p. 126, 130; (1925) Ann. & Bull. Soc. Ent. Belg., LXV, p. 302.

annandalei A. d'Orchymont (1919), Ann. Soc. Ent. Fr., LXXXVIII, p. 130; (1922), Rec. Ind. Museum, VIII, p. 625. — Abor country, Rotung.

TRIBE MEGASTERNINI.

Megasternini A. d'Orchymont (1919), Ann. Soc. Ent. Fr., LXXXVIII, p. 131.

Megasternaires Mulsant (1844), Hist. Nat. Col. Fr. Palp., p. 147, 186.

Megasterni G. H. Horn (1890), Trans. Am. Ent. Soc., XVII, p. 281, 307.

ONTOGENESIS. K.C. Joyce Phillips (1923). The Irish Naturalist, XXXII, p. 109-112, (*Megasternum boletophagum* Marsham.)

KEY TO GENERA.

1. Lateral margin of prothorax not inflexed toward the ventral side. Mesostital tablet pentagonal very much wider than long. Metasternum without femoral crests ansd without defined elevated area. Prostital tablet triangular and small with anterior margin curved and longitudinally carinate in the middle, the prostitum very reduced before anterior coxae. Anterior tibiae not sinuate on outer margin. Series of punctures of eleytra very fine, directed obliquely towards the outer margin. No series of larger punctures on hind border of pronotum. Shape apparently depressed. Of very small size. — 26. *Noteropagus.*

1′. Lateral margin of prothorax inflexed toward the ventral side. Mesostital tablet pentagonal, wider than long if not, the size is very small.* Elytra striate-punctate, with interstices sometimes raised. Metasternum with femoral crests. A series of

* In *Pachysternum gestroi* m., from Africa.

larger punctures on hind margin of pronotum. Antennal cavities closed on inner side by the middle part of prostitum abruptly raised to a plane, more or less emarginate behind and fitting closely to the mesostital process. 2.

2. Prostital middle tablet with a longitudinal carina in the middle, if not the outer margin of anterior tibiae is feebly sinuate* 27. *Pachysternum.*

2′. Prostital tablet without carina. Outer margin of anterior tibiae not sinuate. 28. *Cryptoleurum.*

26. Noteropagus A. d'Orchymont.

Noteropagus, A. d'Orchymont (1919), Ann. Soc. Ent. Fr., LXXXVIII, p. 132—Knisch (1921), Archiv. Naturg. (1919), A. 8, 72.

obliquus A. d'Orchymont (1925), Ann. & Bull. Soc. Ent. Belg., LXV, p. 305. — Penang I.

politus A. d'Orchymont (1919), Ann. Soc. Ent. Fr., LXXXVIII, p. 134[1]—Knisch (1921), Archiv. Naturg. (1919), A. 8, p. 86.[2] — Mentawei IJ,[2] Engano I.[1] Java.

var. *punctatus* A. d'Orchymont (1925), *Ann. & Bull. Soc. Ent. Belg., LXV, p. 293. — Haldwani. — Indo-China. Tonkin.

27. Pachysternum Motschulsky.

Pachysternum Motschulsky (1863), Bull. Soc. Nat. Moscou, XXXVI, I, p. 446—Kuwert (1890), Verh. Naturf. Ver. Brünn, XXVIII, p. 22—Reitter (1909), Fna. Germ., II, p. 367, Jakobson (1911) Käf. Russl. p. 858—Knisch (1921), Archiv. Naturg. (1919), A. 8, p. 72, 86—A. d'Orchymont (1924), Ann. Mus. Civ. St. Nat. Genova, LI, p. 263; (1926), Philipp. Journ. Sci., 30, 3, p. 373. Ann. & Bull. Soc. Ent. Belg., LXVI, p. 223; (1926).

apicatum Motschulsky (1863) Bull. Soc. Nat. Moscou, XXXVI, I, p. 448[1]—A. d'Orchymont (1926), Ann. & Bull. Soc. Ent. Belg., LXVI, p 224.[2] — India.[1, 2,] Tonkin.[2] Malacca.[2], Sumatra. Mentawei.[2], java.[2] Lombok.[2]

* In the African *capense* Mulsant, which may perhaps not belong to *Pachysternum.*

cardoni A. d'Orchymont (1926), Ann. & Bull. Soc. Ent. Belg., LXVI, p. 220, 223. — Barway. Sikkim.

coomani A. d'Orchymont (1926), Ann. & Bull. Soc. Ent. Belg., LXVI, p. 221, 225. — Indo-China. Tonkin.

nigrovittatum Motschulsky (1861), Bull. Soc. Nat. Mosc., XXXIV, t. IX, f. 22; (1863), id., XXXVI, I. p. 447[1]—Sharp (1890), Trans. Ent. Soc. Lond., p. 358[2]—Knisch (1921), Archiv. Naturg. (1919), A. 8, p. 87, 88[3]—A. d'Orchymont (1926), Ann. & Bull. Soc. Ent. Belg., LXVI, p. 222, 227.[4] — Ceylon.[1, 2, 3] India.[4] Tonkin.[4] Hongkong.[4] ? India or.[4]

stevensi A. d'Orchymont (1926), Ann. & Bull. Soc. Ent. Belg., LXVI, p. 220, 225. — Madura., Sikkim.
——————
Tonkin.

28. Cryptopleurum Mulsant.

Cryptopleurum Mulsant (1894), Hist. Nat. Col. Fr., Palp., p. 186, 188, t. f. 21—Redtenbacher (1849), Fna. Austr., p. 135; (1858), ed. 2, p. 117; (1874), ed. 3, II, p. 128—Chenu (1851), Encycl. Hist. Nat. Col., I, p. 255—Murray (1853), Ann. Mag. Nat. Hist. (2) XII, p. 88, t. 4. f.—Lacordaire (1854), Gen. Col., I, p. 471, 475—Fairmaire & Laboulbène (1854), Fne. Ent. Fr., I, p. 256—Leconte (1855), Proc. Ac. Phil., VII, p. 357—Jacquelin Duval (1857), Gen. Col. Eur. I, p. 96, 99—Kuhn (1858). Käf. Südbayer. Flachl., p. 37—Bose (1859), in Gutfleisch, Käf. Deutschl., p. 73—Fuss (1859), Archi. Ver. Siebenb. Landesk. n. F. IV, (3) p. 124—Thomson G.C. (1859), Skand. Col. I, p. 19; (1860), II, p. 111; (1867), IX, p. 123—Leconte (1861), Class Col. North Amer., P. 47—Seidlitz (1872), Fna. Balt., p. 12; (1888), ed. 2, p. 23 (Gatt); (1888), Fna. Transsylv., p. 23 (Gatt)—Dalla Torre (1877), Jahresber Ver. Naturk. Linz (8), p. 67—Provancher (1877), Fne. Ent. Canad., I, p. 221—Bedel (1881), Fne. Col. Bass. Seine, I, p. 335, 340—Sharp (1882), Biol. Centr. Amer. I (2), p. 115—Leconte & Horn (1883), Class. Col. North Amer., p. 73—Rey (1886), Ann. Soc. Linn. Lyon, XXXII, p. 120, 169—Fowler (1887), Col. Brit. Isl., p. 252, 261—Kuwert (1890), Verh. Naturf. Ver. Brünn, XXVIII, p. 22, 173—G. H. Horn (1890), Trans. Am. Ent. Soc., XVII, p. 307, 310—Wickham (1895), Canad. Ent., XXVII, p. 182—

Acloque (1896), Fne. Fr., p. 98—Kliment (1899), Cešti, brouci, p. 119—Everts (1899), Col. Neerl., I, p. 639, 654—Lameere (1900), Man. Fne. Belg., II, p. 335—Stierlin (1900), Col. Helv., p. 214—Ganglbauer (1904), Käf. Mitteleur., IV, 1, p. 267, 285—Reitter (1909), Fna. Germ., II, p. 367, 371—Blatchley (1910), Col. Indian in Bull. Ind. Depart. Geol; i. p. 265, 270—Jacobson (1911), Käf. Russl., p. 858—Kuhnt (1913), Ill. Best. Tab. Käf. Deutschl., p. 419, 439, f. 27 B., 107 A—A. d'Orchymont, (1914), Ent. Mitteil., III, p. 326—C. Schaufuss (1916), in Calwer, Käferb., ed. 6 (1913), p. 347, 352—Des Gozis (1917), Misc. Ent., XXIV, p. 64—Tullgren & Wahlgren (1920), Svenska Ins. (2), p. 235—Knisch (1921), Arhiv. Naturg. (1919), A. 8, p. 72, 87.

coomani A. d'Orchymont (1926), Ann. & Bull. Soc. Ent. Belg., LXVI, p. 229, 231.

Tonkin.

ferrugineum Motschulsky (1863), Bull. Soc. Nat. Mosc., XXXVI, 1, p. 448[1]—Règimbart (1902), Ann. Soc. Ent Fr., LXXI, p. 475; (1903), l. c., LXXII, p. 63[2]—A. d'Orchymont (1926), Ann. & Bull. Soc. Ent. Belg., LXVI, p. 227, 230.[3]

Ceylon.[1, 2], India.[3] Cochin-China.[2] (Saigon). Tonkin.[3], Malacca.[3] Sumatra.[3]

pygmaeum A. d'Orchymont (1913), Suppl. Ent., II, p. 15, 17[1]; (1926), Ann. & Bull. Soc. Ent. Belg., LXVI, p. 232.[2]

Formosa.[1, 2] Tonkin.[2]

sulcatum Motschulsky (1863), Bull. Soc. Nat. Moscou, XXXVI, I, p. 448[1]—A. d'Orchymont (1926), Ann. & Bull. Soc. Ent. Bleg., LXVI, p. 228.[2]

Ceylon.[1], India.[1, 2]

Tonkin.[2], Malacca.[2]

Subfam. HYDROPHILINAE.

Hydrophilinae [—*Limnebius;* Everts (1899), Col. Neerl. I, p. 637—Ganglbauer (1904), Käf. Mitteleur., IV, I, p. 153, 219—Reitter (1909), Süsswasserfna Deutschl., 3-4, p. 52, 64; (1919), Fna. Germ., II, p. 346, 356—Kuhnt (1913), Ill. Best. Tab. Käf. Deutschl., p. 417, 429—C. Schaufuss (1916), in Calwer, Käferb., ed. 6 (1913), p. 321, 333—Tullgren & Wahlgren (1920), Svenska Ins. (2), p. 233, 234]—A. d'Orchymont (1913), Ann. Biol. lac., VI, p. 181; (1916), Ann. Soc. Ent. Fr., LXXXV, p. 105; (1916), Bull. Soc. Ent. Fr., p. 235, f. 2, p. 239—Everts (1922), Col. Neerl., III, p. 285.

Hydrophilini Erichson (—*Limnebius*), (1837), Käf. Mark Brandebg., I, p. 201.

Hydrophilites Laporte de Castelnau (—*Limnebius, Spercheus* (1840), Hist. Nat. Anim. Art. II, p. 48.

Hydrophiliens Brullé (—*Limnebius*) (1835), Hist. Nat. Col., V, 2, p. 247, 269.

Hydrophilida Heer (—*Limnebius*) (1841), Col. Helv., I, p. 480.

Hydrophilides + *Hydrobiides* Lacordaire (—*Limnebius*) (1854), Gen. Col., I, p. 447, 454.

Hydrophilidae G.C. Thomson (1859), Skand. Col. I, P. 16; (1860), II, p. 85.

Hydrophilini + *Hydrobiini* Leconte (1861), Class. Col. North Amer., p. 44, 46.

Hydrophilinae + *Hydrobiinae* Fowler (1912), Fna. Brit. Ind. (Gen. Intr.), p. 129—Richmond (1920), Bull. Amer. Mus. Nat. Hist., XLII, Art. 1, p. 31, 42.

ONTOGENESIS. Lancret & Miger (1805?), Bull. Soc. Philomat., III, p. 229, No. 90—Miger (1809), Ann. Mus. Nat. Hist., XIV, p. 441—Lacordaire (1854), Gen. Col. I, p. 449—Camerano (1883), Atti Accad. Torino, XIX, p. 222—Ganglbauer (1904), Käf. Mitteleur., IV, I, p. 219, 220—A. d'Orchymont (1913), Ann. Biol. lac., VI, p. 181—Richmond *(Hydrophilinae + Hydrobiinae)*, l. c., p. 31, 42.

KEY TO TRIBES.

1. Scutellum no longer or not much longer than its width at basis. Antennae at most 9-jointed (6+3). Eyes not prominent and not particularly convex. Pronotum not narrower behind than hind body and not conspicuously separated therefrom being not wholly received in an emargination or furrow formed by base of elytra, sometimes with imperfect faculty of rolling up but then the middle and hind feet have no swimming hairs on their innerside and the eyes are not divided by a canthus. Prostitum not reduced before anterior coxae, the latter moderately conical and prominent or transverse. Metathoracic epsiterna

as a long rectangle. Larvae (*Hydrobiini, Hydrophilini*) metapneustic. 2.

1. Scutellum a long triangle. Antennae at most 8-jointed (5+3). Pronotum separated from hind body or lunulate by more or less reduction from posterior angles and lateral sides, being able then to be received in a furrow formed by the base of elytra and at rest nearly vertical with head applied more or less on ventral side. Maxillary palpi with last joint longer than penultimate. 4.

2. Meso- and metasternal carina not reunited intimately. 3.

2′. Meso- and metasternal carina reunited and forming only one ridge; anterior coxae more or less conical and prominent. *Hydrophilini.*

3. Five not covered ventral segments; sometimes a more or less prominent 6th segment under the 5th. Anterior coxae more or less conical and prominent. Body without power of rolling up. *Hydrobiini.*

3. The first two ventral segments in♂♀ with a common and grerat twin shaped excavation, containing a hyaline transverse and bilobed mass supported by a fringe of stiff and long yellowish hairs or setae inserted on anterior border of first ventral segment. Labrum prominent. Maxillary palpi with last joint longer than penultimate. Antennae 9-jointed anterior coxae transverse and oblique, moderately prominent, the prostital process very small and triangular. Body with imperfect rolling up power, the pronotum received then in a furrow of base of elytra and nearly vertical, with head applied on ventral side. *Chaetarthriini.*

4. Eyes divided by a conspicuous and complete canthus which reaches the vertex behind. Posterior feet without swimming hairs. Labrum concealed by anterior laminate margin of head. Body with rolling-up power, with lunulated pronotum. No sutural stria on elytra. Metathoracic episterna triangular, very wide anteriorly and gradually narrowed behind. First ventral segment nearly

wholly invaded by posterior coxal cavities, the four following segments lifted up and in a plane at an angle to that of the first. *Amphiopini.*

4′. Eyes very convex and prominent without complete canthus. Posterior feet with long swimming hairs. Labrum not concealed by anterior margin of head. Prostitum very reduced before anterior coxae which are strongly concial and prominent. Metathoracic episterna with parallel sides, tolerably narrow or very narrow and concealed under margin of elytra. Larvae (*Berosus*) apneustic. *Berosini.*

TRIBE HYDROBIINI.

Hydrobiini [Leconte (—*Berosus, Cyllidium*) (1861), Classif. Col. North Amer., I, p. 44, 46]—Bedel (1881), Fne Col. Bass. Seine Col., I, p. 302, 306—G.H. Horn (1890), Trans. Amer. Ent. Soc., XVII, p. 237—Everts (1899), Col. Neerl., I, p. 638—Reitter (1909), in Brauer Süsswasserfna Deutschl., 3-4, p. 65; (1919), Fna. Germ., II, p. 357—Kuhnt (1913), Ill. Best. Tab. Käf. Deutschl., p. 418—C. Schaufuss (1916), in Calwer, Käferb., ed. 6 (1913), p. 333, 338.

Hydrobiares Mulsant (1844), Hist. Nat. Col. Fr. Palp., p. 88, 116.

Hydrophilaires Rey (—*Hydrophilus, Hydrous*) (1885), Ann. Linn. Lyon, XXXI, p. 220.

Hydrobiina Fowler (1887), Coll. Brit. Isl. I, p. 219, 221.

Hydrobiitae Kuwert (1890), Verh. Naturf. Ver. Brünn, XXVIII, p. 17.

Ontogenesis, Lacordaire (1854), Gen. Col. I, p. 454—Ganglbauer (1904), Käf. Mitteleur., IV, I, p. 235. Wesenberg Lund (1915), Insektlivet, p. 292.

KEY TO GENERA.

1. Maxillary palpi robust and short nearly as long as antennae or shorter, their last joint as long or usually longer than penultimate. Elytra with sutural stria, if not, antennae have less than 9-joints. Last joint of tarsi shorter than preceding joints taken together. (Subtribe *Hydrobiae*) 2

1′. Maxillary palpi more slender, longer than antennae, with last joint shorter than penultimate. If last joint is longer than preceding, the antennae have less than 9-joins, the feet are slender and elongate and elytra have no sutural stria (*Pelthydrus*). If maxillary palpi are scarcely longer than antennae and if at the same time their last joint is longer than penultimate, the antennae are nevertheless 9-jointed but elytra have no sutural stria (*Helochares—Agraphydrus*). Prefrons usually broadly and regularly emraginate on anterior margin. (Subtribe *Helocharae*). Last joint of tarsi shorter than preceding joints taken together, tarsi all pentamerous, first joint of posterior very short, second much longer in Indian genera. 8.

2. Second joint of intermediate and posterior tarsi not particularly longer than first, sometimes even of same size. Posterior femora not densely pubescent. Eyes not prominent. Elytra irregularly punctured and with sutural stria. Antennae 9 to 7 jointed. 29. *Paracymus.*

2′. First joint of intermediate and posterior tarsi very short, second joint much longer. 3.

3. Posterior trochanters not enlarged. Posterior tibaie not curved. Only 5 ventral segments normally visible. 4.

3′. Posterior trochanters elongate at apex and separated here from femora. Posterior tibiae curved in Asiatic species. Antennae 8-jointed. First articulation of tarsi shorter than second. No sutural stria on elytra, the latter usually seriate-punctate. Fifth ventral segment emarginate on hind margin, a sixth more or less retractile segment can be seen in the emargination. Intermediate and posterior femora without dense hydrofugal pubescence on underside. 35. *Laccobius.*

4. Elytra regularly striate-punctate. Antennae 9-jointed outer margin of elytra dentate, at least at base. System of larger setigerous punctures on sides of pronotum absent. 5.

4′. Elytra without striae or even series of punctures. At most a sutural stria. 6.

5. False epipleuron (outer, glabrous, shining and smooth inflexed part of inferior side of elytra) less overhanging, the true epipleuron (pubescent and dull) wider and longer, continued far nearer the sutural angle witout being folded in, the two parts together nearly horizontal; metasternum and abdomen thereby not very much bodied in the elytral cavity. Shape more elongate. Striae 5 and 6, exceptionally 6 and 7 shortened and united behind, their confluence surrounded by the neighbouring striae one on each side, which are curved here and more approximate to one another before reaching the extremity (also in *rudesculptus?* no specimen seen, diagnosis mute as to this point). Mesostital process. Absent (*rudesculptus*?) or reduced to a transversal and low crest before intermediate coxae, besides the very small and nearly microscopical carinula between the latter. Alternate interstices of elytra, at least 1st, 3rd and 5th, with coarse sometimes somewhat obsolete punctures*. — 30. *Ametor.**

5′. False epipleuron wider, more overhanging, more vertical than true epipleuron which is not so wide and not so developed hindwards, more quickly in flexed and folded in towards the under-surface of elytra; metasternum and abdomen more deeply embodied in the elytral cavity. Shape more rounded. Elytral striae all continued to the hind margin of elytra. Mesostital process more robust with a subacuminate free extremity, its hollowed posterior face emiting a very slender carinula which connects between the middle legs with an arrow prolongation of metasternum. Alternate interstices of elytra with more or less coarse punctures. Sometimes absent (in *formosa* Knisch) — 31. *Hydrocassis.*Ø

---

* In Semenov's diagnosis: Ist, 3rd and 6th, evidently by an error.

Ø Knisch in his revision of the Genus. *Hydrocyclus* Sharp included in this group *Spercheus* (*Sperchopis* Leconte) *tesselatus* Ziegler from North America. As I do not feel convinced as to the rightness of this attribution I do not adopt it. Further I do not think like Knisch that *Ametor* is the same as *Hydrocassis* (*Hydrocyclus*). It may be observed first that Knisch was wrong in thinking that *Sperchopsis* was an "in litt" name, his error arising probably from the that he referred to a copy of Leconte's "Classification of the Coleoptera of North America" in which the original pages 43 to 48, published in 1861, had not been removed and replaced by others published at the end of the second instalment of Part I, in March 1862. Mr. H.S. Barber, assistant at the U.S. National

(*contd.*)

6. Antennae 9-jointed (6+3) Elytra with sutural stria shortened anteriorly 7.

6′. Antennae 8-jointed (5+3). No sutural stria. Maxillary palpi very short and robust. Middle and posterior femora without dense hydrofugal pubescence on underside. Suture not convex behind. Eyes seen from the above somewhat oblique elongate, somewhat approximate behind. Shape hemispherical recalling *Anacæna.* 34. *Oocyclus.*

7. Eyes not particularly oblique. Maxillary palpi short, their second joint thicker than 3rd. All femora pubescent on underside except knees. Posterior angles of pronotum not prolonged backwards. 32. *Anacaena.*

Museum kindly drew my attention to this point, So Knisch thought probably that *Sperchopsis* appeared for the time, and litt.; In Horn's paper of 1873, p. 133 *Sperchopsis* occurs nevertheless on No. 47 of of those in 1862 published pages, with a short diagnosis in dichotomic key, and although partly erroneous (Like in *Hydrobius* the last joint of maxillary palpi is not equal but truly longer than the third) has evidently right to priority if the name must stand. On the other hand *Sperchcus teselotus* Ziegler and also *Hydrobius scabrous.* G. H. Horn are not very near *Hydrocassis (Hydrocyclus)* and separated therefrom as also from normal *Hydrocylus,* by the nearly horizontal epipleural region of elytra, fitted in a manner very similar to that of same part of body in *Ametor.* So these two species make a considereble approach to this genus, but the aspect of both is very different from *Ametor,* more elongate and narrower in . . . more irregularly convex in tesselatus, without larger punctures on alternate intersitices in the two: these alternate intervals are costiform on the apex fo 3rd on the beginning of 7th and on the whole length of 11th in *scabrosus,* all more or less so at the apex, especially 3rd, also the 7th at base in *tesselatus.* I do not think that their place is with *Hydrobius,* as accepted hitherto by American workers, on account of shape of epipleural region, already alluded to, and of the outer margin of elytra minutely serrate. Moreover the larva of *scabrosus* was found different by Richimmond (1920) from that of *Hydrobius fuscipes* and of *globosus* in which last two, as usual among species of the same genus, no constant differences could be found. It seems at best thus to restore the name of *Sperchopsis* Leconte for *scabrosus* and *tesselatus* and to differentiate this genus from true *Hydrobius* by the eppleural morphology (not observed hitherto I think by former students), by the absence of larger punctures on alternate interstices of elytra, which are partly raised to costæ, and by the presnce of serrate lateral margins of elytra, at least at base. It must be added however that these two species do not seem very related to one another at first sight, but *Sperchosis* presents certainly strong parallelism with *Ametor* and is far more apart from *Hydrocassis.* Finally I would call attention on this occasion to that fact that *scabrosus* was described by Horn as destitute of any trace fo scutellar stria. However I have studied a series of *scalbrosus* in the Dahlem Museum in which this stria is present and on both sides in two specimens present only on one side in two others and absent in the remaining part of material. In *tesselatus* this stria, is always present. Horn had already observed in 1873 that *scabrosus* was somewhat variable in form and sculpture.

As to *Hydrocyclus* Sharp (Japan), as suggested by Knisch, it is really a synonym of *Hydrocassis* Fairmaire (China). I have seen a cotype of *Hydrocyclus lacustris* Sharp and Mr. R. Peschet from has been so kind as to be willing to study for me the unique type of *Hydrocassis scapulata* Fairmaire together with a specimen of new *Hydrocassis* (*Hydrocyclus*)—species which I had sent him for comparison. *Hydrocassis* is more eastern in its distribution than *Ametor.*

7'. Eyes seen from above very oblique, very elongate, and very near one another behind on postfrons. Posterior angles of pronotum very acutely, spinelike, prolonged backwards. Middle and posterior femora with glabrous underside. 33. *Scoliopsis.*

8. Second joint of maxillary palpi with anterior side concave or straight, convex on posterior side, last joint articulated from the outer to the inner towards mouth. Palpi sometimes very long. 9.

8'. Maxillary palpi rather short, never very long, with second joint convex on anterior side and concave posteriorly, the last joint bending outwards. Eyes rounded. A sutural stria at least. Antennae 9-jointed. 38. *Enochrus.*

9. Last joint of maxillary palpi shorter than preceding, if not antennae are 9-jointed, the palpi are very short and the size is small (subg. *Agraphydrus.* Shape of body never navicular of shield-like. Elytra sometimes striate-punctate (*Hydrobaticus*), or with a single, sutural, stria but in the last case maxillary palpi are very elongate (*Chasmogenus*) 36. *Helochares.*

9'. Last joint of maxillary palpi longer than preceding and thickened at extremity. No sutural stria. Antennal joints less than nine in number. Elytra irregularly punctured. Shape of body often navicular. 37. *Pelthydrus*

Subtribe **HYDROBIÆ.**

Hydrobiæ A. d'Orchymont (1919), Ann. Soc. Ent. Fr., LXXXVIII, p. 136.

Hydrobiates Mulsant (1844), Hist. Nat. Col. Fr. Palp., P. 117.

29. Paracymus G.C. Thomson.

Paracymus G.C. Thomson (1867), Skand, Col., IX, p. 119, 120—Sharp (1870), Ent. Mo. Mag., VI, p. 255— G. H. Horn (ex p.) 1873) Proc. Am. Phil. Soc., XIII, p. 135—B del (1881), Fne. Col. Bass. Seine, I, p. 306, 308— Sharp (1884), Trans. Ent. Soc. Lond., p. 466— Rey (1885), Ann. Soc. Linn. Lyon, XXXI, p. 268— Fowler (1877), Col. Brit. Isl., I, p. 221, 225; (1913),

VI, p. 24—Seidlitz (1888) Fna Balt. ed 2, Gatt. p. 22; (1888), Fna Transsylv., Gatt., p. 22—Blackburn (ex p.) (1888), Proc. Linn. Soc. N. S. Wales, (2), III, p. 819—Kuwert (1890), Verh. Naturf. Ver. Brünn, XXVIII, p. 19, 63—G. H. Horn (1890), Trans. Amer. Ent. Soc., XVII, p. 261—Everts (1899), Col. Neerl., I, p. 639, 646—Kliment (1899), Češti brouci, p. 113—Stierlin (1900), Col. Helv., p. 213— Ganglbauer (1904), Käf. Mitteleur., IV, I, p. 236, 241—Reitter (1909), in Brauer, Süsswasserfna Deutschl., 3-4, p. 65, 70; (1909) Fna Germ., II. p. 357, 361—Jacobson (1911), Käf. Russl., p. 852—Kuhnt (1913), Ill Best. Tab. Käf. Deutschl., p. 419, 431, f. 17 E—Schaufuss C (1916), in Calwer, Käferb., ed. 6 (1913), p. 338, 340— Des Gozis (1916), Misc. Ent., XXIII, p. 15—Tullgren & Wahlgren (1920), Svenska Ins. (2), p. 234—A. d'Orchymont (1926), Phillip. Journ. Sc., 30, 3, p. 375-377.

Hydrobius Sharp (nec Leach) (1882), Biol. Centr. Amer., I (2), p. 64.

Creniphilus G. H. Horn (ex p.) (1890), Trans. Am. Ent. Soc., XVII, p. 262, 267.

ONTOGENESIS. Richmond (1920), Bull. Amer., Mus. Nat. Hist., XLII, Art. I, p. 74, 83, 86, 88.

chalceolus Solsky (1874), Fedtsch. Reis. Turkest. II, p. 149[1] (*Hydrobius*)—de Marseul (1882), L'abeille, XX, p. 130 (*Hydrobius*).	Turkestan : Samarkand
evanescens Sharp (1890), Trans. Ent. Soc. London, p. 349 (*Hydrobius*)[1]—H. Scott (1913), Trans. Linn. Soc. London, XVI, 2, p. 102—A. d'Orchymont (1926), Philipp. Journ. Sc., 30, 3, p. 377.[2]	Ceylon.[1] India : Bengal, Genjui, Bihar. ——— Tonkin., Sumatra. Philippine I.[2]
punctillatus Rey (1885) var. *atomus* A. d'Orchymont (1925), Ann. & Bull. Soc. Ent. Belg., LXV, p. 202[1]; (1926) , philipp. Journ. Sco., 30, 3, p. 378.[2]	Philippines I.[1, 2] Singapore.[2] Madagascar.[2] East Africa.[2] Belgian Congo.[2]

30. **Ametor** Semenov*.

Ametor Semenov (1900), Hor. Soc. Ent. Ross., XXXIV, p. 614—Jakobson (1911), Käf. Russl., p. 852.

rudesculptus Semenov (1900), l. c., p. 617. — Turkestan : Kschtut: Artutsch.

rugosus Knisch (1924), Wien. Ent. Zeit., 41, p. 32 (*Hydrocyclus.*) — W. Almora, Soth Garhwal, Chakrata (Khedar Khud) Punjab (Gahan, Simla)

31. **Hydrocassis** Fairmaire.**

Hydrocassis Fairmaire (1878), Ann. Soc. Ent. Fr., (5), VIII, p. 88.

Hydrocyclus Sharp (1884), Trans. Ent. Soc. Lond., p. 451—Knisch (ex p.—*lesselatus* Zeigler) (1921), Entom. Anzeiger, I, p. 99, 100.

32. **Anacaena** G. C. Thomson.

Anacæna G. C. Thomson (1859), Sk. Col., I, p. 18; (1860), II, p. 99; (1867), IX, p. 119—Sharp (187), Ent. Mo. Mag., VI, p. 255—Kiesenwetter (1875), Deuts. Ent. Zeits. XIX, p. 229—Bedel (1881), Fne. Col. Bass Seine, I, p. 307, 308—Fowler (1887), Col. Brit. Isl., I, p. 221, 226—Seidlitz (1888), Fna Balt. ed. II, Gatt., p. 22; (1888), Fna Transsylv., Gatt. p. 22—Kuwert (1890) Verh. Naturf. Ver. Brünn, XXVIII, p. 19, 67—Everts (1899), Col. Neerl. I, p. 639-646—Lameere (1900), Man. Fne Belg., II, p. 330—Stierlin (1900), Col. Helv., p. 213—Ganglbauer. (1904), Käf Mitteleur., IV, I, p. 236, 239—Edwards (1909), Ent. Mo. Mag., XLV, p. 169-171—Reitter (1909) in Brauer, Süsswasser—Fna Deutschl., 3-4, p. 65, 70, f. 48; (1909), Fna Germ., II, p. 357, 361—Jacobson (1911), Käf. Russl., p. 852—Kuhnt (1913) Ill Best. Tab. Käf. Deutschl., p. 419, 431, f. 18—C. Schaufuss (1916), in Calwer, Käferb.,

---

* *Hydrocyclus rugosus* Knisch by its generic characters as pointed out above bleongs to *Ametor*. I possess a third, still unnamed, species from Sikkim.

** Listed here on account of a still underscribed species from Upper Burma (Man. Ton, Mongmit State, 4,200 feet, Ruby Mines district.) Other species: *Formosa* and *imperialis* Knisch (*Hydrocyclus*) from China, *Lacustris* Shharp (*Hydrocyclus*) from Japan and *Scapulata* Fairmaire (genotype) from China. Also a sixth,, also still underscribed species from the latter country.

ed. 6 (1913), p. 338, 440—Tullgren & Wahlgren (1920), Svenska Ins., (2), p. 234.

Brachypalpus Laporte & Castelnau (ex. p.) (1840), Hist. Nat. Anim. Art. Col., II, p. 56 (nec Macquart 1834).

cryniphilus Motschulsky (1845), Bull. Soc. Nat. Moscou, XVII, p. 32 (in litt.).

Greniphilus Des Gozis (1881), Ann. Soc. Ent. Fr. (6) I, Bull., p. (135)—G.H. Horn (ex. p.) (1890), Trans. Amer. Ent. Soc., XVII, p. 262, 267.

Crenephilus Wickham (ex. p.) (1895), Canad. Ent., XXVII, p. 182.

Laccobiellus Abèille de Perrin, (1901), Echange, XVII, p. 60.

ONTOGENESIS. A. d'Orchymont (1913), Ann. Biol. lac., VI, p. 209—Richmond (1920), Bull. Amer. Mus. nat. Hist., XLII, Art. I, p. 71, 83, 86 (*Ancæna*).

advena Sharp (1890), Trans. Ent. Soc. London, p. 349 (*Hydrobius*)[1]. — Ceylon.[1]

minima Sharp (1890), l. c., p. 348 (*Hydrobius*)[1] —A. d'Orchymont (1925), Ann. & Bull. Soc. Ent. Bleg., LXV, p. 65.[2] — Cylon.[1] India : Madura, Shembaganur.[2]

33. **Scoliopsis** A. d'Orchymont

Scoliopsis A. d'Orchymont (1919), Ann. Sco., Ent. Fr., LXXXVIII, p. 138.

spinosa A. d'Orchymont (1919), l. c., p. 141. — Ceylon.

34. **Oocyclus** Sharp.

Oocyclus Sharp (1882), Biol. Centr. Amer I, 2, p. 61—A. d'Orchymont (1919), Bull. Soc. Ent. Fr., p. 226.

Enochroides Kuwert (1890), Verh. Naturf. Ver. Brünn, XXVIII, p. 18, 62, 327.

latus A. d'Orchymont (1919), Ann. Soc. Ent. Fr., LXXXVIII, p. 142, 145. — Ceylon.

35. **Laccobius** Erichson.

Laccobius Erichson (1837), Käf. Mark Brandebg. I, p. 202—Westwood (1840), Introd. Mod. Class. Ins. II, Synopsis p. 10—Mulsant (1844), Hist. Nat. Col. Fr. Palp., p. 129—Redtenbacher (1845), Gatt. Deutsch.

Käferfna, p. 71; (1849), Fna Austr., p. 130; (1858), ed. 2, p. 105; (1874), ed. 3, II, p. 114—Chenu (1851), Encycl. Hist. nat. p. 130—Bach (1851), Käferfna Nord und Mitteldeutschl., I, p. 122, 130—Fairmaire & Laboulbène (1854), Fne Ent. Fr., I, p. 230—Lacordaire (1854), Gen. Col., I, p. 455, 457—Wollaston (1854), Ins. Mader., p. 95—Leconte (1855), Proc. Acad. Philad. VII, p. 357—Jacquelin Duval (1857), Gen. Col. Eur., I, p. 88, 98—Kuhn (1858), Käf. Subayer. Flachl., p. 39—Bose (1859), in Gutfl., Käf. Deutschl., p. 66— Fuss (1859), Archiv. Ver. Siebenb. Landesk. n. F. IV, (3), p. 111—G. C. Thomson (1859), Skand., Col., I, p. 18; (1860), II, p. 92; (1867), IX, p. 119—Leconte (1861) Class. Col. North Amer., p. 46—Seidlitz (1872), Fna Bat. p. 11; (1888), ed. 2, Gatt., p. 22; (1888) Fna Transsylv. Gatt., 22—Paasch (1868), Berl. Ent. Zeitschr., XII, p. 308—G. H. Horn (1873), Proc. Amer. Phil. Soc. XIII, p. 118—Rottenberg (1874) Berl. Ent. Zeits. XVIII, p. 305—Solsky (1876), Fedtsch. Reise Turkestan, II, p. 20—Dalla Torre (1877), Jahresber. Ver. Naturk. Linz (8) p. 66—Provancher (1877), Fne Ent. Canada, I, p. 216—Bedel (1881), Fne. Col. Bass. Seine, I, p. 306, 312—Leconte & Horn (1883), Class. Col. North Amer., p. 72—Rey (1885), Ann. Soc. Linn. Lyon, XXXI, p. 289— Fowler (1887), Col. Brit. Isl., I, p. 221, 228; (1913), VI, p. 26—Kuwert (1890), Verh. Naturf. Ver. Brünn, XXVIII, p. 19, 69—Wickham (1895), Canad, Ent., XXVII, p. 182—Acloque (1896), Fne Fr., p. 92—Everts (1899), Col. Neerl., I, p. 638, 647—Kliment (1899), Cešti brouci, p. 113—Lameere (1900), Man. Fne. Belg., II, p. 329—Stierlin (1900), Col. Helv., P. 212— Ganglbauer (1904), Käf. Mitteleur., IV, I, p. 235, 250—Newbery (1908), Ent. Mo. Mag., (2) XIX, p. 30— Reitter (1909), in Brauer, Süsswasserfna Deutschl., 3-4, p. 66, 74, fig. 51; (1909), Fna Germ., II, p. 357, 364—Blatchley (1910), Col. Indiana, Indiana Depart. Geol., I, p. 257— Jakobson (1911), Käf. Russl., p. 852—Kuhnt (1913), Ill. Best. Tab. Käf. Deutschl., p. 419, 433, f. 11C, 21—Schaufuss C (1916) in Calwer, Käferb., ed. 6 (1913) p. 338, 342—Des Gozis (1916), Misc. Ent. XXIII, p. 14— Tullgren & Wahlgren (1920), Svenska Ins. (2) p. 234, 235—Everts (1922), Col. Neerl., III, p. 289.

Limnebius Brullè (1835), Hist. Nat. Ins. Vol. V, II, p. 286 (nec Leach 1815).

Brachypalpus Laporte de Castelnau (ex. p.) (1840), Hist. Nat. Anim Art., Col., II, p. 56 (nec Macquart, 1834).

Hydroxenus Wollaston (1867), Col. Hesper., p. 40.

ONTOGENESIS. A. d'Orchymont (1913), Am. Biol. Lac., VI, p. 196, fig. 19 p. 209—Richmond (1920) Bull. Amer. Mus. Nat. Hist. XLII, Art. I, p. 43, 85, 87.

Sub-gen. **Laccobius** s. str. Ganglbauer.

Laccobius Ganglbauer (1904), Käf. Mitteleur. IV, I, p. 251—A. d'Orchymont (1914), Journ. As. Soc. Bengal, X, p. 359.

argillaceus J. Sahlberg (1900), Ofvers. Fnisk. Vet. Soc. Forh. XLII, p. 188.	Turkestan. R.Tschu.
discicollis Règimbart (1903), Ann. Soc. Ent. Fr., LXXII, p. 58, 336.	India : Coromandel, Genji, Mahè.
hauserianus Kniz (1914), Verh. Zool. Bot. Ges. Wien, LXIV, Sitzber, p. 116.	Buchara.
hingstoni A. d'Orchymont (1926), Ann. & Bull. Soc. Ent. Belg. LXVI, p. 104.	Tibet.
imperialis Knisch (1924), Wien. Ent. Zeit., XLI, p. 35[1].	Kumaon. W. Almora, Ranikhet.[1] North Garhwal.
litoralis J. Sahberg (1900), Ofvers. Finsk. Vet. Soc. Forh. XLII, p. 189.	Turkestan : L. Issykkul.
orientalis Knisch (1924), Wein. Ent. Zeit., XLI, p. 33.[1]	Kumaon. W. Almora, Ranikhet.[1]
pallidissimus Reitter (1899), Deutsch. Ent. Zeitschr. p. 197.	Afghanistan Sefid Kuh. penschdeh.
patruelis Knisch (1924), Wien. Ent. Zeit., XLI, p. 36.	Kumaon, Bodiar, Haldwani.
rectus Sharp (1890), Trans. Entom. Soc. London, p. 353.	Ceylon.

regalis Knisch (1924), Wien. Ent. Zeit., XLI, p. 34.[1] — Kumaon, (W. Almora, Sunderdhunga). Kotgarh, Gahan, Simla Hills, Punjab. Chakrata, Khedar Kud. Bangar.

roseiceps Règimbart (1903), Ann. Soc. Ent. Fr., LXXII, p. 59. — Indo-China. Annam (thuan An).

rotundatus Règimbart (1903), l. c., p. 59. — Belgaum. Tharrawaddy.

simulans A. d'Orchymont (1923), Mem. Dep. Agr. India, VIII, p. 7. — Yunnan.

———————

Bihar, Kumaon, Kurseong, Khassi Hills.

sublaevis J. Sahlberg (1900), of v. Fnisk. Vet. Soc. Forh. XLII, p. 189. — turkestan. R. Tschu.

zugmayeri Kniz (1910), Zool. Jahrb. Abt. Syst. Geoga. u. Biol. Tiere, XXIX p. 454. — W. Tibet. L. Panggong.

Sub-tribe **HELOCHARÆ.** — Ceylon.[1,2], India.[2]

Helocharæ A. d'Orchymont (1919), Ann. Soc. Ent. Fr., LXXXVIII, p. 147.

Philhydrates Mulsant (1844), Hist. Nat. Col. Fr. Pal. p. 117, 131.

36. Helochares Mulsant.

Helochares Mulsant (1844), Hist. Nat. Col. Fr. Palp. p. 197—Bach (1851), Käferfn. Nord. Mitteldeutschl. I, p. 392—Jacquelin Duval (1857), Ann. Soc. Ent. Fr. (3), V, Bull. p. (54)—Reiche (1857), id, p. 77—Redtenbacher (1858), Fna Austr., ed. 2, p. 104—Fuss (1859), Archiv Ver. Siebenb. Landesk. n. F. (3) p. 111—Thomson G. C. (1859), Skand Col., I. p. 18; (1860), II, p. 98; (1867), Vol. IX, p. 119—Wollaston (1867), Col. Hesperid, p. 43—Sharp (1870), Ent. Mo. Mag., VI, p. 254—Seidlitz (1872), Fna. Balt., p. 10; (1888), ed. 2, Gatt., p. 22; (1888), Fna Transsylv., Gatt., p. 22—Dalla Torre (1877), Jahresber Ver. Naturk. Linz (8) p. 66—Bedel

(1881), Fne Col. Bass Seine, I, p. 307, 311—Rey (1885), Ann. Soc. Lin. Lyon, XXXI, p. 281—Fowler (1887), Col. Brit. Isl., I, p. 221, 227—Wickham (1895), Canad. Ent., XXVII, p. 182—Everts (1899), Col. Neerl., I, p. 638, 643—Kliment (1899), Cešti brouci, p. 111—Lameere (1900), Man. Fne Bleg., II, p. 330—Stierlin (1900), Col. Helv., p. 213—Ganglbauer (1904), Käf. Mitteleur., IV, I, p. 236, 248—Blatchley (1910), Col. Indiana in Bull. Ind. Depart. Geol. I, p. 257—Kuhnt (1913), Ill. Best. Tab. Käf. Deutschl., p. 419, 433 f. 19—Schaufuss, C. (1916) in Calwer, Käferb., ed. 6 (1913), p. 339—A. d'Orchymont (1919), Ann. Soc. Ent. Fr., LXXXVIII, p. 148—Des Gozis (1916), Misc. Ent., XXIII, p. 14, 25—Tullgren & Wahlgren (1920), Svenska Ins. (2), p. 234, 235.

Philydrus Laporte de Castelnau (ex. p. Hist. Nat. Ins., II, p. 52.

Helophilus Mulsant (1844), Hist. Nat. Col., Fr. Palp., p. 132.

Hylophilus Motschulsky (1845), Bull. Soc. Nat. Mosc., XVIII, I, p. 32 (nom. nudum).

Helophygas Motschulsky (1853), Etudes Entom, I, Suppl., Hydrocanthares de la Russie, p. 11 (nom. nud.).

Pyilydhrus Leconte (ex. p.) (1855) Proc. Acad. Nat. Sc., Philad., p. 369.

Stagnicola Montrouzier (1860), Ann. Soc. Ent. Fr. (3), VIII, p. 246.

Sindolus + *Chasmogenus* + *Helochares* Sharp (1882), Biol. Centr. Amer. I (2), p. 72 & seq.—G. H. Horn (1890), Trans.

Amer. Ent. Soc., XVII, p. 238, 240, 250.

Helochares+Helocharimorphus Kuwert (1890) Verh. Naturf. Ver. Brünn., XXVIII, p. 18, 34, 62.

Helochares + *Crepidelochares* Reitter (1909), in Brauer, Süsswasserfna Deutschl., 3-4, p. 66, 73, 74 f. 50; (1909) Fna Germ, II, p. 357, 363 nota.

ONTOGENESIS. Schiödte (1862), Nat. Tidskr., I (2), p. 223—A. d'Orchymont (1913), Ann. Biol. lac., VI,

p. 210—Wesenberg. Lund (1915), Insektlivet, p. 297—Richmond (1920), Bull. Amer. Mus. Nat. Hist., XLII, Art. I, 1920, p. 61, 83, 86, 88.

Sub-gen. **Hydrobaticus** W. M'Leay.

Hdyrobaticus W. M'Leay (1871), Trans. Ent. Soc. N.S. Wales. II, p. 131—A. d'Orchymont (1919), Ann. Soc. Ent. Fr., LXXXVIII, p. 148, 149.

Graphelochares Kuwert (1890), Verh. Naturf. Ver. Brünn, XXVIII, p. 34.

Grapidelochares Ganglbauer (1904), Käf. Mitteleur., IV, I, p. 248.

anchoralis Sharp (1890), Trans. Ent. Soc. London, p. 352[1]—Règimbart (1903), Ann. Soc. Ent. Fr., LXXII, p. 54.[2]

———
Annam.[2]
Cochin-Chin.[2]
Cambodia.[2]
Laos, Tonkin.
F. M. S., Sumatra.[2]
Philippine I.

crenatus Règimbart (1903), Ann. Soc. Ent. Fr., LXXII, p. 54.[1]

Cochin-China.[1]
Cambodia.[1], Tonkin.

———

Pondicheri,[1] genji,[1]
Tetara.[1], Bengal.[1]

———

F. M. S., Sumatra.[1]
Java.[1], Philippine I.

lentus Sharp (1890), Trans. Ent. Soc. Lond. p. 352[1]—Règimbart (1902), Ann. Soc. Ent. Fr., LXXI, p. 473[2]; (1903), LXXII, p. 55[3]—A. d'Orchymont (1913), Suppl. Ent. II, p. 5[4]; (1922), Rec. Ind. Mus., VIII, p. 626.[5]

Ceylon.[1, 3, 4, 6, 8]
India.[2, 3, 7], Nepal.
N. E. Assam.[5]

———

Indo-China.[2, 7]
Cochin-China.[3]

Tokin., F. M. S.
Sunda. Is.[2, 7]
Sumatra.[3]
Java.[3]
Borneo.[3]
Formosa.[4]

densus Sharp (1890), Trans. Ent. Soc. London, p. 352[6]—Règimbart (1902), Ann. Soc. Ent. Fr., LXXI, p. 473[7]; (1903), LXXII, p. 55.[8]

minor A. d'Orchymont (1925), Ann. & Bull. Soc. Ent. Belg., LXV, p. 293.

Tonkin.

salvazai A. d'Orchymont (1919), Ann. Soc. Ent. Belg., LIX, p. 76.

Cambodia.

Subu-gen. **Chasmogenus** Sharp.

Chasmogenus Sharp (1882), Biol. Centr. Amer., I (2), p. 73—G. H. Horn (1890), Trans. Amer. Ent. Soc., XVII, p. 238—A. d'Orchymont (1919), Ann. Soc. Ent. Fr., LXXXVIII, p. 148, 149.

Crephelochares Kuwert (1890), Verh. Naturf. Ver. Brünn, XXVIII, p. 34.

Crepidelochares Ganglbauer (1904), Käf. Mitteleur., IV, I, p. 248.

livornicus Kuwert (1890), Verh. Naturf. Ver. Brünn, XXVIII, p. 28, 327[1] (*Crephelochares*)—Ganglbauer (1904), Käf. Mitteleur., IV, I, p. 248[2]—Reitter (1909), in Brauer, Süsswasserfna Deutschl., 3-4, p. 74; (1909), Fna. Germ., II, p. 363, nota (*Crepidelochares*)—A. d'Orchymont (1925), Ann. & Bull. Soc. Ent. Belg., LXV, p 70.[3]

abnormalis Sharp (1890), Trans. Ent. Soc. London, p. 351 (*Philydrus*)[4]— Scott (1913), Trans. Linn. Soc. London, XVI, 2, p. 205 (*Philydrus*).[5]

mollis Règimbart (1903), Ann. Soc. Ent. Fr., LXXII, p. 32[6]; LIIV, (1906), p. 263 (*Phiillhydrus*).[7]

ferrugatus Règimbart (1903), l. c., p. 57 (*Phylhydrus*).[8]

abnormicollis Zaitzev (1908) Hor. Soc. Ent. Ross., XXXVIII, p. 385 (*Enochrus—Lumetus*) (Catal.).

Italy.· Livorno.[1, 2]
Herzegovina.[2]
Gabun.[3]
B. E. Africa.[7]
Madagascar.[3, 6]
Seychelles.[3, 5.]
Alsabra.[3, 5]
Ceylon.,[3, 4]
Cochin-China.[8]
Cambodia.[3], Tonkin.
Sumatra.[8]
Philippine I.

rutilus (Knisch, 1921, in litt.) A. d'Orchymont (1925), Ann. & Bull. Soc. Ent. Belg., LXV, p. 71.[1]

? nigritulus Règimbart (1903), Ann. Soc. Ent. Fr., LXXII, p. 57 (*Philhydrus*)[2] (nec Kuwert, 1890).

abnormalis Knisch (1921), Archiv. Naturg. (1919) A. 8, p. 68[3] (nec. Sharp 1890).

? règimbarti Knisch (1924), in Cat. Col., Pars 79, p. 195 (in Catal.).

Gabun.[1, 3]
Belgian Cogo.[1]
Indo-China.[2]
Sumatra.[2, 3]

Sub-gen. **Helochares** s. str. Kuwert.

Helochares Kuwert (1890), Verh. Naturf. Ver. Brünn, XXVIII, p. 34—Ganglbauer (1904), Käf, Mitteleur., IV, I, p. 248—A. d'Orchymont (1919), Bull. Soc. Ent. Fr., p. 228; Annales id., LXXXVIII, p. 149.

Neohydrobius Blackburn (1898), Trans. R. Soc. S. Austr., XXII, 2, p. 221.

pallens W. S. Mac Leay (1825), Annul. Javan., p. 35 (*Enhydrus*)[1]—A. d'Orchymont (1926), Ann. & Bull. Soc. Ent. Belg., LXVI, p. 232; (1926), Philipp Journ. Sci, 30, 3, p. 379.[2]	Java.[1], Sumatra. N. Oceania.[4] Philippine I.[2] TropicalAsia.[4]
? *parvulus* Reiche & Saulcy (1856), Ann. Soc. Ent. Fr., (3), IV, p. 358, 359.[3]	F. M. S. Tonkin.
minutissimus Règimbart (1903), Ann. Soc. Ent. Fr., LXXII, p. 27, 54, 335[4]—A. d'Orchymont (1922), Rec. Ind. Museum, VIII, p. 627[5] (nec. Kuwert, (1890).	---------- Bihar. N. E. Assam.[4] ------------
dispar Sharp (1903) Res. Swed. Zool. Exped. White Nile & Egypt, No. 10, p. 7.[6]	Egypt.[4] Tropical Africa.[4] madagascar.[4] Sudan.[6], Syria.[3]
taprobanicus Sharp (1890). Trans. Ent. Soc. London, p. 351[1]—A. d'Orchymont (1923), Treubia, III, 3-4, p. 419.[2]	Ceylon.[1] Indo-China.[2,3] F.M.S. Sumatra.[2,3] Boreno.[2,3]
atropiceus Règimbart (1903), Ann. Soc. Ent. Fr., LXXII, p. 53.[3]	New Guinea. Philippine I.
vitalisi A. d'Orchymont (1919), Ann. Soc. Ent. Belg., LIX, p. 78.	Indo-China. Cambodia.

Sub-gen. **Agraphydrus** Règimbart.

Agraphydrus Règimbart (1903), Ann. Soc. Ent. Fr., LXXII, p. 33—A. d'Orchymont (1919), l. c., LXXXVIII, p. 155 (*Enochrus* sub-gen. *Agraphydrus*); (1927). Journ. F. M. S. Mus. XIII, p. 250.

coomani A. d'Orchymont (1927). Journ F.M.S. Mus. XIII, p. 248.	Tonkin., F. M. S. Kuala Lampur.
kempi A. d'Orchymont (1922), Rec. Ind. Museum, VIII, p. 626 (*Helochares* s. str.)[1].	Abor, (Yembung)[1] Kaligad, Dehra Dun, U. P.

pauculus Knisch (1924), Wien. Ent. Zeits., XLI, p. 36 (*Helochares—Helocharimorphus*). — Kumaon (W. Almora).

pygmaeus Knisch (1924), l. c., p. 38 (*Helochares—Helocharimorphus*). — Kumaon (W. Almora, Bodiar, Haldwani).

37. **Pelthydrus** A. d'Orchymont.

Pelthydrus A. d'Orchymont (1919), Ann. Soc. Ent. Fr., LXXXVIII, p. 150.

championi A. d'Orchymont (1926), Ann. & Bull. Soc. Ent. Belg., LXVI, p. 241. — Haldwani.

inaspectus A. d'Orchymont (1926), l. c., p. 237, 238. — Yunnan.

minutus A. d'Orchymont (1919) Ann. Soc. Ent. Fr., LXXXVIII, p. 153;[1] (1926), Philipp. Journ. Sc., 30, 3, p. 380.[2] — Sumatra.[1] Philippine I.[2] Indo-China. Tonkin.

sculpturatus A. d'Orchymont (1919), Ann. Soc. Ent. Fr., LXXXVIII, p. 153; (1926), Ann. & Bull. Soc. Ent. Belg., LXVI, p. 238. — Yunnan.

vitalisi A. d'Orchymont (1926), l. c., p. 239. — Indo-China. Tonkin.

38. **Enochrus** Zaitzev.

Enochrus Zaitzev (1919), Rev. russe. d'Entom., VIII, (1908), p. 91—A. d'Orchymont (1913), Suppl. Ent. II, p. 6; (1919), Ann. Soc. Ent. Fr., LXXXVIII, p. 154.

Philhydrus Solier (1834), Ann. Soc. Ent. Fr., III, p. 302, 315 (nec. Duftschmid, 1805).

Philhydrus Brullé (1835), Hist. Nat. Ins., V, II, p. 276.

Enochrus+Philhydrus Sharp (1870), Ent. Mo. Mag., VI, p. 254.

Enochrus+Philhydrus Rey (1885), Rey, Ann. Soc. Linn. Lyon, XXXI, p. 247, 251.

Enochrus+Holcophildrus Jakobson (1911), Käf. Russl. P. 852.

ONTOGENESIS Schiödte (1862), Nat. Tidskr. (3) I, p. 223; (1872), p. 215 (*Philydrus*)—Ganglbauer (1904), Käf. Mitteleur., IV. I, p. 243 (*Philydrus*)—Brocher (1913), L'aquarium de Chambre, p. 222, fig. 77 (*Phiydrus*)—A. d'Orchymont (1913), Ann. Biol.

lac., VI, p. 210—Wesenberg—Lund (1915), Insektlivet, p. 293 f. 242-243 a b. (*Philydrus*)—Schaufuss C. (1916), in Calwer, Käferb., ed. 6 (1913), p. 341 (*Philhydrus*)— Richmond (1920), Bull. Amer. Mus. Nat. Hist., XLII, Art. 1, p. 65, 83, 86, 88 (*Philydrus*).

Sub-gen. **Lumetus** Zaitzev.

Lumetus Zaitzev (1909), Revue russe d'Entom., VIII (1908) p. 91—A. d'Orchymont (1913), Suppl. Ent., II, p. 6; (1914), Journ. & Proc. As. Soc. Bengal, X, p. 358; (1919) Ann. Soc. Ent. Fr., LXXXVIII, p. 155.

Philydrus in sp. Kuwert (1888), Deuts. Ent. Zeitschr., p. 274.

Philhydrus in sp. Everts (1899), Col. Neerl., I, p. 644.

crassus Règimbart (1903), Ann. Soc. Ent. Fr., LXXII, p. 55 (*Philhydrus*). — Indo-China. Annam. Cochin-China Tonkin.

fragilis Sharp (1890), Trans. Entom. Soc. Lond, p. 350 (*Philydrus*)—Scott (1913) Linn. Soc. Lond. XVI, 2, p. 204 (*Philhydrus*). — Ceylon.[1]

fuscatus Motschulsky (1861), Bull. Soc. Nat. Mosc., XXXIV, I, p. 109 (*Philhydrus*). — Ceylon.

icterus Knisch (1924) Wien Ent. Zeit., XLI, p. 38. — Kumaon (W. Almora)

iteratus Sharp (1890), Trans. Ent. Soc. London, p. 349 (*Philhydrus*). — Ceylon.

japonicus Sharp (1873), Trans. Ent. Soc. Lond p. 59 (*Philhydrus*)[1]—A. d'Orchymont (1923), Mem. Depart. Agric. Ind., VIII, p. 9.[2] — Japan.[1] Bihar.[2]

malabarensis Règimbart (1903), Ann. Soc. Ent. Fr., LXXII, p. 335 (*Philhydrus*).[1] — Mahè, Calicut.[1] Philippine I.

minusculus A. d'Orchymont (1925), Ann. & Bull. Soc. Ent. Belg., LXV, p. 294. — Cochin-China.

nigropiceus Motschulsky (1861), Bull. Soc. Nat. Mosc., XXXIV, I, p. 109 (*Philhydrus*). — Ceylon.

rubrocinctus Règimbart (1903), Ann. Soc. Ent. Fr., LXXII, p. 56[1] (*Philhydrus*)—A. d'Orchymont (1926), — Indo-China.[3] Cochin-China.[1]

Treubia, VII, p. 132[2]—(1926), Philippine Journ. Sc., 30, 3, p. 383.[3]

Tonkin., Burma.[1]
Sumatra.[1, 3]
Buru I.[2, 3]
Philippine L.[2, 3]

tetraspilus Règimbart (1903), Ann. Soc. Ent. Fr. LXXII, p. 335 (*Philhydrus*) (*cf.* A. d'Orchymont, Journ. As. Soc. Bengal, n. s. X, 1914, p. 358).

mahè, Calicut.

Sub-gen. **Methydrus** Rey.

Methydrus Rey (1885), Ann. Soc. Linn. Lyon, XXXI, p 253—Ganglbauer (1904), Käf. Mitteleur., IV, 1, p. 243—Zaitzev (1908), Ann. Mus. Zoo. Ac. Imp. S. Petersb. XII, (1907), p. 161, nota—Lomnicki (1911), Kosmos, XXXV, p. 265—Kuhnt (1913) Ill. Best. Tab. Käf. Deutschl., p. 432—A. d'Orchymont (1914), Journ. & Proc. Asiat. Soc. Bengal, n. s. X, p. 357—Schaufuss C. (1916) in Calwer Käferb. ed. 6 (1913). p. 341—Des Gozis (1916), Misc. Ent., XXII, p. 35—A. d'Orchymont (1919), Ann. Soc. Ent. Fr., LXXXVIII, p. 155.

Agraphilydrus Kuwert (1888), Deuts. Ent. Zeits., XXXII, p. 280.

Agraphilhydrus Everts (1899) Col. Neerl., I, p. 644.

esuriens Walker (1858), Ann. & Mag. Nat. Hist. (3) II, p. 209[1] (*Philhydrus*).

Ceylon.[1, 2, 3, 4]

escuriens Sharp (1890), Trans. Ent. Soc. Lond., p. 350 (*Ohilydrus*)[2]—Règimbart (1903), Ann. Soc. Ent. Fr., LXXII, p. 58 (*Philhydrus*).[3]

nigriceps Motschulsky (1859), Etud. Entom., VIII, p. 46 (*Pylophilus*)[4]— Redtenbacher (1867), Reise Novara, Zool. II, p. 26 (*Philhydrus*).[5]

India.[3, 4], Nicobar I.[5]
————
Cochin-China.[3]
Tonkin.
Annam.[3], Siam.
F. M. S., Sumatra.[3]
Java.[3], Borneo.[3]
Philippine I.[3]
Australia.
Queensland.

flavicans Règimbart (1903). Ann. Soc. Ent. Fr., LXXII, 56 (*Philhydrus*)[1]—A. d'Orchymont (1913), Suppl. Ent., II, p. 6 (*Lumetus*).[2]

Annam.[1]
————
Bengal.[1], Bihar.
————
Formosa.[2]

parvulus Kuwert (1890), Verh. Naturf. Ver. Brünn., XXVIII, p. 56 (*Philydrus—Agraphilydrus*)[1]—Guillebeau (1896) Bull. Soc. Ent. Fr., p. 226 (*Philydrus*)[2]—Règimbart (1900), l. c., p. 50 (*Philydrus*)[3]; (1903), Ann. Soc. Ent. Fr., LXXII, p. 31 (*Philhydrus*)[4]—Scott (1913), Trans. Linn. Soc. Lond. XVI, 2, p. 203-204 (*Philydrus*)[5] (nec. *Helochares parvulus* Reiche, 1856).
Syria : Beyrout.[1] Tropical and N. Africa. Egypt.[2, 3, 4] Seychelles I.[3, 4, 5] Aldabra I.[3, 4, 5] Coetivy I.[5] Madagascar.[4] India.[3]

pendleburyi A. d'Orchymont (1927). Journ F.M.S. Mus. XIII, p. 250. F. M. S. Pahang.

Tribe **CHAETARTHRIINI** Bedel.

Chaetarthriini Bedel (1881), Fne Col. Bass. Seine, I, p. 302, 314—Everts (1899), Col. Neerl., I, p. 638—Ganglbauer (1904), Käf. Mitteleur., IV, I, p. 220, 225—Kuhnt (1913), Ill. Best. Tab. Käf. Deutschl., p. 418— Shaufuss C. (1916), in Calwer, Käferb., ed. 6 (1913), p. 333, 344.

Cyllidiaires Mulsant (1844), Hist. Nat. Col. Fr. Palp., p. 88, 143.

Chètarthriaires Rey (1885), Ann. Soc. Linn. Lyon, XXXI, p. 310.

Chaetarthriitae Kuwert (1890), Verh. Naturf. Ver. Brünn, XXVIII, p. 20.

Chaetarthriina Zaitzev (1908), Hor. Soc. Ent Ross., XXXVIII, p. 398 (—Catal.) Jakobson (1911), Käf. Russl., p. 840.

KEY TO GENERA.

1. Elytra with ten very narrow and deep striae not distinctly punctured except anteriorly where the striae disappear; they are aprallel to suture from the middle on and remain so on their extremity against outer margin. In consequence the outer stria is very short and exists only in the median part of outer margin at a distance nearly equal to one-fourth of length of elytra. 39. *Thysanarthria.*

1′. Elytra with only a sutural stria and besides, but only sometimes, some vagae punctures arranged here and there in inconspicuous and incomplete series. 40. *Chaetarthria.*

39. **Thysanarthria** A. d'Orchymont.

Thysanarthria A. d'Orchymont (1926), Ann. & Bull. Soc. Ent. Belg, LXVI p. 195, 242.

championi Knisch (1924), Wien. Ent. Zeit., XLI, p. 40 (*Chaetarthria*)[1]—A. d'Orchymont (1926), l. c., p. 195, fig., 243, 244.[2] — Kumaon, W. Almora.[1] Ranikhet. ? Tonkin.[2]

40. **Chaetarthria** Stephens.

Chaetarthria Stephens (1832), Ill. Brit-Mandib., V, p. 401; (1839), Man. Brit Col., p. 88—Westwood (1840), Intr-Mod., Classif. Ins., II, Synopsis, p. 10—Bach (1851), Käferfn. Nord Mitteldeutschl., I, p. 468—G. H. Horn (1873), Proc. Am. Philos. Soc., XIII, p. 118, 124—Redtenbacher (1874), Fna. Austr., ed. 3, II, p. 115—Dalla Torre (1877), Jahresber. Ver. Naturk Linz (8), p. 66—Bedel (1881), Fne Col. Bass. Seine, I, p. 314—Sharp (1882), Biol. Centr. Amer., I, 2, p. 86—Lec. & Horn (1883), Class. Col. North Amer., p. 72—Rey (1885), Ann. Soc. Linn. Lyon, XXXI, p. 310—Fowler (1887), Col. Brit. Isl., I, p. 222, 232—Seidlitz (1888), Fna Balt., ed. 2, Gatt., p. 23; (1888), Fne Transsylv., Gatt., p. 23—Kuwert (1890), Verh. Naturf. Ver. Brünn, XXVIII, p. 20, 106—Wickham (1895), Canad. Ent, XXVII, p. 182—Everts (1899), Col. Neerl., I, p. 639, 650—Kliment (1899), Cešti brouci, p. 115—Lameere (1900) Man. Fne Belg., II, p. 330—Stierlin (1900), Col. Helv. p. 211—Ganglbauer (1904), Käf. Mitteleur., IV, I, p. 256—Reitter (1909), in Brauer, Süsswasserfna Deutschl., 3-4, p. 65, 71; (1909), Fna Germ., II, p. 357, 361—Blatchley (1910), Col. Ind. in Bull. Ind. Depart. Geol. I, p. 257, 259—Kuhnt (1913), Ill. Best Tab. Käf. Deutschl., p. 418, 435, f. 15B—Schaufuss C. (1916) in Calwer, Käferb., ed. 6 (1913), p. 344— Des Gozis (1916), Misc. Ent., XXIII, p. 14—Tullgren & Wahlgren (1920), Svenska Ins. (2), p. 234, 235—Richmond (1920) Bull. Amer. Mus. Nat. Hist., XLII, Art. 1. p. 51.

Cyllidium Erichson (1837), Käf. Mark. Brandebg., I, p. 211.

Chaetarthrias Chenu (1851), Encycl. Hist. Nat. Col. I, p. 251.

Chaethartria G.C. Thomson (1859), Skand Col., I, p. 18; (1860), II. p. 99 (1867), IX, p. 119.

indica A. d'Orchymont (1920), Ann. Soc. Ent. Belg., LX, p. 19[1]; (1926), Philipp. Journ. Sci., 30, 3, p. 383.[2] — Nilgiri Hills.[1] Sikkim-Dikchu.

almorana Knisch (1924), Wien. Ent. Zeit., XLI, p. 39.[3] — Kumaon (W. Almora, Bodiar, Haldwani.[3])

Indo-China. Tonkin.[2] Philippine I.[2]

saundersi A. d'Orchymont (1923), Treubia, III, 3-4, p. 420.[1] — Singapore, Cochin-China. Sumatra.[1]

TRIBE HYDROPHILINI.

Hydrophilini Leconte (1861), Classif. Col. N. Amer., I, p. 44, 46 (1861), 44, 45 (1862)—Bedel (1881), Fne Col. Bass. Seine, I, p. 302, 304—Leconte & Horn (1883), Class. Col. North Amer., p. 71—Everts (1899), Col. Neerl., I, p. 637—Ganglbauer (1904), Käf. Mitteleur., IV, I, p. 220, 226—Reitter (1909), in Brauer., Süsswasserfna Deutschl., 3-4, p. 64; (1909), Fna Germ., II, p. 356—Blatchley (1910), Bull. Indiana Depart. Geol., I, p. 249, 254—Kuhnt (1913), Ill. Best. Tab. Käf. Deutschl., p. 418—Schaufuss C. (1916), in Calwer, Käferb. ed. 6 (1913), p. 333—A. d'Orchymont (1919), Ann. Soc. Ent. Fr., LXXXVIII., p. 156.

Hydrophilaires Mulsant (1844), Hist. Nat. Col. Fr. Palp., p. 103.

Hydrophilii Fairmaire & Laboulbène (1854), Fne Ent. France., I, p. 224.

Hydrophilides Lacordaire (1854), Gen. Col. I, p. 446, 447.

Hydrophilates + Hydrates Mulsant (1856), Opusc. Entom., VII, p. 167.

Idrofilini Camerano (1884), Atti R. Acad. Sc. Torino, XIX, p. 220.

Hydrophilina Fowler (1887), Col. Brit. Isl., I, p. 219.

Hydrophilitæ Kuwert (1890), Verh. Naturf. Ver. Brünn, XXVIII, p. 16.

ONTOGENESIS Mulsant (1844), Hiist Nat. Col. Fr. Palp. p. 105 (*Hydrophilaires*)—Lacordaire (1854), Gen. Col. I, p. 440 (*Hydrophilides*)—Camerano (1883), Atti Ac. Torino XIX, p. 222 (*Indrofilini*)—Ganglbauer (1904), Käf. Mitteleur, IV, I, p. 226—Richmond (1920) Bull. Amer. Mus. Nat. Hist., XLII, Art. 1, p. 31 (*Hydrophilinoe*).

KEY TO GENERA.

1. All femora clothed at base with a silky procumbent and dense pubescence. Antennae with normal club, composed of joints not lunulate and not particularly irregular or ciliate. Middle and posterior tarsi compressed and oarlike. All ventral segments pubescent without shining spot on the middle of posterior margin of fifth.

   Prostital carina ridge-like with an anterior brush of long setae and, sometimes, posteriorly with a small emargination; a small notch on anterior part of mesostital carina with on the ground a bunch of small setae. Maxillary palpi with last joint as long or usually longer than the preceding joint. 41. *Sternolophus.*

1′. The anterior femora only are clothed with pubescence at base, it is wanting or very much reduced on middle and posterior femora. Prostital carina without anterior brush of long setae. 2.

2. Prostital carina cultriform. Mesostital carina with a small notch on anterior part. Claws of anterior tarsi of male hook-like but not very dissimilar. Size smaller. 3.

2′. Prostital carina not cultriform, excavate to receive the anterior part of mesostitum. Prefrons on anterior side widely emarginate, the emargination filled with a short membraneous poorly chitinized sclerite sometimes of yellow colour, the preclypeus.

Ante-penultimate joint of maxillary palpi curved, with concavity directed towards inner side, Anntennal club perfoliate and asymmetrical. Middle and posterior tarsi strongly compressed and oarlike. Claws of all tarsi dentate at base, those of anterior tarsi of ♂ usually unequal and differently shaped. Size great or very great. 44. *Hydrous.*

3. Prefrons on anterior margin widely emarginate, the preclypeus conspicuous in the emargination as in *Hydrous.* Antennal club perfoliate and asymmetrical as in that genus: joints 7-9 excavate on posterior side, the 7th and 8th lunulate as an irregular sickle with ciliated upper end. Labrum with a posterior transversal row of setigerous punctures, interrupted in the middle. Claws of all tarsi strongly denate at base in ♂♀, the basal tooth much more developed and spatuliform on anterior tarsi in the ♂. 42. *neohydrophilus.*

Labrum anteriorly with two isolated pores, not much separated giving insertion to a brush of short and stiff hairs. Maxillary palpi very long, with last joint showing a tendency to equal the 3rd in length; the latter not dilated in ♂ nor compressed on underside, although it is less narrowed in its first half than in ♂. Prostital carina sometimes prolonged as a robust prominent spine, sometimes also the metasternal carina. Middle and posterior tarsi strongly compressed and oar-like

3′. Prefrons not emarginate on anterior margin, truncate and as a nearly straight line behind labrum. 7th and 8th joints of antennae not or only feebly lunulate, only slowly prolonged and not ciliate at upper end. Labrum with a transverse in the middle not interupted row of setigerous punctures, without the two anterior pores and brush-like hairs. Maxillary palpi not extra-ordinarily long, the last joint notably shorter than the penultimate. Posterior spine of prostital carina, when present, never very long. Posterior point of metasternal carina reaching scarcely farther than base of posterior trochanters. Claws of tarsi slender, more or less curved, with a small tooth at base, pointed in ♀, obtuse in ♂. 43. *Hydrophilus.*

41. **Sternolophus** Solier.

Sternolophus Solier (1834), Ann. Soc. Entr. Fr., III, p. 302, 310—Brullé (1835), Hist. Nat. Ins., V, II, p. 279— Laporte de Castelnau (1840), Hist. Nat. Anim. Art., II, p. 54—Chenu (1851), Enc. Hist. Nat. Col. I, p. 247—Lacordaire (1854), Gen. Col. I, p. 450, 453—Mulsant (1856), Opusc. Entom., VII, p. 171—Kuwert (1890), Verh. Naturf. Ver. Brünn., XXVIII, p. 17, 28—Zaitzev (1909), Revue Russe d'Ent (1908), p. 228, 229—A. d'Orchymont (1911), Mem. Soc. Ent. Belg., XIX, p. 53; (1919), Ann. Soc. Ent. Fr., LXXXVIII, p. 156.

Helobius Mulsant (1851) Mem. Ac. Sc. Lyon, (2) I, p. 75.

Sub-gen. **Neosternolophus** Zaitzev.

Neosternolophus Zaitzev (1909), Revue Russe d'Ent. (1908), p. 228—A. d'Orchymont (1911), Mem. Soc. Ent. Belg., XIX, p. 53, 55, 59, nota a; (1919) (emend) Ann. Soc. Ent. Fr., LXXXVIII, 1[a]. 157, 158.

brachyacanthus Règimbart (1902), Ann. Soc. Ent. Fr., LXXI, p. 472[1]—Zaitzev (1909), l. c., p. 229[2]—A. d'Orchymont (1911), l. c., p. 54.[3]

Ceylon.[1, 2], Burma.[1]
Bihar., U. Burma.

Siam.[1]
Indo-China.[2]
Annam.[1]
Cambodia.[3]
Tonkin.[1], F. M. S.
Sumatra., Java.[2]
Philippine I.
Aru I.

tenebricosus Blackburn (1888), Proc. Linn. Soc. N. S. Wales, (2), III, p. 813[1]—A. d'Orchymont (1911), Mem. Soc. Ent., Belg., XIX, p. 54, 55, f. 1, 2;[2] (1919) Ann. Soc. Ent. Fr., LXXXVIII, p. 156[3]; (1923), Treubia, III, p. 420.[4]

oceanicus Zaitzev (1910), Rev. Russe d'Ent., X, p. 225.[5]

Australia North Terr.[1]
Mulgrave R.
Queensland.

Tranvancore.
(Wallardi).[2]
Malabar (Mahè).[2]
Abor.

Java (Bandoeng).[2]
? Natolna I.[3]

Borneo.[3, 5]
Philippine I.
? Timor.[3], Buru I.
Celebes.[4]
New Guinea.[4]

Sub-gen. **Sternolophus** s. str.

Sternolophus in sp. (1909), Rev. Russe d'Ent., Ent., (1908), p. 229—A. d'Orchymont (1919) (3 emend.), Ann. Soc. Ent. Fr., LXXXVIII, p. 158.

decens Zaitzev (1909), l. c., p. 231[1]; A. d'Orchymont (1911), Mem. Soc. Ent. Belg., XIX, p. 54, 58.[2]

Persia.[1], Arabia.[2]
——————
India., Bihar.

rufipes Fabricius (1792), Entom. Syst. I, p. 183; (1801), Syst. Eleuth., I, p. 251 (*Hydrophilus*)—Herbst (1797), Natur Syst. Käf., VII,[1] p. 307[1] (*Hydrophilus*)—Laporte de Castelnau (1840), Hist. Nat. Anim. Art. II, p. 54[2]—Mulsant (1856), Opusc. Ent., VII, p. 173[3]—Zaitzev. (1909), l. c., p. 230[4]—A. d'Orchymont (1911) l. c., p. 54.

mergus Redtenbacher (1844), Hügel Kashm. IV, 2, p. 514 (*Tropisternus*)[5]—Zaitzev (1909), l. c.[6]

fulvipes Motschulsky (1853), Etudes Entom., II, p. 45.[7]

China.[1, 2, 3, 6, 7]
Asia trop.[4], India or [3]
Ceylon., Kashmir.[5, 6]
——————
Indo-China.
F. M. S.
Sumatra.
Java.[2, 6], Formosa.
Japan.[4]
Philippine I.

42. **Neohydrophilus** A. d'Orchymont.

Neohydrophilus A. d'Orchymont (1911). Mem. Soc. Ent. Belg., XIX. p. 59: (1916), Ann. Soc. Ent. Fr., LXXXV, p. 98, f. 6; (1919), l. c., LXXXVIII p. 160.

Hydrous Laporte de Castelnau (ex. p.) (1840), p. 51 (nec. Leach, 1815).

cambodiensis Règimbart (1903), Ann. Soc. Ent. Fr., LXXI, p. 53 (*Hydrophilus*).[1]

Cambodia.[1]
Burma.[1]
Tonkin (Hanoi)

rufiventris Nietner (1856), Journ. As. Soc. Bengal, XXV, p. 537; (1857), Ann. Mag. Nat. Hist. (2), XIX, p. 387 (Hydrous)[1]—A. d'Orchymont (1911), Mem. Soc. Ent. Belg., XIX, p. 67, t. 6 f. 15, 16[2]; (1912), Rev. Zool. Afric., I, p. 332, f. 3[3]; (1913), Ent. Mitteil., II, f. 2. p. 102.

Ceylon.,[1, 3, 4]
Malabar (Mahè).[2, 3]
——————
Sumatra.[3, 4]
Madagascar.[3]

horni Règimbart (1902), Ann. Soc., Ent. Fr., LXXI, p. 471 (*Hydrophilus*).[4]

spinicollis Eschscholtz (1922), Entomographien, I, p. 41 (*Hydrophilus*)[1]—A. d'Orchymont (1923), Mem. Dep. Agr. Ind., VIII, p. 10.[2*] — India Or.[1] Pusa, yellapur, Kasara Brahmaputra R.[2]

var. *elongatus* Règimbart (1902), Ann. Soc. Ent. Fr., LXXI, p. 471 (*Hydrophilus*).[1] — Ceylon.[1], Burma.[1] Philippine I.

43. Hydrophilus O. F. Müller.

Hydrophilus O. F. Müller (1764) (ex. p.) Fna Ins. Fridrichsdalina, p. 16—Leach (1815), in Brewster's Edinb. Encycl., IX, p. 96; (1817), Zool. Misc., III, p. 92, 94—Curtis (1827), Brit. Ent., IV, p. 159—Stephens (1829), Ill. Brit. Ent., Mandib., II, p. 123—Solier (1834), Ann. Soc. Ent. Fr., III, p. 302, 312—Erichson (1837), Käf Brandebg., I, p. 207—Redtenbacher (1849), Fna Austr., p. 131—Bach (1851), Käferfn. Nord Mitteldeutschl., I, p. 131—Chenu (1851), Enc. Hist. Nat. Col., I, p. 244—Kuhn (1858), Käf. Südbayer. Flachl., p. 39—Seidlitz (1888), Fna Balt., ed. 2, Gatt., p. 21; (1888) Fna Transsylv., Gatt., p. 21—Kuwert (1890), Verh. Naturf. Ver. Brünn, XXVIII, p. 17, 29—Bedel (1892), Rev. Entom. Caen, X, p. 308—Kliment (1899), Cešti brouci, p. 110—Semenov (1900), Hor. Soc. Ent. Ross, XXXIV, p. 585—Stierlin (1900), Col. Helv., p. 212—Ganglbauer (1904), Käf. Mitteleur., IV, I, p. 226, 233—Reitter (1909), in Brauer, Süsswasserfna Deutschl., 3-4, p. 65, 68, fig. 47; (1909), Fna Germ., II, p. 356, 359—Zaitzev (1909), Rev. Russe Ent. (1908), p. 229—Jakobson (1911), Käf. Russl., p. 850—A. d'Orchymont (1911), Mem. Soc. Ent. Belg., XIX, p. 58 (*Hydrophilus* s. str.)—Kuhnt (1913), Ill. Best Tab. Käf. Deutschl., p. 418, 430 f. 14—Schaufuss C. (1916), in Calwer, Käferb., ed. 6 (1913), p. 335, 337—Des Gozis (1916), Misc. Ent. XXIII, p. 13 18—A. d'Orchymont (1919), Ann. Soc. Ent. Belg., LIX, p. 81, 82—Tullgren & Wahlgren (1920), Svenska Ins (2), p. 234.

---

* I think Règimbart was wrong in his interpretation of this species. So I do not quote his papers (1903 & 1906, Madagascar & Africa Or.) nor mine after him (1912), Belgian Congo). These specimens belong to a different species now on new study.

Hydrochara Berthold (1827), Latr. Nat. Fam. Thierr., p. 355.

Hydrous Brullè (1835), Hist. Nat. Ins. V, II, p. 275 [nec Leach (1817).]

Hdyrocharis Hope, [1838, in litt.] Westwood (1840), Introd. Med. Classif. Ins., II, Synopsis, p. 9.

Hydroechus Stephens (1839), Man. Brit. Col., p. 88.

Hydrochares Solsky (1876), Fedtsch. Reise Turkestan, II, p. 20.

ONTOGENESIS Thomson (1860), Skand. Col., II, p. 91 (*Hydrous*)— Schiödte (1862), Nat. Tidskr. (3) I, p. 222, 223; (1872), VIII, p. 204—Haase (1884), Zeits. Ent. Bresl. N. F. (9), Vereinsnachr., p. 31 (*Hydrous*) Ganglbauer (1904), Käf. Mitteleur., IV, I, p. 223, 234—Portier (1911-12), Archiv. Zool. Expér. gén (5) VIII, p. 258, f. 41-42 (*Hydrophilides*)—A. d'Orchymont (1913), Ann. Biol. lac., VI, p. 211—Schaufuss C. (1916), in Calwer, Käferb., ed. 6 (1913), p. 337—Richmond (1920), Bull. Amer. Mus. Nat. Hist., XLII, Art. I, p. 31, 83, 86, 88.

? inconspicuus Nietner (1856), Journ. As. Soc. Bengal, XXV, p. 538; (1857) Ann. & Mag. Nat. Hist., (2) XIX, p. 387, cum pupa (*Hydrous*). (Gen. dub.). — Ceylon.

similis A. d'Orchymont (1919), Ann. Soc. Ent. Belg., LIX, p .81; (1919), Ann. Soc. Ent. Fr., LXXXVIII, p. 162. — India Or.

vitalisi A d'Orchymont (1919), Ann. Soc. Ent. Belg., LIX, p. 79, 81, 82[1]; (1923), Mem. Dep. Agric. India, VIII, p. 10.[2] — Indo-China.[1] India.[2]

44. **Hydrous** Leach.

Hydrous Dahl., Amaenitates Academicae, VIII, p. 308 nota (1785)—Leach (1815), Brewster's Edinb. Encycl., IX, p. 96; (1817), Zool. Misc., III, p. 92. 94—Curtis (1828), Brit. Ent., V, p. 239—Stephens (1829) Ill. Brit. Ent. Mandib. II, p. 122—Solier (1834), Ann. Soc.Ent. Fr., III, p. 302, 304 (+*Stethoxus* + *Temnopterus*)—Erich, son (1837), Käf. Mark Brandebg. I, p. 206—Hope (1838) Col. Man., II, p. 125—Redtenbacher (1849), Fna Austr.,

p. 131—Bach (1851) Käfer Nord Mitteldeutschil., I, p. 131—Kuhn (1858), Käf, Südbayer, Flachil, p. 39—Seidlitz (1888), Fna Balt., ed 2, Gatt., p. 21; (1888), Fna Transsylv., Gatt., p. 21—Kuwert (1890), Verh. Naturf. Ver. Brünn. XXVIII, p. 16; (1893), Deuts. Ent. Zeits., p. 81—Kliment (1899), Cešti brouci, p. 109—Stierlin (1900), Col. Helv., p. 212—Ganglbauer (1904), Käf. Mitteleur, IV, I, p. 226—Reitter (1909), in Brauer, Süsswassserfna Deutschl., 3-4, p. 65, 57; (1909), Fna Germ., II, p. 356, 359—Zaitzev (1909), Rev. Russe d'Entom. (1908), p. 228—A. d'Orchymont (1911), Mem. Soc. Ent. Belg., XIX, p. 59, nota 2-Kuhnt (1913), Ill. Best. Tab. Käf. Deutschl., p. 418, 429, f. 12, 13—Schaufuss C. (1916), in Calwer, Käferb., ed. 6 (1913) p. 335—Des Gozis (1916), Misc. Ent., XXIII, p. 13, 15—Tullgren & Wahlgren (1920), Svenska Ins. (2), p. 234.

Hydrophilus Brullé (1835), Hist. Nat. Ins. V, II, p. 273 (nec. Leach, 1817).

Hydrous + *Stethoxus* + *Temnopterus* Solier (1834), Ann. Soc. Ent. Fr., III, p. 302, 304, 307.

Hydrous + *Temnopterus* Kuwert (1890) Verh. Naturf. Ver. Brünn, XXVIII, p. 16, 24, 26.

Stethoxus Bedel (1892), Rev. d'Ent. Caen, X, p. 308.

Stethoxus + *Dibolocelus* Règimbart (1901), Ann. Soc. Ent. Fr., LXX, p. 188.

ONTOGENESIS Thomson G.C. (1860), Skand. Col. II, p. 89 (*Hydrophilus*) Schiödte (1862), Nat. Tidskr. (3) I; (1872) VIII, p. 214—Ganglbauer (1904), Käf-Mitteleur IV, I, p. 229, 231—A. d'Orchymont (1913), Ann. Biol. lac., VI, p. 211—Schaufuss C. (1916), in Calwar, Käferb., ed. 6 (1913), p. 15, t. 2, f. 9 a. b. p. 336—Richmond (1920), Bull. Amer. Mus. Nat. Hist., XLII, Art. 1, p. 39, 83, 86, 88.

Sub-gen. **Hydrous** s. str. Kuwert.

Hydrous in. sp. Kuwert (1890), Verh. Naturf. Brünn. XXVIII, p. 16, 24—Ganglbauer (1904), Käf. Mitteleur., IV, I, p. 227.

Hydrous + *Stethoxus* Solier (1834), Ann. Soc. Ent. France, III, p. 302, 304, 307. *Hydrous* s. str. + *Mesocanthicus* Hope (1838) Col. Man. (2), p. 126.

Hydrophilus s. str. + *Hydrosoma* + *Hydrodema* Laporte de Castelnau (1840), Hist. Nat. Ins., II, p. 49-51.

Stethoxus s. str., Bedel (1892), Rev. d'Ent. Caen, X, p. 309—Règimbart (1901), Ann. Soc. Ent. Fr., LXX, p. 192.

Hydrous s. str. + *Stethoxus* + *Pagipherus* Kuwert (1893), Deuts. Ent. Zeitschr., p. 82.

cashmirensis Redtenbacher (1844), Hügel. Kaschmir, IV, 2, p. 513 Pl. 24, f. 2 (*Hydrophilus caschmirensis*)[1]—Bedel (1892), Rev. Ent. Caen, X, p. 310, 315 (*Stethoxus*)[2]—Kuwert (1893), Deuts. Ent. Zeits., p. 86[3]—Règimbart (1901). Ann. Soc. Ent. Fr., LXX, p. 193, 201[4] (*Stethoxus*)—A. d'Orchymont (1919), Ann. Soc. Ent. Belg., LIX, p. 83[5]—Knisch (1921), Archiv. Naturg. (1919), A. 8, p. 63.[6]

japonicus Sharp (1873), Trans. Ent. Soc. London, p. 57[7] (*Hydrophilus*).

var. *birmanicus* Règimbart (1901), Ann. Soc. Ent. Fr., LXX, p. 202, Pl. 7, f. 13. (*Stethoxus*).[8]

Kashmir.[1, 2, 3, 4]
India.[4], Burma.[2]
Ceylon.[4, 6]
____
Indo-China.
Tonkin.[2, 4]
Cambodia.[5], Annam.[6]
China.[2, 4, 6], Japan.[3, 7]
Kiu-Siou I.[2]
Cagoshima I.[4]
Sumatra.[4], Burma.[8]
Carin, Asciuii
Cheba.

cavisternum Bedel (1892), Rev. Ent. Caen, X, p. 310, 316 (*Stethoxus*)[1]—Kuwert (1893), Deuts. Ent. Zeits., p. 92[2]—Règimbart, Ann. Soc. Ent. Fr., LXX, p. 193, 202, Pl. 7, f. 12, 12a♂ (*Stethoxus*).[3]

Tonkin.[1, 3]
Cochin-China.[3]
Laos, Annam.[3]
Cambodia.
Siam.[1, 2, 3]
Malacca.[1, 3]
Selangor, Kuala.
Java.[2]

hastatus Herbst (1779), Beschäfte, Berl. Ges. Naturf. Freunde, IV, p. 317, Pl. 7, f. 2 A.B.[1] (*Dytiscus*)—Bedel (ex. p.) (1892), Rev. Ent. Caen, X, p. 310, 315[2] (*Stethoxus*)—Kuwert (1893), Deuts. Ent Zeits., p. 92[3]—Règimbart (1901), Ann. Soc. Ent. Fr., LXX, p. 193, 200, Pl. 7, f. 11 (*Stethoxus*).[4]

olivaceus Herbst (1797), Natursyst. Insekt. Käf. VII, p. 295, p. 113, f. 6 (nec Fabricius, 1781)[5] (*Hydrophilus*).

aberrans Kuwert (1893), Deuts. Ent. Zeits., p. 93.[6]

India Or.[1, 5, 6]
India.[4]
Tonkin.[2, 4]
Burma.
China mer.

indicus Bedel (1892), Rev. Ent. Caen, X, p. 310, 316 (*Stethoxus*)[1]—Kuwert (1893), Deuts. Ent. Zeits., p. 88[2]—Règimbart (1901), Ann. Soc. Ent. Fr., LXX, p. 194, 205 (*Stethoxus*).[3]

N. India.[1, 2, 3]

alivaceus Fabricius (1781), Spec. Ins. I, p. 289; (1787), Mant. Ins., I, p. 188; (1792), Ent. Syst., I, p. 182; (1801), Syst. Eleuth., I, p. 250 (*Hydrophilus*)—Linné (1789), Syst. Nat., ed. XIII Gmel. I, IV, p. 1941 (*Hydrophilus*)—Olivier (1792), Encycl. Method., VII, p. 125, t. 183, f. 2; (1795), Entom., III, No. 39, p. 10, Pl. 1, f. 6 A B (*Hydrophilus*)—Latreille (1804), Hist. Nat. Crust. Ins., X, p. 62 (*Hydrophilus*)—Laporte de Castelnau (1840), Hist. Nat. Anim. Art, II, p. 50 (*Hydrophilus*)—Kuwert (1893), Deuts. Ent. Zeits., p. 93—Règimbart (1901), Ann. Soc. Ent. Fr., LXX, p. 193, 200, Pl. 7, f. 2, 2a, 10 (*Stethoxus*)—Fowler (1912), Fna Brit. Ind. (Gen. Intr.), f. 35, p. 128 (*Hydrophilus*)—A. d'Orchymont (1919) Ann. Soc. Ent., Belg., LIX, p. 83.
N. India.

oblongus Olivier (1792), Encycl. Meth., VII, p. 125 (*Hydrophilus*).

hastatus Bedel (ex. p.) (1892), Rev. Ent. Caen, X, p. 310, 315 (*Stethoxus*) (nec. Herbst 1779).

pallidipalpis W. S. M'Leay (1825), Annul. Jav., p. 35;[1] A. d'Orchymont (1911), Mem. Soc. Ent. Belg., XIX, p. 70.[2]
Java.[1, 2], Sumatra.[7] Formosa.[6]

acuminatus Motschulsky (1853), Etud. Entom., II, p. 44[3] (*Hydrophilus*)—Règimbart (1901), Ann. Soc. Ent. Fr., LXX, p. 195, 211, Pl. 7, f. 21[4] (*Stethexus*).
Japan.[4, 5, 6] China.[3, 4, 6], Corea.[4]

Tibet.[4], India.[6] Kashmir.[6] Burma.[4, 6] Kachin Cauri.

cognatus Sharp (1873), Trans. Ent. Soc. London, p. 57[5] (*Hydrophilus*).

viridicollis Bedel (1892), Rev. Ent. C. en, X, p. 312, 318[6] (*Stethoxus*) (nec Redtenbacher, 1844).

sumatræ Kuwert (1893), Deuts. Ent. Zeits. p. 85.

piceus Linné (1758), Syst. Nat. ed. X, p. 411 (*Dytiscus*)—Erichson (1837), Käf Mark Brandebg, I, p. 206 (*Hydrophilus*)—Mulsant (1844), Hist. Nat. Col. Fr., Palp., p. 108 (*Hydrophilus*)—Jacquelin—Duval (1857), Gen. Col. Eur., I, Pl. 29, f. 141 (*Hydrophilus*)—Thomson G. C. (1860), Skand. Col., II, p. 90 (*Hydrophilus*)—Bedel (1881), Fne Col. Bass. Seine, I, p. 305 (*Hydrophilus*); (1892), Rev. d'Ent Caen, X, p. 313, 318 (*Stethoxus*—Camerano (1883), Atti. Accad. Torino, XIX, p. 228, t. 7, f. 1-3, t. 8, 12, 13, 15 (*Hydrophilus*)—Rey (1885), Ann. Soc. Linn.
Palæarctic. Kashmir.

Lyon, XXXI, p. 224 (*Hydrophilus*)—Kuwert (1890), Verh. Naturf. Ver. Brünn, XXVIII, p. 26; (1893), Deuts. Ent. Zeits., p. 85—Everts (1899), Col. Neerl., I, p. 641, f. 60, 62 p. 636 (*Hydrophilus*); (1922), III, p. 285—Lameere (1900), Man. Fne Belg., II, p. 334 (*Hydrophilus*)—Règimbart (1901), Ann. Soc. Ent. Fr., LXX, p. 195, 211, Pl. 7, f. 22 (*Stethoxus*)—Ganglbauer (1904), Käf. Mitteleur, IV, I, p. 231—Reitter (1909), in Brauer, Süsswasserfn. Deutschl., 3-4, p. 68; (1909), Fna. Germ., II, p. 359, T. 79, f. 5—Kuhnt (1913), III. Best. Tab. Käf. Deutschl., p. 430 f. 12—Schaufuss C. (1916), in Calwer, Käferb., ed. 6, (1913), p. 337, t. 6, f. 17—Des Gozis (1916), Misc. Ent., XXIII, p. 17, f.—Tullgren & Wahlgren (1920), Svenska Ins (2), p. 234, f. 169 p. 235—Blunck & Speyer (1925), Zool. Anz., 63, p. 241-249.*

ruficornis De Geer (1774), Mem., IV, p. 371, Pl. 14, f. 1-11 (*Hydrophilus*).

picaeus Fourcroy (1785), Ent. Paris, I, p. 65 (*Hydrous*).

viridicollis Redtenbacher (1844), Hügel Kaschmir., IV, 2, p. 513, Pl. 24, f. 3 (*Hydrophilus*).

niger Eichler (1876), Ent. Nachr., II, p. 168 (*Hydrophilus*).

turkestanus Kuwert (1893) Deuts. Ent. Zeits., p. 85.

rufocinctus Bedel (1892), Rev. d'Ent. Caen, X, p. 309, 317 (*Stethoxus*)—Kuwert (1893), Deuts. Ent. Zeits., p. 90—Règimbart (1901), Ann. Soc. Ent. Fr., LXX, p. 192, 205 (*Stethoxus*). — N. India. N. India.

senegalensis Percheron (1835), Gen. d.Ins., III, Nr. 5, Pl. 9, f. 1 (*Hydrophilus*)[1]—Bedel (1892), Rev. Ent. Caen, X, p. 310, 315 (*Stethoxus*)[2]—Kuwert (1893), Deuts. Ent. Zeits., p. 93[3]—Règimbart (1901), Ann. Soc. Ent. Fr. LXX, p. 192, 199, Pl. 7, f. 9 (*Stethoxus*)[4]—A. d'Orchymont (1919), Ann. Soc. Ent. Fr., LXXXVIII, p. 163. — Africa.[1, 2, 3, 4, 5, 6] N. India.[4, 7] Kurseong, Sangli, Himalaya. Mussoorie.

convexus Laporte de *Castelnau* (1840), Hist. Nat. Anim. Art., II, p. 50 (*Hydrophilus*)[5]

---

* Principal authors, For complete bibliography including Biology see; knisch, Col. Cat. para 79, 1924.

aulinieri Guerin (1842), Icon. Régne Animal VII Ins. (1834), p. 73.[6]

piesbergeni Reitter (1906), Wien. Ent. Zeit., XXV, p. 40.[7]

sternitalis Reitter (1906), Wien. Ent. Zeit., XXV, p. 41, nota. — China., Turkestan.

temnopteroides A d'Orchymont (1913), Ann. Soc. Ent. Belg., LVII, p. 319; (1914), LVIII, p. 47 (erratum). — Bihar.

unguicularis Règimbart (1901), Ann. Soc. Ent. Fr., LXX, p 195, 213, Pl. 7, f. 23 (*Stethoxus*). — Patria ignota. ? India.

TRIBE AMPHIOPINI.

Amphiopini A. d'Orchymont (1919), Ann. Soc. Ent. Fr., LXXXVIII, p. 164.

Amphiopitae Kuwert (1890) Verh. Naturf. Ver. Brünn., XXVIII, p. 20.

Amphiopina Zaitzev. (1908), Hor. Soc. Ent. Ross., XXXVIII, p. 399 (Catal)—Jakobson (1911), Käf. Russl., p. 840.

KEY TO GENUS.

Comprises only the genus *Amphiops* Erichson.

45. **Amphiops** Erichson.

Amphiops Erichson (1843), Arch. f. Naturg., IX, I, p. 229—Chenu (1851), Encycl. Hist. Nat. Col., I, p. 242—Lacordaire (1854), Gen. Col., I, p. 455, 462—Kuwert (1890), Verh. Naturf. Ver. Brünn, XXVIII, p. 20, 120—Blackburn (1898), Trans. R. Soc. S. Austr., XXII, 2, p. 225.

Cyprimorphus Fairmaire (1873), Rev. Mag. Zool., III, 3, p. 334.

annamita Règimbart (1903), Ann. Soc. Ent Fr., LXXI, p. 62. — Annam.

coomani A. d'Orchymont (1926), Ann. & Bull. Soc. Ent. Belg., LXVI, p. 246. — Tonkin.

gibbus Illiger (1801), Mag. Insektenk., I, p. 168 (*Hydrophilus*)[1]—L. von Heyden (1911), Abhand. Senekenb. Naturf. Ges., XXXIII, p. 353.[2] — India.[1], Bengal.[1] Burma.[6] N. E. Assam.[5]

? *mirabilis* Sharp (1890), Trans. Ent. Soc. London, p. 355[3]—Règimbart (1903), Ann. Soc. Ent. Fr., LXXII, p. 62[4]—A. d'Orchymont (1922), Rec. Ind. Museum, VIII, p. 628.[5]	Ceylon.[3, 4] ---------- Indo-China.[4] Tonkin., Sumatra.
? *variolosus* Règimbart (1903), Ann. Soc. Ent. Fr., LXXII, p. 63.[6]	Java.[6] Philippine I., Aru I.[2]
pedestris Sharp (1890), Trans. Ent. Soc. Lond. p 354.[1] var *varians* A. d'Orchymont (1922), Rec. Ind. Mus., VIII, p. 629.[2]	Ceylon.[1], India.[3] Pondicherri. Coromondel.[2] ----------
pedestris Règimbart (1903), Ann. Soc. Ent. Fr., LXXII, p 62.[3]	Indo-China: Salgon.[3] ? Tonkin., Sumatra.[3]
pisiformis Fairmaire (1884), Ann. Soc. Ent Fr. (6) IV, Bull. p. XLVI[1]—Règimbart (1903), Ann. Soc. Ent. Fr., LXXII, p. 61.[2]	Indo-China.[1, 2] Cambodia. ? Sumatra.
simplex Sharp (1890), Trans. Ent. Soc. London, p. 354.	Ceylon.
sumatrensis Règimbart (1903), Ann. Soc. Ent. Fr., LXXII, p. 61.[1]	Sumatra.[2], F. M. S. Selangor. ? Philippine I.

TRIBE BEROSINI.

Berosini Bedel (1881), Fne Col. Bass. Seine, I, p. 302—Everts (1899), Col. Neerl., I, p. 637—Ganglbauer (1904), Käf. Mitteleur., IV, I, p. 220—Reitter (1909), in Brauer, Süsswasserfna Deutschl., 3-4, p. 64; (1909), Fna. Germ., II. p. 356—Kuhnt (1913), Ill. Best. Tab. Käf. Deutschil., p. 418—Schaufuss C. (1916), in Calwer, Käferb., ed. 6 (1913), p. 333—A. d'Orchymont (1919), Ann. Soc. Ent. Fr., LXXXVIII, p. 164.

Berosaires Mulsant (1844), Hist. Nat. Col. Fr., Palp., p. 88, 97.

Berosina Thomson G. O (1859), Skand. Col. I, p. 17; (1860), II, p. 85; (1867), IX, p. 118.

Berositae Kuwert (1890), Verh. Naturf. Ver. Brünn, XXVIII, p. 20.

ONTOGENSIS Ganglbauer (1904), Käf. Mitteleur., IV, I, p. 221.

KEY TO GENERA.

1. Five ventral not retractile segments. Antennae composed of seven joints (4+3). Upper surface never of a uniform, deep and shining black. Shape elongate. Femora at least in part without dense pubescence. Sutures of ventral segments not deepened. Usually a more or less prominent 6th retractile ventral segment. 46. *Berosus.*

1′. Only four not retractile ventral segments, the first one not invaded by the posterior coxal cavities. Antennae of eight joints (5+3). Upper surface always of a uniform, deep and shining or metallic black. Pronotum narrowly applied within the emargination of anterior side of elytra. The fifth ventral segment is retractile and more or less porminent, emarginate in the ♂. 2.

2. Body convex, elongate and compressed on sides. Elytra everywhere striate punctate and pubescent. Metasternum with a simple carina along the middle. Ventral segments crenate on their posterior margin. 47. *Règimbartia.*

2′. Body also very convex but globular, of a short and broad shape. Elytra glabrous with anteriorly only 2-4 short and oblique series of larger punctures on the sides. Metasternum higly carinate along the middle, the carina deeply excised in its median part : it looks as if two short carinae were placed on a line, one before the other. Ventral sutures as deep gaps separating the segments 48. *Globaria.*

46. **Berosus** Leach.

Berosus Leach (1817), Zool. Misc., III, p. 92—Latreille (1829), in Cuv. Règne Anim. n. ed. IV, p. 524—Stephens (1829), III. Brit. Ent. Mandib., II, p. 125; (1839), Man. Brit. Col., p. 88 —Solier (1834), Ann. Soc. Ent. Fr., III, p. 302, 316—Sturm. (1836), Deutschl. Fna. Ins., X, p. 23, t. 228—Brullé (1835), Hist. Nat. Ins., V, 2, p. 284—Erichson (1837), Käf Mark, Brandebg., I, p. 204 (—*Volvulus*)—Westwood (1840), Intr. Mod. Class. Ins., II, Synopsis, p. 10—Laporte de Castelnau (1840), Hist. Nat. Anim. Art., II, p. 55—Heer (1841), Fna. Col. Helv., I, p. 482—Mulsant (1844), Hist. Nat. Col. Fr. Palp., p. 97—

Redtenbacher (1845), Gatt. deutsch. Käferfna, p. 71; (1849), Fna Austr., p. 131; (1858), ed. 2, p. 105; (1874), ed. 3, II, p. 114—Solier (1849), in Gay, Hist. de Chile, Zool., IV, p. 300—Bach (1851), Käferfn. Nord Mitteldeutschl., I, p. 122, 130— Lacordaire (1854), Gen. Col., I, p. 455, 458—Fairmaire & Laboulbène, Fne. Fr., I, p. 227, 231—Leconte (1855), Proc. Acad., Phil., VII, p. 357—Jacquelin Duval (1857), Gen. Col. Eur., I, p. 89, 98—Kuhn (1859), Käf. Südbayer, Flachl., p. 38—Bose (1859), in Gutfl., Käf. Deutschl., p. 67—Fuss (1859), Archiv. Ver. Siebenb. Landesk. n. F. (3) p. 112—Leconte (1861), Class. Col. N. Amer., I, p. 46—G. H. Horn (1873) (—*Volvulus*) Proc. Amer. Phil. Soc., XIII, p. 119 (+ *Enoplurus*) Solsky (1876), Fedtsch. Reise Turkestan, II, p. 21—Dalla Torre (1877), Jahresber. Ver. Naturk. Linz (8) p. 66—provancher (1877), Fne Ent. Canada, I, p. 216—Schlosser Klek (1877), Fna Kornjašah, I, p. 107—Bedel (1881), Fne Col. Bass. Seine, I, p. 302—Sharp (1882), Biol. Centr. Amer., I, 2, p. 79—Lec. & Horn (—*altus*) (1883), Class. Col. North Amer., p. 76—Rey (1885), Ann. Soc. Linn. Lyon, XXXI, p. 337—Fowler (1887), Col. Brit. Isl., I, p. 221, 229—Kuwert (1888), Deutsch. Ent. Zeits. p. 129—Wickham (1895), Canad. Ent., XXVII, p. 182, 185—Acloque (1896), Fne Fr., p. 92—Blackburn (1898), Trans. R. Soc. Austr. XXII, 2, p. 221—Lameere (1900), Man. Fne Belg., II, p. 329—Ganglbauer (1904), Käf. Mitteleur., IV, I, p. 221—Reitter (1909), in Brauer Süsswasserfna Deutschl., 3-4, p. 64, 66, fig. 45; (1909), Fna Germ., II, p. 356, 358—Blatchley (1910), Bull. Indiana Dept. Geol., I, p. 257, 258—Jakobson (1911), Käf. Russl., p. 849—Kuhnt (1913), Ill. Best. Tab. Käf. Deutschl., p. 418, 429, f. 10 A B., 11 B—Schaufuss C. (1916), in Calwer, Käferb., ed 6 (1913), p. 333—Leng & Mutchler (1918), Bull. Amer. Mus. Nat. Hist, XXXVIII, p. 103—A. d'Orchymont (1919), Ann. Soc. Ent. Fr., LXXXVIII, p. 165—Tullgren & Wahlgren (1920), Svenska Ins. (2), p. 234.

ONTOGENESIS Schiödte (1862), Nat. Tidskr. (3), I, p. 222; (1872), (3), VIII, p. 213—Bedel (1881), Fne Col. Bass. Seine, I, p. 303—Ganglbauer (1904), Käf. Mitteleur., IV, I, p. 222, 223—A. d'Orchymont

(1913), Ann. Biol. lac., VI, p. 212—Wesenberg—Lund (1915), Insektlivet, p. 292—Richmond (1920), Bull. Amer. Mus. Nat. Hist., XLII, Art. 1, p. 47, 83, 85, 87.

Subg. **Berosus** s. str. Hope.

Berosus Hope (1838), Col. Man. (2) p. 128—Thomson G. C. (1859), Skand. Col., I, p. 17; (1860), II, p. 86—Horn G. H. (1873), Proc. Am. Phil. Soc., XIII, p. 119—Bedel (1881), Fne Col. Bass. Seine I, p. 88; (1885), Ann. Soc. Linn. Lyon, XXXI, p. 340—Seidlitz (1888), Fna Balt, ed. 2, Gatt. p. 22, 111; (1888), Fna Transsylv., Gatt. p. 22, 118—Everts (1899), Col. Neerl., I, p. 639, 651—Ganglbauer (1904), Käf. Mitteleur., IV, I, p. 223—Reitter (1909), in Brauer., Süsswasserfna Deutschl., 3-4, p. 67; (1909), Fna Germ., II, p. 358—Schaufuss C. (1916). in Calwer Käzferb., ed. 6 (1913), p. 334—Des Gozis (1916), Misc. Ent., XXIII, p. 13.

paraberosus + *Berosus* Kuwert (1890), Verh. Naturf. Ver. Brünn, XXVIII, p. 20, 113, 114.

Euberosus Acloque (1896), Fn. Fr., p. 94.

seneiceps Motschulsky (1861), Bull. Soc. Nat. Mosc., XXXIV, I, p. 110[1]—A. Orchymont (1923), Mem. Dep. Agr. India, VIII, p. 12.[2] — Ceylon.[1] India.[2] Obock.[2,3]

immaculicollis Fairmaire (1892), Rev. Ent. Caen, XI, p. 88.[3]

nigropictus Règimbart (1903), Ann. Soc. Ent. Fr., LXXII, p. 60. — Indo-China : Saigou.

pulchellus W. M'Leay (18250, Annul. Jav., p. 35[1]—A. d'Orchymont (1923), Mem. Depart. Agric. India, VIII, p. 11[2]; (1923), Treubia, III, p. 421.[3] — Java.[1,3], Sumatra.[5] Borneo.[3] Philippine I.[3,4,5] Formosa.[3] Australia.[3,9] Indo-China : Annam.[5] Tonkin.

pubescens Mulsant & Rey (1859), Opusc. Entom. IX, p. 61[4]—Règimbart (1903), Ann. Soc. Ent. Fr., LXXII, p. 60.[5]

decrescens Walker (1859), Ann. Mag. Nat. Hist. (3), III, p. 258[6]—Sharp (1890), Trans. Ent. Soc. Lond. p. 353[7]—Règimbart (1903), Ann. Soc. Ent. Fr., LXXII, p. 61.[8]

India.[2,5,8]

devisi Blackburn (1898), Trans. R. Soc. S. Austr., XXII, 2, p. 225 (*Hygrotrophus*).[9] — Burma.[8] Ceylon.[3, 6, 7, 8]

Subg. **Enoplurus** Hope.

Enoplurus Hope (1838), Col. Man., II, p. 128—G. H. Horn (1873), Proc. Amer. Phil. Soc. XIII, p. 119—Bedel (1881) Fne Col. Bass. Seine, I, p. 303—Rey (1883), Rev. Ent. Caen, II, p. 90; (1885), Ann. Soc. Linn. Lyon, XXXI, p-340—Seidlitz (1888), Fna Balt., ed. 2, Gatt. p. 22; (1888), Fne Transsylv., Gatt., p. 22—Acloque (1896), Fne Fr., p. 94—Everts (1899), Col. Neerl., I, p. 639, 650—Ganglbauer (1904), Käf. Mitteleur., IV, I, p. 223—Reitter (1909), in Brauer, Süsswasserfna Deutschl, 3-4, p. 67; (1909), Fna Germ., II, 358—Kuhnt (1913), Ill. Best. Tab. Käf. Deutschl., p. 429, f. 70—A d'Orchymont (1913), Suppl. Ent., II, p. 2—Schaufuss C. (1916), in Calwer, Käferb., ed. 6 (1913), p. 334—Des Gozis (1916), Mis. Ent., XXIII, p. 13.

Anchialus G. C. Thomson, (1859), Skand. Col., I, p. 17; (1860), II, p. 86, 87; (1868) X, p. 309, 310.

Acanthoberosus Kuwert (1888), Deuts. Ent. Zeits. p. 130; (1890), Verh. Naturf. Ver. Brünn., XXVIII, p. 20, 107.

bowringi A. d'Orchymont (1925), Ann. & Bull. Soc. Ent. Belg., LXV, p. 168. — Fenapore (Dacca).

fairmairei Zaitzev (1908), Hor. Soc. Ent. Ross. XXXVIII, p. 355 (Catal.)[1]—A. d'Orchymont (1913), Suppl. Ent., II, p. 3, 4, t. I, f. 4.[2]

acutispina Fairmaire, (1888), Ann. Soc. Ent. Fr., (6), VIII, p. 336 (nec *acutispina* Fairmaire, (1869).

Indo-China : Tonkin[1, 2, 3] Laos, Annam. Siam. F. M. S. (Selangor). India : Bihar. Sumatra., Java. Borneo. Philippine I.

indicus Motschulsky (1861), Bull. Soc. Nat. Mosc., XXXIV, I, p. 110[1]—Règimbart (1902), Ann. Soc. Ent. Fr., LXXI, p. 473;[2] (1903), LXXII, p. 60[3]—A. d'Orchymont (1913), Suppl. Ent. II, p. 4, T. I, f. Ia, 3 ♀, 4 ♂.[4] — Ceylon.[1, 2, 3] India.[3], Burma.[3]

Siam. Indo-China :[3] Tonkin.

indicus Motschulsky (1861),—(*contd.*)

Laos, Annam. Cambodia.[3], F. M. S. Sumatra., Java.[3] Philippine I. Formosa.[4] China.[3]

47. Règimbartia Zaitzev.

Règimbartia Zaitzev (1908), Hor. Soc. Ent. Ross., XXXVIII, p. 362 (Catal.) —Jakobson (1911), Käf. Russl., p. 849—A. d'Orchymont (1913) Suppl. Ent., II, p. 5, nota 3; (1919), Ann. Soc. Ent. Fr., LXXXIII, p. 166.

Volvulus Brullé (1835), Hist. Nat. Ins., V, 2, p. 282 (nec *Volvulus*) Oken, 1915). *Spheroides* Hope ex. p. (1838), Col. Man., II, p. 168 (—*Globaria*).

Brachygaster Mulsant (1853), Ann. Soc. Linn. Lyon (2), I, p. 381; (1853), Opusc. Entom., II, p. 173 (nec Leach (1817).

attenuata Fabricius (1801), Syst. Eleut., I, p. 253[1] (*Hydrophilus*)—Erichson (1837), Käf. Mark Brandebg., I, p. 205 (*Berosus*)—A. d'Orchymont (1913), Suppl. Ent., II. p. 4.[2]

aenea Brulle (1835), Hist. Nat. Ins., V, 2, p. 288[3] (*Volvulus*).

metallescens Mulsant (1853), Ann. Soc. Linn. Lyon, (2) I, p. 386; (1853), Opusc. Entom, II, p. 178[4] (*Brachygaster*).

indica Mulsant (1853), l. c., p. 386; (1853), l. c. p. 179 (*Brachygaster*).[5]

Profunda Sharp (1873), Trans. Ent. Soc. London, p. 62 (*Volvulus*).[6]

scaphiformis Fairmaire (1878), Journ. Mus. Godeffroy, XIV, p. 83. (*Volvulus*).[7]

Ceylon.[1, 3] India.[4, 5] Indo-China : Annam. Cochin-China. Cambodia. Sumatra. Philippine I. Formosa.[2] Japan.[6] Australia :[7] Queensland.

48. Globaria Latreille.

Globaria Latreille (1829), in Cuvier, Règne Anim., n. ed. IV, p. 521—Laports de Castelnau (1840), Hist. Nat. Anim. Art., II, p. 57—Chenu (1851), Encycl.

Hist. Nat. Col., I, p. 241—Lacordaire (1854), Gen. Col., I, p. 455, 460—A. d'Orchymont (1919) Ann. Soc. Ent. Fr., LXXXVIII, p. 166.

Spheroides Hope ex. p. (1838), Col. Man., II, p. 168 (—*Volvulus*).

? *Allocotocerus* Kraatz (1883), Deuts. Ent. Zeits., XXVII, p. 14.

leachi Hope (1838), Col. Man., II, p. 167, Pl. 3, f. 6 (*Globaria Spheroides*)[1]—Sharp (1890), Trans. Ent. Soc. London, p. 354[2]—A. d'Orchymont (1919), Ann. Soc. Ent. Fr., LXXXVIII, p. 167.[3]	India Or.[1, 4, 5] Ceylon.[2], Bengal.[3] Madras.[3]
striatopunctata Laporte de Castelnau (1840), Hist. Nat. Anim. Art. II, p. 57[4]—Chenu (1851), Encycl. Hist. Nat. Col., I, p. 242 f. 414.[5]	Cambodia., Annam. Tonkin., F. M. S.
mülleri Kirsch (1875) Mittheil. Zool. Mus. Dresden, I, p. 27.[6]	Malacca.[6], Sumatra.

ADDENDA.

[Knisch, A. (1927) Spolia Zeylanica, XIV, 1 pp. 129—133.

15. **Cercyon** [Leach] Mulsant.

aviarius Knisch (1927) *loc. cit.*, p. 129, (*cercyon* s. str.)	Ceylon.

35. **Laccobius** Erichson.

affinis Knisch (1927) *loc. cit.*, p. 132 (*Laccobius* s. str.).	Kumaon, Gori Valley,
championi Knisch (1927) *loc., cit.*, p. 132 (*Laccobius* s. str.) ♂	Kumaon, Shelshel.

C. F. C. B.]

PART XV – CECIDOMYIDAE.

Howlett, Ind. Ins. Life 580-83, Tab. lx, figs. 2-2*b*, [General acct.]; Kieff., Ann. Soc. Sci. Brux., xxix, 143-200, Tab. and 15 t. figs., [Bengal species]; *id., loc. cit.*, xxxiii, [Asiatic species]; Kert., Cat. Dipt., II, 141-154, [species tabulated under host plants]; Will., N. Amer. Dipt., 117-130, figs. 37, 38, [key to N. A. genera]; Kieff., Gen. Ins. Facs. 152, [Revision of genera of World]; Felt, Mem. Dept. Agric. Ind., vii, 16-17, [List of Indian species affecting grasses]; Rao, Proc. Vth Ent. Meetg. Pusa, 272-3, [List of Indian species affecting grasses].

CECIDOMYINAE

1. **Lasioptera** Meigen.

Lasioptera, Mg., Syst. Besch., I, 88, (1818), [Genotype *C. albipennis* Mg.]; Schin., F. Auster., II, 406; Kieff., Gen. Ins. Fasc. 152, 30; Brun., R. I. M., XVII, 8; Will., N. Amer. Dipt., fig. 38, 1, [wing]; Winn., Moneg Gällmuck., 191; O.-S., Mon. N. Amer, Dipt., I, 175.

Dionyza, Meg. in Meig., Syst. Besch., I, 89, (1818), [Genotype *L. picta* Mg.].

bryoniae Schiner.

Lasioptera bryoniæ, Schin., Novara Reise Dipt., 5, Tab. I, 2, 3, (1868)[1]; Wulp, Cat. Dipt. S. Asia, 29[2]; Kert., Cat. Dipt., II, 16[3]; Kieff., Gen. Ins. Fasc. 152, 31[4]; Brun., R. I. M., XVII, 9.[5] — Madras.[1, 2, 3, 4, 5]

eriochloa Felt.

Lasioptera eriochola,, Felt, Mem. Dept. Agric. Ind., IX, 244 (1926).[1] — Coimbatore.[1]

fluitans Felt.

Lasioptera fluitans, Felt, Ent. News, XXVIII, 73, (1917); *id.*, Pusa Bull., 89, 42,[1] [redescr.]; Brun., R. I. M., XVII, 9.[2] — Coimbatore.[1, 2]

textar Kieffer.

Lasioptera textor, Kieff., Ann. Soc. Sci. Brux., XXIX, 155, (1905); *id.,* Gen. Ins. Fasc. 152, 33[1]; Brun., R. I. M., XVII, 9.[2] — Bengal.[1, 2]

2. Panteliola Kieffer.

Panteliola, Kieff., Bull. N. H. S. Metz, XXVIII, 49, (1913). [Genotype *R. haasi* Kieff.]; *id.,* Gen. Ins. Fasc. 152, 46, Tab. VI, 9, Tab. VIII, 3; Brun., R. I. M., XVII, 9.

hassi, Kieffer.

Rhopalomyia haasi, Kieff., Ann. Soc. Sci. Brux., XXIX, 151, (1905); — E. Indies.[1] Bengal.[2]

Panteliola haasi, Kieff., Gen. Ins. Fasc. 152, 47[1]; Brun., R. I. M., XVII, 9.[2]

3. Oligotrophus, Latreille.

Oligotrophus, Latr., Hist. Crust. Ins., XIV, 288, (1805), [Genotype *Tipula juniperinus* L.]; Rubs., Berl. Ent. Zeit., xxxvii, 376; Kert., Cat. Dipt., II, 71; Kieff., Gen. Ins. Fasc. 152, 49; Brun., R. I. M., XVII, 9.

indicus Kieffer.

? *Oligotrophus indianus,* Kieff., Marcellia, VII, 152, (1908),[1] [preocc.]. — E. Indies.[1,2]

Oligotrophus indicus, Kieff., *nom. nov.,* Gen. Ins. Fasc. 152, 50,[2] Brun., R. I. M., XVII, 9.

mangiferæ Kieffer.

Oligotrophus mangiferæ, Kieff., Marcellia, VII, 15 (1908); *id.,* Gen. Ins. Fasc. 152, 50[2]; Brun., R. I. M., XVII, 9. — E. Indies.[2]

quadrilobatus Kieffer.

Oligotrophus quadrilobatus, Kieff., Marcellia, VII, 151, (1908); *id.,* Gen. Ins. Fasc. 152, 50[2]; Brun., R. I. M., XVII, 9. — E. Indies.[2]

saligneus de Nicéville (?).

Oligotrophus saligneus, Nice., Ind. Mus. Notes, V, 115, Tab. xvi, 1, (1903)[1]; Howl., Ind. Ins. Life, 582, fig. 377; Kieff., Gen. Ins. Fasc. 152, 50[2]; Brun., R. I. M., XVII, 9. — W. Himalaya (9,000 ft.) E. Indies.[2]

tenuispatha Kieffer.

Oligotrophus tenuispatha, Kieff., Margcellia, XV, 150, (1908); *id.,* Gen. Ins. Fasc. 152, 50[2]; Brun., R. I. M., XVII, 9. — E. Indies.[2]

4. Trichoperrisia Kieffer.

Trichoperrisia, Kieff., R. I. M., IX, 199, (1913), [Genotype *T. pipericola* Kf.]; *id.,* Gen. Ins. Fasc. 152, 65; Brun., R. I. M., XVII, 10.

pipericola Kieffer.

Trichoperrisia, pipericola, Kieff., R. I. M., IX, 199, (1913)[1]; *id.,* Gen. Ins. Fasc. 152, 65[2]; Brun., R. I. M., XVII, 10.[2] — Peradeniya.[1, 2] E. Indies.[2]

5. Harpomyia Felt.

Harpomyia, Felt, Can. Ent., XLVIII, 401, (1916), [Genotype *H. indica* Ft.]; Brun., R. I. M., XVII, 10.

indica Felt.

Harpomyia indica Felt, Can. Ent., XLVIII, 401, (1916)[1]; Brun., R. I. M., XVII, 10.[2] — ? Pusa.[1, 2]

6. Dasyneura Rondani.

Dasineura, Rond., Sopra alcuni nuovi generi d'Insetti Ditteri, Parma, p. 17, (1840), [Genotype *D. obscura* Rond.].

Dasyneura, Rond., Nuov. Ann. Sci. Nat. Bologna, (2) VI, 12, [emend.]; Aldr., Cat. N. Amer. Dipt., 155; Kieff., Gen. Ins. Fasc. 152, 86; Brun., R. I. M., XVII, 10.

gossypii Felt.

Dasyneura gossypii, Felt, Can. Ent., XLVIII, 29, (1916)[1]; Brun., R. I. M., XVII, 10[2]; Pusa Bull., 89, 42. — S. India.[1, 2] Coimbatore.[3]

Contarinia sp., Fletcher, S. Ind. Ins., 363-64, fig. 223.[3]

mangiferae Felt.

Dasyneura mangiferae, Felt, Ind. Agr. Ent. Mem. X. 1 (1927).[1] — Coimbatore.[1]

7. Schizomyia Kieffer.

Schizomyia, Kieff., Ent. Nachtr., XIV, 183, (1889), [Genotype *S. galii* Kieff.]; *id.,* Gen. Ins. Fasc. 152, 88; Brun., R. I. M., XVII, 10.

assamensis Felt.

Schizomyia assamensis, Felt. Mem. Dept. Agric. Ind., VII, 3, (1920).[1] — Shillong.[1]

incerta Kieffer.

Schizomyia incerta, Kieff., Marcellia, VII, 153, (1908); Brun., R. I. M., XVII, 10.[2] — E. Indies.[2]

indica Kieffer.

Schizomyia indica, Kieff., Marcellia, VII, 154, (1908); *id.,* Gen. Ins. Fasc. 152, 89[2]; Brun., R. I. M., XVII, 10. — E. Indies.[2]

mæruæ Felt.

Schizomyia maeruae, Felt, Mem. Dept. Agric. Ind., IX, 243, (1926).[1] — Coimbatore.[1]

8. **Asphondylia** Loew.

Asphondylia, Lw., Dipt. Beitr., IV, 21, (1850), [Genotype *Cec. sarothamni* Lw.]; Winn., Monog. Gallmück., 187; O.-S., Monog. N. Amer. Dipt., I, 176; *id.,* Trans. Am. Ent. Soc., (1869), 209; Schin., F. Austr., II, 395; Aldr., Cat. N. Am. Dipt., 156; Kief., Gen. Ins. Fasc. 152, 91; Brun., R. I. M., XVII, 10.

Cylindrocera, Lioy, Atti. Inst. Veneto, (3) IX, 503, (1863).

Phyllophaga, Rond., Dipt. Ital. Prodr., I, 199, (1856).

ipomoeæ, Felt.

Asphondylia ipomoeoe Felt, Mem. Dept. Agric Ind., IX, 243, (1926).[1] — Coimbatore.[1]

lantanæ Felt.

Asphondylia lantanoe, Felt, Mem. Dept. Agric. Ind., VII, 2, (1920).[1] — Kallar (Nilgiris).[1]

osbeckiæ Senior-White.

Asphondylia osbeckiœ, S.-W., Mem. Dept. Agric. Ind., VII, 107, Tab. XV, 1, p. 83, [note], (1922). — Shillong[1]

phyllanthi Felt.

Asphondylia phyllanthi, Felt, Mem. Dept. Agric. Ind., VII, 2, (1920).[1] — Maymyo.[1]

pongamiae Felt.

Asphondylia pongamioe, Felt, Mem. Dept. Agric Ind., VII, 24, (1921).[1] — Coimbatore.[1]

sesami Felt.

Asphondylia sesami, Felt, Can. Ent. XLVIII, 31, (1916)[1]; Brun., R. I. M., XVII, 11[2]; Pusa Bull. 89, 41. — S. India.[1, 2] Coimbatore.[2]

'Gingelly Gall Fly', Fletcher, S. Ind. Ins., 364, figs. 224-5.[3]

9. Daphnephila Kieffer.

Daphnephila, Kieff., Ann. Soc. Sci. Brux., XXIX, 144, (1905), [Genotype *D. haasi* Kieff.]; *id.,* Gen. Ins. Fasc. 152, 96; Brun., R. I. M., XVII, 11.

glandifex Kieffer.

Daphnephila glandifex, Kieff., Ann. Soc. Sci. Brux. XXIX, 144, (1905); *id.,* Gen. Ins. Fasc. 152, 96[2]; Brun., R. I. M., XVII, 11.[3] — Bengal.[2, 3]

haasi Kieffer.

Daphnephila haasi, Kieff., Ann. Soc. Sci. Brux., XXIX, 144, (1905)' *id.,* Gen. Ins. Fasc. 152, 96[2]; Brun., R. I. M., XVII, 11.[3] — Bengal.[2, 3]

linderae Kieffer.

Daphnephila linderoe, Kieff., Ann., Soc. Sci. Brux., XXIX, 144, (1905); *id.,* Gen. Ins. Fasc. 152, 96[2]; Brun., R. I. M., XVII, 11.[3] — Bengal.[2, 3]

10. Hallomyia Kieffer.

Hallomyia, Kieff., Spol. Zeyl., VIII, 25, (1912), [Genotype *H, iris* Kieff.]; *id.,* Gen. Ins. Fasc. 152, 106; Brun., R. I. M., XVII, 11.

iris Kieffer.

Hallomyia iris, Kieff., Spol. Zeyl., VIII, 25, figs. 1, 2, (1912)[1]; *id.,* Gen. Ins. Fasc. 152, 106[2]; Brun., R. I. M., XVII, 11.[3] — Peradeniya.[1, 3] Ceylon.[2]

11. Hormomyia Loew.

Hormomyia, Lw., Dipt. Beitr., IV, 20, 31, (1850), [Genotype *H. dubitata* Rub.]; Kieff., Gen. Ins. Fasc. 152, 137, Tab. VI, 7; Will., N. Amer. Dipt., fig. 34, 4, [wing.]

Angelinia, Rond., Atti. Soc. Sci. Milan, II, 290, (1860).

ischaemi Kieffer.

Oligotrophus ischoemi, Kieff., Kew Bull., III, 71, (1910); *id.,* Gen. Ins. Fasc. 152, 50[2]; Brun., R. I. M., XVII, 9, 271. — E. Indies.[2] Coimbatore.[2] Bellary.[4]

Hormomyia ischoemi, Pusa Bull. 89, 46-89, [redescr.], 56[3]; Rao, Proc. Vth Ento. Meetg. Pusa, 270-1, [notes], 272.[4]

subaptera Felt.

Hormomyia subaptera, Felt, Mem. Dept. Agric. Ind., IX, 244 (1926).[1] — Nilgiris (8,300 ft.).[1]

12. **Chrysodiplosis** Kieffer.

Chrysodiplosis, Kieff., Trans. Linn. Soc. Lond., XIV, 318, (1911), [Genotype *C. pulchricornis* Kieff.]; *id.,* Gen. Ins. Fasc. 152, 145, Tab. VIII, 11.

squamatipes Kieffer.

Chrysosdiplosis squamatipes, Kieff., Spol., Zeyl., VIII, 28,[1] (1912); Brun., R. I. M., XVII, 12.[2] — Peradeniya.[1, 2]

13. **Horidiplosis** Felt.

Horidiplosis, Felt, Mem. Dept. Agric. Ind., VII, 10, (1920), [Genotype *H. fici* Ft.].

fici Felt.

Horidiplosis fici, Felt, Mem. Dept. Agric. Ind., VII, 10, (1920).[1] — Pusa.[1]

14. **Orseolia** Kieffer & Massalongo.

Orseolia, Kf. & Mas., Marcellia, I, 56, (1902), [Genotype *O. cynodontis* Kf. & Mas.]; Kieff., Gen. Ins. Fasc. 152, 151.

cynodontis Kieffer & Massalongo.

Orseolia cynodontis, Kf. & Mas., Marcellia, I, 57, (1902); Kieff., Gen. Ins. Facs. 152, 152[2]; Pusa Bull. 89, 54[3]; Rao, Proc. Vth. Ento. Meetg. Pusa, 272.[4] — Italy.[2] Algeria.[2] S. India. (general).[3] Coimbatore.[4]

15. Orseoliella Kieffer.

Orseoliella, Kieff., Marcellia, XI, 231, (1912), [Genotype *O. javanica* Kieff.]; *id.,* Gen. Ins. Fasc. 152, 152, Tab. VIII, 13, Tab. X, 8, 9; Brun., R. I. M., XVII, 15.

apludæ Felt.

Orseoliella apludœ, Felt, Mem. Dept. Agric. Ind., VII, 8, (1920)[1]; Rao, Proc. Vth. Ento. Meetg. Pusa, 272.[2] — Coimbatore.[1] S. India.[2]

graminis Felt.

Orseoliella graminis, Felt, Mem. Dept. Agric. Ind., VII, 22, (1921)[1]; Rao, Proc. Vth. Ento. Meetg. Pusa, 272. — Kistna D.[1]

javanica Kieffer & Leeuw.

Orseolia javanica, Kieff., Marcellia, VIII, 125,[1] (1909). — Java.[1, 2, 3]

Orseoliella javanica, Kieff., Marcellia, XI, 231; *id.,* Gen. Ins. Fasc. 152, 152[2]; Brun., R. I. M., XVII, 15[3]; Felt, Mem. Dept. Agric. Ind., VII, 16[4]; Rao, Proc. Vth. Ento. Meetg., 272. — Kistna D.[4]

16. Feltodiplosis Kieffer.

Feltodiplosis, Kieff., Gen. Ins. Fasc. 152, 154, (1913), [Genotype *F. abdominalis* Felt]; Brun., R. I. M., XVII, 12.

Endaphis, Felt, Ent. News, XXII, 128-224, (1911), [preoecc.].

hirta Felt.

Endaphis hirta, Felt, Ent. News, XXII, 224, (1911).[1] — Ceylon.[1, 2, 3]

Feltodiplosis hirta, Kieff., Gen. Ins. Fasc. 152, 155[2]; Brun., R. I. M., XVII, 12.[3]

17. Arthrocnodax Rübsaamen.

Arthrocnodax, Rübs., Wien. Ent. Zeit., XIV, 189, (1895), [Genotype *A, vitis* Rüb.]; Kieff., Gen. Ins. Fasc. 152, 155, Tab. V, 8, 9; Brun., R. I. M., XVII, 12.

rutherfordi Felt.

Arthrocnodax rutherfordi, Felt, Jo. N. Y. Ent. Soc., XXIII, 180[1], (1915); Brun., R. I. M., XVII, 12[2]. — Ceylon.[1, 2]

walkeriana Felt.

Arthrocnodax walkeriana, Felt, Jo. N. Y. Ent. Soc., XXIII, 181,[1] (1915); Brun., R. I. M., XVII, 12[2]. — Ceylon.[1, 2]

18. **Plutodiplosis** Kieffer.

Plutodiplosis Kieff., Spol. Zeyl., VIII, 27, (1912); [Genotype *P. magnifica* Kieff.]; *id.*, Gen. Ins. Fasc. 152, 166; Brun., R. I. M., XVII, 12.

magnifica Kieffer.

Plutodiplosis magnifica, Kieff.: Spol. Zeyl., VIII, 27, (1912)[1]; *id.*, Gen. Ins. Fasc. 152, 166[2]; Brun., R. I. M., XVII, 12.[3] — Peradeniya.[1, 3] Ceylon.[2]

19. **Indodiplosis** Felt.

Indodiplosis, Felt, Can. Ent., XLVIII, 403, (1916), [Genotype *I. mangiferoe* Ft.]; Brun., R. I. M., XVII, 13.

mangiferae Felt.

Indodiplosis mangiferoe, Ft., Can. Ent., XLVIII, 403, (1916)[1]; Brun., R. I. M., XVII, 13.[2] — Pusa.[1, 2]

20. **Streptodiplosis** Felt.

Streptodiplosis, Ft., Can. Ent., XLVIII, 405, (1916), [Genotype *S. indica* Ft.]; Brun., R. I. M., XVII, 13.

indica Felt.

Streptodiplosis indica, Ft., Can. Ent., XLVIII, 405, (1916)[1]; Brun., R. I. M., XVII, 13.[2] — Kusti (Wynaad).[1, 2]

21. **Dentifibula** Felt.

Dentifibula, Ft., Bull. N. Y. Mus., CXXIV, 389, (1908), [Genotype *Con. viburni* Ft.]; Kieff., Gen. Ins. Fasc. 152, 174; Brun., R. I. M., XVII, 13.

ceylanica Felt.

Dentifibula ceylanica, Ft., Jo. N. Y. Soc., XXIII, 175, (1915)[1]; Brun., R. I. M., XVII, 12.[2] — Ceylon.[1, 2]

obtusilobae Felt.

Dentifibula obtusiloboe, Felt, Jo. N. Y. Ent. Soc., XXIII, 176, (1915)[1]; Brun., R. I. M., XVII, 13.[2] — Ceylon.[1, 2]

22. **Contarinia** Rondani.

Contarinia, Rond., Atti. Soc. Sci. Nat. Milan, II, 287, (1860), [Genotype, *C. loti* DeG.]; Kert., Cat. Dipt., II, 91; Kieff., Gen. Ins. Fasc. 152, 178; Brun., R. I. M., XVII, 270.

Eudiplosis, Kieff., Bull. Soc. Ent. Fr., (1894), 28 (1894).

andropoginis Felt.

Contarinia andropoginis, Felt, Mem. Dept. Agric. Ind., VII, 25, (1921)[1]; Ballard, *loc. cit.*, VII, i, ii, figs., [bionomics]; *id.*, Proc. IVth. Ento. Meetg. Pusa, 22, Tab. I, [all stages]. — Coimbatore.[1]

caudata Felt.

Contarinia caudata, Ft., Mem. Dept. Agric. Ind., VII, 4, (1920).[1] — Coimbatore.[1]

23. **Stictodiplosis** Kieffer.

Stictodiplosis Kieff., Bull. Soc. Ent. Fr., LXIII, 28, (1894), [Genotype *S. nubilipennis* Kf.]; *id.*, Gen. Ins. Fasc. 152, 183, Tab. IV, 14; Brun., R. I. M., XVII, 13.

pulcherrima Kieffer.

Contarinia pulcherrima, Kieff., R. I. M., III, 130, figs.[1] — Kurseong.[1, 3]

Stictodiplosis pulcherrima, Kieff., Gen. Ins. Fasc. 152, 183,[2] [*pulchripennis, laps.*]; Brun., R. I. M., XVII, 13.[3] — E. Indies.[2]

24. **Procontarinia** Kieffer & Cecconi.

Procontarinia, Kf. & Cec., Marcellia, V, 135, (1906), [Genotype *P. matteiana* Kf. & Cec.]; Kieff., Gen. Ins. Fasc. 152, 183; Brun., R. I. M., XVII, 13.

matteiana Kieffer & Cecconi.

Procontarinia matteiana, Kf. & Cec., Marcellia, V. 135, (1906); Kieff., Gen. Ins. Fasc. 152, 183[2]; Brun., R. I. M., XVII, 13.[3] — India.[2, 3]

25. **Oribremia** Kieffer.

Oribremia, Kieff., R. I. M., IX, 199, (1913), [Genotype *O. multifida* Kf.]; *id.*, Gen. Ins. Fasc. 152, 194, Tab. VIII, 9; Brun., R. I. M., XVII, 14.

multifida Kieffer.

Oribremia multifida, Kieff., R. I. M., IX, 199, (1913)[1]; *id.,* Gen. Ins. Fasc. 152, 194[2]; Brun., R. I. M., XVII, 14.[3] — Bhowali (Kumaon).[1,3] C. Himalaya.[2]

26. Lestodiplosis Kieffer.

Lestodiplosis, Kieff., Bull. Sec. Ent. Fr., LXIII, 28, (1894), [Genotype *L. alternans* Kieff.]; *id.,* Gen. Ins. Fasc. 152, 195; Will., N. Amer. Dipt., fig. 38, 14, [wing]; Brun., R. I. M., XVII, 14.

Coprodiplosis, Kieff., Feuille Jeune Natural., XXIV, 84, Tab. XIV, 13, (1894);

Hemidiplosis, Kieff., Ann. d. Sci. Nat., I, 9, (1894).

ceylanicus Kieffer.

Lestodiplosis ceylanicus, Kieff., Spol. Zeyl., VIII, 28[1]; (1912); Brun., R. I. M., XVII, 14.[2] — Peradeniya.[1,2]

27. Diplecus Kieffer.

Diplecus, Kieff., Bull. N. H. S. Metz. XXVIII, 55, (1913), [Genotype *Cop. inconspicuus* Kf.]; *id.,* Gen. Ins. Fasc. 152, 201; Brun., R. I. M., XVII, 14.

inconspicuus Kieffer.

Coprodiplosis inconspicuus, Kief., Spol. Zeyl., VIII, 29,[1] (1912); — Peradeniya.[1,3] Ceylon.[2]

Diplecus inconspicuus, Kieff., Gen. Ins. Fasc. 152, 201[2]; Brun., R. I. M., XVII, 14.[3]

28. Dyodiplosis Rübsaamen.

Dyodiplosis, Rüb., Zeit. wiss. Insek., VIII, 49, (1912), [Genotype *D. arenarioe* Rüb.]; Kief., Gen. Ins. Fasc. 152, 205; Brun., R. I. M., XVII, 14; Felt, Mem. Dept. Agric. Ind., VII, 17, 18, [key to Indian spp.].

andropogonis Felt.

Dyodiplosis andropogonis, Felt, Ent. News, XXVIII, 369, (1917); Pusa Bull. 89, 49, 55[2]; Felt, Mem. Dept. Agric. Ind., VII, 15[3]; Rao, Proc. Vth. Ento. Meetg. Pusa, 272, 273[4]. — Adoni (Ballary D.).[2,3] Coimbatore.[3] Samalkota.[3] Kistna Dist.[3] S. India.[4]

cornea Felt

Pseudhormomyia cornea, Felt, Ent. News, XXVIII, 76, (1917)[1]; Brun., R. I. M., XVII, 15[2]; Pusa Bull. 89, 45, [redescr.].[3] — Coimbatore.[1, 2, 3, 4]

Dyodiplosis cornea, Pusa Bull. 89, 54; Brun., R. I. M., XVII, 271; Rao, Proc. Vth Ento. Meetg. Pusa, 272.[4]

fluvialis Felt.

Pseudhormomyia fluvialis, Felt, Ent. News, XXVIII, 74, (1917)[1]; Brun., R. I. M., XVII, 15,[2] [*fluviatis laps.*]. — Coimbatore.[1, 2, 3, 4]

Dyodiplosis fluvialis, Brun. R. I. M., XVII, 271; Pusa Bull. 89, 43, 53, [redescr.][3]; Rao, Proc. Vth, Ento. Meetg. Pusa, 273.[4]

generosi Felt.

Dyodiplosis generosi, Felt, Jo. N. Y. Ent. Soc., XXIII, 183, (1915)[1]; Brun., R. I. M., XVII, 15.[2] — Ceylon.[1, 2]

indica Felt.

Dyodiplosis indica, Felt, Mem. Dept. Agric. Ind., VII, 19, (1921).[1] — Ballary D.[1]

monticola Felt.

Dyodiplosis monticola, Felt, Mem. Dept. Agric. Ind., VII, 20, (1921)[1]; Rao, Proc. Vth. Ento. Meetg. Pusa, 272. — Ballary D.[1]

plumosa Felt.

Dyodiplosis plumosa, Felt, Mem. Dept. Agric. Ind., VII, 21, (1921)[1]; Rao, Proc. Vth. Ento. Meetg. Pusa, 272.[2] — Samalkota.[1, 2] Kistna D.[1, 2]

29. **Löwiola** Kieffer.

Löwiola, Kieff., Miscell. Ent., V, 5, (1896), [Genotype *L. centaureœ* Lw.]; *id.,* Gen. Ins. Fasc. 152, 206, Tab. XIII, 8; Brun., R. I. M., XVII, 15.

costata Felt.

Löwiola costata, Felt, Jo. N. Y. Ent. Soc., XXIII, 182 (1915)[1]; Brun., R. I. M., XVII, 15.[2] — Ceylon.[1, 2]

30. **Cecidomyia** Meigen.

Cecidomyia, Mg., Illig. Mag., II, 261, (1803), [Genotype *Tipula pini* DeG.]; *id.,* Klassif., 1804; O.-S., Mon. N.

Amer. Dipt., I, 175; *id.*, Ent. Mo. Mag., (2) XII, 40; Karsch, Rev. Gallmück., 19; Rubs., Berl. Ent. Zeit., XXXVII, 324; Will., N. Amer. Dipt., fig. 38, 3; Aldr., Cat. N. Amer. Dipt., 159; Sch., F. Austr., II, 569; Kieff., Gen. Ins. Fasc. 152, 211, Tab. XIII, 12; Brun., R. I. M., XVII, 15; Kert., Cat. Dipt., II, 106.

Cryptodiplosis, Kieff., Bull. Soc. Ent. Fr., LXIV, 194, (1895);

Diplosis, Lw., Dipt., Beitr., IV, 20, (1850).

artocarpi Felt.

Cecidomyia artocarpi, Felt, Mem. Dept. Agric. Ind., VII, 27, (1921)[1]; — Godavari.[1]

malabarensis

Cecidomyia malabarensis, Felt, Ind. Agr. Ent. Mem. X. 2-3 (1927).[1] — Taliparamba.[1]

penniseti Felt.

Cecidomyia penniseti, Felt, Mem. Dept. Agric. Ind., VII, 11, (1920).[1] — Coimbatore.[1]

31. Raodiplosis Felt.

Raodiplosis, Felt, Mem. Dept. Agric. Ind., VII, 6, (1920), [Genotype *R. orientalis* Felt].

orientails Felt.

Raodiplosis orientalis, Felt, Mem. Dept. Agric. Ind., VII, 7, (1920).[1] — Thaton (Burma).[1]

32. Pachydiplosis Kieffer.

Pachydiplosis, Kieff., Bull. N. H. S. Metz., XXVIII, 108, (1913), [Genotype *P. apricans* Kf.]; *id.,* Gen. Ins. Fasc. 152, 224; Brun., R. I. M., XVII, 15.

ceylanica Kieffer.

Clinodiplosis ceylanica, Kf., Spol. Zeyl., VIII, 26, fig. 3, (1912).[1] — Peradeniya.[1,3] Ceylon.[2], Pusa.[3]

Pachydiplosis ceylanica, id., Gen. Ins. Fasc. 152, 224[2]; Brun., R. I. M., XVII, 16.[3]

graminicola Kieffer & Massalongo.

Clinodiplosis graminicola, Kf. & Mass., Marcellia, VIII, 127, (1909)[1]; — Java.[1,2,3] Kistna D.[4,5]

Pachydiplosis graminicola, Kieff., Gen. Ins. Fasc. 152, 224[2]; Brun., R. I. M., XVII, 16[3]; Rao, Proc. Vth. Ento. Meetg. Pusa, 272[4]; Felt, Mem. Dept. Agric. Ind., VII, 16.[5]

oryzæ Wood-Mason.

Cecidomyia oryzæ, W.-M., Amer. Nat., 1881, 149, (1881); *id.,* Ind. Mus. Notes, I, 103, Tab. VI, 6[1]; Wulp., Cat. Dipt. S. Asia, 29; Howl., Ind. Ins. Life, 581; Kert., Cat. Dipt., II, 115[2]; Kieff., Gen. Ins. Fasc. 152, 217.[3]

Monghyr.[1], E. India.[2] Bengal.[6, 3, 2] Kistna D.[4], Cuttack.[5] Balasore.[5]

Pachydiplosis oryzæ, Felt, Mem. Dept. Agric. Ind., VII, 16[4]; Rao, Jo. A. S. B., n. s., XIII, 184; Misra, Mem. Dept. Agric. Ind., V, 231[5]; Brun., R. I. M., XVII, 16[6]; Proc. IId. Ento. Meetg., Pusa, 169, Tab.[7]; Ballard; *loc. cit.,* 21[8], IVth. Meetg., 21,[8] [thermotropism]; Ghosh, *loc. cit.,* 115, [life-history]; Rao, *loc. cit.,* Vth. Meetg., 270, 273, [notes]; Ayyar, *loc. cit.,* IVth. Meetg., 53, Tab. VII, [dates in Madras].[9]

Ranchi.[7] Coimbatore.[8] N. Kanara.[9] N. Arcot.[9] Tanjore. D.[9] S. E. Madras.[9]

33. **Clinodiplosis** Kieffer.

Clinodiplosis, Kieff., Feuille Jeune Natur., Paris, XXIV, 121, (1894), [Genotype *C. cilicrus* Kieff.]; *id.,* Gen. Ins. Fasc., 152, 237, Tab. V, 6, 18, Tab. XIII, 1, Tab. XV, 14; Brun., R. I. M., XVII, 16.

artemisiarum Kieffer.

Clinodiplosis artemisiarum, Kieff., Marcellia, VII, 156, (1908); *id.,* Gen. Ins. Fasc., 152, 237[2]; Brun., R. I. M., XVII, 16.[3]

E. Indies.[2, 3]

cellularis Kieffer.

Clinodiplosis cellularis, Kieff., Marcellia, VII, 156, (1908); *id.,* Gen. Ins. Fasc. 152, 238[2]; Brun., R. I. M., XVII, 16.[3]

E. Indies.[2, 3]

nodifex Kieffer.

Clinodiplosis nodifex, Kieff., Mercellia, VII, 155, (1908); *id.,* Gen. Ins. Fasc. 152, 238[2]; Brun., R. I. M., XVII, 16.[3]

E. Indies.[2, 3]

34. **Androdiplosis** Felt.

Androdiplosis, Felt, Jo. N. Y. Ent. Soc., XXIII, 182, (1915). [Genotype *A. coccidivora* Felt.]; Brun., R. I. M., XVII, 16.

coccidivora Felt.

Androdiplosis coccidivora, Felt, Jo. N. Y. Ent. Soc., XXIII, 183, (1915)[1]; Brun., R. I. M., XVII, 16.[2] — Ceylon.[1, 2]

35. Diadiplosis Felt.

Diadiplosis, Felt, Jo. N. Y. Ent. Soc., XIX, 54, (1911), [Genotype *D, coccidivora* Felt]; Kieff., Gen. Ins. Fasc. 152, 240; Brun., R. I. M., XVII, 14.

coccidivora Felt.

Diadiplosis coccidivora, Felt, Entom., XLVII, 86, (1914)[1]; Rutherfd., Trop. Agric., XLIII, 50[2]; Brun., R. I. M., XVII, 14.[3] — Ceylon.[1, 2, 3]

indica Felt.

Diadiplosis indica, Felt, Mem. Dept. Agric. Ind., VII, 6, (1920).[1] — Pusa.[1]

35A. Schizobremia.*

Schizobremia,

malabarensis, Felt.

Schizobremia malabarensis, Felt, Ind. Agr. Ent. Mem. X, 1-2 (1927).[1] — Taliparamba.[1]

36. Mycodiplosis Rübsaamen.

Mycodiplosis, Rüb., Ent. Nacht., XXI, 186, (1895), [Genotype *D. coniophaga* Winn.]; Kieff., Gen. Ins. Fasc. 152, 241, Tab. V, 17; Brun., R. I. M., XVII, 16.

indica Felt.

Mycodiplosis indica, Felt, Mem. Dept. Agric. Ind., VII, 5, (1920).[1] — Coimbatore.[1]

36A. Lopesiella Tavares.

Lopesiella, Tav. Broteria, vii, 145, (1908), [Genotype, *L. combreti* Tav.]; Kieff., Gen. Ins. Fasc. 152, 252.

polliniae, Felt.

Lopesiella polliniae, Felt, Ind. Agr. Ent. Mem., X, 2-3 (1927)[1] — Taliparamba.[1]

---

* We are unable to trace this generic name, which does not appear in the *Zoological Record* up to 1925, the latest volume received, *T. B. F.*

37. **Xiphodiplosis** Felt.

Xiphodiplosis, Felt, Jo. N. Y. Ent. Soc., XXIII, 180, (1915); [Genotype *X. fulva* Felt]; Brun., R. I. M., XVII, 16.

fulva Felt.

Xiphodiplosis fulva, Felt, Jo. N. Y. Ent. Soc., XXIII, 180, (1915)[1] Brun., R. I. M., XVII, 16.[2] — Ceylon.[1, 2]

38. **Itonida** Meigen.

Itonida, Mg., Nouv. Class., 19, (1800); Kieff., Gen. Ins. Fasc. 152, 258.

This genus is invalid. To accept it involves accepting all the other genera erected in the same work, which in its turn means upsetting some of the best known genera in Diptera. It has had no genotype designated, and would appear to contain only a few recent species by Felt, probably referable to *Cecidomyia.*

penniseti Felt.

Itonida penniseti Felt, Mem. Dept. Agric. Ind., VII, 9, (1920).[1] — Coimbatore.[1]

seminis Felt.

Itonida seminis, Felt, Mem. Dept. Agric. Ind., VII, 26, (1921).[1] — Coimbatore.[1]

39. **Colpodia** Winnertz.

Colpodia, Winn., Linn. Ent. Stett., VIII, 188, (1853), [Genotype *C. angustipennis* Winn.]; Kieff., Gen. Ins. Fasc. 152, 268; Will., N. Amer. Dipt., fig. 38, 10, [wing]; Brun., R. I. M., XVII, 17.

fletcheri Felt.

Colpodia fletcheri, Felt, Can. Ent., XLVIII, 400, (1916)[1]; Brun., R. I. M., XVII, 17.[2] — Pusa.[1, 2]

40. **Dicroneurus** Kieffer.

Dicroneurus, Kieff., Ent. Nacht., XXI, 122, (1895), [Genotype *Ep. lineatus* Kf.]; *id.,* Gen. Ins. Fasc. 152, 270, Tab. XI, 25, Tab. XIV, 11; Brun., R. I. M., XVII, 16.

indicus Kieffer.

Epidosis indicus, Kieff., R. I. M., IX, 200, (1913)[1];	Kalighat. (6,000 ft.).[1, 3]
Dicroneurus indicus, Kieff., Gen. Ins. Fasc. 152, 270[2]; Brun., R. I. M., XVII, 17.[3]	E.Indies.[2]

41. **Misocosmus** Kieffer.

Misocosmus, Kieff., Bull. N. H. S. Metz, XXVIII, 55, (1913), [Genotype *E. ceylanicus* Kieff.]; *id.,* Gen. Ins. Fasc. 152, 276; Brun., R. I. M., XVII, 17.

ceylanicus Kieffer.

Epidosis ceylanicus, Kieff., Spol. Zeyl., VIII, 29, (1912).[1]	Peradeniya.[1, 3]
Misocosmus ceylanicus, Kieff., Gen. Ins. Fasc. 152, 276[2]; Brun., R. I. M., XVII, 17.[3]	Ceylon.[2]

42. **Camptomyia** Kieffer.

Camptomyia, Kieff., Ann. Soc. Ent. Fr., LXIII, 323, (1894); *id.,* Gen. Ins. Fasc. 152, 277; Tab. VII, 9, Tab. XIV, 3, Tab. VI, 1.

No genotype appears to have been indicated.

hibisci Felt.

Camptomyia hibisci, Felt, Mem. Dept. Agric. Ind., IX, 241, (1926).[1]	Coimbatore.[1]

morindae Felt.

Camptomyia morindoe. Felt, Mem. Dept. Agric. Ind., IX, 241, (1926).[1]	Coimbatore.[1]

ricini Felt.

Camptomyia ricini, Felt, Mem. Dept. Agric. Ind., VII, 23, (1921).[1]	Coimbatore.[1]

LESTREMINÆ.

Kieffer, Gen. Ins. Fasc. 152, 285, [Table of genera].

43. **Peromyia** Kieffer.

Peromyia, Kieff., Bull. Soc. Ent. Fr., 63, (1894). [Genotype *P. leveillei* Kf.]; *id.,* Gen. Ins. Fasc. 152, 292; Brun., R. I. M., XVII, 17.

bengalensis Kieffer.

Peromyia bengalensis, Kieff., Ann. Soc. Sci. Brux., XXIX, 158, fig. 4,[1] (1905); *id.,* Gen. Ins. Fasc. 152, 292[2]; Brun., R. I. M., XVII, 17.[3]	Bengal.[1, 3] E. Indies.[2]

44. **Lestremia** Macquart.

Lestremia, Macq., Rec. Soc. Sci. Ag. Lille, 173, (1826), [Genotype *L. cinerea* Macq.]; Kieff., Gen. Ins. Fasc. 152, 307; Will., N. Amer. Dipt., fig. 38, 6, [wing]; Kert., Cat. Dipt., II, 12; Brun., R. I. M., XVII, 17, 18.

Cecidogona, Lw., Stett. Ent. Zeit., V, 324, (1844).

Furcinerva, Rond., Nuov. Ann. Soc. Nat. Bologna, (2) VI, 369, (1846).

Mimosciara, Rond., Nuov. Ann. Soc. Nat. Bologna, (2) VI, 10, (1846); *id.,* Mem. la Ditt. Ital., 25, (1840).

Molobroea, Rond., Atti. Soc. Sci. Nat. Milan., II, 287, (1860).

Yposatoea, Rond., Dipt. Ital. Prodr., I, 198, (1856).

ceylanica Kieffer.

Lestremia ceylanica, Kieff., Spol. Zeyl. VIII, 29, (1912)[1]; Brun., R. I. M., XVII, 18.[2]	Peradeniya.[1, 2]

indica Kieffer.

Lestremia indica, Kieff., R. I. M., III, 129, fig. 1, (1909)[1]; *id.,* Gen. Ins. Fasc. 152, 308[2]; Brun., R. I. M., XVII, 18.[3]	Calcutta.[1, 3] Kurseong.[1, 3] E. Indies.[2]

PART XVI – COSMOPTERYGIDÆ.

Elachistidæ [*partim*], Stainton, Ins. Brit. Tineina, pp. 224-226 (1854) Stainton, Manual II 393-394 (1859); Meyrick, Handbook, pp. 638-640 (1895); Meyrick, Proc. Linn. Soc. N. S. W. XXII 297-299 (1897).

Lavernidae, Wocke, Hein. Schmett. Deutschl. II, ii, 415 (1876); Forbes, Lep. New York, pp. 318-322, figs. 180-194 (1924).

Batrachedrae, Wocke, Hein. Schmett. Deutschl. II ii. 524 (1876).

Momphidae, Spuler, Schmett. Eur. II 381 (1910).

Cosmopterygidae, Meyrick. P. E. S. 1909. 17 (1909); *id.*, Exot. Micr. I 339; *id.*, Tr. N. Z. Inst. XLVII 208 (1915).

1. Cosmopterix, Hübner 1926.

***Cosmopterix*, Hb., Verz. p. 424 (1826)**

Cosmopteryx, Zeller, Isis XXXII 210 (1839) [redescr.]; Dup., Cat. Meth. Lep. Eur. pp. 373-374 (1845) [redescr.]; Herr-Schäff., Schmett. Eur. V 54 (1853) [non-descr.]; Stainton, Ins. Brit. Tin. pp. 228-229, t. 7 ff. 4*a-h* (1854) [redescr.]; Stt., Manual II, 395 (1859) [redescr.]; Stt., Nat. Hist. Tin. XII, 2–9 (1870) [general remarks, biology, etc.]; Snellen, Vlind. Nederl. Microlep. pp. 867-868 (1882) [redescr.]; Wlsm., T. E. S. 1891. 124 (1891); Meyr., Handb. p. 662, fig. [neur] (1895) [redescr.]; Wlsm., P. Z. S. 1897, 105-107 (1897); Meyr., P. Linn. Soc. N. S. W. XXII 338 (1897) [redr.]; Dyar, Bull U.S. Nat. Mus. LII 534-535 (1902); Busck, Proc. U.S. Nat. Mux. XXX 707-713 (1906) [revision of American spp.]; Wlsm., P. Z. S. 1907. 964-967 (1908); Meyr., P. Z. S. 1908. 733-734 (1908); Meyr., B. J. XIX. 411-412 (1909) [Tabulation of Indian spp.]; Wlsm., Biol. Centr. Am., Het. IV 4 (1909); Spuler, Schmett. Eur. II 389, f. 152 [neur., *druryella* Z.] (1910) [redescr.] Forbes, Lep. N. York p. 331, f. 185 (1924) [redescr.]

Type: *zieglerella*, Hb. [Europe].

aculeata, Meyr. 1909.

Cosmopteryx aculeata, Meyr., B. J. XIX 419-420 (1909) ♂ ♀.[1]

Maskeliya[1]
Fort Stedman[1] (Burma)
Khasis[1]
Cherrapunji.
Shillong.

anadoxa, Meyr. 1909.

Cosmopteryx anadoxa, Meyr., B.J. XIX 412 (1909) ♀[1]; Meyr., Exot. Micr. II 567 (1922) [descr. revised].[2]

Nilgiris (3,500 ft.)[1]
N. Coorg.[2]
Kodaikanal.[2]

ancalodes, Meyr. 1919.

Cosmopteryx ancalodes, Meyr., Exot. Micr. II 283 (1919) ♂[1]; *id., t. c.* 411 (1921) [descr. ♂♀ extended].[2]

Shillong.[1]

artemidora, Meyr. 1909.

Cosmopteryx artemidora, Meyr., B. J. XIX 415 (1909)♂[1]

Pollibetta (S. Coorg).
N. Coorg.[1]
N. Kanara.
Shillong.

artifica, Meyr. 1909.

Cosmopteryx artifica, Meyr., B. J. XIX 421 (1909). ♂[1]

Diyatalawa.[1] (Ceylon; 4000 ft.)
Dharwar.

asiatica, Stainton 1859.

Cosmopteryx asiatica, Stainton, T. E. S. (n. s.) V 122-123 (1859)[1]; Cotes and Swinlioe Cat. Moths India, p. 722, No. 4937 (1889)[2]; Meyr., T. E. S. 1894. 23 (1894)[3]; Meyr., B. J. XIX 411 (1909)[4]; Meyr., Rec. Ind. Mus. V 223 (1910)[5].

Calcutta.[1, 5]
Fort Stedman[3] (Burma).
Shillong.

bambusae, Meyr. 1917.

Cosmopteryx bambusae, Meyr., Ent. Mo., Mag. LIII 258 (1917) ♂♀[1]; Fletcher, Rept., third Entl. Meeting I 150 (Nov. 1920)[2]; Fletcher, Ind. Agr. Ent. Mem. VI 102-102, t. 23. f. 1 (Jan. 1921).[3]

Pusa.[1, 3]

Larva descr.[3] ; mining leaves of bamboo.[1-3]

basilisca, Meyr. 1909.

Cosmopteryx basilisca, Meyr., B. J. XIX, 416-417 (1909) ♂[1]; Meyr., Rec. Ind. Mus. V 223 (1910).[2]

Puttalam.[1], Sukna.[2] (E. Him.; 500ft.)

basilisca, Meyr. 1909.—

Gurdaspur.

———

Philippines.

belonacma, Meyr. 1909.

Cosmopteryx belonacma, Meyr., B. J. XIX, 416 (1909) ♂[1].

Khasis.[1]
Shillong., Pusa.

calypso, Meyr. 1919.

Cosmopteryx calypso, Mery., Exot. Micr. II 283-284 (1919) ♀[1].

Cherrapunji.[1]

catharacma, Meyr. 1909.

Cosmopteryx catharacma, Meyr., B. J. XIX, 415-416 (1909) ♂♀[1].

Peradeniya.[1]
Galle.

crassicervicella, Chrétien 1896.

Cosmopteryx crassicervicella, Chrét., Naturaliste, p. 105 (1896)[1]; Rebel, Stdgr. Cat. Lep. Pal. II 185, No. 3556 (1901)[2]; Chrét., Ann. S. E. Fr. 1916. 490 [biol.][3]

Maskeliya.

———

N. Africa.[1, 3]

cyclopæa, Meyr. 1909.

Cosmopteryx cyclopaea, Meyr., B. J, XIX-413 (1909). ♂[1]

N. Coorg.[1]
Pollibetta (S. Coorg).

erethista, Meyr. 1909.

Cosmopteryx erethista, Meyr., B. J. XIX, 413-414 (1909). ♂♀[1]

Khasis.[1], Shillong.
Cherrapunji.

flavofasciata, Wollaston 1879.

Cosmopteryx flavofasciata, Woll, A.M.N. H. (5) III 438 (1879)[1]; Woll, Lep. St. Helena, p. 53 (1879)[1a]; Wlsm., P.Z.S. 1907. 965-966 (1908), [*partim*][1b].

Cosmopteryx mimetis, Meyr., Proc. Linn. Soc. N. S. W. XXII 339 (1897) ♂♀[1c]; Meyr., B. J. XIX 417 (1909) [♂♀ redescr.][2]; Meyr., T. E. S. 1910. 372 (1910)[3]; Meyr., Tr. Linn. Soc. (2) XIV 282 (1911)[4]; Meyr., T. E. S. 1915. 205 (1915)[5]; Meyr., Voyage Alluaud Afrique Orient. II 80 (1920)[6]; Fletcher, Ind. Agr. Ent. Mem. VI 102 (1921)[7]; Meyr., T. E. S. 1923. 552 (1924)[8]; Rebel, Bull. Soc. Ent. Egypte 1926, 190 (1927) [*minetis*–error][9]; Meyr., Ins. Samoa III 84 (1927).[10]

Puttalam.[2]
Maskeliya.[2]
Nilgiris.[2]
Bassein Fort.[7] Pusa.[7]
Peshawar.[7]
Bankipur.

———

E. Australia.[1c, 2, 10]
New Guinea.[2, 10]
Borneo.[2, 10]
Tonkin.[10], Samoa.[10]

Larva descr.[7]; mining leaves of *Cyperus rotundus*.[6]

Fiji.[10] Seychelles.[4, 10]
Mauritius.[8, 4, 10]
Rodriguez.[8, 10]
Br. E. Africa.[6]
Egypt.[9, 10]
St. Helena,[1, 1a, 1b]
Bermuda.[10]
Brit. Guiana.[5]
Ceylon.[1]

hamifera, Meyr. 1909.

Cosmopteryx hamifera, Meyr., B. J. XIX 420-421 (1909)♂[1]; Meyr., Rec. Ind. Mus. V-223 (1910).[2]

(prob. low-country).
Sukna.[2]
(E. Him.; 500 ft.)
Pusa.

S. Africa.

hieraspis, Meyr. 1924.

Cosmopteryx hieraspis, Meyr., Exot. Micr. III 89-90 (1924). ♂♀[1].

Pusa.[1]

Larva mining leaves of *Saccharum fuscum*.[1]

holophracta, Meyr. 1909.

Cosmopteryx holophracta, Meyr., B.J. XIX 414 (1909). ♂ ♀[1]

Khasis.[2,] Shillong.
Cherrapunji.

ingeniosa, Meyr. 1909.

Cosmopteryx ingeniosa, Meyr., B. J. XIX 421 (1909) ♂♀[1]; Meyr., Tr. Linn. Soc. (2) XIV-282 (1911).[2]

Khasis.[1], Cherrapunji.
Shillong.

Seychelles.[2]

iphigona, Meyr. 1915.

Cosmopteryx iphigona, Meyr., Exot. Micr. I-323 (1915).♀[1]

N. Coorg.[1]

laetifica, Meyr. 1909.

Cosmopteryx laetifica, Meyr., B. J. XIX 418 (1909). ♂♀[1]

Diyatalawa.[1]
Madulsima.
Nilgiris (3500 ft.)[1]

licnura, Meyr. 1909.

Cosmopteryx licnura, Meyr., B. J. XIX 414-415 (1909). ♂[1]

Khasis.[1]

ligyrodes, Meyr. 1915.

Cosmopteryx ligyrodes, Meyr., Exot. Mir. I 322-323 (1915). ♂[1]

Kharwar.[1]
"Bengal."[1] [Pusa].

manipularis, Meyr. 1909.

Cosmopteryx manipularis, Meyr., B. J. XIX 419 (1909). ♂♀[1] — Maskeliya.[1] N. Coorg[1] Pollibetta (S. Coorg).

neodesma, Meyr. 1915.

Cosmopteryx neodesma, Meyr., Exot. Micr. I-323 (1915).♂[1] — N. Coorg[1] Dharwar., Shillong. Pyinmana.

paltophanes, Meyr. 1909.

Comoopteryx paltophanes, Meyr., B. J. XIX 417 (1909).♂[1] — Khasis.[2] Cherrapunji

panopla, Meyr. 1909.

Cosmopteryx panopla, Meyr., R. J. XIX 412-413 (1909).♂[1] — Hakgala.[1] (Ceylon; 6000 ft.)

phaeogastra, Meyr. 1917.

Cosmopteryx phaeogastra, Meyr., Ent., Mo. Mag. LIII 257-258 (1917) ♂♀[1]; Fletcher, Ann. Rept. Impl. Entom. 1917-1918, p. 101 (1918)[2]; Fletcher, Rept. Third Entl. Meeting I 150 (Nov. 1920)[3]; Fletcher, Ind. Agr. Ent. Mem. VI 103-104, t. 23 f. 2 (Jan. 1912).[4] — Pusa[1, 4] Coimbatore[4]

Larva described,[4] mining in bean leaves.[1-4]

plesiasta, Meyr. 1919.

Cosmopteryx plesiasta, Meyr., Exot. Micr. II-284 (1919) ♂[1]; *id., t. c.* p. 567 (1922) [♀ antenna].[2] — Castle Rock.[1, 2]

semnota, Meyr. 1914.

Cosmopteryx semnota, Meyr., B. J. XXII-775 (1914).♂[1] — Nilgiris.[1] (7000 ft.)

ipiculata, Meyr. 1909.

Cosmopteryx spiculata, Meyr.,. B. J. XIX 420 (1900). ♂[1] — Maskeliya.[1]

vexillaris, Meyr. 1909.

Cosmopteryx vexillaris, Meyr., B. J. XIX 418 (1909). ♂[1] — Khasis.[1]

zathea, Meyr. 1917.

Cosmopteryx zathea, Meyr., Exot. Micr. II 36 (1917). ♂♀[1] — Pollibetta.[1] (S. Coorg.)

2. Labdia, Walker 1864.

Labdia, Wlk., Cat. XXIX 823 (1864); Turner, Proc. R. Soc. Vict. XXXVI 68 (1923). [characters] — Type: *deliciosella,* Wlk. (Australia).

acroplecta, Meyr. 1915.

Pyroderces acroplecta, Meyr., Exot. Micr. I 313-314 (1915).♂♀[1].

Pattipola.[1] (Ceylon; 6000 ft.)
Maskeliya.[1]

albilineella, van Deventer 1904.

Pyroderces albilineella, Dev., Tijds. V. Ent. XLVII 33-34, t. 2 f. 5 (1904)[1]; Meyr., Exot. Micr. I 310 (1915)[2]; *id., i. c.* 566 (1916)[3]; Fletcher, Rept. third Entl. Meeting I 150 (Nov. 1920)[4]; Fletcher, Ind. Agr. Ent. Mem. VI 100 (Jan. 1921).[5]
Larva boring in pods of *Cassia corymbosa,*[3] in indigo pods[5].

Ceylon.[2]
Madulsima.
Kegalle., Coimbatore[5]
S. Coorg.[5]

Java.[1]
Kei Isds.[2]

amphi terna, Meyr. 1917.

Pyroderces amphipterna, Meyr., Exot. Micr. II 36-37 (1917). ♂[1]

N. Coorg.[1]

anthinopa, Meyr. 1917.

Pyroderces anthinopa, Meyr., Exot. Micr. II 37 (1917).♂[1]

Maskeliya.[1]

arachnitis, Meyr. 1907.

Stagmatophora arachnitis, Meyr., B.J. XVII-744 (1907). ♂♀[1]

Maskeliya.[1]

callibrocha, Meyr. 1915.

Pyroderces callibrocha, Meyr., Exot. Micr. I-312 (1915).[1]

Pusa.[1]

callistrepa, Meyr. 1917.

Pyroderces callistrepta, Meyr., Exot. Micr. II 38-39 (1917) ♂♀[1]; Fletcher, Ind. Agr. Ent. Mem. VI 101, t. 22 (Jan. 1921).[2]

Larva mining leaves of teak *(Tectona grandis).*[1, 2] Early stages descr.[2]

Pusa.[1, 2]

catapneusta, Meyr. 1917.

Pyroderces catapneusta, Meyr., Exot. Micr. II-39 (1917).♂[1]

Madulsima.[1]
Hakgala.[1]

caudata, Meyr. 1917.

Pyroderces caudata, Meyr., Exot. Micr. II 37 (1917)♀[1]

Colombo[1]

citracma, Meyr. 1915.

Pyroderces citracma, Myer., Exot. Micr. I 310-311 (1915). ♂♀[1]

Khasis[1], Cherrapunji.
Sadiya (N. Assam).

clopaea, Meyr. 1917.

Pyroderces clopaea, Meyr., Exot. Micr. II 40 (1917).♂[1] — Galle.[1]

cremasta, Meyr. 1915.

Pyroderces cremasta, Meyr., Exot. Micr. I 311 (1915).♀ — Khasis.[2]

emphanopa, Meyr. 1922.

Labdia, emphanopa, Meyr., Exot. Micr. II-570 (1922).♀[1] — Khasis.[1]

incompta, Meyr. 1917.

Pyroderces incompta, Meyr., Exot. Micr. II-40 (1917).♀[1] — Pollibetta.[1] (S. Coorg.)

inodes, Meyr. 1922.

Labdia inodes, Meyr., Exot. Micr. II-569 (1922).♀[1] — Shillong.[1]

ioxantha, Meyr. 1915.

Pyroderces ioxantha, Meyr., Exot. Micr. I-313 (1915).♂[1] — Maskeliya.[1]

lampropeda, Meyr. 1917.

Pyroderces lampropeda, Meyr., Exot. Micr. II-39 (1917).♀[1] — N. Kanara.[1]

liolitha, Meyr. 1922.

Labdia liolitha, Meyr., Exot. Micr. II-569-570 (1922).♀[1] — N. Kanara.[1]

microglena, Meyr. 1915.

Pyroderces microglena, Meyr., Exot. Micr. I-312 (1915).♂[1] — Maskeliya.[1] Hakgala.

molybdaula, Meyr. 1915.

Pyroderces molybdaula, Meyr., Exot. Micr. I 309-310 (1915).♂ ♀[1] — N. Coorg.[1] Nilgiris (7000 ft.)[1]

Labdia molybdaula, Meyr., Exot. Micr. III 58 (1923).[2]

notochorda, Meyr. 1907.

Stagmatophora notochorda, Meyr., B. J. XVII 744-745 (1907).♂ ♀[1] — Puttalam.[1] (Ceylon).

oxyleuca, Meyr. 1915.

Pyroderces oxyleuca, Meyr., Exot. Micr. I-315 (1915).♂[1] — Maskeliya.[1]

paropis, Meyr. 1915.

Pyroderces paropis, Meyr., Exot. Micr. I-310 (1915).♂ ♀[1] — Maskeliya[1]

planetopa, Meyr. 1915.

Pyroderces planetopa, Meyr., Exot Micr. I-309 (1915).♂[1]

Maskeliya.[1]

promacha, Meyr. 1897.

Pyroderces promacha, Meyr., Proc. Linn. Soc. N. S. W. XXII-351 (1897).♂[1]

Stagmatophora promacha, Meyr., Entom. Mitteil. Suppl. No. III, p. 54 (1914)[2]

Pyroderces promacha, Fletcher, Rept. third Entom. Meeting I 150 (Nov. 1920)[3]; Fletcher, Ind. Agr. Ent. Mem. VI 100-101 (Jan. 1921).[4]

Labdia promacha, Meyr., Ins. Samoa III-90 (1927).[5]

Larva descr.[4], in stem of *Cajanus indicus*[4], mining *Phaseolus mungo* leaves. [?].[4]

Ceylon.[5], Colombo.
Peradeniya., Matale.
Haputale.
Madulsima.
Coimbatore.[4]
N. Coorg.
Karwar., Pusa.[4]
Peshawar.[4]

Sydney.[1]
Formosa.[2, 5]
Java[5]
Samoa[5]

pyrrhota, Meyr. 1915.

Pyroderces pyrrhota, Meyr., Exot. Micr. I-314 (1915).♂[1]

Maskeliya[1]

redimita, Meyr. 1917.

Pyroderces redimia, Meyr., Exot. Micr. II-38 (1917).♂[1]

N.Coorg.[1]

sarcodryas, Meyr. 1922.

Labdia sarcodryas, Meyr., Exot. Micr. II 570 (1922).♂[1]

Surat.[1]

selenopis, Meyr. 1905.

Pyroderces selenopis, Meyr., B. J. XVI-606-607 (1905).♂[1]

Pundaluoya.[1]
Maskeliya.

semicoccinea, Stainton 1859.

Cosmopteryx ? semicoccinea, Stt., T.E.S. (n. s.) V-123 (1859).[1]

Cosmopteryx semicoccinea, Cotes and Swinhoe, Cat. Moths India, p. 722, No. 4938 (1889).[2]

[*Cosmopteryx semicoccinea*] Meyr., B. J. XIX-411 (1909).[3]

Pyroderces semicoccinea, Meyr., Exot. Micr. I 310 (1915)[4]; Fletcher, Rept. third Entom. Meeting I 150 (Nov. 1920[5]); Fletcher, Ind. Agr. Ent. Mem.VI-100 (Jan. 1921).[6]

Labdia semicoccinea, Meyr., Exot. Micr. III-58 (1923).[7]

Biology: reared from stems of *Cajanus indicus,*[6] from old bean stem.

Puttalam., Galle.
Shevaroys.[6]
Pollibetta.[6]
N. Coorg.
N. Kanara.
Calcutta.[1]
Pusa.[6]

Java.

Note. Australian references are to *Labdia cosmangela,* Meyr.]

sirenia, Meyr. 1917.

Pyroderces sirenia, Meyr., Exot. Micr. II-37-38 (1917).♀[1]

Msakeliya.[1]
Galle.[1]

sophista, Meyr. 1917.

Pyroderces sophista, Meyr., Exot. Micr. II 36 (1917).♂[1]

Madulsima.[1]
Ceylon; 4000 ft.)

sphenoclina, Meyr. 1922.

Labdia sphenoclina, Meyr., Exot. Micr. II 570 (1922).♀[1]

N. Coorg.[1]

terenopa, Meyr. 1917.

Pyroderces terenopa, Meyr., Exot. Micr. II 39-40 (1917).♀[1]

Namunakuli.[1]
(Ceylon; 6000 ft.)

3. Spiroterma, Meyr. 1915.

Spiroterma, Meyr., Exot. Micr. I-324 (1915).

Type: *caranaea,* Meyr. [Ceylon.]

caranaea, Meyr. 1915.

Spiroterma caranaea, Meyr., Exot. Micr. I-324-325 (1915) ♂♀[1]; Meyr., Ins. Samoa III 93 (1927).[2]

Galle.[1], Maskeliya.[1]
Sidapur (Coorg.).

Samoa.[2]
New Hebrides.[2]

4. Pyroderces, Herrich-Schaffer 1853.

Pyroderces, H. S. Schmett. Eur. V 47, t. 13 ff. 29, 30 (1853); Wocke, Hein. Schmett. Deutschl., Kleinschmett. II ii 425-426 (1876) (redescr.); Meyr., Proc. Linn. Soc. N. S. W. XXII 341 (1897) [redescr.]; Spuler, Schmett. Eur, II 388-389, f. 150 (neur., *argyrogrammos,* Z.] (1910) [redescr.]; Durrant, Bull. Ent. Res. III 208 (1912); Meyr., Tr. N. Z Inst. XLVII 208 (1915) [redescr.]; Busck, Jl. Agric. Res. IX 363, f. 5 [neur. *rileyi*] (1917) [redescr.]; Heinrich, Jl. Agric. Res. XX 820-881, figs. (1921) [*rileyi,* Wlsm.]; Turner, Proc. R. Soc. Vict. XXXVI 66-68 (1923); Forbers, Lep. N. York, p. 327 (1924) [redescr.]

Type: *argyrogrammos,* Z. [Europe.]

Syntomactis, Meyr., Tr. N.Z. Inst. 1888. 173; Meyr., Proc. Linn. Soc. N.S.W. XXII 378-379 (1897) [redescr.]; Meyr., Tr. R. Soc. S. Austr. XXX 53 (1906); Meyr., Tr. N.Z. Inst. XLVII 210 (1915) [redescr.]

Type: *deamatalla,* Wlk. [New Zealand.]

Anatrachyntis, Meyr., Exot. Micr. I-325 (1915).

Type: *falcatella,* Stt. [India].

amphisaris, Meyr., 1922.

Pyroderces amphisaris, Meyr., Exot. Micr. II-572 (1922).♂[1]

Maskeliya.[1]

Malaya.

calefacta, Meyr., 1922.

Pyroderces calefacta, Meyr., Exot. Micr. II-571-572 (1922).♀[1]

Calcutta.[1]
N. Kanara[1]

centrophanes, Meyr. 1915.

Anatrachyntis, centrophanes, Meyr., Exot. Micr. I-325 (1915).♂ ♀[1]

N. Coorg.[1]
Nilgiris (3500 ft.).[1]
Khasis.[1]

Fiji.

exagria, Meyr. 1915.

Anatrachyntis exagria, Meyr., Exot. Micr. I-326 (1915).♂[1]

N. Coorg.[1]
Virajpet (Coorg.).

falcatella, Stainton 1859.

Gracilaria ? falcatella, Stt., T. E. S. (n. s.) V-121 (1859).[1]

Gracilaria falcatella, Cotes and Swinhoe, Cat. Moths India, p. 720, No. 4925 (1889).[2]

Pyroderces spodochtha, Meyr., B. J. XVI-607 (1905).♂ ♀[3]

.. .. *falcatella,* Meyr., B. J. XVIII 817 (1908).[4]

Pyroderces spodochtha, Meyr., Exot. Micr. I-280 (1914).[5]

Anatrachyntis falcatella, Meyr., Exot. Micr. I-325 (1915)[6]; Fletcher, Rept. Second Entom. Meeting, p. 114 (1917)[7]; Fletcher, Rept. Third Ent. Meeting I-149 (1920)[8]; Fletcher, Ind. Agr. Ent. Mem. VI 99 (1921)[9]; Misra, Rept. Fourth Ent. Meeting, pp. 244-252, t. 44 (Dec. 1921).[10]

Pyroderces dendrophaga, Meyr., Exot. Micr. II-318 (1920).♂ ♀[11]

Pyroderces falcatella, Turner, Proc. R. Soc. Vict. XXXVI 66 (1923).[12]

Larva descr.[10], in resinous masses of Lac Coccid,[3] on dead female Lac Insects,[10] on cotton shoots,[9] on *Dactylopius* on cotton,[9] bred from *Xanthorrhæa arborea* (Liliaceae).[11]

Pupa, descr.[5,9,10]

Kandy.[3], Puttalam.
Coimbatore.
N. Coorg.
Calcutta.[1]
Pusa.[9, 10]
Shillong.[9]

Java.
Queensland.[11, 12]
Hawaii.

hemipelta, Meyr. 1917.

Anatrachyntis hemipelta, Meyr., Exot. Micr.II 44-45 (1917). ♂[1]

N. Coorg.[1]

mesoptila, Meyr. 1897.

Pyroderces mesoptila, Meyr., Proc. Linn. Soc. N. S. W. XXII 347-348 (1897). ♀[1]

N. Kanara.[2]

Anatrachyntis, mesoptila, Meyr., Exot. Micr. I-325 (1915).[2]

Queensland.[1]

Pyroderces mesoptila, Turner, Proc. R. Soc. Vict. XXXVI-66 (1923).[3]

mythologica, Meyr. 1917.

Anatrachyntis mythologica, Meyr., Exot. Micr. II 45-46 (1917).♂ ♀[1]

Maskeliya.[1]
Madulsima.[1]

nephelopyrrha, Meyr. 1917.

Anatrachyntis nephelopyrrha, Meyr., Exot. Micr. II 45 (1917).♂[1]

Pusa.[1]

rhizonympha, Meyr., 1924.

Pyroderces rhizonympha, Meyr., Exot. Micr. III 90 (1924).♂[1]

Pusa.[1]

Larva feeding among surface roots of bamboo.[1]

simplex, Walsingham 1891.

Pyroderces simplex, Wlsm., T. E. S. 1891. 119-120, t. 6 f. 58 (1891).♂ ♀[1]

Batrachedra coriacella, Snellen, Tijds. V. Ent. XLIV 95, t. 6 f. 7 (1901).[2]

Stagmatophora gossypiella, Wlsm., A. M. N. H. (7) XVIII 178-179 (1906)[3]; Morstatt, Pflanzer VII 253 (1912).[4]

Stagmatophora coriacella, Lefroy, Ind. Ins. Life, p. 536 (1909)[5]; Meyr., T. E. S. 1910. 372 (1910).[6]

Pyroderces simplex Durrant, Bull Ent. Res. III 206-207, f. 2 (1912)[7]; Lamborn, Bull. Ent. Res. V 201 (1914).[8]

Pyroderces coriacella, Fletcher, S. Ind. Ins., pp. 458-459, f. 334 (Sept. 1914)[9]; Meyr. Entom. Mitteil., Suppl. III, p. 52 (1914).[10]

Anatrachyntis coriacella, Meyr., Exot. Micr. I 325 (1915).[11]

Ceylon.[15]
Galle.
Myingyan.[17]
Amherst.[5]
Coimbatore.[17]
Chidambaram.[17]
Hagari (Bellary).[17]
Dharwar.
Belgaum.[17]
Baroda., Surat.[17]
Calcutta.[17]
Pusa.[17]
Ferozepur.[17]
Rohtak.[17]
Sialkote.[17]
Shillong.
Lyallpur.

Pyroderces simplex, Willcocks, Ins. Pests Egypt I i 317-320, t. 7 ff. 7-9 (1916)[12]

———————
Java.[2]

Anatrachyntis simplex, Fletcher, Rept. Second Entom. Meeting, p. 114 (Dec. 1917)[13]; Fletcher, Rept. Third Entom. Meeting I 149 (Nov. 1920).[14]

Malay Islands.
Formosa.[10]
Mesopotamia.

Pyroderces coriacella, Meyr., Voyage Alluaud Afrique Orient. II 81 (1920);[15] Myer., Exot. Micr. II 412 (1921).[16]

Mauritius.[6]
Br. E. Africa.[15]
s. Africa.

Anatrachyntis simplex, Fletcher, Ind. Agr. Ent. Mem. VI 97-99 (Jan. 1921);[17]

Gambia.[1], Barbados.[1]
N. America.

Larva in cotton-seed[2-9,12,] cotton bolls,[17] cotton-buds,[17] cotton seed-heads,[15] *Hibicus esculentus*[17] rotten peach fruit,[17] dry fallen *Ficus glomerata* fruit,[17] rotten bambo stalk[17], *Cajanus indicus* stem[17], dry *Vigna catjang* pods,[17] wheat flour,[17] castor fruits[17], maize cobs,[17] *Sorghum* earsheads,[17] rotten pomegranate fruit[17]; "mining species of mallow."[1]

Biology[12,17]

strangalota, Meyr. 1922.

Pyroderces strangalota, Meyr., Exot. Micr. II 572 (1922).♂[1]

Nilgiris (3500 ft.)[1]

5. Dorodoca, Meyr. 1915.

Dorodoca, Meyr., Exot. Micr. I-324 (1915).

Type: *chrysomochla,* Meyr. [India]

chrysomochla, Meyr. 1915.

Dorodoca chrysomochla, Meyr., Exot. Micr. I 324 (1915). ♂♀[1]; Meyr., Exot. Micr. III-90 (1924).[2]

Larva in dried spadix of *Borassus flabelliformis* (Palmaceae).[2]

Pusa.[1, 2]
Cuddapah.[1](4000 ft.)
Chapra.

leucomochla, Meyr. 1922.

Dorodoca leucomochla, Meyr., Exot. Micr. II 572-573 (1922).♂[1]

Thaton.[1]
(Burma).

6. Stagmatophora, Herrich-Schäffer 1853.

Stagmatophora, H.S., Schmett. Eur. V 49, t. 13 ff. 27, 28 (1853); Wocke, Hein. Schmett Deutschl., Kleinschmett. II ii 423 (1876) [redescr.]; Wlsm., E. M. M. 1906, 196-197; Meyr., Tr. R. Soc. S. Asutr. XXX 53 (1906); Wlsm.,

Type: *heydeniella,* F. R. [Europe.]

Fauna Hawaii I 514 (1907); Meyr., B. J. XIX 410 (1909); spuler; Schmett. Eur. II 382, f. 138 [neur. *heydeniella*, F. R.] (1910) [redescr.]; Turner, Proc. R. Soc. Vict. XXXVI 68 (1923); Forbes, Lep. N. York, P 328, f. 189 (1924) [redescr.]

drosophanes, Meyr. 1921.

Stagmatophora drosophanes, Meyr., Exot. Micr. II 412-413 (1921)♂♀[1]; Meyr., Sarawak Mus. Jl. II 159 (1926).[2] — Shillong.[1], Muktesar (Kumaon; 7000 ft.)[1]
——————
Java., Borneo.[2]

faceta, Meyr. 1917.

Pyroderces faceta, Meyr., Exot. Micr. II 38 (1917).♂♀[1] — Puttalam., Matalé.[1] Haputale.[1] N. Coorg.[1], Pusa.[1]

peristrepta, Meyr. 1917.

Pyroderces peristrepta, Meyr., Exot. Micr. II 39 (1917).♂[1] — Peshawar.[1]

7. Persicoptila, Meyr. 1886.

Persicoptila, Meyr., T. E. S. 1886 295 (1886); Meyr., Proc. Linn. Soc. N. S. W. XXII 329 (1897) [redescr.] — Type: *erythrota*, Meyr., [New Hebrides.]

acrostigma, Meyr. 1915.

Persicoptila acrostigma, Meyr., Exot. Micr. I-320 (1915).♂[1] — Nilgiris.[1] (7000 ft.)

chiasta, Meyr. 1909.

Persicoptila chiasta, Meyr., B. J. XIX 421-422 (1909).♂[1] — Ohiya.[1] (Ceylon, 6000 ft.)

libanotris, Meyr. 1906.

Persicoptila libanotris, Meyr., B. J. XVII 411 (1906). ♀[1] — Maskeliya.[1]

meliteucta, Meyr. 1915.

Persicoptila meliteucta, Meyr., Exot. Micr. I 319-320 (1915). ♂♀[1] — N. Coorg.[1]

oriaula, Meyr. 1915.

Persicoptila, oriaula, Meyr., Exot. Micr. I 321 (1915).♂[1] — Palnis (6000 ft.)[1]

petrinopa, Meyr. 1915.

Persicoptila petrinopa, Meyr., Exot. Micr. I 320 (1915). ♂♀[1] — Nilgiris (7000 ft.)[1]

picrodes, Meyr. 1915.

Persicoptila picrodes, Meyr., Exot. Micr. I 322 (1915).♂[1] — Palnis (6000 ft.)[1]

scholarcha, Meyr. 1915.

Persicoptila scholarcha, Meyr., Exot. Micr. I 321 (1915)♂[1] — Palnis.[1]

8. Trissodoris, Meyr. 1914.

Trissodoris, Meyr., B. J. XXII 775 (1914). — Type: *honorariella,* Wlsm.[Hawaii; Ceylon].

honorariella, Walsingham 1907.

Stagmatophora honorariella, Wlsm., Fauna Hawaii, I 515-516, t. 15 f. 21 (1907).♂[1]

Stagmatophora quadrifasciata, Wlsm., Fauna Hawaii I 516, t. 15 f. 22 (1907).♀[2]; Swezey, Proc. Hawaii. Ent. Soc. II 139 (1910).[3]

Trissodoris honorariella Meyr., B. J. XXII 776 (1914)[4]; Meyr., Ins. Samoa III 92 (1927).[5]

Biology[3]; Larva on dead leaves of *Pandanus.*[5]

Ceylon.[5]
Barberyn. Isd.[4]
Ambalangoda.[4]
Weligama.[4]
____
New Guinea.[4, 5]
Pitcairn Isd.[5]
Fiji.[5], Samoa.[5]
Hawaii.[1, 2]

9. Schendylotis, Meyr. 1910.

Schendylotis, Meyr., Rec. Ind. Mus. V 225-226 (1910). — Type: *chrysota,* Meyr., (Sikkim).

chrysota, Meyr. 1910.

Schendylotis chrysota, Meyr., Rec. Ind. Mus. V 226 (1910).♀[1] — Kurseong.[1]

10. Apothetodes, Meyr. 1919.

Apothetodes, Meyr., Exot. Micr. II 233 (1919). — Type: *dialectica,* Meyr. [India].

dialectica, Meyr. 1919.

Apothetodes dialectica, Meyr., Exot. Micr. II 233 (1919). ♀[1] — Dharwar.[1]

11. Limnæcia, Stainton 1851.

Limnæcia, Stt., Suppl. Cat. Brit. Tin. p. 4 (1851); Wocke, Hem. Schmett. Deutschl., Kleinschmett. II ii. 421 (1876) [redescr.]; Rebel, Stdgr. Cat. Lep. Pal. II 187(1901); Spuler, Schmett. Eur., pp. 384-385, f. 142 [neur., *phragmitella*] (1910) [redescr.]; Forbes, Lep. N. York, p. 324, f. 181 (1924). [redescr.] — Type: *phragmitella,* Stt. [Europe; N. & S. Africa; E. Australia; N. Zealand.]

Laverna [*part.*]; St. Ins. Brit. Tin. pp. 236, 238-239 (1854) [*phragmitella*]; Stt., Manual II 399 (1859) [*phragmitella*].

Limnoecia, Meyr., Tr. N. Z. Inst. 1888. 172 [redescr.]; Meyr., Handb., p. 674 (1895) [redescr.]; Meyr., Proc. Linn. Soc. N. S. W. XXII 358 (1897) [redescr.]; Meyr., Tr. N. Z. Inst. XLVII 209 (1915) [redescr.]

Anybia, Stainton, Ins. Brit. Tin. pp. 244-245, t. 7 ff. 11^{a-c} (1854); Stt., Manual II 401 (1859) [redescr.]; Wlsm., P. Z. S. 1891. 537 (1892); Meyr., Hand b. p. 675 (1895) [redescr.]; Spuler, Schmett. Eur. II 384 (1910) [redescr.].
Type: [*epilobiella,* Röm.=] *langiella,* Stt. [Europe.]

Atremnaea, Staudinger, Berlin. Ent. Zeits. XIV 317 (1870).
Type: *lonchoptera,* Stdgr.

Ptilochares, Meyr., Proc. Linn. Soc. N. S. W. XI 1046 (1886).
Type: *trissodesma,* Meyr., [Victoria.]

Opsizyga, Lower, Tr. R. Soc. S. Asutr. XXVII 230 (1903).
Type: *eugramma,* Low. [Australia].

Erechthiodes, Meyr., Ann. Transv. Mus. IV 195 (1914).
Type: *audax,* Meyr. [Transvaal].

Callixestis, Meyr., Exot. Micr. II 41 (1917).
Type: *cassandra,* Meyr. [Ceylon].

Thalerostoma, Meyr., Exot. Micr. II 42 (1917).
Type: *orthocentra,* Meyr., [India].

amblopa, Meyr. 1915.

Limnoecia amblopa, Meyr., Exot. Micr. I 319 (1915).♂[1]
Nilgiris.[1] (7000 ft.)

antidesma, Meyr. 1913.

Limnoecia antidesma, Meyr., Ann. Transv. Mus. II 308 (1913).♂[1]; Meyr., Exot. II 42 (1917).[2]
Trichinopoly. Kanara.[2], Surat.
____
Transvaal.[1]

asterodes, Meyr. 1915.

Limnoecia, asterodes, Meyr., Exot. Micr. I 319 (1915).♂[1]
Khasis.[1]

asteroleuca, Meyr. 1917.

Limnoecia asteroleuca, Meyr., Exot. Micr. II 43 (1997).♂[1]
Peshawar.[1] Kharagodha (Gujrat).

auximena, Meyr. 1915.

Limnoecia auximena, Meyr., Exot. Micr. I 318 (1915).♂[1]
N. Coorg.[1]

callicosma, Meyr. 1915.

Limnoecia callicosma, Meyr., Exot. Micr. I 318 (1915).♂[1]	Maskeliya.[1]

cassandra, Meyr., 1915.

Xestocasis cassandra, Meyr., Exot. Micr. I 317 (1915).♂[1]	Colombo.[1]
Callixestis cassandra, Meyr., Exot. Micr. II 41 (1917).[2]	

chromaturga, Meyr. 1915.

Xestocasis chromaturga, Meyr., Exot. Micr. I 316-317 (1915).♂♀[1]	Galle., Peradeniya.[1] Matalé.[1]
Callixestis chromaturga, Meyr., Exot. Micr. II 41 (1917).[2]	Nilgiris (3,500 ft.)[1] N. Coorg.[1] N. Kanara.[1] Margherita.[1], (Assam).

chrysidota, Meyr. 1917.

Callixestis chrysidota, Meyr., Exot. Micr. II 41 (1917).♂[1]	Pusa.[1]

crocodelta, Meyr. 1915.

Xestocasis crocodelta, Meyr., Exot. Micr. I 316 (1915).♂[1]	N. Kanara.[1]

enclista, Meyr. 1921.

Limnoecia enclista, Meyr., Zool. Mededeelingen VI 168-169 (1921).♂[1] ♀	Ceylon. ——— Java.

leucomelanella, Rebel, 1916.

Leucomelanella, Rebel, Denkschr. Kais. Akad. Wiss. Wien XCIII 445, f. 1 (1916).[1]	Maskeliya.[2] Coimbatore.[2] Pusa., Surat Peshawar. ——— Java.[1]
Limnoecia heterosticta, Meyr., Exot. Micr. II 42 (1917).♂ [2]	N. E. Africa.

lubricata, Meyr. 1917.

Callixestis lubricata, Meyr., Exot. Micr. II 41-42 (1917).♂[1]	Maskeliya.[1]

magica, Meyr. 1905.

Pyroderces magica, Meyr., B. J. XVI 607 (1905).♂[1]	Maskeliya.[1] Pattipola. (Ceylon; 6,000 ft.)
Limnoecia magica, Meyr., Exot. Micr. I 318 (1915).[2]	

metacypha, Meyr. 1914.

Limnoecia metacypha, Meyr., Exot. Micr. I 203-204 (1914). ♂[1]; Fletcher, Ind. Agr. Ent. Mem. VI 101 (Jan. 1921).[2] *Cocoon* descr.[1] — Peradeniya.[1]

orthocentra, Meyr. 1917.

Thalerostoma orthocentra, Meyr., Exot. Micr. II 42 (1917).♂ ♀[1] — Nilgiris. (7,000 ft.)

peronodes, Meyr. 1915.

Limnoecia peronodes, Meyr. Exot. Micr. I 318 (1915)♂; Fletcher, Ind. Agr. Ent. Mem. VI 101-102 (Jan. 1921).[2] — Pusa.[1, 2]

Larva feeding in leaf-sheaths of bamboo,[1] on Coccidae on bamboo.[2]

proclina, Meyr. 1907.

Limnoecia proclina, Meyr., B. J. XVII 745 (1907).♀[1] — Maskeliya.[1]

psalidota, Meyr. 1917.

Limnoecia psalidota, Meyr., Exot. Micr. II 43 (1917).♂ ♀[1] — Kegalle.[1], (Ceylon.) Dharwar[1], Pusa.[1] Abbottabad.[1] Peshawar.

Java.

scaeosoma, Meyr. 1905.

Pyroderces scaeosoma, Meyr., B. J. XVI 607-608 (1905).♂[1] — Peradeniya.[1], Kandy. Maskeliya.

Java.

stabilita, Meyr. 1915.

Limnoecia stabilita, Meyr., Exot. Micr. I 319 (1915).♂ ♀[1] — N. Coorg.[1], Yercand.

trissodelta, Meyr. 1922.

Limnoecia trissodelta, Meyr., Exot. Micr. II 573 (1922).♂ ♀[1] — Shillong.[1]

tyriarcha, Meyr. 1919.

Limnoecia tyriarcha, Meyr., Exot. Micr. II 234 (1919).♀[1] — Bhim Tal.[1] (Kumaon).

12. Eumenodora, Meyr. 1906

Eumenodora, Meyr., Tr. R. Soc. S. Austr. XXX 55 (1906). — Type: *encrypta*, Meyr. [Queensland].

tetrachorda, Meyr. 1924.

Eumenodora tetrachorda, Meyr., Exot. Micr. III 91 (1924). ♂♀[1] — Khurda[1] (Puri distr.)

Larva on needles of *Casuarina*.[1]

13. Microcolona, Meyr. 1897.

Microcolona, Meyr., Proc. Linn. Soc. N. S. W. XXII 370-371 (1897); Meyr., Tr. N. Z. Inst. XLVII 209 (1915) [redescr.]. — Type: *characta*, Meyr. [E. Australia; N. Zealand].

autotypa, Meyr. 1922.

Microcolona autotypa, Meyr., Exot. Micr. II 575 (1922). ♂[1] — Kodaikanal.[1]

citroplecta, Meyr. 1917.

Microcolona citroplecta, Meyr., Exot. Micr. II 49-50 (1917) ♂[1]; Fletcher, Ind. Agr. Ent. Mem. VI 104 (1921).[2] — N. Coorg.[1] Pusa.[1]

Biology: the foodplant is perhaps *Eugenia jambolana*.[1]

cricota, Meyr. 1917.

Microcolona cricota, Meyr., Exot. Micr. II 50 (1917).♂[1] — N. Coorg.[1], Khasis.[1]

emporica, Meyr. 1917.

Microcolona emporica, Meyr., Exot. Micr. II 50-51 (1917).♀[1] — Maskeliya.[1]

eriptila, Meyr. 1915.

Microcolona eriptila, Meyr., Exot. Micr. I 334 (1915).♂[1] — Shevaroys.[1] Bababudin Hills.

phalarota, Meyr. 1915.

Microcolona phalarota, Meyr., Exot. Micr. I 335 (1915).♂♀[1] — Khasis.[1] Cherrapunji.[1]

porota, Meyr. 1917.

Microcolona porota, Meyr., Exot. Micr. II 50 (1917).♂♀[1] — Khasis.[1]

pycnitis, Meyr. 1915.

Microcolona pycnitis, Meyr., Exot. Micr. I 334 (1915). ♂♀[1] — Maskeliya.[1] Namunakuli.[1] (Cylon; 6,000 ft.) N. Coorg.

14. Stilbosis, Clemens 1860.

Stilbosis, Clem., Proc. Acad. Nat. Sci. Philad., p. 170 (May 1860); Stainton, Tin. N. Amer., p. 129 (1872); Forbes, Lep. N. York, p. 325, f. 188 (1924) [redescr.]. — Type: *tesquella,* Clem. [N. & C. America.

synclista, Meyr. 1917.

Stilbosis synclista, Meyr., Exot. Micr. II 49 (1917).♂[1] — Khasis.[1]

turrigera, Meyr. 1921.

Stilbosis turrigera, Meyr., Exot. Micr. II 414 (1921).♂[1] — Simla.[1]

15. Meneptila, Meyr. 1915.

Meneptila, Meyr., Exot. Micr. I 333 (1915). — Type: *praedonia,* Meyr. [India].

praedonia, Meyr. 1915.

Meneptila praedonia, Meyr., Exot. Micr. I 333 (1915).♂ ♀[1] — Khasis.[1]

16. Trachydora, Meyr. 1897.

Trachydora, Meyr., Proc. Linn. Soc. N.S.W. XXII 390 (1897). — Type: *illustris, Meyr.* [E. Australia].

Anataractis, Meyr., Exot. Micr. I 565 (1916). — Type: *plumigera,* Meyr. [India; Java].

plumigera, Meyr. 1916.

Anataractis plumigera, Meyr., Exot. Micr. I 565-566 (May 1916) ♂ ♀[1]; Fletcher, Entl. Note 83 (June 1916)[2]; Fletcher, Rept. Second Entl. Meeting, p. 81 (1917)[3]; Fletcher, Rept. Third Entl. Meeting I 149 (1920)[4]; Fletcher, Ind. Agr. Ent. Mem. VI 99-100 (1921).[5] — Coimbatore.[1, 5] Dindigul., Dharwar. Pusa.[1, 5], Chapra. ———— Java.

Larva in galls in stem of *Indigofera linifolia.*[3, 5]

17. Leptozestis, Meyr. 1924.

Leptozestis, Meyr., Exot. Micr. III 91 (1924). — Type: *parascia,* Meyr. [W. Australia].

epochaea, Meyr. 1917*.

Syntomactis epochaea, Meyr., Exot. Micr. II 51 (1917).♀[1] — N. Kanara.[1]

18. Chrysoclista, Stainton 1854.

Chrysoclista, Stainton, Ins. Brit. Tin., pp. 240-241, t. 7 f. 9a-c (1854); Stt., Manual II 400 (1859) [redescr.]; Wocke, — Type: *linneella,* Clk. [Europe].

* This species, only known from a single female specimen, many perhaps be referable to *Pyroderces.*

Hein. Schmett. Deutschi., Kleinschmett. II ii 432 (1876) [redescr.); Snellen, Vlind. Nederl. Microlep., pp. 860-861 (1882) [redescr.]; Meyr., Handb. pp. 676-677 (1895) [redescr.]; Spuler, Schmett. Eur. II 386, f. 145 [neur., *linneella*] (1910) [redescr.].

Blastodacna, Wocke, Hein. Schmett. Deutschl., Kleinschmett. II ii 428-429 (1876); Rebel, Stdgr. Cat. Lep. Pal. II 186 (1901); Spuler, Schmett. Eur. II 387, f. 147 [neur., *putripennella,* Z.] (1910) [redescr.]; Frobes, Lep. N. York, p. 329 (1924) [redescr.]. — Type: *hellerella,* Dup. [Europe].

Spuleria, Hofmann, Iris X 230 (1897); Spuler, Schmett. Eur. II 386, f. 146 [neur., *aurifrontella,* Hb.] (1910) [redescr.] — Type: *aurifrontella,* Hb. [C. & S. Europe].

thrypsiphila, Meyr. 1917.

Chrysoclista, thrypsiphila, Meyr., Exot. Micr. II 52 (1917).♂[1] — Matalé[1] (Ceylon).

19. Aganoptila, Meyr. 1915.

Aganoptila, Meyr., Exot. Micr. I 333-334 (1915). — Type: *phanarcha,* Meyr., (Ceylon).

durata, Meyr. 1922.

Aganoptila durata, Meyr., Exot. Micr. II 576 (1922).♂[1] — Cherrapunji.[1] (Khasis; 4,500 ft.).

phanarcha, Meyr. 1915.

Aganoptila phanarcha, Meyr., Exot. Micr. I 334 (1915)♂♀[1]; Fletcher, Ind. Agr Ent. Mem. VI 104 (1921).[2] — Pattipola[1] (Ceylon; 6,000 ft.).

Biology: Bred from galls on undertermined tree.[1]

20. Ascalenia, Wocke 1876.

Ascalenia, Wocke, Heinemann's Schmett. Deutschl., Kleinschmett. II ii 421-422 (1876); Rebel, Stdgr. Cat. Lep. Pal. II 187 (1901); Spuler, Schmett. Eur. II 384, f. 140 [neur., *vanella,* Frey] (1910) [redescr.]; Meyr., Ann. Transv. Mus. VIII 96 (1921) [=*Cholotis,* Meyr.]. — Type: *vanella,* Frey [Europe].

Cholotis, Meyr., Trans. Linn. Soc. (2) XIV 284 (1911); Meyr., Exot. Micr. I 328 (1915) [neuration]. — Type: *semnostola,* Meyr. [Australia].

archaica, Meyr. 1917.

Cholotis archaica, Meyr., Exot. Micr. II 44 (1917). ♂[1] — Maskeliya.[1]

crypsiloga, Meyr. 1915.

Cholotis crypsiloga, Meyr., Exot. Micr. I 329-330 (1915)♀[1]; Fletcher, Ind. Agr. Ent. Mem. VI 104 (1921).[2] — Puttalam (Ceylon). Coimbatore.[1]

Larva on *Acacia.*[1]

epicrypta, Meyr. 1915.

Cholotis epicrypta, Meyr., Exot. Micr. I 327 (1915).♂♀[1] — N. Coorg.[1] Dharwar.

icriota, Meyr. 1915.

Cholotis icriota, Meyr., Exot. Micr. I 328-329 (1915).♂♀[1] — N. Coorg.[1] Pollibetta (S. Coorg).

molifera, Meyr. 1915.

Cholotis molifera, Meyr., Exot. Micr. I 327 (1915).♂♀[1] — Peradeniya.[1] N. Coorg.[1]

pachnodes, Meyr. 1917.

Cholotis pachnodes, Meyr., Exot., Micr. II 44 (1917)♂♀[1]; Fletcher, Ind. Agr. Ent. Mem. VI 104 (1921).[2] — Pusa.[1, 2]

Larva in twigs of *Tamarix gallica.*[1,2]

spermatica, Meyr. 1915.

Cholotis spermatica, Meyr., Exot. Micr. I 328 (1915).♂♀[1] — Khasis.[1] Cherrapunji.

thoracista, Meyr. 1915.

Cholotis thoracista, Meyr., Exot. Micr. I 327-328 (1915). ♂♀[1] — Peradeniya.[1] Kandy.[1] Madulsima.[1] N. Coorg.[1] N. Kanara.[1] Pusa., Khasis.[1]

Fiji.

21. Aeronectris, Meyr. 1917.

Aeronectris, Meyr., Exot. Micr. II 35 (1917). — Type: *euacta,* Meyr. (India).

euacta, Meyr. 1917.

Aeronectris euacta, Meyr., Exot. Micr. II 35-36 (1917).♂[1] — N. Coorg.[1]

22. Batrachedra, Stainton 1854.

Batrachedra, Stainton, Ins. Brit. Tin., pp. 230-231, t. 7 ff. 5^{a-c} (1854); H. S., Schmett. Eur. V 54, t. 9 ff. 18-21 — Type: *praeangustata,* Haw, [Europe].

(1853) [redescr.]; Stt., Manual II 396 (1859) [redescr.]; Wocke, Hein. Schmett. Deutschl., Kleinschmett. II ii 524 (1876) [redescr.]; Snellen, Vlind. Nederl. Microlep., pp. 862-863 (1882) [redescr.]; Meyr., Tr. N.Z. Inst. 1888. 180 [redescr.; remarks]; Meyr., Handb. p. 661 (1895) [redescr.]; Meyr., Proc. Linn. Soc. N. S. W. XXII 299-301 (1897) [redescr.]; Rebel, Stdgr. Cat. Lep. Pal. II 185 (1901); Wlsm., Fauna Hawaii. I 508 (1907) [redescr.]; Spuler, Schmett. Eur. II 388, f. 148 [neur., *praeangustata,* Hw.] (1910) [redescr.]; Durrant, Bull. Ent. Res. III 208 (1912); Wlsm. Biol. Centr. Amer., Het. IV 320 (1914); Meyr., Tr. N. Z. Inst. XLVII 228-229 (1915) [redescr.]; Meyr., Exot. Micr. II 30 (1916); Forbes, Lep. N. York, p. 330, f. 186 (1924) [redescr.].

Tetanocentria, Rebel, Berl. Ent. Zeit. XLVII 107 (1902); Spuler, Schmett. Eur. II 387-388 (1910) [redescr.].

Type: *gelechiella,* Rebel [Greece].

Eustaintonia, Spuler, Schmett. Eur. II 388, f. 149 [neur., *pinicolella,* Dup.] (1910).

Type: *pinicolella* Dup. [Europe].

amydraula, Meyr. 1916.

Batrachedra amydraula, Meyr., Exot. Micr. II 30 (1916) ♂ ♀[1]; Ramachandra Rao, Rept. Fourth Entl. Meeting, pp. 167-168 (1921)[2]

Larva boring young fruits of Date-palm.[2]

Shillong., Pusa.[1]

Mesopotamia.[2]
Baghdad.[1]
Egypt.

arenosella, Walker 1864.

Gracilaria arenosella, Wlk., Cat. XXX 857 (1864).♀[1]

Batrachedra arenosella, Meyr., Tr. N. Z. Inst. 1888. 181[2]; Meyr., Proc. Linn. Soc. N. S. W. XXII 302-303 (1897)[3]; Meyr., Exot. Micr. II 30-31(1916)[4]; Fletcher, Ind. Agr. Ent. Mem. VI 104 (1921).[5]

Batrachedra psilopa, Meyr., B. J. XII 982 (1907).♂ ♀[6]

Larva amongst seeds of *Juncus*[3], on flowers of *Cocos nucifera.*

Maskeliya.[4, 6]
Kodaikanal.
N. Coorg.[4]
Bangalore.[4]
S. Kanara.
Calcutta.[4]
Pusa.[5,] Khasis.[4]

New Zealand.[1, 3]
Tasmania.[3]
S. Australia.[3]
N.S. Wales.[3]
Queensland.[3]
Solomon Isds.[1]
Brit. Guiana.[4] Peru.

epimyxa, Meyr. 1916.

Batrachedra epimyxa, Meyr., Exot. Micr. II 32 (1916).♂[1] N. Coorg.[1]

macroloncha, Meyr. 1916.

Batrachedra macroloncha, Meyr., Exot. Micr. II 32 (1916).♀[1] Maskeliya.[1]

orinarcha, Meyr. 1917.

Batrachedra orinarcha, Meyr., Exot. Micr. II 33 (1917).♂[1] Nilgiris.[1] (3,500 ft.)

pacabilis, Meyr. 1922.

Batrachedra pacabilis, Meyr., Exot. Micr. II 583 (1022).♂♀[1] Murree[1] (Punjab; 7.500 ft.).

promylaea, Meyr. 1917.

Batrachedra promylaea, Meyr., Exot. Micr. II 35 (1917).♂[1] Khasis.[1]

rhysodes, Meyr. 1917.

Batrachedra rhysodes, Meyr., Exot. Micr. II 34 (1917).♂ ♀[1] Khasis.[1]

sacrata, Meyr. 1921.

Batrachedra sacrata, Meyr., Exot. Micr. II 415 (1921).♂[1] Garo Hills.[1]

Biology: Bred from *Shorea robusta* probably refuse-feeder).[1]

scapulata, Meyr. 1917.

Batrachedra scapulata, Meyr., Exot. Micr. II 34 (1917).♀[1] Maskeliya.[1]

silvatica, Meyr. 1917.

Batrachedra silvatica, Meyr., Exot. Micr. II 35 (1917). ♂ ♀[1]; Fletcher, Ind. Agr. Ent. Mem. VI 105 (1921).[2] Almora.[1] (Kumaon; 6,000 ft.)

Biology: Bred from twigs of *Pinus longifolia;* probably a feeder on dry refuse.[1]

substrata, Meyr. 1916-1917.

Batrachedra substrata, Meyr., Exot. Micr. II 32 (Oct. 1916)–33 (Oct. 1917).♀[1] Maskeliya.[1]

verax, Meyr. 1917.

Batrachedra verax, Meyr., Exot. Micr. II 33 (1917).♂[1] Madulsima. (Ceylon; 4,000 ft.).[1]

xanthocrena, Meyr. 1917.

Batrachedra xanthocrena, Meyr., Exot. Micr. II 34 (1917).♂[1] — N. Coorg.[1]

23. Clemmatista, Meyr. 1921.

Clemmatista, Meyr., Exot. Micr. II 414 (1921). — Type: *metacirrha,* Meyr. [India].

metacirrha, Meyr. 1921.

Clemmatista metacirrha, Meyr., Exot. Micr. II 414-415 (1921).♂[1] — Surat.[1]

24. Rhadinastis, Meyr. 1897.

Rhadinastis, Meyr., Proc. Linn. Soc. N. S. W. XXII 311 (1897). — Type: *microlychna,* Meyr., [E. Australia].

loraria, Meyr. 1917.

Rhadinastis loraria, Meyr., Exot. Micr. II 43-44 (1917). ♂ ♀[1] — Khasis.[1]

phoenicopa, Meyr. 1907.

Rhadinastis phoenicopa, Meyr., B. J. XVII 982 (1907).♂♀[1] — Maskeliya.[1]

---

162 Species.

PART XVII – YPONOMEUTIDAE.

Yponomeutidæ, Stainton, Ins. Brit. Tineina, pp. 53-54 (1854); *id.*, Manual II 304-305 (1859); Heinemann, Schmett. Deutschl., Kleinschmett. II. i. 100-101 (1870); Spuler, Schmett. Eur. II 441 (1910).

Plutellidæ (*partim*), Meyrick, Handbook, pp. 692-694 (1895); *id.*, Proc. Linn. Soc. N.S.W. XXXII 47-51 (1907).

Yponomeutidæ (*pa tim*), Forbes, Lep. New York, pp. 337-338 (1924).

1. Argyresthia, Hübner 1826.

Argyresthia, Hb., Verz. p. 422 (1826); Zeller, Isis XXXII 204 (1839) [redescr.]; Zeller, Linn. Ent. II 234-241 (1847) [redescr.]; H.S., Schmett. Eur. V 52, t. 13 ff. 48-52, 54, 55 (1853) [redescr.]; Stainton, Ins. Brit. Tin., p. 181, t. 6 ff. 6 (1854) [redescr.]; Stt., Manual II 369 (1859) [redescr.]; Wocke, Hein. Schmett. Deutschl., Kleinschmett. II ii 647 (1876) [redescr.]; Snellen, Vlind. Nederl. Mcirolep. pp. 523-524 (1882) [redescr.]; Meyr., Handb. p. 761 fig. [neur., head, *nitidella*] (1895) [redescr.]; Spuler, Schmett. Eur. II 446, fig. [neur., *goedartella,* Linn.] (1910) [redescr.]; Forbes, Lep. N. York, p. 344. f. 202 (1924) [redescr.].

Type: *gœdartella,* Linn. [Europe; N. America.]

Oligos, Treitschke, Schmett. Eur. VIII 299 (1830) [*non-descr.*]

Ederesa, Curtis, Entom. Mag. I 191 (1833); Stephens, Ill. Brit. Entom., Haust. IV 397 (1835); Westwood, Synopsis Brit. Ins., p. 112 (1840) [redescr.].

Type: *pruniella,* Linn. [Europe].

Argyrosetia, Stephens, Cat. Brit. Ins. II 205 (1829) [*non-descr.*]; Steph., Ill. Brit. Entom. Haust. IV 251-252 (1834); Westwood, Synops. Brit. Ins., p. 112 (1840) [redescr.; type cited as *goedartella*].

Type: *gœdartella,* Linn.

Ismene, Stephens, Ill. Brit. Entom., Haust. IV 247-248 (1834); *id., t.c.* p. 397 (1835) [Syn. of *Ederesa,* Curtis].

Type: [*nitidella,* Fabr. =] *pruniella,* Steph [Europe].

Blastotere, Ratzeburg, Forst. Ins. II 240 (1840). — Type: [*illuminatella, Z=*] *bergiella,* Rtz. [Europe].

dislocata, Meyr. 1914.

Argyresthia dislocata, Meyr., B. J. XXIII 123 (1914) ♂♀.[1] — Khasis.[1]

icterias, Meyr. 1907.

Argyresthia, icterias, Meyr., B. J. XVII 984 (1907)♂♀.[1] — Maskeliya.[1] (Ceylon; 4000 ft.) N. Coorg. Cotacamund.

iopleura, Meyr. 1918.

Argyresthia iopleura, Meyr., Exot. Micr. II 187 (1918)♂[1]; Fletcher, Ind. Agr. Ent. Mem. VI 132 (Jan. 1921).[2] — Almora.[1] (Kumaon; 6,000 ft.)

Biology: Bred from twigs of *Pinus longifolia,* probably feeding in the shoots.

leuculias, Meyr. 1914.

Argyresthia leuculias, Meyr., B. J. XXIII 123 (1914)♂♀.[1] — Maskeliya.[1]

semitrunca, Meyr. 1914.

Argyresthia semitrunca, Meyr., B. J. XXIII 123 (1914)♀.[1] — Khasis.[1]

2. Lycophantis, Meyr. 1914.

Lycophantis, Meyr., B. J. XXIII 122-123 (1914). — Type: *chalcoleuca,* Meyr. [India].

chalcoleuca, Meyr. 1914.

Lycophantis chalcoleuca, Meyr., B. J. XXIII 123 (1924) ♂♀.[1] — Khasis.[1], Shillong. Cherrapunji.

3. Zelleria, Stainton 1849.

Zelleria, Stainton, Cat. Brit. Tin. Pter., p. 22 (1849); Stt., Ins. Brit. Tin., pp. 191-192, t. 6 ff. 9[a-c] [*hepariella*] (1854) [redescr.]; Stt., Manual II 374 (1859) [redescr.]; Stt, Nat. Hist. Tin. XI 88-99 (1870) [general remarks; biology, etc.]; Meyr., Proc. Linn. Soc. N.S.W. XVII 580-581 (1892) [redescr.]; Meyr., Handb. p. 767 (1895) [redescr.]; Meyr., Proc. Linn. Soc. N.S.W. XXXII 69 (1907); Spuler, Schmett. Eur. II 445-446, fig. [neur. FW. *phillyrella*] (1910) [redescr.]; Meyr., Tr. N. Z. Inst. XLVII 227 (1915) [redescr.]. — Type: *hepariella,* Stt. [Europe; Asia Minor].

Kessieria, Nowieki, Micr. spec. nov., p. 13 (1864). Type: *zimmermanni,* Now. [Alps.]

Hofmannia, Wocke, Hein. Schmett. Deutschl., Kleinschmett. II ii 644 (1876); Meyr., Tr. N. Z. Inst. 1888. 162 [redescr.] Type: *saxifragae,* Stt. [N. & C. Europe].

[nec *Hofmannia,* Pagenstecher 1900, which is a synonym of *Triscaedecia,* Hampson– Orneodidae.]

Circostola, Meyr., Tr. N. Z. Inst. XXI 163 (1889); Meyr., Tr. N. Z. Inst. XLVII 227 (1915) [syn. of *Zelleria.*]. Type: *copidota,* Meyr., [New Zealand]

petrias, Meyr. 1907.

Zelleria petrias, Meyr., B. J. XVII 746 (1907)♂♀;[1] Meyr., Rec. Ind. Mus. V. 228 (1910).[2] Simla[1, 2], Kasauli. Dalhousie.,Bhim Tal. (Kumaon; 4,500 ft.). Kanara., N. Coorg. Ootacamund.

strophaæ, Meyr. 1914.

Zelleria strophaæ, Meyr., B. J. XXIII 124 (1914)♂♀.[1] Puttalam.[1] Maskeliya[1] Numunakuli.[1] (Cylon; 6,000 ft.)

4. Paradoxus, Stainton 1869.

Paradoxus, Stainton, Tin. South Europe, p. 167, fig. (1869); Spuler, Schmett. Eur. II 442 (1910) [redescr.]; Meyr., B. J. XXIII 124 (1914) [note on structure]. Type: *osyridellus,* Stt. [S. Europe; Asia Minor].

deformis, Meyr. 1914.

Paradoxus deformis, Meyr., B. J. XXIII 124 (1914)♀.[1] Kegalle.[1], (Cylon.)

5. Xyrosaris, Meyr. 1907.

Xyrosaris, Meyr., Proc. Linn. Soc. N. S. W. XXXII 71 (1907). Type: *dryopa,* Meyr. [E. Australia].

campsiptila, Meyr. 1914.

Xyrosaris campsiptila, Meyr., B. J. XXIII 124 (1914)♂♀.[1] Puttalam.[1] N. Coorg.[1]

lichneuta, Meyr. 1918.

Xyrosaris lichneuta, Meyr., Exot. Micr. II 188 (1918)♂♀.[1] Khasis.[1], Kodaikanal.

maligna, Meyr. 1907.

Xyrosaris, maligna, Meyr., B. J. XVII 984 (1907)♂♀.[1] Puttalam.[1], Maskeliya.[1]

5. Xyrosaris, Meyr. 1907.—

Kegalle., Haputale.
N. Coorg.
N. Kanara.

6. Orthosaris, Meyr. 1914.

Orthosaris, Meyr., B. J. XXIII 126 (1914)

Type: *strictulata,* Meyr. [Ceylon].

strictulata, Meyr. 1914.

Orthosaris, strictulata, Meyr., B. J. XXIII 126 (1914)♂.[1]

Matale.[1]
(Ceylon; 1,800 ft.)

7. Prays, Hübner 1826.

Prays, Hübner, Verz. p. 413 (1826); Zeller, Isis XXXII 191 (1839) [redescr.]; Stt., Ins. Brit. Tin., p. 64, t. 3, ff. 4[a-c] (1854) [redescr.]; Stt., Manual II 310 (1859) [redescr.]; Wlk., Cat. XXVIII 540 (1863); Hein., Schmett. Deutschl., Kleinschmett. II i 111-112 (1870) [redescr.]; Snellen, Vlind. Nederl. Microlep., pp. 517-518 (1882) [redescr.]; Meyr., Handb., p. 694 (1895) [redescr.]; Rebel, Stdgr. Cat. Lep. Pal. II 133 (1901); Meyr., Proc. Linn. Soc. N. S. W. XXXII 74-75 (1907) [redescr.]; Wlsm., Fauna Hawaii. I 651 (1907) [citations]; Spuler, Schmett. Eur. II 442, fig. [neur., *curtisellus,* Don.] (1910) [redescr.]; Forbes, Lep. N. York, p. 338 (1924) [redescr.].

Pepilla, Guenée, Ann. Soc. Ent. France (2) III 344 (1845) [*non-descr.*]; Gn., Eur. Micr. Index, p. 106 (1846) [*non-descr.*].

Type: *[curtisellus,* Don.=].

cœnobitella, Hb. [Europe]

acmonias, Meyr. 1914.

Prays acmonias, Meyr., B. J. XXIII 125-126 (1914)♀.[1]

Rawalpindi.[1] [?]*
Murree.

citri, Millière 1873.

Acrolepia, citri, Mill., Pet. Nouv. Ent. V 310 (1873).[1]

Prays citri, Mill., Icon., p. 150, ff. 17-20[2]; Rebel, Stdgr. Cat. Lep. Pal. II 133 (1901)[3]; Spuler, Schmett. Eur. II 442 (1910) [redescr.][4]; Grandi, Disp Entom. Agraria, pp. 287-288, f. 285[5]; Essig, Calif. Mthly. Bull. II 722-723, f. 389 (1913)[6]; Meyr., B. J. XXIII 125 (1914)[7]; Wlsm., P. Z. S. 1907, 974 (1908)[10]; Meyr., Ann. Transv. Mus.

Colombo.[7]
Maskeliya.[7, 16]
Madulsima.[7]
N. Coorg.[7]
Pusa.[14, 16]

Corsica.[1-4, 10.]
S. France.[10]

* Specimens of *P. acmonias* in Mr. Meyrick's collection were sent to him by the late Colonel Manders labelled as from Rawalpindi, but it seems probable, in this and a few other similar cases, that they were actually obtained in the adjacent Hills.

VI 56 (1918)[11]; Fletcher, Entom. Note 89, f. 16 (1916)[12]; Quayle, U. S. A. Agric. Dept. Bull. 134, p. 22 (1914)[13]; Fletcher, Rept. Second Entl. Meeting, pp. 17, 212 (1917)[14]; Fletcher, Rept. Third Entl. Meeting I 160 (1920)[15]; Fletcher, Ind. Agr. Ent. Mem. VI 132, t. 32, f. 1 (1921)[16]; Mañalac San Juan, Philipp. Agriculturist XII 339-348, tt. 1-2 (1924)[17]; Meyr., Ins. Samoa III 104 (1927).[18] — Sicily.[3, 4, 10] Canary Isds.[10] Transvaal.[11] Java.[18] Philippines.[6, 7, 17] N. S. Wales.[8] Fiji.[18] Samoa.[18]

Prays nephelomima, Meyr., Proc. Linn. Soc. N.S.W. XXXII 76 (1907).[8]

Larva on *Citrus,* a serious pest of the orange in South Europe and the Philippines.

Biology.[1, 5, 6, 13, 16, 17]

curulis, Meyr. 1914.

Prays curulis, Meyr., B. J. XXIII 125 (1914)♀.[1] — Khasis.[1], Shillong.

Larva mining leaves of *?Symplocos.*

ducalis, Meyr. 1914.

Prays ducalis, Meyr., B. J. XXIII 125 (1914)♀.[1] — Namunakuli.[1] (Ceylon; over 6,000 ft.). Nilgiris.

erebitis, Meyr. 1914.

Prays erebitis, Meyr., B. J. XXIII 124-125 (1914)♂.[1] — Ootacamund.[1]

peperitis, Meyr. 1907

Prays peperitis, Meyr., B. J. XVII 984-985 (1907)♂♀.[1] — Maskeliya.[1] Kodaikanal. (Palnis; 7,000 ft.)

temulenta, Meyr. 1910.

Yponomeuta temulentus, Meyr., Rec. Ind. Mus. V 228 (1910)♂.[1] — Theog.[1] (Simla Hills; 8,000 ft.)

Prays temulenta, Meyr., Cat. Hyponomeut., p. 15 (1913).[2] — Muktesar. (Kumaon; 7,500 ft.)

8. Diaphragmistis, Meyr. 1814.

Diaphragmistis, Meyr., B. J. XXIII 126 (1914). — Type: *macroglena,* Meyr., [india].

macroglena, Meyr. 1914.

Diaphragmistis macroglena, Meyr., B. J. XXIII 126 (1914)♂.[1] — Khasis.[1]

9. Yponomeuta, Latreille 1802.

Yponomeuta, Latr., Précis caract. gen., p. 146 (1796) [descr. without associated species]; Latr., Hist. nat. Crust. Ins. III 417-418 (1802); Latr., H. N. Crust. Ins. IV 222; Latr., H. N. Crust. Ins. XIV 250-251 (1805); Latr., Consid. générales... Ins., p. 364 (1810) [redescr.; no spp.]; *id., t.c.* p. 441 (1810) [*evonymellus,* Fb., quoted as tyep]; Oken, Lehrbuch III i 655 (1815) [charact.]; Leach, Brewster's Edinb. Encycl. IX 133 (1815) [redescr.; *evonymella* sole sp.]; Stephens, Ill. Brit. Entom., Haust. IV 241-242 (1834) [redescr.; *Nygmia,* Hb., is synon.]; Zeller, Isis. XXXII 194 (1839) [redescr.]; Westwood, Synopsis Brit. Ins., p. 111 (1840) [redescr.]; type cited as *padella,* L.]; Duponchel, Cat. Méth. Lep. Eur., pp. 327-328 (1845) [redescr.]; Herrich-Schäffer, Schmett. Eur. V 28, t. 3, ff. 25-27, t. 9, f. 13 (1853) [redescr.]; Meyr., Handb., p. 695, fig. [neur., *plumbellus*] (1895) [redescr]; Rebel, Stdgr. Cat. Lep. Pal. II 132 (1901); Meyr., Proc. Linn. Soc. N. S. W. XXXII 77 (1907) [redescr.]; Forbes, Lep. N. York, p. 339, f. 199 (1924) [redescr.].

Type: *evonymella,* L. [Europe].

Hyponomeuta, Zeller, Isis, pp. 199-209, tab. (1844) [redescr.; revision]; Stt., Ins. Brit. Tin., pp. 58-59, t. 3, ff. $2^{a\text{-}c}$ (1854) [redescr.]; Stt., Manual II 307 (1859)[redescr.]; Heinemann, Schmett. Deutschl., Kleinschmett. II i 107-108 (1870) [redescr.]; Snellen, Vlinders Nederl. Microlep., pp. 504-506 (1882) [redescr.]; Turner, Proc. Linn. Soc. N. S. W. XXVIII 77 (1903) [redescr.]; Spuler, Schmett. Eur. II 443-444. fig. [neur., *cognatellus*] (1910) [redescr.]; Meyr., Cat. Hyponom., p. 16 (1914); Wlsm., Biol. Centr. Amer., Het. IV 324 (1914) [synon.]; Meyr., Tr. N. Z. Inst. XLVII 227 (1915) [redescr.].

Hyphantes, Hübner, Tentamen, p. 2 (1806) [*non-descr.; "evonymella"* sole species cited].

Erminea, Haworth, Lep. Brit. III 512 (1811); Haw., Lep. Brit. IV 512 (1828) [redescr.].

Type: *evonymella,* Linn.

Nygmia, Hübner, Verz. p. 412 (1826).

Type: *evonymella,* Linn.

Teinoptila, Sauber, Semper's Schmett. Philipp. II 701, t. 66, f. 16 (1902).

Type: *interruptella*, Saub. [Philippines; Papua; N. Australia].

alienellus, Walker 1863.

Hyponomeuta alienellus, Wlk., Cat. XXVIII 532 (1863)♂[1]; Moore, Lep. Ceylon III 504 (1887)[2]; Cotes and Swinhoe, Cat. Moths India, p. 707, No. 4830 (1889).[3]

Ceylon.[1-3]
Maskeliya.
Pattipola.

brunnescens, Moore 1887.

Hyponomeuta brunnescens, Mo., Lep. Atk., p. 282 (1887).[1]; Cotes and Swinhoe, Cat. Moths India, p. 707, No. 4831 (1889)[2]; Meyr., Cat. Hyponomeut., p. 19 (1913).[3]

Darjiling.[1, 2]
Assam.[3], Khasis.
Bhutan., Pusa.

calculosa, Meyr. 1914.

Hyponomeuta calculosa, Meyr., B. J. XXIII 127 (1914)♂♀.[1]

Khasis.[1]

corpuscularis, Meyr. 1907.

Yponomeuta corpuscularis, Meyr., B. J. XVIII 153 (1907)♂.[1]

Matalé., Hakgala.
Nilgiris.[1], (6,000 ft.).
Shevaroys.

corusca, Meyr. 1914.

Hyponomeuta corusca, Meyr., B. J. XXIII 127 (1914)♀.[1]

Khasis.[1]

lapidellus, Walsingham 1880.

Hyponomeuta lapidellus, Wlsm., P. Z. S. 1880-86, t. 12. f. 1 (1880)[1]; Cotes and Swinhoe, Cat. Moths India, p. 708, No. 4832 (1889)[2]; Fletcher, Ind. Agr. Ent. Mem. VI 133 (1921).[3]

Dharmsala.[1]

Larva descr.[1,3], on "soongroo"(wild *Salvia*).[1]

melanaster, Meyr. 1907.

Yponomeuta melanaster, Meyr., B. J. XVII 746 (1907) ♂♀.[1]

Puttalam.[1], Karwar.

Hyponomeuta delicata, Schultze, Philipp. Jl. Sci. III 36-37, t. 1 f. 14 (1908)♂.[2]

Philippines.[2]

millepunctatellus, Warren 1888.

Hyponomeuta, millepunctatellus, Warren, P. Z. S. 1888, 338 (1888)[1]; Cotes and Swinhoe, Cat. Moths India, p. 708, No. 4835 (1889).[2]

Kala Pani.[1, 2]
(Murree Hills.)
Shillong.

minuellus, Walker 1863.

Hyponomeuta, minuellus, Wlk., Cat. XXVIII 533 (1863)♀[1]; Cotes and Swinhoe, Cat. Moths India, p. 708, No. 4836 (1889).[2]

Note—The Indian records are perhaps a mistaken identification for *millepunstatellus*, Warren.

Nepal.[1, 2]
—————[1, 2]
N. China.[1, 2]
Shanghai.[1], Ningpo.
Tientsin.

munda, Meyr. 1921.

Hyponomeuta munda, Meyr., Exot. Micr. II 438 (1921) ♂.[1] — Kharaghoda.[1] (Gujarat.)

numerosa, Meyr., 1921.

Hyponomeuta numerosa, Meyr., Exot. Micr. II 438-439 (1921)♀.[1][Probably a synonym of *millepunctatella,* War. ren.—T.B.F.]. — Murree.[1] (Punjab Hills; 7,500 ft.)

praetincta, Meyr. 1914.

Hyponomeuta, praetincta, Meyr., B. J. XXIII 126-127 (1914)♂♀.[1] — Khasis.[1]

10. Calamotis, Meyr. 1918.

Calamotis, Meyr., Exot. Micr. II 188 (1918). — Type: *prophracta,* Meyr., [India].

prophracta, Meyr. 1918.

Calamotis, prophracta, Meyr., Exot. Micr. II 189 (1918)♂.[1] — N. Kanara.[1] Bhim Tal. (Kumaon; 4,500 ft.).

11. Atteva, Walker 1854.

Atteva, Wlk., Cat. II 526 (1854); Berg., Ann. Soc. Ci. Argent. X 99-100 (1880); Wlsm., P.Z.S. 1897 112 (1897); Wlsm., Swinhoe's Cat. Het. Oxford Mus., II 558 (1900); Dyar, List N. Amer. Lep., p. 490 (1902); Turner, Proc. Lin. Soc. N.S. W. XXVIII 78 (1903) [redescr.]; Meyr., Proc. Linn. Soc. N.S.W. XXXII 78 (1907); Wlsm., Biol. Centr. Amer., Het. IV 327-328 (1914) [synon., citations]; Forbes, Lep. N. York. p. 340 (1924) [redescr.]. — Type: *niveigutta,* Wlk. [India].

Pæciloptera, Clemens, Proc. Acad. Nat. Sci. Philad. XII 546-547 (1860) [*nec* Latreille 1829, Hemiptera]. — Type: *aurea,* Fitch [N. America].

Amblothridia, Wallengren, Resa Eugenies II, Zool. Ins., p. 385 (1861). — Type: *niveigutta,* Wlk.

Corinea, Walker, Cat. XXVIII 542 (1863). — Type: [*fabriciella, Swed.*=] *niveiguttella,* Wlk. [India].

Oeta, Grote, Proc. Ent. Soc. Philad. V. 230-231 (1865). — Type: *aurea,* Fitch.

Carthara, Walker, Cat. XXXV 1871-1872 (1866) [*nec* Walker 1865, Lepidoptera]. — Type: *flavivitta*, Wlk. [S. America].

Synadia, Walker, Cat. XXXV 1984 (1866). — Type: *flavivitta*, Wlk.

Scintilla, Guenee, Ann. Soc. Ent. France (5) IX 287 (1879). [*nec* Desh. 1858—Mollusca]. — Type: *pustulella*, Fb.

Syblis, Guenee, Ann. Soc. Ent. France (5) IX 288-289 (1879). — Type: *fulviguttata*, Z. [W. Indies].

emissella, Walker 1863.

Corinea, emissella, Wlk., Cat. XXVIII 543 (1863)♂.[1]

Atteva emissella, Meyr., Cat. Hyponomeut., p. 22 (1913).[2]

Bhutan.[2]

Borneo.[1,2]
Sarawak.[1]

fabriciella, Swederus 1787.

Phalaena Tinea fabriciella, Swederus, Kongl. Svensk. Vet. Akad. nya Handl. VIII 277, No. 28 (1787)[1]; Gmelin, Linn. Syst. Nat. (ed. xiii) I (50) p. 2608, No. 1533 (1788-1793).[2]

Corinea niviguttella, Walker, Cat. XXVIII 542-543 (1863) ♂ ♀[3]; Zeller, Horae Soc. Ent. Ross. XIII 225 (1877).[4]

Atteva niveiguttella, Wlsm., P.Z.S. 1885, 883 (1886)[5]; Cotes and Swinhoe, Cat. Moths India, p. 718, No. 4916 (1889).[6]

Atteva fabriciella, Wlsm., Swinhoe's Cat. Het. Oxford Mus. II 559, No. 3660 (1900)[7]; Meyr., Catal. Hyponomeut., p. 22 (1913)[8]; Fletcher, South Ind. Ins., pp. 461-463, f. 338 (Sept. 1914)[9]; Fletcher, Rept. Third Entl. Meeting I 160 (Nov. 1920)[10]; Fletcher, Ind. Agr. Ent. Mem. VI 133-134 (Jan. 1921).[11]

Larva descr,[9], on *Ailanthus excelsa*.[9, 10, 11]

Ceylon.[3]
Coimbatore.[7, 11]
Trimulgheri.[11]
Nilgiris.[7], Bombay.[7]
Deesa.
Ahmedabad.[11]
Nagpur.[11]
Arrah.

Java.[2]
Borneo.[8]

impariguttata, Zeller 1877.

Oeta impariguttata, Zeller, Horae Soc. Ent. Ross. XIII 224-225, t. 3 f. 67 (1877) ♀.[1]

Atteva impariguttata, Moore, Lep. Ceylon III 505, t. 209 f. 2 (1887)[2]; Cotes and Swinhoe, Cat. Moths India, p. 718, No. 4914 (1889).[3]

Ceylon.[1-3]
Puttalam.
Kalutara.
Kandy.

niveigutta, Walker 1854.

Atteva niveigutta, Wlk., Cat. II 526-527 (1854);[1] Moore, P.Z.S. 1867, 669[2]; Cotes and Swinhoe, Cat. Moths India, p. 718, No 4915 (1889)[3]; Wlsm., in Swinhoe's Cat. Het. Oxford Mus. II 558 (1900)[4]; Fletcher, Rept. Third Entl. Meeting I 160 (1920)[5]; Fletcher, Ind. Agr. Ent. Mem. VI 134 (1921).[6]

Larva on *Ailanthus excelsa.*[2, 6]

Sylhet.[1, 4], Khasis. Cherrapunji. Bhutan., Sikkim.[6] Madras.

China.[4], Szechuan. Hunan.

pulchella, Moore 1887.

Atteva pulchella, Mo., Lep. Atk., p. 282 (1887)[1]; Cotes and Swinhoe, Cat. Moths India, p. 719, No. 4917 (1889).[2]

Loc. ?[1] India.[2]

sciodoxa, Meyr. 1908.

Atteva sciodoxa, Meyr., Rec. Ind. Mus. II 398-399 (1908) ♂♀.[1]

Dawna Hills.[1] (S. Burma).

Penang.[1], Labuan.[1] Tonkin.[1]

subaurata, Durrant 1900.

Atteva subaurata, Durrant, Swinhoe's Cat. Het. Oxford Mus. II 560, No. 3662 (1900)♂♀.[1]; Snellen, Tijds. v. Ent. XLVI 30 (1903).[2]

Atteva modesta, Snellen, Tijds. v. Ent. XLIV 79, t. 5. f. 5 (1901)♂♀.[3]

Cherrapunji.[1] Khasis. Margherita.[1] Assam.[2, 3], Sikkim.[1]

12. Pronomeuta, Meyr. 1905.

Pronomeuta, Meyr., B. J. XVI 608 (1905).

Type: *sarcopis,* Meyr. [India; Ceylon.]

sarcopis, Meyr. 1905.

Pronomeuta sarcopis, Meyr., B. J .XVI 608 (1905)♂♀.[1]

Heneratgoda.[1] (Ceylon). Kandy. Matalé., Karwar., (N. Kanara).

13. Nosymna, Walker 1863.

Nosymna, Wlk., Cat. XXIV 831 (1863); Wlsm., Swinhoe's Cat Het. Oxford Mus. II 566 (1900).

Type: *repletella,* Wlk. [Borneo].

Androgyne, Wlsm., Swinhoe's Cat. Het. Oxford Mus. II 565-566 (1900).

Type: *punctata,* Wlsm., [India].

lapillata, Meyr. 1914.

Nosymna lapillata, Meyr., B. J. XXIII 127 (1914) ♀.[1] — Darjiling.[1]

punctata, Walsingham 1900.

Androgyne punctata, Wlsm., Swinhoe's Cat. Het. Oxford Mus. II 566-567 (1900)♂ ♀.[1]

Nosymna punctata, Meyr., Cat. Hyponomeut., p. 23 (1913).[2]

Nosymna repletella [nec Wlk.], Moore, Jl. A. S. Bengal LIII, pt. 2, No. 3, p. 237 (1884);[3] Cotes and Swinhoe, Cat. Moths India, p. 716, No. 4897 (1889).[4]

Sikkim.[1] (1000-4000 ft.) Assam.[1], Khasis. Nemotha.[3, 4] (Cachar).

14. Aetherastis, Meyr. 1909.

Aetherastis, Meyr., B. J. XIX 422 (1909). — Type: *uranias,* Meyr. [Ceylon].

circulata, Meyr. 1918.

Aetherastis circulata, Meyr., Exot. Micr. II 190 (1918)♂[1]; Fletcher, Rept. Third Entl. Meeting I 160 (1920)[2]; Fletcher, Ind. Agr. Ent. Mem. VI 134 (1921)[3]. — Trivandrum.[1]

Larva on *Eugenia, jambolana* (Myrtaceae)[1-3]

constellata, Meyr. 1914.

Aetherastis constellata, Meyr., B. J. XXIII 129 (1914)♂.[1] — Wellawaya.[1] (Ceylon).

uranias, Meyr. 1909.

Aetherastis uranias, Meyr., B. J. XIX 422 (1909) ♀.[1] — Peradeniya.[1] Kandy.

15. Comocritis, Meyr. 1894.

Comocritis, Meyr., T.E.S. 1894, 24 (1894). — Type: *olympia,* Meyr. [India; Burma; China].

cyanobactra, Meyr. 1922.

Comocritis cyanobactra, Meyr., Exot. Micr. II 552 (1922) ♂.[1] — Assam.[1], Sylhet.[2]

Comocritis pieria var., Antam, Ind. Tea Assocn. Bull. 5, p. 17, tab. 9 f. 2 (1907).[2]

Larva damaging bark of Tea bushes.[2]

enneora, Meyr. 1914.

Comocritis enneora, Meyr., B. J. XXIII 130 (1914) ♂.[1] — Khasis.[1]

nephelista, Meyr. 1914.

Comocritis nephelista, Meyr., B. J. XXIII 129-130 (1914) ♂♀.[1] — Kandy.[1], Maskeliya.[1]

olympia, Meyr. 1894.

Comocritis olympia, Meyr., T.E.S. 1894, 24 (1894), ♂.[1]; Meyr., Cat. Hyponomeut., p. 24 (1913).[2] — Koni.[1] (Upper Burma.) Assam.[2], Khasis.

China.

pieria, Meyr. 1906.

Comocritis, pieria, Meyr., B. J. XVII 416 (1906) ♂ ♀[1]; Antram, Ind. Tea Assocn. Bull. 5, pp. 16-17, t. 9 ff. 1, 3, 4 (1907)[2]; Meyr., Cat. Hyponomeut., P. 24 (1913)[3]; Fletcher, Rept. Third Entl. Meeting I 160 (Nov. 1920)[4]; Senior White, Rept. Third Entl. Meeting III 872-875 (Dec. 1920) [biology][5]; Fletcher, Ind. Agr. Ent. Mem. VI 134, t. 32 f. 2 (Jan. 1921).[6] — Neboda.[1], (Ceylon.) Matalé.[5, 6] Peradeniya. Assasm.[2, 3] Cachar.[2, 6] Sylhet.[2, 6] Duars.[2]

Larva Described[1, 2, 5, 6,] in galleries on bark of Para Rubber (*Hevea brasiliensis*) [1, 5, 6,] on bark of Tea.[2, 6]

præcolor, Meyr. 1914.

Comocritis, præcolor, Meyr., B. J. XXIII 129 (1914) ♂.[1] — Trincomali.[1]

thespias, Meyr. 1909.

Comocritis thespias, Meyr., B. J. XIX 423 (1909) ♂.[1] — Kandy.[1]

16. Ethmia, Hübner 1822.

Ethmia, Hübner, Verz. p. 163 (1822); Zeller, Isis XXXII 194 (1839) [redescr. as Subg. of *Psecadia*]; Meyr., Tr. R. Soc. S. Austr. XXX 49 (1906) [*Psecadia* is synon.]; Wlsm., Fauna Hawaii. I 507 (1907) [Type stated to be *aurifluella,* Hb.=*pyrausta,* Hb. *nec* Pallas]; Wlsm., Biol. Centr. Amer., Het. IV 143-144 (1912) [synon; Type again stated to be *aurifluella,* Hb.]. — Type: *pyrausta,* Pallas*. [Europe; N. Asia.]

Psecadia Hübner, Verz., p. 412 (1826); Zeller, Isis XXXII 194 (1839) [redescr.]; Duponchel, Cat. Meth. Lep. Eur., p. 328 (1845) [redescr.]; Stainton, Nat. Hist. Tin. XIII 238-251 (1873) [general account; biology, etc.]; Meyr., — Type: *decemguttella,* Hb. [C. & S. Europe].

---

* The sole species quoted by Hübner is "Pyrausta Pall: Reis. I Anh, 20," which must therefore be the genotype of *Ethmia.* There is no reference to *Bombyx Pyrausta,* Hb. 260, which is a synonym of *aurifluella,* Hb., a species distinct from *pyrausta,* Pallas.

Handb., p. 630 (1895) [redescr.]; Rebel, Stdgr. Cat. Lep. Pal. II 167 (1901); Meyr., Tr. R. Soc. S. Austr. 165 (1902) [redescr.]; Meyr., Tr. R. Soc. S. Austr. XXX 49 (1906) [syn. of *Ethmia*]; Meyr., Cat. Hyponomeut., p. 25 (1914) [syn. of *Ethmia*].

Anesychia, Hübner, Verz. p. 413 (1826); Stephens, Ill. Brit. Ent., Haust. IV 238-239 (1834) [redescr.]; Westwood, Synopsis Brit. Ins., p. 111 (1840) [redescr.; type cited as *pusiella*]; Stainton, Ins. Brit. Tin., pp. 61-62, t. 3 ff. 3^{a-c} (1854) [redescr.]; Stt., Manual II 308 (1859) [redescr.]; Meyr., Cat. Hyponomeut, p. 25 (1913) [syn. of *Ethmia*].

Type: *pusiella*, Rœmer. [C. & S. Europe; W. C. Asia].

Disthymnia, Hübner, Verz. p. 413 (1826).

Type: *funerella*, Fb. [C. & S. Europe].

Melanoleuca, Stephens, Cat. Brit. Ins. II 202 (1829) [*non-descr.*]. Steph., Ill. Brit. Ent., Haust. IV 239 (1834) [sunk under *Anesychia*, Hb.].

Aedia, Duponchel, Hist. Nat. Lep. France X 296, 305-316 (1836); Dup., Cat. Meth. Lep. Eur., p. 327 (1845) [redescr.]. [nec *Aedia*, Hübner 1825].

Type: *bipunctella*, Fb. [C. & S. Europe, N. Africa].

Chalybe, Duponchel, Hist. Nat. Lep. France X 296 (1836); Dup., Cat. Meth. Lep. Eur., p. 329 (1845) [redescr.]; Stainton, Ins. Brit. Tin., p. 64 (1854) [redescr.]; Stt., Manual II 309 (1859) [redescr.].

Type: *aurifluella*, Hb. [Europe].

Azinis, Walker, Cat. XXVIII 541 (1863); Wlsm., Fauna Hawaii. I 508 (1907).

Type: *hilarella*, Wlk. [Ceylon].

Tamarrha, Walker, Cat. XXIX 816 (1864).

Type: *nivosella*, Wlk. [Antilles].

Theoxenia, Walsingham, Moore's Lep. Ceylon III 506 (1887).

Type: [*hilarella*, Wlk. ♀=] *penicillata*, Wlsm. ♂ [Ceylon.]

Babaiaxa, Busck, Jl. N. Y. Ent. Soc. X 95 (1902).

Type: *delliella*, Fernald. [N. America].

acontias, Meyr. 1906.

Ethmia acontias, Meyr., B. J. XVII 409-410 (1906) ♂.[1]

Larva on leaves of "Datranga" (Hindi).

Puttalam.[1]
Hambantota.
(S. E. Ceylon).
N. Kanara.
Bhagalpur.

assamensis, Butler 1879.

Hyponomeuta assamensis, Butler, T.E.S. 1879, 6 (1879).[1]

Azinis assamensis, Warren, P.Z.S. 1888, 338 (1888)[2]; Cotes and Swinhoe, Cat. Moths India, p. 719, No. 4918 (1889).[3]

Ethmia assamensis, Durrant, P. Z. S. 1906, 496 (1906)[4]; Fletcher, Ind. Agr. Ent. Mem. VI 134-135 (Jan. 1921).[5]

Psecadia, hockingella, Wlsm., P. Z. S. 1880, 90-91, t. 12, ff. 8, 9, 9^a (188)[6]; Cotes and Swinhoe, Cat. Moths India, p. 718, No. 4912 (1889).[7]

Larva descr.[5, 6,] on "Poonah" [probably *Ehretia serrata*].[6]

Cachar.[1, 3]
Naga Hills.[4]
Khasis., Sadiya.
Bhutan.
Darjiling.
Khatmandu.[5]
Dharmsala.[4, 7]
Kulu.[7]
Kashmir.[4]
Masuri.

crocosoma, Meyr. 1914.

Ethmia crocosoma, Meyr., Exot. Micr. I 173-174 (1914)♀.[1]

Darjiling.[1]

duplicata, Meyr. 1914.

Ethmia duplicata, Meyr., B. J. XXIII 130 (1914) ♂♀.[1]

Puttalam.[1]
Maskeliya.[1]
Pattipola.[1]

ermineella, Walsingham 1880.

Psecadia ermineella, Wlsm., P. Z.S. 1880, 90, t. 12, f. 7 (1880)[1]; Cotes and Swinhoe, Cat. Moths India, p. 718, No. 4911 (1889).[2]

Ethmia ermineella, Meyr., Cat. Hyponomeut., p. 27 (1913).[3]

Dharmsala.[1]
Rawalpindi [?][3]*
Kurseong.
Darjiling distr.

euarithma, Meyr.

Ethmia euarithma, Meyr., Exot. Micr. III 120 (1924) ♂.[1]

Teng-yueh-ting.[1]
(W. China).

hilarella, Walker 1863.

Azinis hilarella, Walker, Cat. XXVIII 542 (1863)♀[1]; Moore, Lep. Ceylon III 506; t. 209 f. 3 (1887) [predescr.][2]; Cotes and Swinhoe, Cat. Moths India, p. 719, No. 4919 (1889)[3]; Wlsm., Fauna Hawaii., I 508 (1907).[4]

Ethmia hilarella, Meyr., Cat. Hyponomeut., p. 26 (1913).[5]

Ceylon.[1–6]
Puttalam., Eppawela.
Pussellawa.
Dimbula., Maskeliya.
Passara.
South India.[5]
N. Coorg.

* A specimen in Mr. Meyrick's collection is labelled as collected at Rawalpindi by the late Colonel Manders. As in other similar cases, it seems more probable that this specimen was taken in the Adjacent Hills.

Theoxenia penicillata, Wlsm., Mo. Lep. Ceylon III 506, t. 209, f. 4 (1887) ♂.[6]	Shembaganur (Palnis; 6,000 ft.). Karwar. (N. Kanara.)

Note—Several other species have been wrongly recorded as *hilarella,* Wlk. *Psecadia hilarella,* Meyr., T.E.S. 1894, 22, is *systematica,* Meyr., (*q.v.*). *Psecadia hilarella,* Meyr., Tr. R. Soc. S. Austr. 1902, 166, is *exhilarella,* Durrant, an Australian species, *Azinis hilarella,* Semper, Schmett. Philipp. II 700, t. V, ff. 12, 13 (1902), is probably, *praeclara,* Meyr., also known from Lombok and Borneo.

lineatonotella, Moore 1867.

Hyponomeuta lineatonotella, Moore, P.Z.S. 1867, 669, t. 33, f. 18 (1867)[1]; Cotes and Swinhoe, Cat. Moths India, p. 708, No. 4833 (1889).[2]	Darjiling.[1] Kurseong. ——— China.

pagiopa, Meyr. 1918.

Ethmi, pagiopa, Meyr., Exot. Micr. II 189-190 (1918)♀.[1]	Kashmir.[1], (6,000 ft.).

systematica, Meyr. 1922.

Ethmia systematica, Meyr., Exot. Micr. II 552 (1922) ♂ ♀.[1]	Fort Stedman.[1, 2] (Shan States, Burma Silchar.[1]
Psecadia hilarella [nec Wlk.], Meyr., T.E.S. 1894, 22 (1894).[2]	Bhagalpur., Chapra.

zelaea, Meyr. 1906.

Ethmia zelaea, Meyr., B. J. XVII 409 (1906)[1].	N.C. Prov., Ceylon.[1]

17. Rœslerstammia, Zeller 1839.

Rœslerstammia, Zeller, Isis XXXII 202-203 (1839); Duponhel, Cat. Meth. Lep. Eur., p. 365 (1845) [redescr.]; Herrich-Schäffer, Schmett. Eur. V 32, t. 11, ff. 25-26 (1853) [redescr.]; Stainton, Ins. Brit. Tin., p. 172, t.5, ff.12[a-c] (1854) [redescr.] [*Röslerstammia*]; Stainton, manual II 363 (1859) [redescr.] [*Roslerstammia*]; Heinemann, Schmett. Deutschl., Kleinschmett. II i 98 (1870) [redescr.]; Meyr., Handb., p. 770 (1895) [redescr.]; Meess, in Spuler's Schmett. Eur. II 454, F. 204 [neur., *erxlebella,* Fb.] (1910) [redescr.].	Type: *erxlebella,* Fb. [Europe].
Chrysitella, Zeller, Isis XXXII 203 (1839).	Type: [*erxlebella,* Fb.=] *erxlebeniella,* Z.

hemiadelpha, Meyr. 1922.

Rœslerstammia, hemiadelpha, Meyr., Exot. Micr. II 553 (1922)♂♀.[1] — Murree.[1] (7,500 ft.).

metaplastica, Meyr. 1921.

Rœslerstammia metaplastica, Meyr., Exot. Micr. II 439 (1921) ♂.[1] — Murree.[1]

18. Lamyristis, Meyr. 1911.

Lamyristis, Meyr., B. J. XXI 131 (1911). — Type: *leucopselia,* Meyr. (Ceylon].

laucopselia, Meyr. 1911.

Lamyristis leucopselia, Meyr., B. J. XXI 131 (1911)♂♀.[1] — Maskeliya.[1]

19. Anticrates, Meyr. 1905.

Anticrates, Meyr., B. J. XVI 612 (1905); Meyr., Proc. Linn. Soc. N.S.W. XXXII 84-84 (1907) [redescr.]; Meyr., B. J. XXIII 129 (1914). — Type: *chrysantha,* Meyr. [Ceylon].

Pyrozela, Meyr., B. J. XVIII 414 (1906); Meyr., Cat. Hyponomeut., p. 32 (1914) [syn. of *Anticrates*]. — Type: *xanthomima,* Meyr. [Ceylon].

? *Epopsia,* Turner, Proc. Linn. Soc. N. S. W. XXVIII 89-90 (1903). — Type: *metreta,* Turner [N. Queensland].

anticlina, Meyr. 1907.

Pyrozela anticlina, Meyr., B. J. XVII 747-748 (1907)♂♀.[1] — Sikkim (1,800 ft.).[1] Bhutan (1,800 ft.).[1] Darjiling distr.

autobrocha, Meyr. 1922.

Anticrates autobrocha, Meyr., Exot. Micr. II 553 (1922) ♀.[1] — Gauhati.[1]

———————

Borneo.
New Guinea.

chrysantha, Meyr. 1905.

Anticrates chrysantha, Meyr., B. J. XVI 612 (1905)♂.[1] — Maskeliya.[1]

decaplaca, Meyr. 1914.

Anticrates decaplaca, Meyr., B. J. XXIII 127-128 (1914)♀.[1] — Anuradhapura.[1] Hambantota. South India. Coimbatore.

erythromima, Meyr. 1906.

Pyrozela erythromima, Meyr., B. J. XVII 415 (1906)♂♀.[1] — Maskeliya.[1] Namunakuli. (Ceylon 6,000 ft.)

lucifera, Meyr. 1914.

Anticrates lucifera, Meyr., B.J. XXIII 128-129 (1914) ♂ ♀[1]; Fletcher, Ind. Agr. Ent. Mem. VI 135 (1921).[2] — Karwar.[1] (N. Kanara.)

Larva on *Sideroxylon tomentosum* (Sapotaceae).

miltochorda, Meyr. 1914.

Anticrates miltochorda, Meyr., B. J. XXIII 128 (1914)♀.[1] — Khasis.[1]

sanguinicornis, Walsingham 1900.

Mieza ? sanguinicornis, Wlsm., Swinhoe's Cat. Het. Oxford Mus. II 562-563, No. 3667 (1900)♂.[1] — Ceylon.[1] Maskeliya.

thermastris, Meyr. 1914.

Anticrates thermastris, Meyr., B. J. XXIII 129 (1914)♀.[1] — Nilgiris (3,500 ft.).[1] N. Kanara.[1]

tridelta, Meyr. 1907.

Pyrozela, tridelta, Meyr., B. J. XVII 747 (1907) ♂♀.[1] — Khasis.[1]

———

China.

venatrix, Meyr. 1914.

Anticrates venatrix, Meyr., B. J. XXIII 128 (1914)♀.[1] — Nilgiris.[1], (7,000 ft.).

xanthomima, Meyr., 1906.

Pyrozela xanthomima Meyr., B. J. XVII *415* (1906)♂♀.[1] — Maskeliya.[1] N. C. Prov., Ceylon.

20. Litaneutis, Meyr. 1913.

Litaneutis, Meyr., Exot. Micr. I 143 (1913). — Type: *sacrifica,* Meyr., [India; Australia; Samoa].

sacrifica, Meyr. 1913.

Litaneutis sacrifica, Meyr., Exot. Micr. I 143-144 (1913) ♂[1]; Meyr., Ins. Samoa III 105 (1927)♀.[1] — Khasis.[1]

———

Queensland.[1]
Samoa.[2]

21. Thecobathra, Meyr. 1922.

Thecobathra, Meyr., Exot. Micr. II 553 (1922). Type: *acropercna,* Meyr. [India].

acropercna, Meyr. 1922.

Thecobathra acropercna, Meyr., Exot. Micr. II 553-554 (1922)♂♀.[1] Cherrapunji.[1] (Khasis; 4,500 ft.)

22. Tarphyscelis, Meyr. 1913.

Tarphyscelis, Meyr., Exot. Micr. I 144 (1913). Type: *palaeota,* Meyr., [India].

palaeota, Meyr. 1913.

Tarphyscelis palaeota, Meyr., Exot. Micr. I 145 (1913) ♂♀.[1] Khasis.[1]

23. Metharmostis, Meyr. 1921.

Metharmostis, Meyr., Exot. Micr. II 439 (1921). Type: *asaphauta,* Meyr. [India].

asaphaula, Meyr. 1921.

Metharmostis asaphaula, Meyr., Exot. Micr. II 439 (1921) ♂♀.[1] Nasik.[1] (w. Ghats.)

Species omitted.

Yponomeuta malinellus, Zeller; Wlsm., P.Z.S. 1885 883; Cotes and Swinhoe, Cat. Moths India, p. 708, No. 4834 (1889).

The above European species was recorded from Poona in the references quoted above, but it seems probable that these records were based on misidentification and *malinellus* seems better omitted from the Indian List until confirmed.

---

84 Species